The Analysis of
Linear Circuits

Harbrace Series in Electrical Engineering

Glen Wade, EDITOR
UNIVERSITY OF CALIFORNIA, SANTA BARBARA

The Analysis of Linear Circuits

Charles M. Close
RENSSELAER POLYTECHNIC INSTITUTE

 *Harcourt, Brace & World, Inc.*
New York / Chicago / San Francisco / Atlanta

To *Ann, Doug, and Kim*

Preface

This book was written for a one-year course in linear circuit analysis beginning in the sophomore or junior year. It is an integrated treatment of passive and active circuit theory, with emphasis on the significance of the analysis techniques and on the circuit's effect upon the waveshape and frequency spectrum of an incoming signal. It contains separate chapters on such topics as the analysis of nonelectrical systems and linear electronic circuits and provides a foundation for subsequent courses in electronic circuits, system components, control systems, and communication systems.

Although many students will have studied electric and magnetic phenomena and possibly differential equations, the only prerequisite for this textbook is a knowledge of differential and integral calculus. Mathematical techniques that are appropriate to the increasing maturity of today's undergraduate have been used throughout, but the necessary background material has been summarized at suitable places: differential equations in Section 4.1, complex algebra in Section 5.1, the Fourier series in Section 9.1, complex-variable theory in Section 10.4, and matrices in Section 12.4. At the same time, every effort has been made to emphasize the physical significance of the steps in the solution of circuits.

There has been a clear trend during the past decade toward a more unified treatment of the subject matter in the first three years of the engineering curricula. For example, in the electrical engineering curriculum, separate courses in resistive circuits, steady-state analysis, and transients have largely disappeared. There still exists, however, an artificial barrier between active and passive circuit theory, and many students have difficulty extending the techniques learned in a course on passive networks to linear active networks. For example, a student who has mastered Thévenin's theorem for passive networks is frequently unable to apply it to the models of transistor circuits. Rather than ignore active circuits or confine their discussion to an isolated chapter near the end of the book, I have chosen to introduce models with controlled sources in the very first chapter and to include them throughout the book. The analysis techniques that are developed are sufficiently general to apply to both active and passive circuits.

Another, perhaps artificial, barrier raised in many textbooks on circuits is represented by the absence of a discussion of nonelectrical systems. Certainly, it is important for the student to realize that all the techniques developed for circuits can be applied to the analysis of any linear system, and this is the central theme of Chapter 11.

Finally, to illustrate the application of the techniques developed earlier to a particular subclass of problems, I consider in Chapter 12 some specific electronic circuits. The first part of this chapter, which the student can read early in the course if he chooses to do so, also gives a much more complete treatment of the need for and the development of incremental models with controlled sources than does the brief discussion in Section 1.5.

Besides attempting to reduce unnecessary division of the subject, I have tried to show the advantages, limitations, significance, and interrelationships of different analysis techniques. A number of examples throughout the book are solved by more than one method, and there is frequent repetition of the more important concepts, e.g., the resolution of signals into sets of elementary functions, the free and forced responses, the frequency discrimination of circuits.

A major decision facing the author of a book on circuits is where to put the development and application of the Laplace transform. I have chosen to consider it in Chapter 10, for several reasons. First, a proper foundation and motivation can be provided for this topic by the earlier discussions of complex frequency, response to exponential signals, and Fourier methods. Thus, the Laplace transform can be introduced as a generalization of previous techniques and not as an arbitrary device to reduce integral-differential equations to algebraic equations. Second, the advantages and physical significance of other techniques can be emphasized before the Laplace transform is introduced. It is not uncommon for students who have been introduced to the Laplace transform in a mechanical way at the beginning of their study to transform routinely every equation (even for a circuit that contains only one capacitance and a few resistances) and to overlook any physical significance of the terms in the solution. Finally, the applications of complex-variable theory in Sections 10.4 and 10.5 put the Laplace transform on an even firmer foundation and better prepare the student for advanced courses.

In writing the individual chapters, I have attempted to provide motivation and continuity, to explain what is to be done next, and to help the student obtain mathematical and physical insight into the different techniques. I have made liberal use of illustrative examples and figures and have included summaries at the end of the chapters. The majority of the problems (some of which have many parts) have been used during the past five years at Rensselaer for quizzes or homework assignments; about one-half of the answers are either contained in the problem statements or are given at the end of the book.

The nomenclature used is fairly standard, with lower-case letters denoting functions of time and capital letters representing phasors and transformed quantities. Boldface type is used only for phasors, in order to emphasize the special character of a phasor (a complex quantity that conveys the essential information about a real function of time). The boldface type would lose its effectiveness if it were used for all complex quantities, such as the complex frequency s and transformed quantities. Because instructors cannot write easily in boldface, I have not made the typeface essential to the proper interpretation of the voltage and current symbols. Thus, $|\mathbf{E}|$ and not E is used for the magnitude of the voltage phasor $\mathbf{E}$.

The order in which the topics are presented is not the only one that could be followed. For example, for some courses it might be convenient to follow Chapter 5 with Chapter 8 or to follow Chapter 6 with Chapters 9 and 10. A number of individual sections, including Sections 2.4, 3.5, 3.6, 6.5, 8.4, 8.6, 10.4, and 10.5, may be omitted without any serious lack of continuity, if the instructor chooses.

Among the many people who have provided invaluable assistance and advice are a dozen colleagues, over a thousand students, nearly a dozen reviewers who offered constructive comments at various stages, and two patient typists (Sandra Elliott and Rosana Laviolette). In addition, I am indebted to those who provided the inspiration and encouragement that were essential ingredients in the completion of the book.

CHARLES M. CLOSE

Troy, New York

Contents

The Analysis of
Linear Circuits

1

Preliminary Concepts

The two subjects that provide the theoretical basis for the major part of electrical engineering are circuit theory and field theory, the first of which is introduced in this book. *Field theory* is needed to explain the behavior of transistors, inductors, and other devices, and to analyze the propagation of energy through space. *Circuit theory* is concerned with the effects of interconnecting components whose characteristics are already known, and *not* with the reasons behind the behavior of the individual components. The discussion of circuit theory in this book not only provides a strong foundation for future electrical engineering courses, but develops many techniques that may also be applied to other engineering disciplines. In this preliminary chapter, we indicate the general nature and scope of the problems that will be considered, and present many of the definitions and laws upon which succeeding chapters are based.

Figure 1.1-1a is a symbolic representation of a *system* (a collection of interconnected components) that is subjected to an *input* (stimulus or excitation) and that produces an *output* (response). In general, the input x and the output y will both be functions of time t, which is implied by the symbols $x(t)$ and $y(t)$. A specific mechanical system is shown in Fig. 1.1-1b, where the input is the applied force. The output in Fig. 1.1-1b might be the resulting velocity of the mass or the potential energy stored in the spring. The term output refers to the result that is to be observed, so its choice depends upon the individual observer. If several results are to be observed, the system will have several outputs.

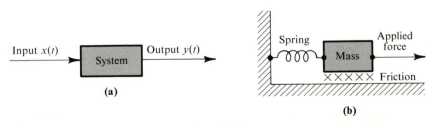

(a)

(b)

Figure 1.1-1

In some systems, the *form* of the output may be more important than its size. A loudspeaker, for example, changes electrical signals (the input) into sound waves (the output). An artificial earth satellite may measure temperature and radiation (the inputs), and encode this information into a high-frequency radio signal (the output). In both of these examples, the shape of the output signal conveys information about the input. The exact size of the output is less important, as long as it can be easily detected. In other examples, such as the distribution network of an electric utility, the output is *not* used to convey information, but to provide power in a suitable form and in sufficient quantity.

An existing system may be observed experimentally, and conclusions may be drawn from the data. It is important, however, to have available analytical methods for examining the input-output relationship of a proposed system. These methods can be used to predict the chances of success, they can provide valuable insight into system operation, and through them ways of overcoming deficiencies can be found. For example, it certainly would not be possible to build and launch an earth satellite successfully without a lengthy stage of theoretical design.

The development of analytical methods involves several considerations. First, a mathematical description of the individual components must be formulated. This description must approximate the observed behavior of the physical component under consideration, but should otherwise be as simple as possible, so that the calculations do not become needlessly difficult. Second, a mathematical description of the effect of interconnecting different components must be found by postulating interconnection laws. A given system can then be represented by a *model* (or idealized system), which is based upon the mathematical description of the components and their interconnection. Then, the input-output relationship can be determined by applying known mathematical techniques. The ultimate test of the validity of this procedure is whether the analytical results consistently agree with experimental observations. If there is not sufficiently close agreement, then the mathematical description of either the components or their interconnection must be revised.

Some knowledge of elementary physics is very helpful when the idealized elements and the interconnection laws are postulated. In most of this book, however, we are content to present, without detailed physical arguments, the elements and laws that have been found to agree with experimental results. In the mathematical analyses, we assume that the student has a good understanding of differential and integral calculus; any additional mathematics that may be needed is summarized at appropriate places.

1.1

Analysis, Linearity, and Circuits

Each of the three principal words in the title of this book indicates a limitation on its scope. In an *analysis* problem, the system and the input are given, and the output must be found. In contrast, the task in a *synthesis* problem is to design a system when the input and the desired response are specified. Still other engineering problems may require the determination of the input necessary to produce a desired response from a given system; or they may be research problems where very few constraints are placed upon the input, system, and output. Although the analysis problem is, in itself, the least important type of engineering problem, it must be mastered before more challenging problems can be solved.

Input signals, such as the rectangular pulse shown in Fig. 1.1-2a, are real functions of time. (Figure 1.1-2a might represent an applied force of constant size for $0 < t < 1$ in the mechanical system of Fig. 1.1-1b.) The response to such an input signal will be another real function of time and will depend upon both the input

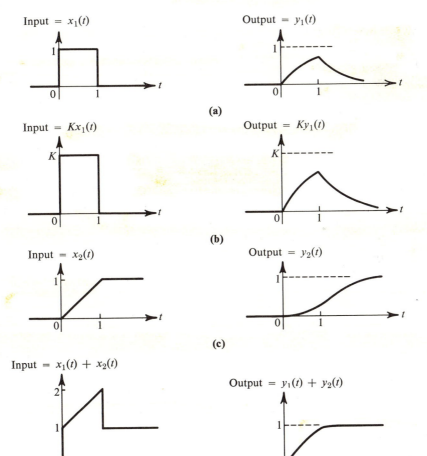

(a)

(b)

(c)

(d)

Figure 1.1-2

and the characteristics of the system. In later chapters, it is sometimes convenient to assume, as an intermediate step in the analysis, that the input and output are complex functions. Nevertheless, the final solution represents the response to a real input signal.

The input-output relationship is one of cause and effect; therefore, the value of $y(t)$ at some time t_0 cannot be influenced by the values of $x(t)$ for $t > t_0$; i.e., the present output can depend upon the past and present input, but not upon the future input. If the response of a proposed system to all real inputs is not real, or if the response violates the normal cause-and-effect relationship, the system is said to be *unrealizable*.

Systems are further classified by placing additional restrictions on the input-output relationship.[†] A system is said to be *linear* if it obeys the following rules, which are illustrated in Fig. 1.1-2.

1. When the input is increased by a multiplying factor K, the output is also multiplied by K, as shown in Fig. 1.1-2b.

2. If two inputs are applied simultaneously (whether at the same or different points in the system), then the total response is the sum of the individual responses to each one separately (Fig. 1.1-2d).[††]

In summary, if $y_1(t)$ and $y_2(t)$ denote the responses to two independent inputs $x_1(t)$ and $x_2(t)$, then a system is linear if and only if the response to

is

$$x(t) = K_1 x_1(t) + K_2 x_2(t) \tag{1.1-1}$$

$$y(t) = K_1 y_1(t) + K_2 y_2(t) \tag{1.1-2}$$

for all inputs and for all values of the constants K_1 and K_2. Note that linearity does not *necessarily* imply that the output function of time has the same shape as the input.

In practice, most systems are to some extent nonlinear, but in many cases the effect of the nonlinearity is so slight that it may be neglected. Quite often, the non-linearity of a component becomes evident only when the inputs are large. For example, a metallic shaft under tension can be described by the curve in Fig. 1.1-3. If the elongation and applied force are chosen as the output and input, respectively, the component is linear only as long as the force does not exceed the elastic limit f_c. For very large forces, however, doubling the force does *not* double the elongation. Similarly, many electrical components, such as transistors and vacuum tubes, behave in a linear way only for small inputs. This book is restricted to components that can be assumed to be linear. Any system which is composed of only linear components is itself linear.

[†] The analysis of a much broader class of systems than those considered in this book can be found in P. M. DeRusso, R. J. Roy, and C. M. Close, *State Variables for Engineers* (New York: Wiley, 1965).

[††] The second part of our definition of linearity, which is illustrated in Fig. 1.1-2d, is known as the *superposition principle*.

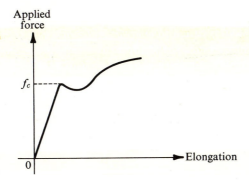

Figure 1.1-3

Example 1.1-1. A differentiator is a device characterized by the relationship $y = dx/dt$. Is the device a linear system?

Solution. Let $y_1(t)$ and $y_2(t)$ denote the responses to the inputs $x_1(t)$ and $x_2(t)$, respectively, so that

$$y_1(t) = \frac{dx_1(t)}{dt} \quad \text{and} \quad y_2(t) = \frac{dx_2(t)}{dt}$$

The response to the input $K_1 x_1(t) + K_2 x_2(t)$ is

$$\frac{d}{dt}\left[K_1 x_1(t) + K_2 x_2(t) \right] = K_1 \frac{dx_1(t)}{dt} + K_2 \frac{dx_2(t)}{dt} = K_1 y_1(t) + K_2 y_2(t)$$

which agrees with Eq. 1.1-2; therefore, the system is linear. A circuit that acts like a differentiator is discussed in Example 1.4-4.

Example 1.1-2. Determine whether or not the following systems are linear.

(a) A system characterized by $y = x^2$.

(b) A system characterized by $y = t(dx/dt)$.

(c) A system characterized by $y = x(dx/dt)$.

Solution. The first and third systems are nonlinear, since doubling the input quadruples (not doubles) the output. The second system is linear, since the response to $K_1 x_1(t) + K_2 x_2(t)$ is $K_1 y_1(t) + K_2 y_2(t)$.

Circuits (or *networks*) are systems composed of electrical components, where the observed signals (including the input and output) are currents and voltages. Nonelectrical systems are treated briefly in Chapter 11. It is pointed out there that many of the concepts and mathematical techniques which are developed for the analysis of circuits can be carried over to mechanical, hydraulic, thermal, acoustical, and other types of systems, and so have very wide application.

The circuits analyzed in this book are not only linear; they also are time invariant and lumped. A system is called *time invariant* (or *fixed*) if the input-output relationship is independent of time. If the input to a time-invariant system is delayed by t_0 seconds, the output is similarly delayed, but its size and shape are not affected. Figure 1.1-4a shows a typical relationship between the input $x(t)$ and output $y(t)$. Delaying a function of time by t_0 corresponds to shifting the curve representing it to the right, and to replacing every t by $t - t_0$ in the equation describing it. For a time-invariant system, the response to $x(t - t_0)$ must therefore be $y(t - t_0)$, as shown in Fig. 1.1-4b.

A *time-varying* system is a system whose characteristics change with time. Since the mass of a rocket decreases as its fuel is ejected, the mass of the rocket is a time-varying component. Or, the properties of a circuit might change if the surrounding temperature changes significantly. In time-varying systems, the form and size of the response do depend upon when the input is applied.

Example 1.1-3.   Are the systems described in Examples 1.1-1 and 1.1-2 time invariant?

Solution.   The differentiator $y = dx/dt$ is time invariant, since the output waveshape is simply the slope of the input function. As illustrated in Fig. 1.1-5, the size and shape

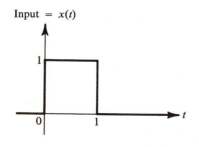
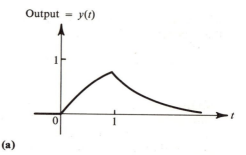

(a)

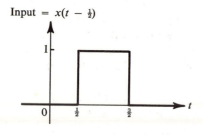
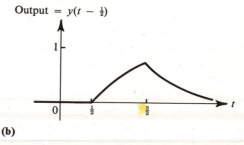

(b)

Figure 1.1-4

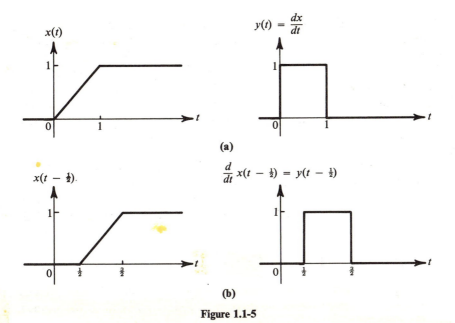

Figure 1.1-5

of the response is unaffected by the time at which the input is applied. Mathematically, the response to an input $x(t - t_0)$ is

$$\frac{dx(t - t_0)}{dt} = \frac{dx(t - t_0)}{d(t - t_0)} \frac{d(t - t_0)}{dt} = [y(t - t_0)](1 - 0) = y(t - t_0)$$

In Example 1.1-2, the system characterized by $y = t(dx/dt)$ is time varying. Because of the multiplying factor of t on the right, the output to an input applied later in time will be larger than to the same one applied earlier. The other systems in Example 1.1-2 are time invariant, since, in each case, the size and shape of the response depend only upon the size and shape of the input.

The distinction between *lumped* and *distributed* systems is illustrated in Fig. 1.1-6 by the vibrating string stretched between two posts. The mass of the string in Fig. 1.1-6a is uniformly distributed along its entire length. Each infinitesimal particle in the string can move with a different velocity, so there are an infinite number of points at which an output velocity could be measured. This is one example of a *distributed system*. Another is an electrical transmission line, where the voltage varies continuously along the line and can be measured at an infinite number of positions.

A *lumped system* has a finite number of points at which *different* output measurements may be made. Although the velocity of the moving mass in Fig. 1.1-1b could be measured at any point on its surface, there is *one* unique velocity associated with this lumped component. Very often a distributed system can be approximated by a lumped system. Figure 1.1-6b shows several equally spaced masses joined by

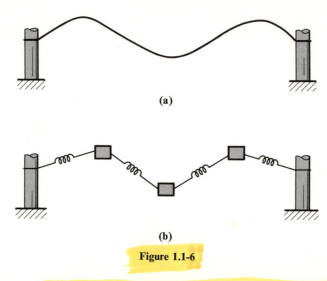

(a)

(b)

Figure 1.1-6

idealized springs. The velocities (outputs) of the masses can be measured, and together they will completely describe the motion. As the number of masses used is increased, the behavior of the lumped system approaches that of the distributed system. In a somewhat similar way, the electrical transmission line can be approximated by a finite number of lumped electrical components.

The restriction to the analysis of linear, time-invariant, lumped circuits might appear to be a severe limitation. Many important problems do, however, fall into such a classification. Furthermore, an understanding of the material presented in this book is essential when such restrictions are removed in more advanced courses.

1.2

Current, Voltage, Power, and Energy

The mks system of units is used throughout this book, and the units for the electrical quantities discussed in this section are summarized in Table 1.2-1.[†] All five quantities in the table are, in general, functions of time t, which can be emphasized by adding a t in parentheses to the symbols, e.g., by denoting the current by $i(t)$ instead of i.

Whenever a number is multiplied or divided by a large power of ten, the prefixes in Table 1.2-2 may be used. Thus 2×10^{-6} seconds (sec) is 2 microseconds (μsec) and 8×10^3 watts is 8 kilowatts (kw). (The prefixes kM, mμ, and $\mu\mu$ were commonly used in the past, but are no longer officially accepted in engineering publications.)

[†] The units in circuit theory are the same whether the "rationalized" or "unrationalized" mks units are used.

TABLE 1.2-1

Quantity	Symbol	mks units
Charge	q	coulombs
Current	i	amperes
Voltage	e	volts
Energy	w	joules
Power	p	watts

It is assumed that the student has some acquaintance, from elementary physics, with the concepts of charge, current, and voltage. The two kinds of charge, called positive and negative, are carried by protons and electrons, respectively, and have the units of *coulombs*. Let q denote the amount of charge passing a given point on a conducting wire. The rate at which charge moves past the point, expressed in units of *coulombs per second* or *amperes*, is defined as the *current i*.

$$i = \frac{dq}{dt} \qquad\qquad (1.2\text{-}1)$$

Since the effect of the current depends upon the kind and direction of the moving charge, this equation is not adequate unless it is accompanied by a statement of explanation (or by a symbol that is equivalent to such a statement).

The line in Fig. 1.2-1a represents a conducting wire, while the reference arrow indicates an assumed positive direction for the current $i_1(t)$. The current is defined as *positive* if positive charge is moving in the direction of the arrow, or if negative charge is moving in the opposite direction. Since these two possibilities produce the same electrical effects, there is no need to determine which one is actually taking place. The current is *negative* if positive charge is flowing in the direction opposite to that of the arrow, or equivalently, if negative charge moves with the arrow.

A current is a function of time and, in general, is positive during some periods of time and negative during others. A sketch of a current versus time, such as that in Fig. 1.2-1b, is called the current *waveshape* (or *waveform*). If $i_1(t)$ has the

TABLE 1.2-2

Multiple or submultiple	Prefix	Symbol
10^9	giga (kilomega)	G (kM)
10^6	mega	M
10^3	kilo	k
10^{-3}	milli	m
10^{-6}	micro	μ
10^{-9}	nano (millimicro)	n (mμ)
10^{-12}	pico (micromicro)	p ($\mu\mu$)

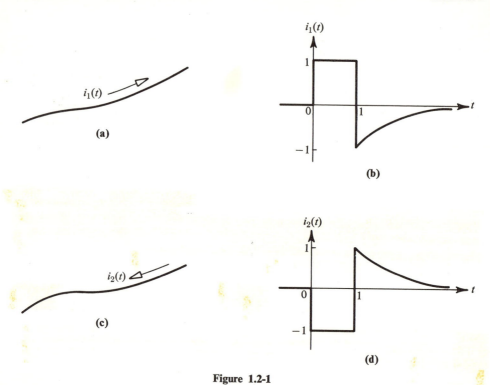

Figure 1.2-1

waveshape shown in Fig. 1.2-1b, positive charge moves to the right at a constant rate for $0 < t < 1$ and moves to the left at a decreasing rate for $t > 1$.[†] Figure 1.2-1c shows the same conducting wire, but with a different reference arrow. If the physical movement of charge is the same in (a) and (c), then $i_2(t) = -i_1(t)$ and $i_2(t)$ must have the waveshape in (d). It should be emphasized that the reference arrow may be drawn either way, as long as the sketch (or equation) is consistent with the arrow direction selected.

The circuit elements that are defined in Section 1.3 have two accessible terminals, as shown in Fig. 1.2-2a. Net charge cannot accumulate inside these elements, so any current entering one terminal must leave the other. The movement of charge through an element is usually associated with the absorption or generation of *energy*, which is denoted by w and which has units of *joules*. The *voltage* or *potential difference* between the two terminals is denoted by e, has units of *volts*, and is indirectly defined in the following paragraph.

[†] As mentioned in the previous paragraph, $i_1(t)$ may also be interpreted in terms of negative charge moving in the opposite directions. It is now known that electrons (negatively charged particles) and not protons (positively charged particles) move along a conductor, though positively charged particles may move in other media. Historically, however, current was assumed to be the flow of positive charges, and it has become customary in circuit theory to interpret current in this way, no matter what kind of particles are actually involved.

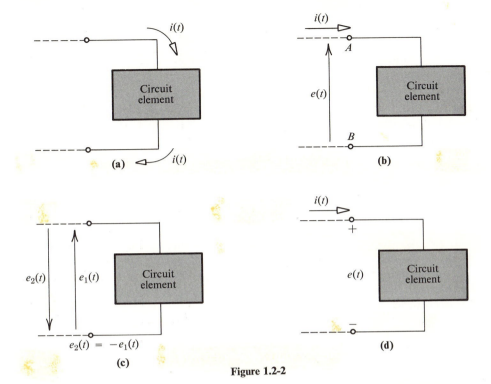

Figure 1.2-2

The energy *absorbed* by the circuit element in Fig. 1.2-2b when a differential amount of charge dq is moved through the element from A to B, i.e., from the head to the tail of the voltage reference arrow, is

$$dw = e \, dq$$

A negative value of dw indicates that energy is actually being *supplied* by the element (presumably to some part of the circuit which is not shown). The sign of $e(t)$ is positive if and only if energy is absorbed by the circuit element when positive charge is moved from the head to the tail of the voltage arrow.

In elementary physics, the electric potential of a point is commonly defined as the work required to move a unit of positive charge from a reference point to the point in question. This definition is equivalent to the definition in the previous paragraph, if $e(t)$ represents the difference in the potentials of points A and B. If $e(t)$ is positive at some instant of time, the potential at the head of the arrow is more positive than that at the tail. If at some instant $e(t)$ is negative, then the converse is true.[†] Reversing the direction of the arrow changes the sign of the voltage, as shown in

---

[†] Some textbooks differentiate between a "voltage rise" that is synonymous with our definition of e, and a "voltage drop," which is sometimes denoted by v and which differs by a minus sign. Although the voltage-drop concept is not used here, the student may encounter this and still other notations when consulting other books.

Fig. 1.2-2c, so that any curve or equation representing a voltage must refer to a particular direction of the reference arrow.

The plus-minus notation of Fig. 1.2-2d is sometimes used in place of the voltage arrow. The plus and minus signs replace the head and the tail of the arrow, respectively. At times, when several voltages are indicated on the same diagram, the proper pairing of the plus and minus signs may be confused; therefore, they will not be used in this book. As an aid in distinguishing voltage and current arrows, however, the current arrowheads are closed and the voltage arrowheads are open.

Since, by Eq. 1.2-1, $dq = i\,dt$, the energy absorbed in dt seconds by the circuit element in Fig. 1.2-2b is $dw = ei\,dt$, and the power absorbed by it is

$$p = \frac{dw}{dt} = ei \tag{1.2-2}$$

where p is expressed in *watts* (joules per second). In Eq. 1.2-2 it is assumed that the current arrow enters the terminal toward which the voltage arrow is directed, for otherwise a minus sign would be required. Figure 1.2-3 shows four possible combinations for the reference arrows. Since $e_2(t) = -e_1(t)$ and $i_2(t) = -i_1(t)$, the four expressions for power give identical results.

Since $dw = p\,dt = ei\,dt$, the total net energy absorbed by an element is

$$w(t) = \int p(t)\,dt = \int e(t)i(t)\,dt \tag{1.2-3}$$

An indefinite integral requires the evaluation of a constant of integration. In order to avoid this step in the solution of circuit problems, it is better to interpret such relationships in terms of definite integrals. The total energy absorbed from time t_1 to t_2 is

$$\int_{t=t_1}^{t=t_2} dw(t) = \int_{t_1}^{t_2} p(t)\,dt$$

or

$$w(t_2) - w(t_1) = \int_{t_1}^{t_2} p(t)\,dt \tag{1.2-4}$$

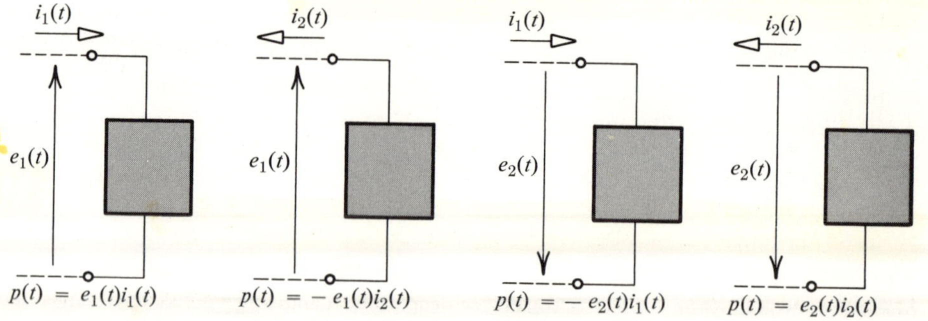

$$p(t) = e_1(t)i_1(t) \qquad p(t) = -e_1(t)i_2(t) \qquad p(t) = -e_2(t)i_1(t) \qquad p(t) = e_2(t)i_2(t)$$

Figure 1.2-3

If we let $t_1 = t_0$ and $t_2 = t$, the total energy received up to time t is

$$w(t) = w(t_0) + \int_{t_0}^{t} p(\lambda)\, d\lambda \qquad (1.2\text{-}5)$$

In the last integrand, t has been replaced by the dummy variable λ, to avoid confusion between the upper limit and the variable of integration. The variable of integration disappears after the limits are inserted, so the symbol used does not affect the result.

An alternative interpretation assumes that at some time sufficiently far in the past (denoted by $t \to -\infty$) no power had yet been delivered to the element. Then

$$w(t) = \int_{-\infty}^{t} p(\lambda)\, d\lambda \qquad (1.2\text{-}6)$$

which is a symbolic way of saying that the energy at time t depends upon all the power that has ever flowed into or out of the element from its birth up to t. If the integral is broken into two parts,

$$w(t) = \int_{-\infty}^{t_0} p(\lambda)\, d\lambda + \int_{t_0}^{t} p(\lambda)\, d\lambda = w(t_0) + \int_{t_0}^{t} p(\lambda)\, d\lambda$$

it agrees with Eq. 1.2-5. This entire discussion about the interpretation of integrals in circuit theory will be applied to the circuit elements discussed in Section 1.3.

1.3

Circuit Elements

The relationship between the input and output of a circuit depends upon the nature and size of the components, and the manner in which these components are connected. In this section the components are discussed, and in Section 1.4 the laws governing their interconnection will be presented. The components considered first have two accessible terminals, as shown in Fig. 1.2-2a.

Suppose that voltage and current measurements are made on a two-terminal component, when it is used in circuits subjected to a wide variety of inputs. From these measurements, it may be possible to formulate a simple equation that approximates the observed voltage-current relationship under most circumstances. An ideal element could then be defined by this equation and could be used as a model for the physical device. All the elements defined in this section are ideal and only approximate the behavior of their physical counterparts.

Circuit theory is concerned only with quantities that may be measured at the terminals of a component and not with the physical phenomena occurring inside it. Of course, it is always more satisfying if the model that is chosen can be related to the internal phenomena. Furthermore, an understanding of the physical principles, obtainable from physics and field theory, may make the task of devising a satisfactory model easier. Nevertheless, detailed explanations of the reasons behind the observed behavior of components are left to other textbooks.

The mathematical description of the ideal elements that are finally adopted should be as simple as possible, and the number of different kinds of elements should not be excessive. Consequently, some physical components cannot be adequately represented by a single ideal element; but they can be represented by several interconnected elements, as discussed in Section 1.5.

The Passive Circuit Elements. A circuit element is said to be *passive* if the total net energy delivered to it from the rest of the circuit (calculated from Eq. 1.2-5 or 1.2-6) is always nonnegative. Power delivered to a passive circuit element may be

TABLE 1.3-1

Circuit element	Symbol	Defining equation
Resistance: R ohms (Ω) Conductance: G (mhos) $= 1/R$	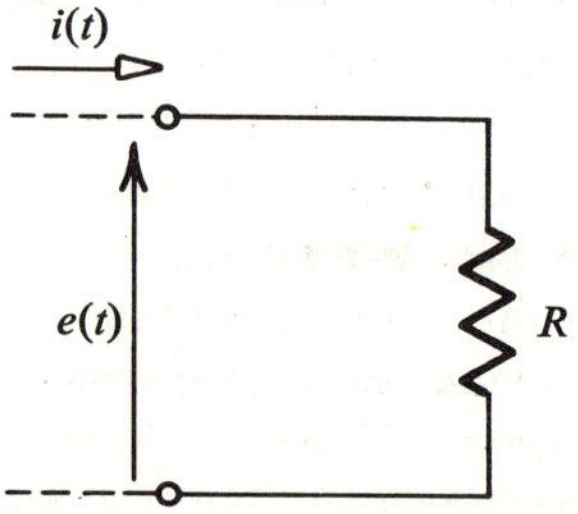	$e(t) = Ri(t)$ $i(t) = \dfrac{e(t)}{R} = Ge(t)$
Capacitance: C farads (f) Elastance: S (darafs) $= 1/C$	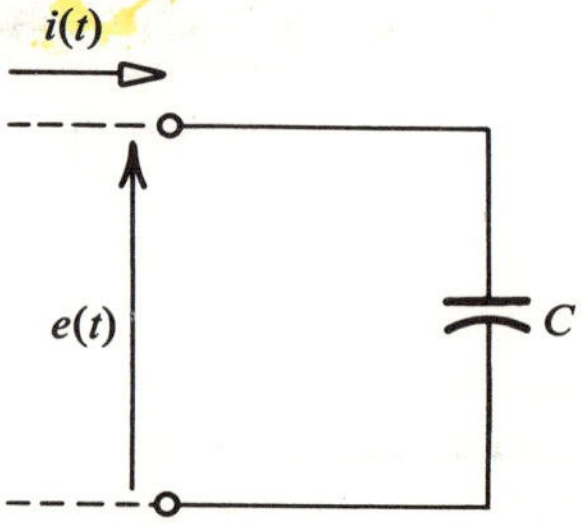	$i(t) = C\dfrac{de}{dt}$ $e(t) = \dfrac{1}{C}\int i\,dt$
Self-inductance: L henrys (h) Inverse inductance: Γ (inverse henrys) $= 1/L$	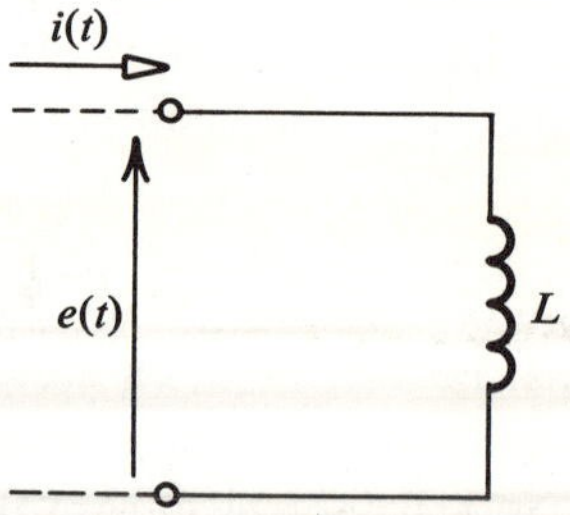	$e(t) = L\dfrac{di}{dt}$ $i(t) = \dfrac{1}{L}\int e\,dt$

Important →

dissipated as heat, in which case it is irretrievably lost, or it may be stored in the surrounding electric or magnetic fields. An *active* element, on the other hand, may supply energy to the rest of the circuit.

Among the components found in common circuits are the resistor, capacitor, and inductor. The elements used to approximate these components are resistance (R), capacitance (C), and inductance (L), respectively. Their symbols, definitions, and units are summarized in Table 1.3-1. Unless otherwise stated, R, C, and L are real, nonnegative constants. The units used for these quantities have been derived from the names of three nineteenth-century workers: Georg Simon Ohm (who proposed the law taken as the definition of resistance) and Michael Faraday and Joseph Henry (who discovered independently that voltage could be produced by a changing magnetic field). The symbols Ω, f, and h are used for ohm, farad, and henry, respectively.

The definition of a *resistance*, $e(t) = Ri(t)$, indicates that the current and voltage are directly proportional to each other. Thus, curves of $e(t)$ and $i(t)$ plotted versus t always have exactly the same shape, differing only in the scale used for the vertical axis, as is illustrated in Fig. 1.3-1a. In fact, the waveshape of a current is often determined experimentally by examining the voltage waveshape across some resistor through which it passes. The *conductance* is defined as $G = 1/R$, and its unit is the *mho*, obtained by spelling ohm backwards.

In Table 1.3-1, the voltage arrow is consistently directed toward the terminal entered by the current arrow. If one of the reference arrows is reversed, a minus sign is required in the equations, as illustrated in Fig. 1.3-2.

From Eq. 1.2-2, the power delivered to the resistance in Fig. 1.3-2a is

$$p = ei = Ri^2 = Ge^2 \tag{1.3-1}$$

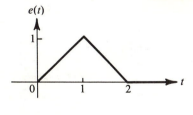
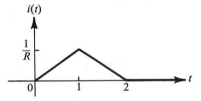

(a)

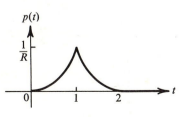
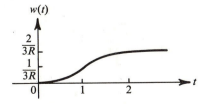

(b)

Figure 1.3-1

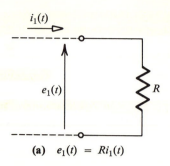

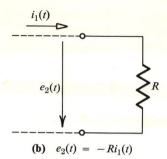

(a) $e_1(t) = Ri_1(t)$

(b) $e_2(t) = -Ri_1(t)$

Figure 1.3-2

which is always nonnegative if R is assumed to be nonnegative. Hence, the resistance can never be a source of power, but can only absorb it and dissipate it as heat; consequently, the resistance is a passive element. The total energy supplied to a resistance can be calculated by Eq. 1.2-5. Figure 1.3-1b shows the power and energy corresponding to the voltage and current waveshapes of Fig. 1.3-1a. In plotting $w(t)$ in Fig. 1.3-1b, it is assumed that no power has been delivered before $t = 0$.

A resistance of value zero is called a *short circuit*, and a conductance of value zero—i.e., an infinite resistance—is called an *open circuit*. These two extreme cases are shown in Fig. 1.3-3. In the first case, $e(t)$ is constrained to be zero, while in the second $i(t) = 0$.

Although all metallic conductors exhibit *some* resistance, *resistors* are devices that have been constructed to make resistance the dominant effect. We shall assume that the resistance of a wire connecting two components is either negligible or that it is represented by a separate element in the model of a circuit. All the connecting wires shown in the figures should be interpreted as short circuits having no potential difference between the two ends.

A *capacitor* consists of two metal plates separated by a thin layer of insulating material, as shown in Fig. 1.3-4. The electrical charges on the two plates must be of equal magnitude but of opposite sign and must produce in the insulating layer an

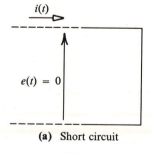

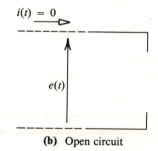

(a) Short circuit

(b) Open circuit

Figure 1.3-3

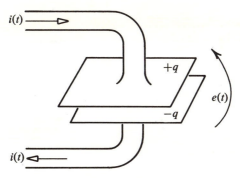

Figure 1.3-4

electrostatic field that is capable of storing energy. [The mechanism by which the current $i(t)$ "passes through" the capacitor is discussed in field-theory textbooks.] *Capacitance* is defined by the equation $q = Ce$, or, differentiating with respect to time and using Eq. 1.2-1, $i = C(de/dt)$.

The expression $i = dq/dt = C(de/dt)$ may be rewritten as $e = q/C = (1/C) \int i\, dt$. The last equation is more easily interpreted, however, as a definite integral, by arguments similar to those used in developing Eqs. 1.2-5 and 1.2-6. Since $i\, dt = C\, de$,

$$\int_{t_1}^{t_2} i\, dt = \int_{t=t_1}^{t=t_2} C\, de$$

$$= C[e(t_2) - e(t_1)]$$

Or if t_1, t_2, and t are replaced by 0, t, and λ, respectively,[†]

$$e(t) = e(0) + \frac{1}{C} \int_0^t i(\lambda)\, d\lambda \qquad (1.3\text{-}2)$$

An alternative but equivalent equation for the voltage across a capacitance is

$$e(t) = \frac{1}{C} \int_{-\infty}^t i(\lambda)\, d\lambda \qquad (1.3\text{-}3)$$

where $\int_{-\infty}^t i(\lambda)\, d\lambda$ symbolizes the net charge delivered to the capacitance by all the current that has ever flowed into or out of it from its birth up to the time t. This integral may be broken up into two parts as follows:

$$e(t) = \frac{1}{C} \int_{-\infty}^0 i(\lambda)\, d\lambda + \frac{1}{C} \int_0^t i(\lambda)\, d\lambda = e(0) + \frac{1}{C} \int_0^t i(\lambda)\, d\lambda$$

which is the same as Eq. 1.3-2. In practice, the entire previous history of a capacitance is never known, but its voltage at $t = 0$ may be known; consequently, Eq. 1.3-2 is normally the more useful expression.

The waveshapes of voltage and current for a capacitance are not, in general, the

---

† Equation 1.3-2 is still valid if 0 is changed to some other time reference t_0.

ImportanT :   For Capacitance !

same. The current waveshape may be found by differentiating the waveshape of the voltage. Conversely, the voltage waveshape may be determined by integrating the current waveshape.

When a given function consists of straight lines, the differentiation or integration can be easily done graphically. The derivative of a function is the slope of the curve, while $\int_0^t f(\lambda)\, d\lambda$ is the net area underneath the curve from time zero to time t. In Fig. 1.3-5, the student should immediately be able to draw the functions in one column when those in the other column are given. Notice that integration tends to smooth out a waveshape, while differentiation tends to introduce discontinuities.

Figure 1.3-6a shows a possible pair of waveshapes for the voltage and current of a capacitance. If one of the two is given, and if $e(0) = 0$, the other immediately follows.

Equations 1.2-2 and 1.2-5 can be used to plot curves of the power and energy, and

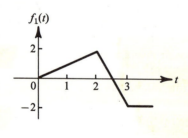
$f_1(t) = voltage, e(t)$

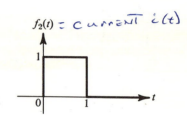
$f_2(t) = current\ i(t)$

(a)

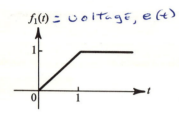
$f_1(t)$

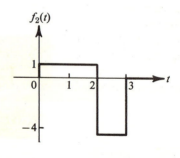
$f_2(t)$

(b)

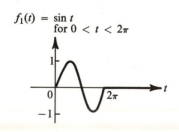
$f_1(t) = \sin t$ for $0 < t < 2\pi$

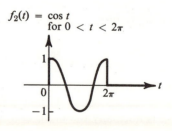
$f_2(t) = \cos t$ for $0 < t < 2\pi$

(c)

$$f_1(t) = \int_0^t f_2(\lambda)\, d\lambda \qquad\qquad f_2(t) = \frac{df_1(t)}{dt}$$

Figure 1.3-5

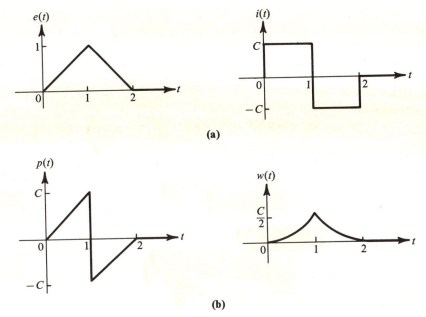

Figure 1.3-6

to derive a general expression for the energy supplied to a capacitance. Plots of power and energy for the waveshapes in Fig. 1.3-6a are given in Fig. 1.3-6b, if $w(0)$ is zero. For $1 < t < 2$, the sign of $p(t)$ is negative, so that power is being delivered *from* the capacitance. Note, however, that $w(t)$ is always nonnegative. Physically, energy is received by the capacitance and stored in its electric field for $0 < t < 1$, and this energy is returned to the rest of the circuit for $1 < t < 2$.

The power delivered to a capacitance can be written as

$$p(t) = e(t)i(t) = Ce(t)\frac{de(t)}{dt}$$

It will be assumed that at some time sufficiently far in the past (denoted by t_0), no voltage, current, or energy had yet been delivered, so $e(t_0) = w(t_0) = 0$. By Eq. 1.2-5, the total energy received by the capacitance up to time t is

$$w(t) = \int_{t_0}^{t} Ce(\lambda)\frac{de(\lambda)}{d\lambda}\,d\lambda = C\int_{\lambda=t_0}^{\lambda=t} e(\lambda)\,de(\lambda)$$

$$= \tfrac{1}{2}C\Big[e^2(\lambda)\Big]_{\lambda=t_0}^{\lambda=t} = \tfrac{1}{2}Ce^2(t)$$

since the lower limit $e^2(t_0)$ vanishes. If the t in parentheses in the final result is omitted, the equation becomes

$$w = \tfrac{1}{2}Ce^2 \tag{1.3-4}$$

Since C is assumed to be nonnegative, the total net energy supplied to it up to some time t is also nonnegative; therefore, the element is passive.

An inductor is a physical device that consists of a number of turns of coiled, conducting wire. A current passing through the wire produces a magnetic field, which is capable of storing energy. If the magnetic field changes with time, it induces a voltage in the coils. The external behavior of this device is approximated by the element *self-inductance* (or *inductance*), which is defined by $e = L(di/dt)$. Inductance is said to be the *dual* of capacitance, since the defining equations are essentially the same, except that the roles of $e(t)$ and $i(t)$ are interchanged.

The integral $i(t) = (1/L) \int e(t)\, dt$ is interpreted as

$$i(t) = \frac{1}{L} \int_{-\infty}^{t} e(\lambda)\, d\lambda \tag{1.3-5}$$

or, equivalently and more clearly, as

$$i(t) = i(0) + \frac{1}{L} \int_{0}^{t} e(\lambda)\, d\lambda \tag{1.3-6}$$

The quantity $\int e(t)\, dt$ represents flux linkages and is discussed in Chapter 7. Figure 1.3-7a shows a consistent pair of voltage and current waveshapes for an inductance if $i(0) = 0$. Although the form of the integrals in Table 1.3-1 is the one most frequently seen—because it is the simplest to write—the integrals should be interpreted as in Eqs. 1.3-5 and 1.3-6.

Power and energy calculations for the inductance are carried out by using Eqs. 1.2-2 and 1.2-5. Figure 1.3-7b shows the curves of power and energy for the waveshapes in Fig. 1.3-7a if $w(0) = 0$. When a general expression for the energy

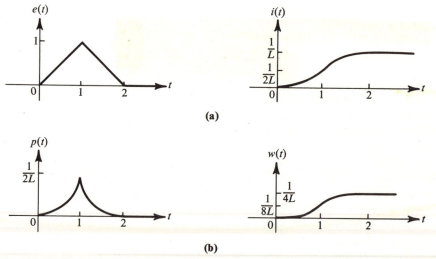

(a)

(b)

Figure 1.3-7

stored in an inductance is derived, it is assumed that at some time t_0, $w(t)$ and $i(t)$ are both zero. Then

$$w = \tfrac{1}{2}Li^2 \tag{1.3-7}$$

The details of this derivation are similar to the arguments used in the derivation of Eq. 1.3-4. Since L is a real, positive constant, $w \geq 0$, and the inductance is passive.

All three of the elements defined in Table 1.3-1 are linear and time invariant. For the resistance, doubling the current doubles the voltage, and if the current consists of two components—e.g., if $i(t) = i_1(t) + i_2(t)$—the voltage $e(t) = R[i_1(t) + i_2(t)] = Ri_1(t) + Ri_2(t)$ is the sum of the voltages produced by each component considered separately. The resistance is time invariant, since delaying the current by t_0 seconds delays the voltage waveshape but does not affect its size or shape. The relationships $i = C(de/dt)$ and $e = L(di/dt)$ are similar (except for a multiplying constant) to the equation $y = dx/dt$, which was discussed in Examples 1.1-1 and 1.1-3.

As has been discussed already, the elements in Table 1.3-1 are passive if R, C, and L are positive constants. A resistance immediately dissipates any energy received, but a capacitance and an inductance store the energy received (in the electric and magnetic fields associated with the physical devices) and can later return it to the circuit.

Inductances are often given in millihenrys (mh) or microhenrys (μh), while capacitances are commonly expressed in microfarads (μf) or picofarads (pf). A 1 f capacitance, for example, is very rare. However, to prevent the numerical work from obscuring the significance of the steps in the solution, values of L and C in the neighborhood of 1 h and 1 f are frequently used in the problems. Furthermore, as discussed in Chapter 11, scaling techniques can always be used to convert circuits with elements of practical size into ones with element sizes that are easier to handle mathematically.

The discussion of another passive circuit element, mutual inductance, is deferred until Chapter 7. Unlike the circuit elements in Table 1.3-1, mutual inductance cannot be characterized by a two-terminal model. Instead, it is one of three parameters characterizing a four-terminal component known as the *transformer*.

Sources. Each element in Table 1.3-1 is defined by a relationship between $e(t)$ and $i(t)$, so that when one of these variables has been determined the other may be calculated. The behavior of many components, however, cannot be described by the elements discussed so far. *Sources* are two-terminal elements for which there is no direct voltage-current relationship. Therefore, when one of the two variables is given, the other cannot be found without a knowledge of the rest of the circuit. (Once again, the sources defined here are ideal elements that only approximate physical devices.)

Independent sources are elements for which either the voltage or current is always given. The independent *voltage source* shown in Fig. 1.3-8a has a voltage $e(t)$ that is a specified function of time and that is independent of any external connections. The current $i(t)$ does depend upon the external connections and may flow in either direction. If at some instant the values of $e(t)$ and $i(t)$ are both positive, the source

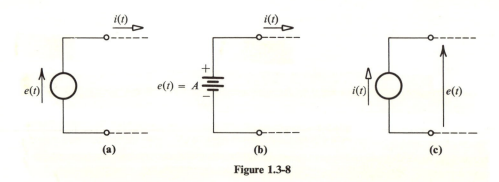

Figure 1.3-8

is supplying and not absorbing power (since the current arrow leaves the terminal toward which the voltage arrow is drawn). The current may take on any value, so the voltage source is theoretically capable of supplying an unlimited amount of power and energy to the rest of the circuit.

A battery is an independent voltage source that provides a nearly constant voltage, regardless of the current drawn from it, so it can be approximated by a voltage source of constant value. When $e(t)$ is a constant and not a function of time, the source is sometimes drawn as in Fig. 1.3-8b, where the short lines suggest the plates of a battery. A voltage source of value zero is equivalent to the short circuit shown in Fig. 1.3-3a.

The independent *current source* shown in Fig. 1.3-8c produces a specified current $i(t)$ regardless of the external connections. Since the voltage across the terminals

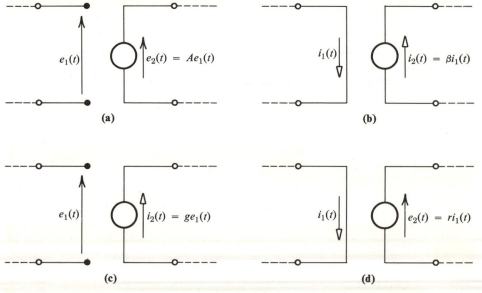

Figure 1.3-9

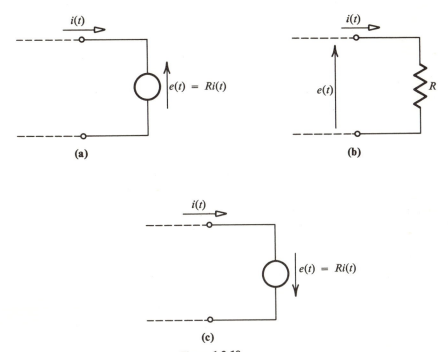

(a)   **(b)**

(c)

Figure 1.3-10

depends upon the rest of the circuit and may have any value, the independent current source can also supply an unlimited amount of power and energy. The student may be more familiar with voltage than with current sources, but many devices are represented by a current source in combination with some passive elements. A current source of value zero is equivalent to the open circuit shown in Fig. 1.3-3b.

The term *dead source* is often used to denote a source of value zero. Thus, a dead voltage source is a short circuit and a dead current source is an open circuit.

A *controlled source* is a source whose value is not independent of the rest of the circuit, but is instead a known function of some other voltage or current. Such sources are particularly important in constructing models of electronic devices. For example, Figs. 1.3-9a and 1.3-9b are very simple models for certain vacuum-tube and transistor amplifiers, respectively. Figure 1.3-9c shows a current source $i_2(t)$ that is controlled by the voltage $e_1(t)$, and Fig. 1.3-9d represents a current-controlled voltage source. In one sense, a controlled source is not a true two-terminal component, since the voltage or current upon which it depends (which may be located anywhere) must also be shown in order to completely characterize it.

The controlled source shown in Fig. 1.3-10a is equivalent to the resistance in Fig. 1.3-10b, since both have the same terminal voltage and current. If the "polarity" of the controlled source in Fig. 1.3-10a is reversed, as in Fig. 1.3-10c, the element corresponds to a negative resistance and supplies power to the rest of the circuit.

1.4 *Kirchhoff's Laws*

MON.
(Feb. 12)

Figure 1.4-1a shows a circuit composed of five two-terminal elements (or *branches*). The junction of two or more elements is called a *node* (or *vertex*). Normally, however, the solid dots in the four corners of the circuit are used only at the junction of three or more wires, as in Fig. 1.4-1b (which is exactly equivalent to the original circuit). Since all the connecting wires are assumed to be short circuits, all points on the bottom line in Fig. 1.4-1b are at the same potential, and the bottom line is treated as a single node.

When a circuit is drawn on a plane surface, it is sometimes necessary to distinguish between wires that meet at a junction and wires that appear to cross but yet do not make an electrical connection. By conventional notation, the wires in Fig. 1.4-2a do have a common junction, while those in Fig. 1.4-2c do not. Most people regard Fig. 1.4-2b as equivalent to Fig. 1.4-2a, but others as equivalent to Fig. 1.4-2c, so we shall avoid using it.

The effect of interconnecting different elements can be described by two laws proposed by Gustav Robert Kirchhoff, a German physicist. Kirchhoff assumed

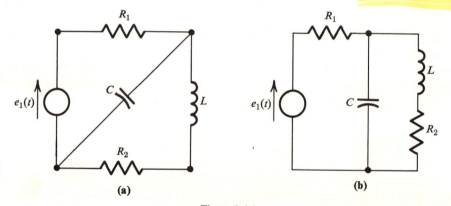

(a)   (b)

Figure 1.4-1

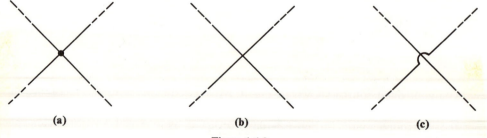

(a)   (b)   (c)

Figure 1.4-2

that no net charge can accumulate within an element (as already discussed in connection with Fig. 1.2-2a) or at a node, and that charge must be conserved. Thus, any charge entering a node must immediately leave by some other path; therefore, the total current entering the node must equal the total current leaving it. In Fig. 1.4-3a, for example,

$$i_1(t) + i_4(t) = i_2(t) + i_3(t)$$

As discussed in Section 1.2, a reference arrow has been drawn for each current to indicate the assumed positive direction. At a given instant, however, some or all of the currents may be negative, indicating that positive charge is actually moving in the direction opposite to that of the arrow.

A current $i_1(t)$ entering a junction is the same as a current $-i_1(t)$ leaving the junction, so (a) and (b) of Fig. 1.4-3 are equivalent. The sum of the four currents in Fig. 1.4-3b must be zero; therefore,

$$-i_1(t) + i_2(t) + i_3(t) - i_4(t) = 0$$

which is, of course, the same as the previous equation. In summary, *Kirchhoff's current law* states that the *algebraic* sum of the currents leaving any node is zero, which is indicated symbolically by

$$\sum i(t) = 0 \qquad\qquad (1.4\text{-}1)$$

In applying Eq. 1.4-1, a current entering the node is to be interpreted as a negative current leaving.

Each of the gray rectangles in Fig. 1.4-4 represents a circuit element. If Eq. 1.4-1 is applied successively to each of the four nodes,

$$-i_1(t) - i_2(t) + i_3(t) = 0$$
$$-i_3(t) - i_4(t) + i_5(t) = 0$$
$$i_1(t) - i_5(t) - i_6(t) = 0$$
$$i_2(t) + i_4(t) + i_6(t) = 0$$

In order that the last equation may be satisfied, at least one of the three currents $i_2(t)$, $i_4(t)$, and $i_6(t)$ must be negative.

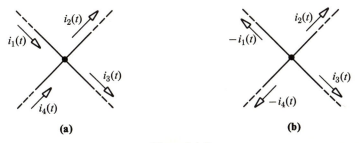

(a)

(b)

Figure 1.4–3

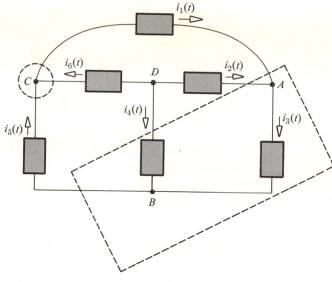

Figure 1.4-4

Kirchhoff's current law may also be applied to a closed surface surrounding any part of a circuit.[†] Since there is no place for any charge to accumulate within the surface, the *algebraic* sum of the currents leaving it must be zero. Applied to the dashed rectangle in Fig. 1.4-4, the law yields

$$-i_1(t) - i_2(t) - i_4(t) + i_5(t) = 0$$

Summing the currents leaving the dashed circle in the figure is equivalent to summing the currents leaving node C.

From the definition of voltage in Section 1.2, the work required to move a unit

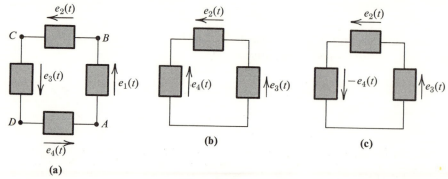

Figure 1.4-5

[†] The closed surface cannot pass *through* an element, since in circuit theory the elements are assumed to be indivisible.

of positive charge from A to B in Fig. 1.4-5a is $e_1(t)$. The work required to move the charge from B to C is $e_2(t)$, and that required to move it completely around the loop and back to the starting point is $e_1(t) + e_2(t) + e_3(t) + e_4(t)$. If energy is conserved, this sum must be zero.[†] In general, *Kirchhoff's voltage law* states that the algebraic sum of the voltages around any closed path is zero, which is written as

$$\sum e(t) = 0 \qquad (1.4\text{-}2)$$

From Fig. 1.2-2c, we observe that reversing a reference arrow reverses the sign of the corresponding voltage; therefore, Figs. 1.4-5b and 1.4-5c are equivalent. If a voltage arrow is directed in opposition to the direction in which the closed path is traversed, the corresponding term in Eq. 1.4-2 is therefore negative.[††] For path $BADB$ in Fig. 1.4-6,

$$e_3(t) + e_2(t) - e_4(t) = 0$$

while for $BDCB$ and $ACDA$,

$$e_4(t) - e_6(t) + e_5(t) = 0$$

and

$$e_1(t) + e_6(t) - e_2(t) = 0$$

Among the other closed paths that may be selected are $BACB$ and $BDACB$, for which

$$e_3(t) + e_1(t) + e_5(t) = 0$$

$$e_4(t) - e_2(t) + e_1(t) + e_5(t) = 0$$

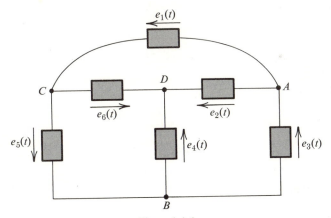

Figure 1.4-6

[†] In field theory, the work required to move a unit charge around a closed loop in space in the presence of a time-varying magnetic field is not necessarily zero because the mechanism that produced the field may be a source of energy. In circuit theory, it is assumed that there are no sources of energy that have not been explicitly represented by a lumped element in the circuit.

[††] Equations 1.4-1 and 1.4-2 certainly remain valid if the sign of every term is reversed. Reversing the sign of every term corresponds to associating positive signs with currents *entering* a node and with voltage arrows directed *opposite* to the direction in which a closed path is traversed, instead of the procedure followed in this section.

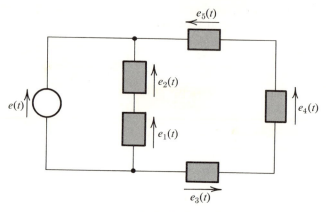

Figure 1.4-7

As a corollary to Eq. 1.4-2, the voltage between any two points is uniquely defined. Since $e_1(t) + e_2(t) - e(t) = 0$ for the left-hand loop in Fig. 1.4-7,

$$e(t) = e_1(t) + e_2(t)$$

For the outside loop, $e_3(t) + e_4(t) + e_5(t) - e(t) = 0$, and

$$e(t) = e_3(t) + e_4(t) + e_5(t)$$

Thus, the source voltage equals the algebraic sum of the voltages in the middle leg, as well as the algebraic sum of the voltages in the right-hand leg. Notice that $e_2(t) = e(t) - e_1(t)$, a result that follows from the above equations, but which the student should learn to deduce directly from inspection of the circuit.

Both Kirchhoff's laws and the definitions of Table 1.3-1 are needed to solve circuit problems. Several simple yet practical circuit problems are solved in the following examples. The student is not expected to have had any previous acquaintance with the devices represented in these problems. Furthermore, although he should thoroughly understand all the steps in the given solutions, he may not yet be able to solve such problems efficiently without some guidance. In the succeeding chapters we discuss in detail some useful, systematic methods of attack.

Example 1.4-1. The I and R in Fig. 1.4-8a represent the inductance and resistance of the horizontal deflection coil in an oscilloscope. If $L = 5$ h and $R = 1 \Omega$, and if the current has the waveshape shown in Fig. 1.4-8b, what must be the waveshape of the source voltage?

Solution. From Kirchhoff's voltage law and the definitions of e_R and e_L,

$$e(t) = e_R(t) + e_L(t) = Ri + L\frac{di}{dt}$$

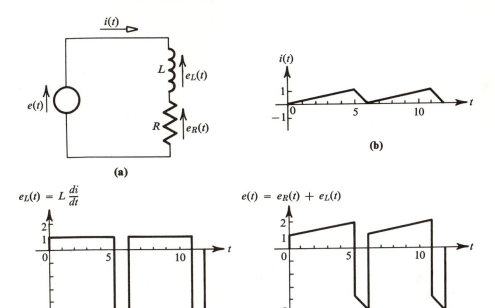

Figure 1.4-8

Then, since $R = 1\,\Omega$ and $L = 5$ h,

$$e(t) = i + 5\,\frac{di}{dt}$$

The voltage $e_L = 5(di/dt)$ is shown is Fig. 1.4-8c, and the required source voltage $e(t)$ is plotted in (d).

Example 1.4-2. Let the voltage $e(t)$ in Fig. 1.4-8a be zero for $t < 0$ and unity for $t > 0$, as plotted in Fig. 1.4-9a. Assume that the current waveshape is a continuous function of time and that $i(t) = 0$ for $t \le 0$. Find and sketch $i(t)$ for $t > 0$.

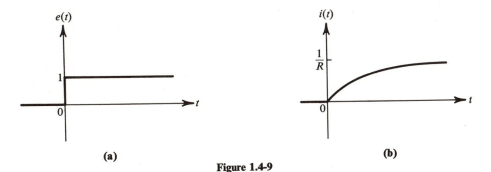

Figure 1.4-9

Solution. The equation $e(t) = Ri + L(di/dt)$ becomes

$$L\frac{di}{dt} + Ri = 1$$

for $t > 0$. The systematic solution of this first-order differential equation is discussed in Chapter 3. For the present, we merely note that the expression

$$i(t) = \frac{1}{R}(1 - K\epsilon^{-(R/L)t})$$

does satisfy the above equation. Since the current is assumed to be zero at $t = 0$ and to be a continuous function, $K = 1$ and

$$i(t) = \frac{1}{R}(1 - \epsilon^{-(R/L)t})$$

for $t > 0$. The current waveshape is sketched in Fig. 1.4-9b. (Whenever the given circuit contains inductances or capacitances, it may be necessary to solve a differential equation in order to find the desired response.)

Example 1.4-3. The part of the circuit enclosed by the dashed line in Fig. 1.4-10 is a model of a negative impedance converter. Find the current $i_1(t)$.

Solution. By Kirchhoff's current law, $i_3(t) = -i_1(t)$, so $e_1(t) = e_3(t) = Ri_3(t) = -Ri_1(t)$, and $i_1(t) = -e_1(t)/R$. Notice that this same current would result if a negative resistance of $-R$ ohms were placed directly across the voltage source. If the resistance R in the figure is replaced by a capacitance C or inductance L, a similar proof shows that the voltage source sees a negative capacitance or inductance, respectively.

Example 1.4-4. Find a differential equation that relates the input voltage $e_1(t)$ and the output voltage $e_o(t)$ in Fig. 1.4-11. To what does this result reduce when the

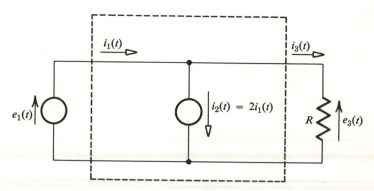

Figure 1.4-10

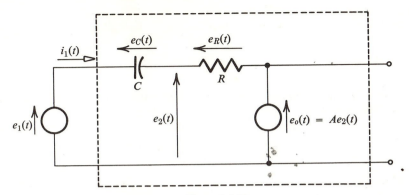

Figure 1.4-11

"amplification factor" A of the voltage-controlled voltage source approaches infinity?
Suggest a name for the circuit enclosed by the dashed line.

Solution.   The current through the capacitance is

$$i_1(t) = C\,\frac{de_C}{dt} = C\,\frac{d(e_1 - e_2)}{dt}$$

Since the same current flows through the resistance,

$$i_1(t) = \frac{e_R(t)}{R} = \frac{1}{R}\,[e_2(t) - e_o(t)]$$

Equating these expressions and eliminating $e_2(t)$ by using $e_2(t) = e_o(t)/A$, we obtain

$$C\,\frac{de_1}{dt} - \frac{C}{A}\frac{de_o}{dt} = \frac{e_o(t)}{AR} - \frac{e_o(t)}{R}$$

If A approaches infinity, $C(de_1/dt) = -e_o(t)/R$, or

$$e_o(t) = -RC\,\frac{de_1}{dt}$$

For sufficiently large values of A, the output voltage is proportional to the derivative
of the input, and the circuit is therefore called a differentiator.
   Notice that the circuit behaves like a differentiator when

$$\left|\frac{C}{A}\frac{de_o}{dt}\right| \ll \left|\frac{e_o(t)}{R}\right| \quad \text{and} \quad \left|\frac{e_o(t)}{AR}\right| \ll \left|\frac{e_o(t)}{R}\right|$$

The second inequality is always satisfied if $A \gg 1$, but the first involves the rate of
change of $e_o(t)$. If the input and output voltages change rapidly with time, the magnitude of de_o/dt becomes large. Thus, the circuit may differentiate some inputs very
well, but may not be a satisfactory differentiator for rapidly changing inputs.

1.5

Representing Physical Devices by Models

There are two basic steps used in the analyses of physical circuits. First, a model that approximates the behavior of the physical system is devised from the ideal elements defined in Section 1.3. Second, this model is solved by means of the mathematical definitions of the elements and Kirchhoff's laws. Several examples of the second of these two steps were given at the end of Section 1.4. In this section, the selection of a model for physical devices is discussed, and an explanation given of why the following chapters do not treat this important subject in any depth.

Inductors, resistors, capacitors, and sources can sometimes be approximated by one of the ideal elements defined in Section 1.3. In other applications, however, even these devices should be represented by the interconnection of several elements, as discussed in the following examples. The models for many other components, such as transistors, nearly always include several elements.

The choice of a model may be based upon the movement of electrons and ions within the device. In physical electronics, for example, the models for transistors and vacuum tubes are commonly related to the internal physical phenomena. Unfortunately, many students reading this book will not yet have sufficient experience to be able to do this. Another approach to the modelling problem is to observe empirically the external behavior of the physical device and to devise a model having similar characteristics. This method requires considerable appreciation for the effect of connecting elements in different ways, and hence cannot be very effective until *after* the student has completed a course in circuit theory.

The external behavior of several different models may be identical, in which case any one of them could be used for a particular problem. A knowledge of the most general model for a physical component with a given number of external terminals can be very helpful when a model for a new device is selected. These matters are discussed later in the book.

For the foregoing reasons, we are now unable to justify adequately the models presented in this book. In the remaining chapters, the model to be used will be included as part of the problem statement. After the student has completed this and related courses and obtained some laboratory experience, he will be in a much better position to attack the modelling problem.

In studying the examples given below, several things should be kept in mind. If the model selected is too simple, it may not adequately approximate the behavior of the physical device, and the calculated response of the circuit may differ appreciably from the observed output. An unnecessarily complicated model will, however, increase the difficulty of obtaining an analytical solution. Furthermore, the form of the solution may be so complicated that its salient features may not be readily recognized. A compromise in the complexity of the model selected usually must be made.

The choice of a model is made even more difficult by the fact that a model that is suitable under some circumstances may not be satisfactory when the device is used under different conditions. Its suitability depends not only upon the desired accuracy of the calculations, but also upon the external connections to the device and upon the nature of the input. Some parts of the model may be unnecessary for one input, but essential when the input is changed.† Although it is dangerous to generalize at this time, many devices require a more complicated model when the input changes rapidly with time.

Example 1.5-1. Discuss the models for an inductor and a capacitor.

Solution. An inductor usually consists of a number of turns of closely coiled, conducting wire. A simple model that often is satisfactory is the single inductance shown in Fig. 1.5-1a. Since any wire has some resistance, a more accurate model is the one shown in Fig. 1.5-1b. Finally, since there is some capacitance between adjacent turns, an even more accurate model is the one shown in Fig. 1.5-1c.

The resistance of the insulating layer (or *dielectric*) between the plates of the capacitor in Fig. 1.3-4 is very large, but still finite. Although the single capacitance in Fig. 1.5-2a usually is a satisfactory model, the resistance of the insulation can be accounted for by the leakage resistance R_1 included in (b). The two wires connected to the plates have some resistance. Furthermore, even a straight, conducting wire has some inductance associated with it. Thus, an even more complete model of a capacitor is the one shown in Fig. 1.5-2c.

The inexperienced student is often surprised to discover that under extreme conditions a device labelled "inductor" may behave more like a capacitance, and a device marked "capacitor" more like an inductance. The models in Figs. 1.5-1c and 1.5-2c indicate why this is sometimes possible.

Example 1.5-2. Discuss the model for a battery, whose purpose is to produce a constant voltage of A volts between its two terminals.

Solution. The basic model is an independent voltage source of constant value, as shown in Figs. 1.3-8b and 1.5-3a. In most batteries, however, the terminal voltage

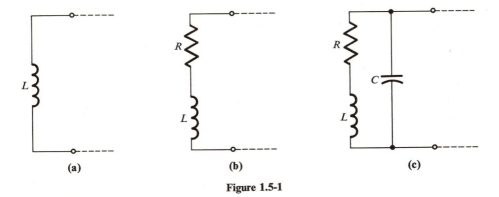

(a)   **(b)**   **(c)**

Figure 1.5-1

---

† One good illustration of this is given in Example 6.5-7.

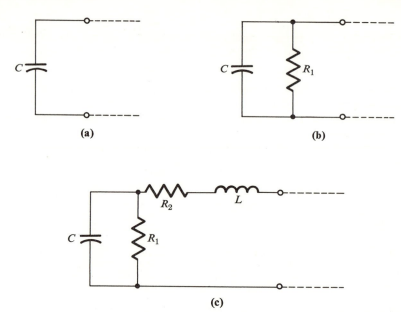

Figure 1.5–2

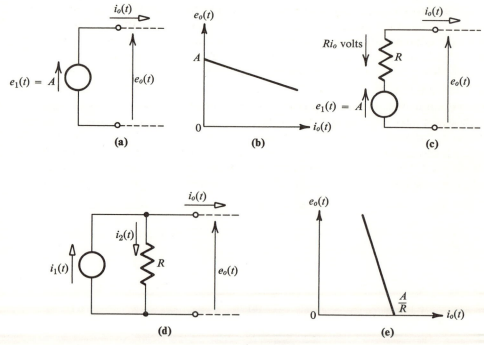

Figure 1.5-3

decreases when the external current increases, as indicated in Fig. 1.5-3b. For the
model in (c), the terminal voltage is

$$e_o(t) = A - Ri_o(t)$$

which agrees with this observed behavior. The resistance R is called the *internal
resistance* of the battery.

The model in Fig. 1.5-3c is not the only one that has the characteristics plotted in
(b). In the circuit in Fig. 1.5-3d, $i_2(t) = i_1(t) - i_o(t)$ and $e_o(t) = Ri_2(t) = Ri_1(t)$
$- Ri_o(t)$. If the value of the independent current source is A/R amperes,

$$e_o(t) = A - Ri_o(t)$$

Since the models in Figs. 1.5-3c and 1.5-3d then give the same relationship between
$e_o(t)$ and $i_o(t)$, they may be used interchangeably. If the terminal voltage is almost
constant—i.e., if the line in (b) is almost horizontal—the model in (c) is more common.
For a device that produces an almost constant current, as suggested by (e), the circuit
in (d) would probably be used. Nevertheless, the two models are equally valid, and
one may always be replaced by the other when $e_o(t)$ and $i_o(t)$ are calculated. This is
true even if the sources are functions of time, as long as $i(t) = e(t)/R$.

The connection of idealized elements in certain ways may lead to circuits that
contain a contradiction and that have no solution. Suppose, for example, that the
two sources in Fig. 1.5-4 are connected by closing the switch K and that $A \neq B$.
The value of an independent source is unaffected by the external connections, but
according to Kirchhoff's voltage law, the voltage across each source must be the
same. The contradiction arises because the model is too crude for the corresponding
physical situation, even if there were some reason for connecting dissimilar batteries.
If the resistance of the connecting wires and the internal resistance of the physical
sources are considered, the contradiction disappears.

Some physical devices have more than two external terminals. Among the most
common three-terminal components are the vacuum tube and the transistor. The
symbol for a triode vacuum tube is shown in Fig. 1.5-5a, where the letters g, p, and
k denote the grid, plate, and cathode terminals. In many applications, the tube acts
like a linear device and can be represented by a model composed of the elements in
Section 1.3. Usually, the current entering the grid terminal is negligibly small, so a
crude model need not contain any elements connected to the grid. The voltage of

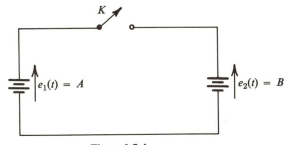

Figure 1.5-4

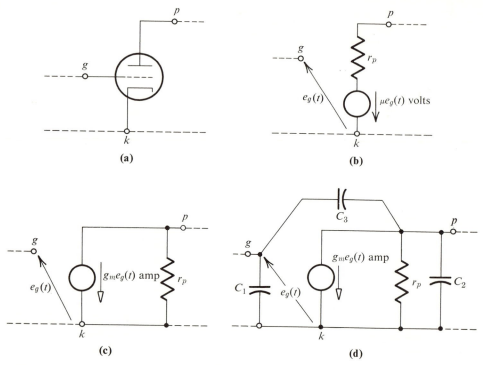

Figure 1.5-5

the cathode with respect to the plate is observed to be roughly proportional to the cathode-to-grid voltage $e_g(t)$, but it decreases as the current flowing into the plate increases. Based on Example 1.5-2, a combination somewhat similar to Fig. 1.5-3c or 1.5-3d might be expected between the cathode and plate terminals. The source is not an independent source, however, but is controlled by the cathode-to-grid voltage. Two models that are often used are shown in Figs. 1.5-5b and 1.5-5c. The symbol μ, a dimensionless constant, is called the *amplification factor*. The plate-to-cathode voltage is μ times the cathode-to-grid voltage *if* the plate current is zero. A typical value of μ for a triode is 20. The symbols r_p and g_m denote the plate resistance and the *transconductance*, respectively. Typical values are $r_p = 10$ kilohms and $g_m = 2$ millimhos. (The use of a lower case rather than a capital letter for the plate resistance is discussed in Chapter 12.) From the discussion in Example 1.5-2,

$$g_m = \frac{\mu}{r_p}$$

in order for the models in Figs. 1.5-5b and 1.5-5c to be equivalent.

The grid, plate, and cathode are metallic electrodes placed fairly close together within an evacuated envelope. Consequently, there is a small but measurable capacitance between any two electrodes. This effect may be represented by the three

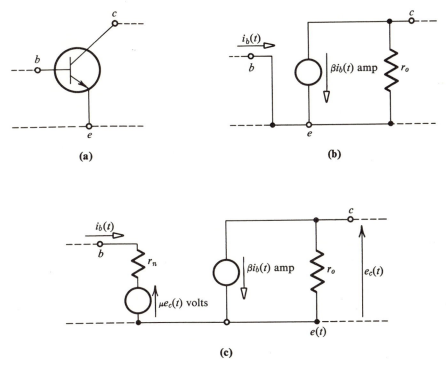

Figure 1.5-6

capacitances in the more complicated model of Fig. 1.5-5d. Any capacitance that may exist between the connecting wires should also be included in the values of C_1, C_2, and C_3.

One symbol for the transistor is given in Fig. 1.5-6a, where b, e, and c denote the base, emitter, and collector terminals. Like the vacuum tube, this component may be operated in either a linear or nonlinear manner. In the linear range, the size of the collector current is roughly proportional to the base current, but also depends upon the emitter-to-collector voltage. If the voltage between the emitter and base, which is usually small, is assumed to be zero, the model in Fig. 1.5-6b may be used. A typical value for the amplification factor β of the current-controlled current source is 50. The small voltage between the emitter and base depends upon the base current $i_b(t)$ and, to a lesser extent, upon the emitter-to-collector voltage. A more exact model that includes these effects is shown in (c). The transistor also possesses some internal capacitance, which may have to be included in some applications, but which is not shown in the figure.

The most commonly used models for the inductor, capacitor, triode, and transistor are those in Figs. 1.5-1a, 1.5-2a, 1.5-5b and 1.5-5c, and 1.5-6c [with the controlled source $\mu e_c(t)$ often omitted]. The additional elements in the more complicated models represent phenomena that are not, in most cases, desirable, but which cannot be entirely avoided. These additional elements cannot always be ignored and are sometimes called *parasitic elements*.

In devices whose models contain controlled sources, the power associated with the output signal is frequently larger than the power of the input signal, and the device continually supplies a net power to the rest of the circuit. The controlled source is an active circuit element and can account for this net power. The mechanism that provides this power for transistors and vacuum tubes and a much more complete justification of their models is found in Chapter 12.

1.6

Summary

The fundamental electrical quantities are current, voltage, power, and energy. In this textbook the calculation of these quantities for circuit models that are composed of idealized elements is considered. The passive elements of resistance, capacitance, and self-inductance approximate the behavior of some common two-terminal devices. The models for transistor and vacuum-tube circuits always include controlled sources and are more fully discussed in Chapter 12. Externally applied signals can be represented by independent sources.

In Chapter 7 an additional element, mutual inductance, is introduced. However, in the latter part of that chapter, it is shown that the effect of mutual inductance can be represented by a combination of the elements defined in Section 1.3.

Models that are composed of the above-mentioned elements are linear and time invariant. Multiplying the input signal by a constant multiplies the output by the same constant. If several inputs are applied, the total response is the sum of the individual responses to each input considered separately.

The effect of interconnecting elements is described by Kirchhoff's laws. Together with the definitions of the elements, they are sufficient for the analysis of any model and can be used to explain and predict the behavior of circuits. If the analytical solution does not agree with experimental observations, the model selected must have been inappropriate.

The remainder of this book is based upon the definitions and laws presented in this chapter. We now wish to study the application of these laws to the analysis of any linear and time-invariant circuit, the systematic solution of the resulting equations, and the significance of the solutions. The next six chapters are primarily concerned with the circuit's effect on the size and shape of incoming signals. Extensive power calculations are deferred until Chapter 8.

PROBLEMS

1.1   Figure 1.1-3 shows a plot of the properties of a metallic shaft under tension, but the properties of the shaft under compression are not included. Is the range of linearity for a typical shaft given by $0 < f < f_c$ or $|f| < f_c$ or neither?

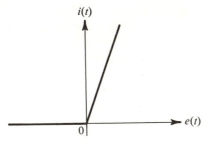

Figure P1.3

1.2   Are the three systems described by the following equations linear and time invariant? Explain your answers. Assume that the systems are initially at rest.

(a) $y = \dfrac{dx}{dt} + 2x$ (b) $\dfrac{dy}{dt} = 3x$ (c) $y\dfrac{dy}{dt} = x$

1.3   A diode is an electrical component, the approximate characteristics of which are shown in Fig. P1.3. If the voltage is allowed to take on both positive and negative values, is the diode a linear device? If the voltage is restricted to only positive values or only negative values, what circuit element does the diode resemble?

1.4   Find the net energy supplied to the network N in Fig. P1.4 during the time interval $0 < t < 4$.

1.5   If $e(t) = 2 \cos 10t$ and $i(t) = 4 \cos 10t$ in Fig. 1.2-2b, give an expression for the power supplied to the circuit element. How much energy is supplied for $0 < t < 2\pi/10$? What is the average power supplied during this time? Repeat the problem with $i(t) = 4 \sin 10t$.

1.6   The capacitance of a parallel-plate capacitor is given by $C = \epsilon A/d$, where A and d denote the area and the separation of the plates. The permittivity (or dielectric constant) ϵ equals $10^{-9}/36\pi$ farads per meter for air and $10^{-9}/6\pi$ farads per meter for mica. The dielectric strength (the maximum field intensity the dielectric can withstand without breakdown) is approximately 3×10^6 volts per meter for air and 200×10^6 volts per meter for mica. Design and discuss a 1 μf parallel-plate capacitor that can withstand 1 kv. Repeat for a 1 f capacitor.

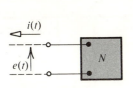 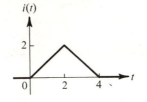 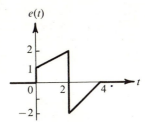

Figure P1.4

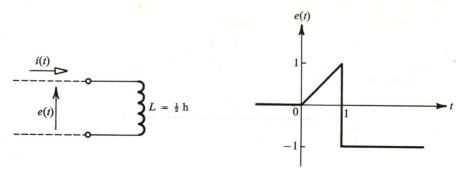

Figure P1.8

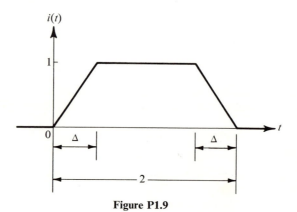

Figure P1.9

Figure P1.10

1.7 Find and sketch the current $i(t)$ for a capacitance C whose voltage for $t > 0$ is

(a) $e(t) = \sin 10t$ (b) $e(t) = t$ (c) $e(t) = te^{-t}$

Assume that the reference arrows have the same direction as in Table 1.3-1 and that the voltage and current are zero for $t < 0$. Repeat for an inductance L.

1.8 Draw a curve of $i(t)$ versus t for the circuit in Fig. P1.8. Also sketch curves of the power delivered to the inductance and the net energy supplied.

1.9 Figure P1.9 shows the current $i(t)$ flowing through an inductance L. Sketch $e(t)$ when $\Delta = \frac{1}{2}$ and when $\Delta = \frac{1}{4}$. Discuss what happens if the dimension Δ approaches zero. (This limiting case is discussed more fully in Chapter 3.)

1.10 The switch K in the circuit in Fig. P1.10 closes at $t = 0$. If all voltages and currents are zero for $t < 0$, sketch $e_2(t)$ and $e_1(t)$ for $t > 0$.

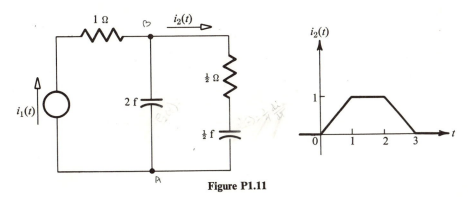

Figure P1.11

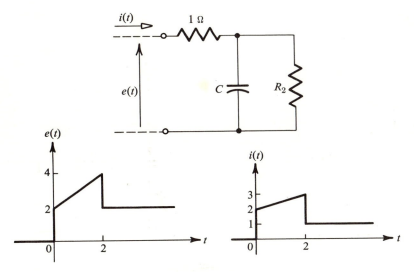

Figure P1.12

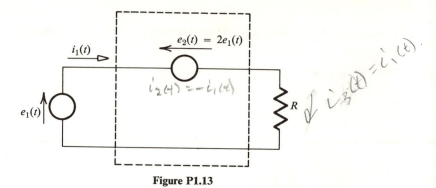

Figure P1.13

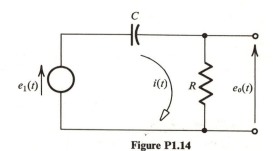

Figure P1.14

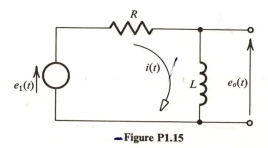

Figure P1.15

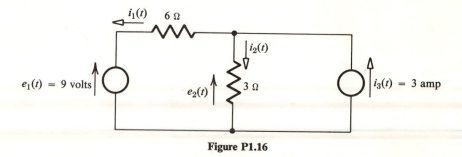

Figure P1.16

1.11 What must be the waveshape of the current $i_1(t)$ in Fig. P1.11 in order to produce the waveshape given for $i_2(t)$?

1.12 Given the waveshapes of $e(t)$ and $i(t)$ in Fig. P1.12, what must be the values of C and R_2?

1.13 Find $e_1(t)/i_1(t)$ for the circuit in Fig. P1.13, and suggest a name for the portion enclosed by the dashed line.

1.14 For the circuit in Fig. P1.14, find an expression for $i(t)$ and then for $e_1(t)$ in terms of $e_o(t)$. If an integral sign appears, differentiate the equation term by term in order to remove it. Under what condition will the circuit act like a differentiator? Can you list any advantages or disadvantages of this circuit compared with the circuit in Fig. 1.4-11?

1.15 Repeat Problem 1.14 for the circuit in Fig. P1.15.

1.16 Express the currents $i_1(t)$ and $i_2(t)$ in Fig. P1.16 in terms of the source voltage $e_1(t)$ and the unknown voltage $e_2(t)$. Apply Kirchhoff's current law to the upper right-hand node, and solve for $e_2(t)$. Find the power supplied by each source.

1.17 Find a differential equation relating the input voltage $e_1(t)$ and the output voltage $e_o(t)$ in Fig. P1.17. If A is allowed to approach infinity, suggest a name for the portion of the circuit enclosed by the dashed line.

1.18 Under what conditions is the relationship between $e_1(t)$ and $e_o(t)$ in Fig. P1.18 similar to that between $e_1(t)$ and $e_o(t)$ in Fig. P1.17?

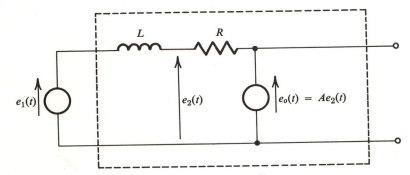

Figure P1.17

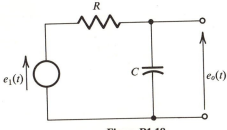

Figure P1.18

1.19 For the problem described in Example 1.4-2, find the net energy supplied by the source for $0 < t < 5$ sec. Assume that $L = 5$ h and $R = 1\ \Omega$. How much energy is stored in the magnetic field of the inductance at $t = 5$ sec? How much energy must have been dissipated as heat in the resistance for $0 < t < 5$? Verify the last answer by direct calculation.

1.20 Suppose that $e(t) = 0$ for $t \geq 0$ in the circuit in Fig. 1.4-8a, but that because of some previous input voltage the current is $1 - 1/\epsilon = 0.632$ amp at $t = 0$. If $L = 5$ h and $R = 1\ \Omega$, find an expression for $i(t)$ for $t > 0$.

1.21 In the circuit in Fig. 1.4-8a, let the source voltage be zero for $t < 0$, 1 volt for $0 < t < 5$ sec, and zero for $t > 5$ sec. If $L = 5$ h and $R = 1\ \Omega$, calculate the total energy dissipated in the resistance for $t > 5$ sec. Show that this equals the energy stored in the inductance at $t = 5$ sec.

36 volts **1.22** If $i_o(t) = 5$ amp in Fig. P1.22, calculate the power supplied to each resistance. Also find the voltage $e(t)$ and the power supplied by each source.

1.23 If a constant 10 volt source is placed across a 2 h inductance, how long will it take to supply 100 joules of energy to the inductance?

1.24 A plot of $i(t)$ versus t for $t > 0$ for the circuit in Fig. P1.24a is given in Fig. P1.24b. If $e_o(t) = -4$ volts at $t = 0$, find and sketch the energy stored in the capacitance for $t > 0$. Also calculate the net energy supplied by the source for $0 < t < 2$.

1.25 Find a differential equation relating $e(t)$ and $i(t)$ for each of the circuits in Fig. P1.25.

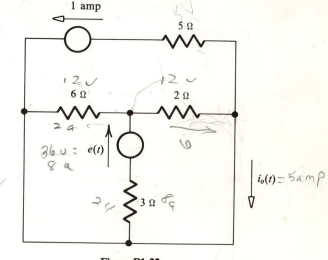

Figure P1.22

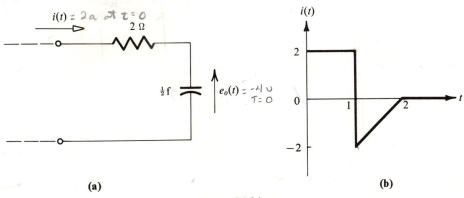

(a)

(b)

Figure P1.24

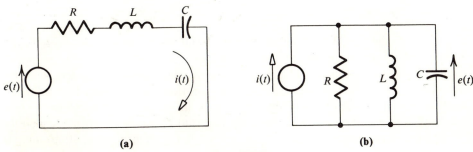

(a)

(b)

Figure P1.25

2

Resistive Circuits

Any circuit can be described by a set of simultaneous equations, which are obtained from Kirchhoff's current law, Kirchhoff's voltage law, and the defining equations for the individual circuit elements. This set of equations must then be solved for the desired voltage or current. Kirchhoff's laws and the definition of resistance are algebraic equations, but inductance and capacitance are defined in terms of derivatives and integrals and lead to integral-differential equations, as in Example 1.4-2.

The only circuit elements considered in most of this chapter are resistances and various types of sources. Because inductance and capacitance are temporarily excluded, the circuits can be described by algebraic equations, which are relatively easy to solve. One reason for considering this special case first is that the mathematical manipulations are less apt to obscure the significant steps in the analysis procedures. Also, the analysis of any circuit containing only one kind of passive element—e.g., only inductances or only capacitances—is very similar to the analysis of resistive circuits.

Many important theorems and procedures are concerned only with the manner in which the elements are interconnected, and not with the nature of the elements. Once these techniques are developed for resistive circuits, they can be applied to general circuit analysis. For example, in later chapters it is shown that the steady-state behavior of any circuit whose sources have a constant value or have a sinusoidal waveshape can be found by the same methods that are developed for resistive circuits. With the Laplace transform of Chapter 10, these techniques can be extended to still more general problems.

In circuits containing capacitance and inductance, the size as well as the shape of the response depends not only upon the size of the input but also upon its waveshape. Recall, however, that the voltage and current of a resistance have similar waveshapes. In a resistive circuit, all of the voltage and current waveshapes are similar to that of the input, and their relative size is unaffected by the form of the input.[†] There is no loss of generality, therefore, if the independent sources in the

---

[†] If there are several inputs with different waveshapes, the superposition theorem of Section 2.2 may be used.

examples are assigned some numerical value instead of a function of time. The sources in resistive circuits are very frequently ones of constant value (d-c sources), but the student should realize that the solution is no more difficult if they are functions of time.

In the first three sections of this chapter we exploit the consequences of the linear nature of our circuit models, and make use of the fact that certain combinations of elements can be replaced by a simpler combination. In Sections 2.4 through 2.6 some systematic methods for analyzing resistive circuits of any degree of complexity are developed.

2.1

Equivalent Resistance

Two circuits are said to be *equivalent* if they cannot be distinguished from one another by voltage and current measurements made at their accessible terminals. The circuits within the boxes labelled A and B in Fig. 2.1-1 are equivalent if and only if $i_A(t) = i_B(t)$ for all possible choices of the voltage source $e(t)$. One example of equivalent circuits was given in Figs. 1.3-10a and 1.3-10b. Any combination of resistances (or resistances and controlled sources) that has only two external terminals is equivalent to a single resistance. The equivalent resistance is defined by $R_{eq} = e(t)/i(t)$, as indicated in Fig. 2.1-2.

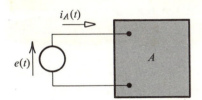 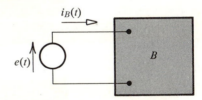

Figure 2.1-1

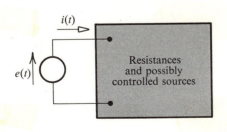 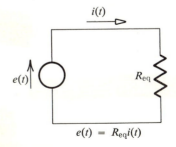

$$e(t) = R_{eq}i(t)$$

Figure 2.1-2

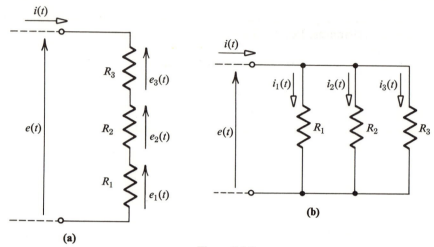

Figure 2.1-3

The series and parallel combinations of resistances shown in Fig. 2.1-3 are considered first. Elements are said to be connected in *series* if the same current passes through all of them, as in Fig. 2.1-3a. For this connection,

$$e(t) = e_1(t) + e_2(t) + e_3(t) = R_1 i(t) + R_2 i(t) + R_3 i(t) = (R_1 + R_2 + R_3)i(t)$$

so the equivalent resistance is

$$R_{eq} = \frac{e(t)}{i(t)} = R_1 + R_2 + R_3 \qquad (2.1\text{-}1)$$

Elements are connected in *parallel* if the same voltage appears across each of them, as in Fig. 2.1-3b. Then

$$i(t) = i_1(t) + i_2(t) + i_3(t) = G_1 e(t) + G_2 e(t) + G_3 e(t) = (G_1 + G_2 + G_3)e(t)$$

The equivalent conductance $G_{eq} = 1/R_{eq} = i(t)/e(t)$ is

$$G_{eq} = G_1 + G_2 + G_3 \qquad (2.1\text{-}2)$$

Since conductance is the reciprocal of resistance, the last equation also can be written

$$\frac{1}{R_{eq}} = \frac{1}{R_1} + \frac{1}{R_2} + \frac{1}{R_3} \qquad (2.1\text{-}3)$$

As illustrated in the examples that follow, these equations may be used several times in the same problem.

For two resistances in parallel, the R_3 term in Eq. 2.1-3 may be disregarded, and

$$R_{eq} = \frac{R_1 R_2}{R_1 + R_2} \qquad (2.1\text{-}4)$$

which is often called the *product-over-sum rule*. The student should be careful when trying to generalize the product-over-sum rule to several resistances in parallel. For three parallel resistances, Eq. 2.1-3 gives

$$R_{eq} = \frac{R_1 R_2 R_3}{R_1 R_2 + R_1 R_3 + R_2 R_3} \qquad (2.1\text{-}5)$$

Somewhat similar results may be obtained for inductances and capacitances in series and in parallel. In Fig. 2.1-4a,

$$e(t) = e_1(t) + e_2(t)$$

$$= \left[e_1(0) + \frac{1}{C_1} \int_0^t i(\lambda)\, d\lambda \right] + \left[e_2(0) + \frac{1}{C_2} \int_0^t i(\lambda)\, d\lambda \right]$$

$$= e_1(0) + e_2(0) + \left(\frac{1}{C_1} + \frac{1}{C_2} \right) \int_0^t i(\lambda)\, d\lambda$$

If the voltages on all the capacitances are known to be zero at $t = 0$, then this equation describes a single capacitance C_{eq} given by

$$\frac{1}{C_{eq}} = \frac{1}{C_1} + \frac{1}{C_2}$$

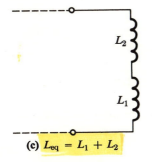

(a) $\dfrac{1}{C_{eq}} = \dfrac{1}{C_1} + \dfrac{1}{C_2}$

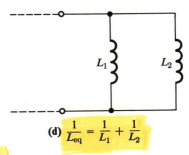

(b) $C_{eq} = C_1 + C_2$

(c) $L_{eq} = L_1 + L_2$

(d) $\dfrac{1}{L_{eq}} = \dfrac{1}{L_1} + \dfrac{1}{L_2}$

Figure 2.1-4

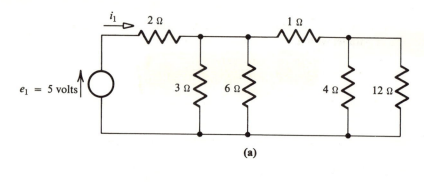

(a)

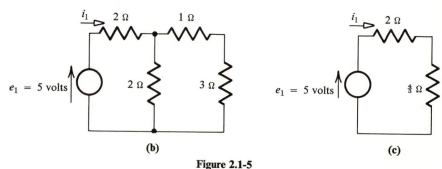

(b)

(c)

Figure 2.1-5

By Eq. 1.3-4, requiring the voltages to be zero initially is the same as requiring the initial energy stored in the capacitances to be zero.

The derivation of the other results given in Fig. 2.1-4 is investigated in Problem 2.5 at the end of the chapter. In the equations given, it is assumed that there are no currents in the inductances at $t = 0$, i.e., that they contain no initial stored energy. It should be emphasized that there is not, in general, any way of combining elements of different kinds that are connected in series or parallel.

Example 2.1-1. Find the equivalent resistance seen by the source in Fig. 2.1-5a and the power supplied by the source.

Solution. The 3 and 6 Ω resistances are connected in parallel, as are the 4 and 12 Ω resistances.[†] By Eq. 2.1-4, these combinations may be replaced by

$$\frac{(3)\,(6)}{3+6} = 2\,\Omega \quad \text{and} \quad \frac{(4)\,(12)}{4+12} = 3\,\Omega$$

respectively, and the circuit may be redrawn as in Fig. 2.1-5b. The two right-hand

---

[†] The student should not make the mistake of assuming that the 2 and 3 Ω resistances (or the 1 and 4 Ω ones) are connected in series. Even though they share one common junction, the same current does not flow through both.

resistances are now in series and may be combined into a single 4 Ω resistance, which in turn is in parallel with a 2 Ω resistance. Then

$$\frac{(4)\,(2)}{4+2} = \frac{4}{3}\;\Omega$$

as in Fig. 2.1-5c. Finally,

$$R_{\text{eq}} = 2 + \frac{4}{3} = \frac{10}{3}\;\Omega$$

and $i_1 = e_1/(\tfrac{10}{3}) = \tfrac{3}{2}$ amp. The power supplied by the source is $p = e_1 i_1 = \tfrac{15}{2}$ watts.

Example 2.1-2. The circuits in Fig. 2.1-6 are called *voltage* and *current dividers,* respectively, and are frequently encountered in subsequent problems. Derive expressions for e_2 and i_2.

Solution. The resistance seen by the voltage source in Fig. 2.1-6a is $R_1 + R_2$, so $i = e/(R_1 + R_2)$ and

$$e_2 = iR_2 = \frac{R_2}{R_1 + R_2}\,e \qquad\qquad (2.1\text{-}6)$$

Since

$$e_1 = iR_1 = \frac{R_1}{R_1 + R_2}\,e$$

$$\frac{e_1}{e_2} = \frac{R_1}{R_2}$$

and the source voltage divides across R_1 and R_2 in proportion to their resistances.

In Fig. 2.1-6b, the resistance seen by the source is $1/(G_1 + G_2)$, so $e = i/(G_1 + G_2)$ and

$$i_2 = G_2 e = \frac{G_2}{G_1 + G_2}\,i = \frac{R_1}{R_1 + R_2}\,i \qquad\qquad (2.1\text{-}7)$$

where the last form is obtained by using $G_1 = 1/R_1$ and $G_2 = 1/R_2$. Since

$$i_1 = \frac{G_1 i}{G_1 + G_2} = \frac{R_2 i}{R_1 + R_2}$$

the current divides in proportion to the conductances, or in *inverse* proportion to the resistances.

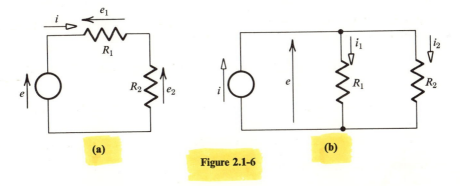

(a)

(b)

Figure 2.1-6

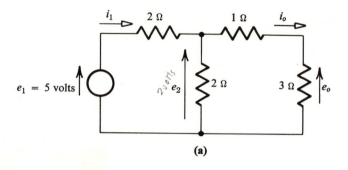

(a)

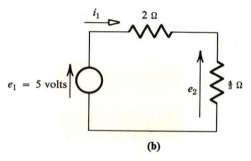

(b)

Figure 2.1-7

Example 2.1-3.   Find the output voltage $e_o(t)$ in Fig. 2.1-7a.

Solution.   The right-hand part of the circuit may be replaced by a $\frac{4}{3}\,\Omega$ resistance, as in Figs. 2.1-5c and 2.1-7b. From Eq. 2.1-6,

$$e_2 = \frac{\frac{4}{3}}{2 + \frac{4}{3}}\, e_1 = 2 \text{ volts}$$

Since the individual identity of resistances that have been combined is lost, it is necessary to refer to Fig. 2.1-7a in order to calculate e_o. Since e_2 is the voltage across the 1 and 3 Ω series combination,

$$e_o = \frac{3}{1+3}\, e_2 = \frac{3}{2} \text{ volts}$$

As an alternative solution, note that $i_1 = \frac{3}{2}$ amp, as in Example 2.1-1. By Eq. 2.1-7,

$$i_o = \frac{2}{2+4}\, i_1 = \frac{1}{2} \text{ amp}$$

and $e_o = 3i_o = \frac{3}{2}$ volts.

Example 2.1-4.   Find the voltage e_o in Fig. 2.1-8.

Solution.   By Eq. 2.1-6,

$$e_2 = (\tfrac{2}{3})(6) = 4 \text{ volts}$$

$$e_3 = (\tfrac{2}{6})(6) = 2 \text{ volts}$$

$$e_o = e_2 - e_3 = 2 \text{ volts}$$

The equivalent resistance of some resistive combinations cannot be found by combining elements in series and in parallel. No two resistances in Fig. 2.1-9, for example, are connected in series or in parallel, so Eqs. 2.1-1 through 2.1-7 are not applicable. Nevertheless, the equivalent resistance of the circuit enclosed by the dashed line can be found by calculating the ratio e/i, which is R_{eq}. Determining the current produced by the voltage source in Fig. 2.1-9 is fairly difficult, so the solution of this particular example is deferred until Section 2.3.

The equivalent resistance for combinations of resistances and controlled sources must also be found from the definition $R_{\text{eq}} = e/i$. The following examples are easily solved in this manner by a direct application of Kirchhoff's laws. As illustrated by Example 1.4-3, the equivalent resistance may be negative in some circuits that contain controlled sources.

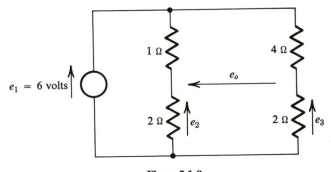

Figure 2.1-8

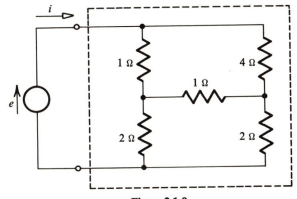

Figure 2.1-9

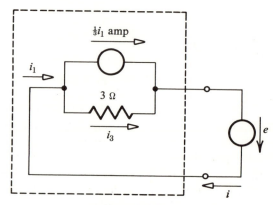

Figure 2.1-10

Example 2.1-5.   Find R_{eq} for the circuit enclosed by the dashed line in Fig. 2.1-10.

Solution.   Connect an external voltage source as shown in the figure, and note that $i = i_1$. By Kirchhoff's current law, $i_3 = \frac{2}{3}i_1$, so the voltage across the resistance (which is the same as the external voltage) is $e = 3i_3 = 2i_1 = 2i$, and $R_{eq} = e/i = 2\,\Omega$.

Example 2.1-6.   Find the equivalent resistance for the circuit enclosed by the dashed line in Fig. 2.1-11.

Solution.   When an external source is connected to the circuit, the same current i flows through every element. By Kirchhoff's voltage law,

$$e_k - \mu e_g + e_p = e$$

or

$$(R_k i) - \mu(-R_k i) + (r_p i) = e$$

from which

$$R_{eq} = \frac{e}{i} = R_k(1 + \mu) + r_p$$

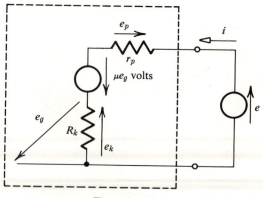

Figure 2.1-11

2.2

Some Consequences of Linearity

Controlled sources and the passive circuit elements are linear components, and any circuit composed of these elements is itself linear. The techniques developed in this and the following section often provide a simpler solution than could otherwise be obtained. They are, however, only the first of many useful procedures that follow directly from the fact that the circuits considered are linear. Other important consequences of linearity appear throughout the book, especially in Chapters 3, 9, and 10.

The inputs to the circuits in this chapter are independent sources. (It is shown in later chapters that any energy initially stored in a capacitance or inductance also can be regarded as an input.) We first consider circuits that are excited by only one independent source. By the definition of linearity, if the value of the source is multiplied by the constant A, so is the response.

Example 2.2-1. Find e_o when $e_1 = 5$ volts in Fig. 2.2-1 by first calculating the source voltage required to produce an output voltage $e_o = 3$ volts.

Solution.   If $e_o = 3$ volts,

$$i_o = \tfrac{3}{3} = 1 \text{ amp}$$

$$e_2 = e_o + 1i_o = 3 + 1 = 4 \text{ volts}$$

$$i_2 = \tfrac{4}{2} = 2 \text{ amp}$$

$$i_1 = i_o + i_2 = 3 \text{ amp}$$

$$e_1 = e_2 + 2i_1 = 4 + 6 = 10 \text{ volts}$$

In practice, the above calculations are so simple that they need not be written down but merely indicated on the figure. The only reason for choosing $e_o = 3$ volts instead of $e_o = 1$ volt or some other arbitrary value was to make the current i_o have a convenient value. The given value of the source voltage was $e_1 = 5$ volts. If the input is multiplied by $\tfrac{1}{2}$, all the other voltages and currents are multiplied by the same constant. Thus, for $e_1 = 5$ volts, $i_1 = \tfrac{3}{2}$ amp and $e_o = \tfrac{3}{2}$ volts, which are the same as the solutions in Examples 2.1-1 and 2.1-3.

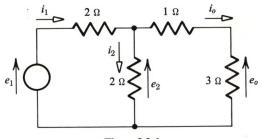

Figure 2.2-1

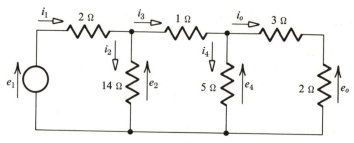

Figure 2.2-2

Example 2.2-2. Find e_o and i_1 for the circuit shown in Fig. 2.2-2.

Solution. Temporarily assume that $e_o = 2$ volts. Then $i_o = 1$ amp, $e_4 = 5$ volts, $i_4 = 1$ amp, $i_3 = 2$ amp, $e_2 = 7$ volts, $i_2 = \frac{1}{2}$ amp, $i_1 = \frac{5}{2}$ amp, and $e_1 = 12$ volts. For the actual source voltage of $e_1 = 2$ volts, $e_o = (\frac{1}{6})(2) = \frac{1}{3}$ volt and $i_1 = \frac{5}{12}$ amp. Figures 2.2-1 and 2.2-2 are examples of *ladder networks*, which are easily solved by the method used here.

If a linear circuit is excited by two or more inputs, then the total response is the sum of the individual responses to each one separately, as in Eqs. 1.1-1 and 1.1-2. When applied to circuits, this property is often called the *superposition theorem* and is rephrased as follows: The response to several independent sources is the sum of

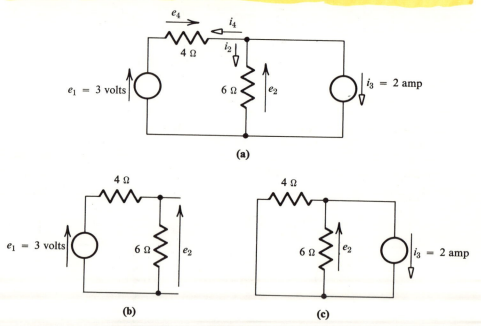

(a)

(b)   (c)

Figure 2.2-3

the responses to each independent source with the remaining independent sources dead.†

In the interpretation of the theorem, the student should recall that a dead source is one whose value is zero, and that dead voltage and current sources are equivalent to short and open circuits, respectively. Notice that the theorem concerns only independent and not controlled sources. In contrast to an independent source, the value of a controlled source depends upon other voltages and currents in the circuit, and should therefore not be regarded as an external input. Controlled sources are *not* made dead when applying the theorem.

Example 2.2-3.   Find the voltage e_2 in Fig. 2.2-3a, using the superposition theorem.

Solution.   We first make the independent current source dead by replacing it by an open circuit, as in Fig. 2.2-3b. By Eq. 2.1-6, the response to the voltage source is

$$e_2 = \frac{6}{4 + 6} e_1 = 1.8 \text{ volts}$$

In (c), the dead voltage source is replaced by a short circuit. The two resistances are

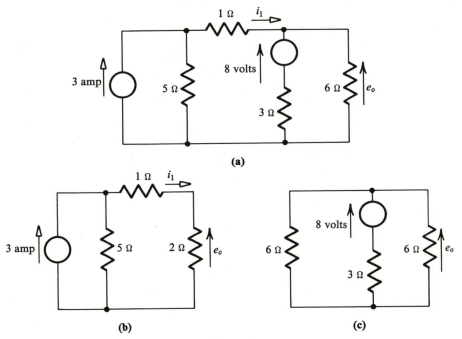

(a)

(b)   (c)

Figure 2.2-4

---

† The application of the theorem to circuits where energy has been initially stored in a capacitance or inductance is considered in Section 3.3.

now in parallel and have an equivalent resistance of $(4)(6)/(4 + 6) = 2.4\ \Omega$. The response to the current source is

$$e_2 = -2.4i_3 = -4.8 \text{ volts}$$

where the minus sign is required because the current arrow does not enter that terminal of the resistance toward which the voltage arrow is directed. Finally, the total response is

$$e_2 = 1.8 - 4.8 = -3 \text{ volts}$$

This answer may be checked by a direct analysis of Fig. 2.2-3a. Since $i_4 + i_2 + i_3 = 0$, $(e_4/4) + (e_2/6) + 2 = 0$. If the substitution $e_4 = e_2 - e_1 = e_2 - 3$ is made,

$$\frac{e_2}{4} - \frac{3}{4} + \frac{e_2}{6} + 2 = 0$$

from which $e_2 = -3$ volts, which is the same as the previous answer.

Example 2.2-4.   Find the voltage e_o across the 6 Ω resistance in Fig. 2.2-4a.

Solution.   When the independent voltage source is replaced by a short circuit, the 3 and 6 Ω resistances are in parallel and are equivalent to a single 2 Ω resistance, as in Fig. 2.2-4b. Then, by Eq. 2.1-7, $i_1 = (\frac{6}{8})(3) = \frac{18}{8}$ amp, and $e_o = (\frac{18}{8})(2) = \frac{18}{4}$ volts.

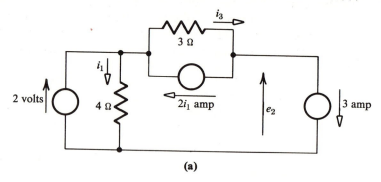

(a)

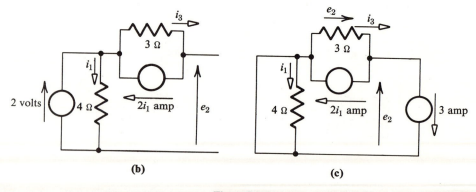

(b)   **(c)**

Figure 2.2-5

When the current source is replaced by an open circuit, the 5 and 1 Ω resistances can be combined, as in Fig. 2.2-4c. The two 6 Ω resistances are then in parallel and are equivalent to a single 3 Ω resistance. By Eq. 2.1-6, $e_o = (\frac{3}{8})(8) = 4$ volts. Finally, when both sources are present, $e_o = 4 + \frac{15}{4} = \frac{31}{4}$ volts.

Example 2.2-5. Find e_2 in the circuit in Fig. 2.2-5a, which contains a current-controlled current source.

Solution. With the independent current source dead as in Fig. 2.2-5b, $i_1 = \frac{2}{4} = \frac{1}{2}$ amp, $i_3 = 2i_1 = 1$ amp, and $e_2 = 2 - 3i_3 = -1$ volt. When the independent voltage source is dead as in (c), $i_1 = 0$ since there is no voltage across the 4 Ω resistance. Then $i_3 = 3$ amp and $e_2 = -3i_3 = -9$ volts. When both independent sources are present, $e_2 = -1 - 9 = -10$ volts.

Thévenin's and Norton's Theorems

Any combination of resistances and controlled sources that has only two external terminals may be reduced to a single resistance R_{eq}, as discussed in Section 2.1. To calculate the external voltages and currents, the entire combination may be replaced by R_{eq}, no matter what is connected to the two terminals. In this section, we

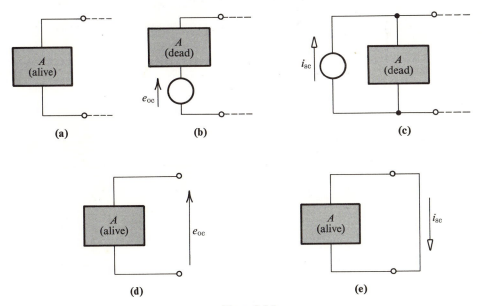

Figure 2.3-1

seek an equivalent circuit to replace a combination which again has only two external terminals, but which contains within it one or more independent sources. As might be expected, such a combination cannot, in general, be replaced by a single element.

Consider the two-terminal network labelled A in Fig. 2.3-1a, where the notation "alive" implies that the circuit contains independent sources. When no external connections are made, so that no current can flow from network A, the voltage across the terminals is denoted by e_{oc}, as in Fig. 2.3-1d. The subscripts are intended to suggest the "open-circuit" termination. The current drawn from the circuit under "short-circuit" conditions, i.e., when the external terminals are directly connected so that the external voltage is forced to be zero, is denoted by i_{sc}, as in Fig. 2.3-1e. The notation "A (dead)" in (b) and (c) means that all the independent sources (but not the controlled sources) inside network A have been made dead. If A does not contain capacitances or inductances, the dead network may be replaced by a single resistance R_{eq}, as in Section 2.1.

Thévenin's theorem states that Fig. 2.3-1a may be replaced by the equivalent circuit in Fig. 2.3-1b, when calculating *external* voltages and currents, no matter what is connected to the two terminals. *Norton's theorem* states that Fig. 2.3-1a is also equivalent to Fig. 2.3-1c. Both theorems are a direct consequence of the linear

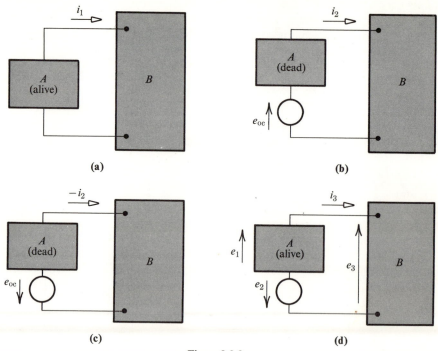

Figure 2.3-2

nature of the circuits considered in this book and may be proved by using the super-position theorem, as follows.[†]

In Fig. 2.3-2, the external connections to network A are represented by network B, which may consist of any combination of elements. In order to simplify the proofs of the theorems, however, we shall initially assume that B contains no sources of any kind, a restriction that is not necessary and that can be removed later. If B contains no sources, then the currents and voltages inside it are completely determined by the incoming current; thus, to prove Thévenin's theorem, it is sufficient to show that i_1 in Fig. 2.3-2a and i_2 in Fig. 2.3-2b are equal.

Since e_{oc} is the only independent source in the circuit in Fig. 2.3-2b, reversing its direction will reverse the directions of all other voltages and currents, as illustrated in Fig. 2.3-2c. Finally, consider Fig. 2.3-2d, which contains an independent voltage source e_2 in addition to the sources within the original network A. If e_2 is adjusted so that $e_2 = e_1$, then $e_3 = 0$. Since there are no sources within B and there is no voltage across its terminals, i_3 must be zero, which in turn means that no current is

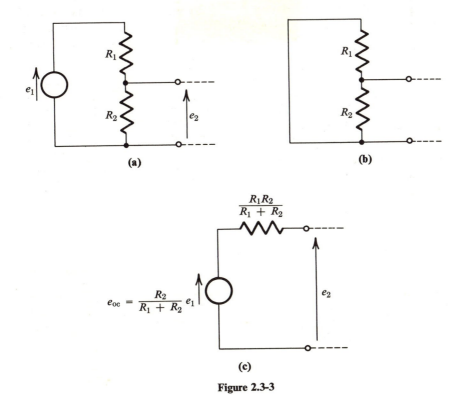

(a)

(b)

(c)

Figure 2.3-3

[†] The application of these theorems to circuits where energy has been initially stored in a capacitance or inductance is considered in Sections 3.3 and 4.3.

drawn from A, and $e_2 = e_1 = e_{oc}$. (Network A does not know that anything is connected to it.) By the superposition theorem, the current i_3 is the sum of the current produced by $e_2 = e_{oc}$ and that produced by the independent sources in A, so $i_3 = i_1 + (-i_2)$. Since i_3 is zero, $i_1 = i_2$, which proves the theorem.

Before proving Norton's theorem or extending the above proof to cases where network B contains sources, we present several examples of Thévenin's theorem. Since controlled sources are not made dead in the application of the superposition theorem, upon which the above proof is based, they should be left alone when forming the dead network A.

Example 2.3-1.   Find the Thévenin equivalent circuit for the circuit in Fig. 2.3-3a. (This result is useful in the analysis of the transistor bias circuits discussed in Chapter 12.)

Solution.   When the external terminals are left open circuited, the same current flows through R_1 and R_2, and the voltage divider rule of Eq. 2.1-6 may be used:

$$e_{oc} = \frac{R_2}{R_1 + R_2} e_1$$

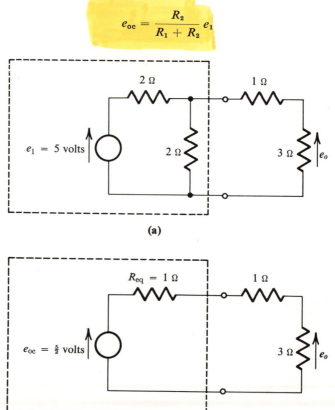

(a)

(b)

Figure 2.3-4

The dead network in Fig. 2.3-3b consists of two resistances in parallel, so

$$R_{eq} = \frac{R_1 R_2}{R_1 + R_2}$$

The complete Thévenin equivalent circuit is shown in Fig. 2.3-3c.

Example 2.3-2.   Use Thévenin's theorem to find e_o in Fig. 2.3-4a.

Solution.   The results of the previous example may be used to find the Thévenin equivalent circuit for the network within the dashed line, which is shown in Fig. 2.3-4b. By the voltage divider rule, $e_o = (\frac{3}{2})(\frac{5}{3}) = \frac{3}{2}$ volts, which is the same as the result obtained in Examples 2.1-3 and 2.2-1.

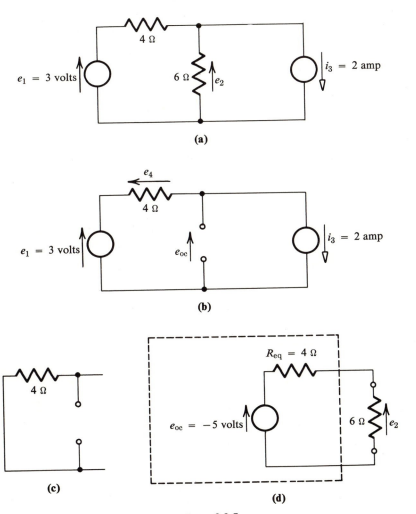

Figure 2.3-5

Example 2.3-3. Find e_2 in Fig. 2.3-5a by replacing everything except the 6 Ω resistance by a Thévenin equivalent circuit.

Solution. The open-circuit voltage is shown in Fig. 2.3-5b, where $e_4 = (2)(4) = 8$ volts and $e_{oc} = e_1 - e_4 = -5$ volts. The dead network A is shown in Fig. 2.3-5c and can be replaced by $R_{eq} = 4\,\Omega$. The complete Thévenin equivalent circuit appears within the dashed line in Fig. 2.3-5d, so $e_2 = (\frac{6}{10})(-5) = -3$ volts, which agrees with Example 2.2-3.

Example 2.3-4. Find e_4 and i_4 in Fig. 2.3-6a.

Solution. Let network A include everything except the 1 Ω resistance whose voltage is to be determined. Under open-circuit conditions, A is identical to Fig. 2.1-8, and so from Example 2.1-4 $e_{oc} = 2$ volts. The dead network in Fig. 2.3-6b can be redrawn as in Fig. 2.3-6c, since all points on the heavy line are at the same potential. Thus

$$R_{eq} = \frac{(1)(2)}{1+2} + \frac{(4)(2)}{4+2} = 2\,\Omega$$

From the Thévenin equivalent circuit enclosed by the dashed line in Fig. 2.3-6d, $i_4 = \frac{2}{3}$ amp and $e_4 = \frac{2}{3}$ volt.

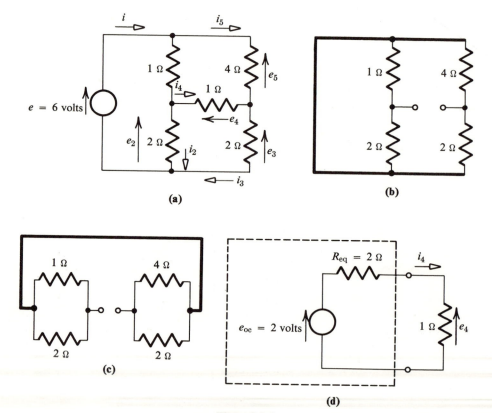

Figure 2.3-6

It is interesting to notice that Figs. 2.3-6a and 2.1-9 are identical. Since we have found the voltage and current for one of the resistances, it may now be possible to find without undue effort the ratio e/i, which is the equivalent resistance seen by the voltage source. In Fig. 2.3-6a, $i_4 + i_5 = i_3$, so $\frac{2}{3} + e_5/4 = e_3/2$. Since $e_5 = 6 - e_3$,

$$\frac{2}{3} + \frac{6}{4} - \frac{e_3}{4} = \frac{e_3}{2}$$

from which $e_3 = \frac{26}{9}$ volts and $i_3 = e_3/2 = \frac{13}{9}$ amp. Then $e_2 = e_3 + e_4 = \frac{26}{9} + \frac{2}{3} = \frac{32}{9}$ volts, and $i_2 = e_2/2 = \frac{16}{9}$ amp. Finally, $i = i_2 + i_3 = \frac{29}{9}$ amp, and

$$\frac{e}{i} = \frac{6}{\frac{29}{9}} = \frac{54}{29}\ \Omega$$

The ratio e/i does not depend upon the value of the voltage source, since multiplying e by any constant will cause all the currents to be multiplied by the same constant.

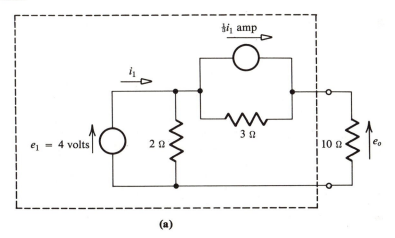

(a)

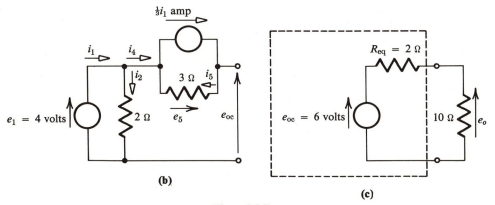

(b)

(c)

Figure 2.3-7

Example 2.3-5.   Find e_o in Fig. 2.3-7a by applying Thévenin's theorem to the part of the circuit within the dashed line.

Solution.   The open-circuit voltage is found from Fig. 2.3-7b, where $i_5 = \frac{1}{3}i_1$ and $i_4 = 0$. Since $i_1 = i_2 = e_1/2 = 2$ amp, $e_5 = 3i_5 = (3)(\frac{2}{3}) = 2$ volts, and $e_{oc} = e_1 + e_5$ $= 6$ volts. The dead network is formed by replacing the independent voltage source by a short circuit, but leaving the current-controlled current source alone. Then the $2\ \Omega$ resistance may be omitted, since there is no voltage across it, and hence no current through it. The dead network is the same as the circuit within the dashed line in Fig. 2.1-10, and from Example 2.1-5 it is equivalent to a $2\ \Omega$ resistance. From the Thévenin equivalent circuit in Fig. 2.3-7c, $e_o = (\frac{10}{12})(6) = 5$ volts.

Example 2.3-6.   Find an expression for e_o/e_1 in Fig. 2.3-8a.

Solution.   Under the open-circuit condition shown in Fig. 2.3-8b, all currents are zero, so there is no voltage across either resistance. Then $e_g = e_1$ and $e_{oc} = -\mu e_g$ $= -\mu e_1$. When the independent source is dead, the circuits within the dashed lines

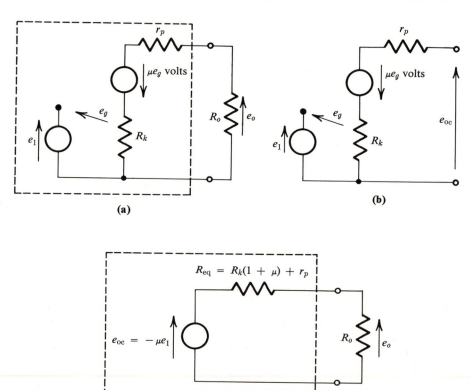

(a)

(b)

(c)

Figure 2.3-8

in Figs. 2.3-8a and 2.1-11 are identical and, from Example 2.1-6, are equivalent to the resistance $R_k(1 + \mu) + r_p$. From the Thévenin equivalent circuit of Fig. 2.3-8c,

$$e_o = \frac{R_o}{R_o + R_{eq}} e_{oc} = \frac{-\mu R_o}{R_k(1 + \mu) + r_p + R_o} e_1$$

Norton's theorem, which states that the external behavior of Figs. 2.3-1a and 2.3-1c is identical, may be proved by referring to Fig. 2.3-9. We first assume that network B contains no sources, and prove the theorem by showing that $i_1 = i_4$. Reversing the direction of the independent current source i_{sc} in Fig. 2.3-9b reverses the direction of the current into B, as in Fig. 2.3-9c. The configuration in Fig. 2.3-9d includes a current source i_5 as well as the sources within the original network A. If i_5 is adjusted so that $i_5 = i_6$, then $i_7 = 0$, and since there are no sources inside B, $e_7 = 0$. Because the voltage across A is now zero, the current drawn from it is $i_6 = i_{sc}$. By the superposition theorem and the fact that $i_5 = i_6 = i_{sc}$, $i_7 = i_1 + (-i_4)$; hence, since i_7 is zero, $i_1 = i_4$.

Example 2.3-7. Solve Examples 2.3-2 through 2.3-6 by using Norton's theorem.

Solution. Figures 2.3-10 through 2.3-14 indicate the solution by Norton's theorem for the circuits in Figs. 2.3-4 through 2.3-8, respectively. Part (a) of each figure shows the network A with a short circuit connected between the external terminals. In Figs. 2.3-10a and 2.3-11a, the calculation of the short-circuit current should be evident

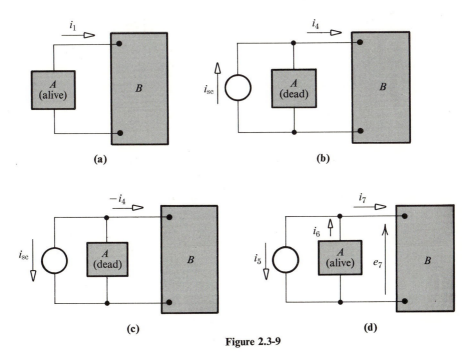

Figure 2.3-9

from the figure. To determine i_{sc} for Fig. 2.3-12a, notice that the total resistance seen by the source is

$$\frac{(1)\,(4)}{1+4} + \frac{(2)\,(2)}{2+2} = \frac{9}{5}\,\Omega$$

so $i = \frac{30}{9}$ amp. By the current divider rule, $i_1 = (\frac{4}{5})(\frac{30}{9}) = \frac{24}{9}$ amp and $i_2 = (\frac{2}{4})(\frac{30}{9})$ $= \frac{15}{9}$ amp, so $i_{sc} = i_1 - i_2 = 1$ amp.

In Fig. 2.3-13a, $i_2 = 2$ amp and $i_3 = \frac{4}{3}$ amp, since the voltage across each resistance is 4 volts. If Kirchhoff's current law is applied to the upper left-hand node, $i_1 = i_2 + i_3 + \frac{1}{3}i_1$, which gives $i_1 = \frac{3}{2}(i_2 + i_3) = 5$ amp. Then

$$i_{sc} = i_3 + \tfrac{1}{3}i_1 = \tfrac{4}{3} + \tfrac{5}{3} = 3 \text{ amp}$$

In Fig. 2.3-14a, $r_p i_{sc} + \mu e_g + R_k i_{sc} = 0$. Since $e_g = e_1 - e_k = e_1 + R_k i_{sc}$,

$$(r_p + \mu R_k + R_k)i_{sc} = -\mu e_1$$

The equivalent resistance of the dead network in each problem was found when the Thévenin equivalent circuits were constructed. The Norton equivalent circuits, consisting of a current source in parallel with R_{eq}, are given within the dashed lines in part (b) of Figs. 2.3-10 through 2.3-14. When the external resistances are attached, the output voltages are easily found and are seen to agree with the previous answers.

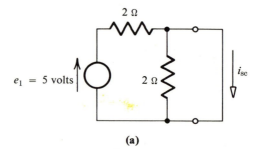

(a)

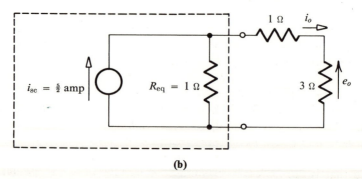

(b)

Figure 2.3-10

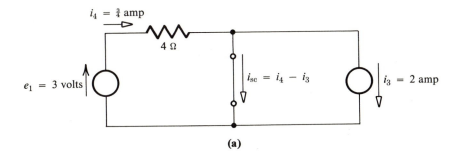

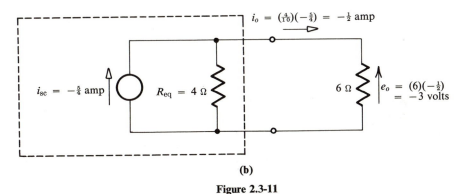

(b)

Figure 2.3-11

A comparison of Figs. 2.3-4 through 2.3-8 with Figs. 2.3-10 through 2.3-14 indicates that

$$i_{sc} = \frac{e_{oc}}{R_{eq}} \qquad\qquad (2.3\text{-}1)$$

This relationship is always true, because the Thévenin and Norton equivalent circuits in Figs. 2.3-15a and 2.3-15b are equally valid, no matter what external connections are made. If the terminals are shorted, the external current in Fig. 2.3-15a is e_{oc}/R_{eq} and in Fig. 2.3-15b is i_{sc}. Equation 2.3-1 also follows from the discussion associated with Figs. 1.5-3c and 1.5-3d, and indicates that a voltage source in series with a resistance may always be replaced by a current source in parallel with the same resistance. In some problems, Eq. 2.3-1 provides the easiest method of finding the equivalent resistance of the dead network.

In the previous proofs of Thévenin's and Norton's theorems, we assumed that network *B* contained no sources. This assumption permitted us to say, for example, that if $e_3 = 0$ in Fig. 2.3-2d then $i_3 = 0$. If *B* contains *controlled* sources, this statement and all other steps in the previous proofs remain valid, unless *B* contains a source that is controlled by a voltage or current in *A*. In the latter event, the theorems cannot be used. In circuit analysis, however, a controlled source is never

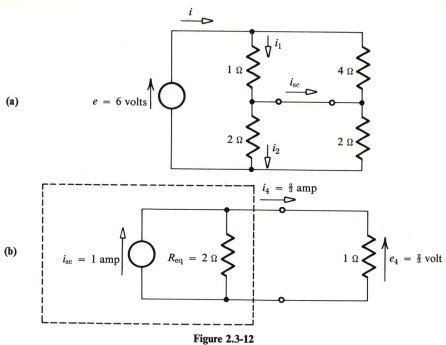

(a)

(b)

Figure 2.3-12

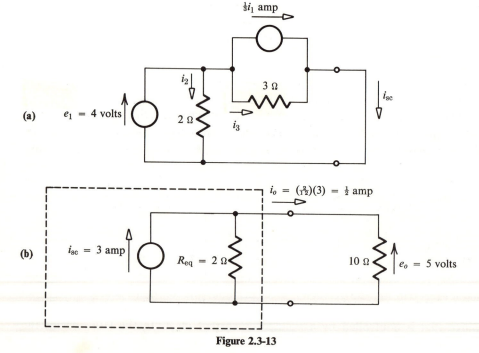

(a)

(b)

Figure 2.3-13

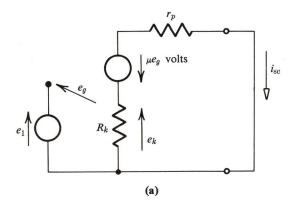

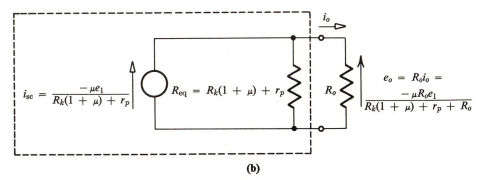

(b)

Figure 2.3-14

isolated from the voltage or current that controls it, so this exception to the theorems is usually not mentioned.

The proof that Thévenin's theorem is valid even when circuit B contains independent sources follows from Fig. 2.3-16, where it is sufficient to show that $i_1 = i_2$. By the superposition theorem, $i_1 = i_3 + i_4$ and $i_2 = i_4 + i_5$. Since the previous

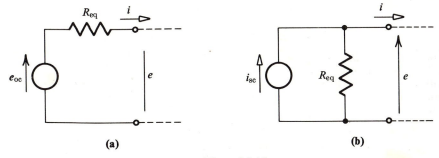

Figure 2.3-15

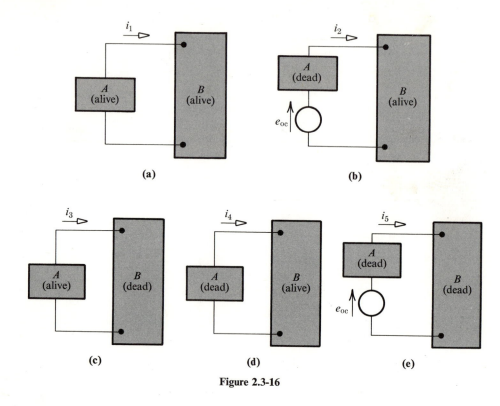

Figure 2.3-16

proof of the theorem demonstrated that $i_3 = i_5$, $i_1 = i_2$ as required. The proof of Norton's theorem can be extended to cases where circuit B contains independent sources by a similar argument.

Example 2.3-8. Find the current through the 2 Ω resistance in Fig. 2.3-17a.

Solution. Thévenin's theorem may be used more than once in the same problem. Replacing the parts of the circuit within the dashed lines by their Thévenin equivalent circuits produces Fig. 2.3-17b, from which $i = \frac{5}{7}$ amp.

The use of Thévenin's or Norton's theorem does not always significantly simplify the solution of a given problem. For example, Thévenin's theorem is of considerable help in finding e_4 and i_4 in Fig. 2.3-6a, but applying Norton's theorem to Fig. 2.3-8a actually increases the required work. The theorems are particularly useful when the external connections to a given two-terminal network are varied, because the same Thévenin or Norton equivalent circuit can be used throughout the analysis. One important example of such a situation is the problem of delivering maximum power to a variable load, which is considered in Problem 2.16 and in Chapter 8.

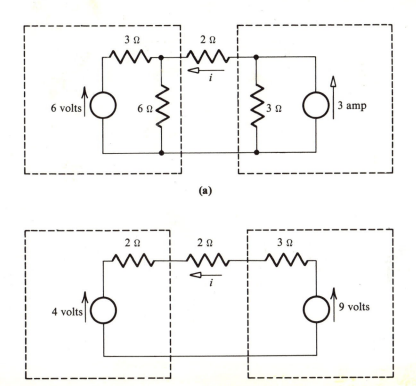

(a)

(b)

Figure 2.3-17

2.4

Network Topology

In Section 2.1 we showed that many, but not all, circuits can be solved by the rules for combining resistances in series and in parallel. The linearity theorems of Sections 2.2 and 2.3 often facilitate the solution of series-parallel circuits, as well as additional problems such as Example 2.3-4. Although simple circuits usually may be most easily analyzed by the methods of the previous sections, it is desirable to develop general and systematic procedures for solving circuits of any degree of complexity. The only equations that are needed are the element laws and Kirchhoff's laws, but writing and solving these equations in a haphazard order often leads to an unnecessarily long and tedious solution.

Two common techniques for finding the output of a circuit are the node-equation method of Section 2.5 and the loop-equation method of Section 2.6. This section provides useful background material and some insight into how each of these methods systematically includes and simultaneously solves the element laws and

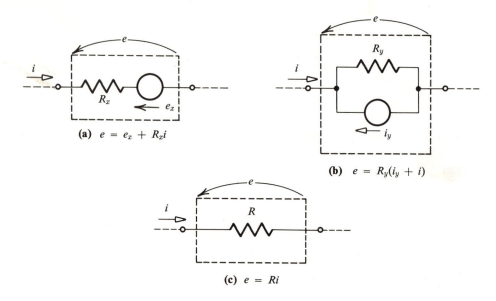

(a) $e = e_x + R_x i$

(b) $e = R_y(i_y + i)$

(c) $e = Ri$

Figure 2.4-1

Kirchhoff's laws. It also introduces some concepts and terminology that may be useful in advanced courses.

Any two-terminal combination of resistances and sources may be represented by a Thévenin or Norton equivalent circuit similar to those in Figs. 2.4-1a and 2.4-1b. A combination that does not contain any independent sources may be represented by a single resistance, as in Fig. 2.4-1c. In each case, there is a relationship between the external voltage and current.

A typical circuit consists of several two-terminal parts (or branches), each of which is characterized by a known voltage-current relationship. The external voltage and current of each such branch is unknown in advance, so if there are B branches there are a total of $2B$ unknowns. In order to solve for these unknowns, $2B$ independent equations are needed. A set of equations is independent if none of them can be formed from the remaining ones, i.e., if every equation contains some information that cannot be deduced from the others. Similarly, a set of n variables is independent only if specifying any $n - 1$ of them does not determine the remaining variable. There are two fundamental questions which will be answered in this section. How can the $2B$ independent equations be found, and how can they easily and systematically be solved?

Consider first the circuit in Fig. 2.4-2a, which can be represented by the five branches in Fig. 2.4-2b.† Current and voltage reference arrows are arbitrarily

---

† The simplest method of solution would be to reduce the circuit to three branches, as in Example 2.3-8. In order to better illustrate the general procedure, however, we shall retain all five branches.

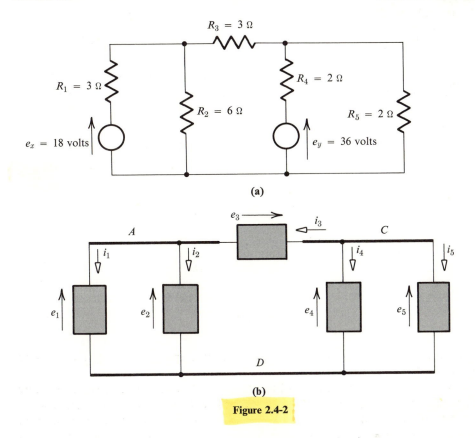

(a)

(b)

Figure 2.4-2

selected for each branch, as shown. Although it is often convenient to direct the voltage arrows opposite to the current arrows, it is not necessary, as was discussed in Section 1.3. The voltage-current relationships for the individual branches in Fig. 2.4-2b are

$$e_1 = e_x + R_1 i_1 = 18 + 3i_1 \quad \text{or} \quad i_1 = -6 + \tfrac{1}{3}e_1$$
$$e_2 = R_2 i_2 = 6i_2 \qquad\qquad \text{or} \quad i_2 = \tfrac{1}{6}e_2$$
$$e_3 = R_3 i_3 = 3i_3 \qquad\qquad \text{or} \quad i_3 = \tfrac{1}{3}e_3 \qquad\qquad (2.4\text{-}1)$$
$$e_4 = e_y + R_4 i_4 = 36 + 2i_4 \quad \text{or} \quad i_4 = -18 + \tfrac{1}{2}e_4$$
$$e_5 = R_5 i_5 = 2i_5 \qquad\qquad \text{or} \quad i_5 = \tfrac{1}{2}e_5$$

which constitute five of the ten required independent equations.

The remaining equations, describing the interconnection of the branches, must be obtained from Kirchhoff's laws. A current law can be written at each of the three nodes indicated by the heavy lines. Summing up the currents leaving nodes A and C, respectively, we obtain

$$i_1 + i_2 - i_3 = 0$$
$$i_3 + i_4 + i_5 = 0 \qquad\qquad (2.4\text{-}2)$$

which certainly are independent equations, since each contains a current not included in the other one. At node D,

$$-i_1 - i_2 - i_4 - i_5 = 0$$

which is *not* another independent equation, since it is the negative sum of the previous two. By Kirchhoff's voltage law,

$$e_1 - e_2 = 0$$

$$e_2 + e_3 - e_4 = 0 \qquad\qquad \text{(2.4-3)}$$

$$e_4 - e_5 = 0$$

where each equation again contains a quantity not included in the other two. Summing the voltages around the perimeter of the figure, we get

$$e_1 + e_3 - e_5 = 0$$

which is just the sum of the previous three equations. Still other voltage-law equations may be written, but they too will be found to be not independent. Fortunately, we already have ten independent equations in Eqs. 2.4-1 through 2.4-3, which are sufficient to solve for the ten unknowns. Two systematic methods of solving the simultaneous equations are of great general importance and are considered next.

With Eqs. 2.4-3, three of the five branch voltages may be expressed in terms of the other two, *provided* that the latter two are independent. Thus,

$$e_1 = e_2$$

$$e_3 = e_4 - e_2 \qquad\qquad \text{(2.4-4)}$$

$$e_5 = e_4$$

where the voltages are expressed in terms of e_2 and e_4. The voltages could equally well be expressed in terms of e_1 and e_3, specifically

$$e_2 = e_1$$

$$e_4 = e_3 + e_2 = e_3 + e_1$$

$$e_5 = e_4 = e_3 + e_1$$

or in terms of several other combinations of two voltages. They cannot, however, be expressed in terms of e_1 and e_2, no matter what algebraic manipulations are used. This should be expected, because e_1 and e_2 are not independent, since $e_1 = e_2$. Finally, although Eqs. 2.4-4 have been obtained by solving Eqs. 2.4-3 algebraically for e_1, e_3, and e_5, the student should be able to write these results down by an inspection of the circuit.

All the branch currents can be now expressed in terms of e_2 and e_4 by substituting Eqs. 2.4-4 into Eqs. 2.4-1.

$$i_1 = -6 + \tfrac{1}{3}e_2$$

$$i_2 = \tfrac{1}{6}e_2$$

$$i_3 = \tfrac{1}{3}(e_4 - e_2) \qquad (2.4\text{-}5)$$

$$i_4 = -18 + \tfrac{1}{2}e_4$$

$$i_5 = \tfrac{1}{2}e_4$$

Next, we substitute these equations into Eqs. 2.4-2:

$$(-6 + \tfrac{1}{3}e_2) + \tfrac{1}{6}e_2 + \tfrac{1}{3}(e_2 - e_4) = 0$$

$$\tfrac{1}{3}(e_4 - e_2) + (-18 + \tfrac{1}{2}e_4) + \tfrac{1}{2}e_4 = 0 \qquad (2.4\text{-}6)$$

or, when like terms are collected,

$$\tfrac{5}{6}e_2 - \tfrac{1}{3}e_4 = 6$$

$$-\tfrac{1}{3}e_2 + \tfrac{4}{3}e_4 = 18 \qquad (2.4\text{-}7)$$

Solving these two equations simultaneously, we obtain

$$e_2 = 14, \qquad e_4 = 17$$

The values of the other voltages, obtained from Eqs. 2.4-4, are

$$e_1 = 14, \qquad e_3 = 3, \qquad e_5 = 17$$

while the currents follow from Eqs. 2.4-1 or 2.4-5. Equations 2.4-6 or 2.4-7 are known as *node equations* and are discussed further in Section 2.5. There, procedures that enable the student to write down these equations from an inspection of the circuit will be developed.

Another systematic solution of Eqs. 2.4-1 through 2.4-3 begins with a consideration of Eqs. 2.4-2. Any two of the five currents can be expressed in terms of the other three, provided that the latter currents are independent. If the equations are solved for i_2 and i_4 in terms of i_1, i_3, and i_5,

$$i_2 = -i_1 + i_3$$

$$i_4 = -i_3 - i_5 \qquad (2.4\text{-}8)$$

These results also can be written down directly from the circuit. All the currents could instead be expressed in terms of i_2, i_3, and i_4, or in terms of several other current combinations. No amount of manipulation, however, will enable the currents to be expressed in terms of only i_1, i_2, and i_3, since by the first of Eqs. 2.4-2 these are not independent variables.

Equations 2.4-8 are substituted into Eqs. 2.4-1 in order to express all the branch voltages in terms of i_1, i_3, and i_5.

$$e_1 = 18 + 3i_1$$
$$e_2 = 6(i_3 - i_1)$$
$$e_3 = 3i_3 \qquad\qquad\qquad\text{(2.4-9)}$$
$$e_4 = 36 - 2(i_3 + i_5)$$
$$e_5 = 2i_5$$

If these equations are in turn substituted into Eqs. 2.4-3,

$$(18 + 3i_1) + 6(i_1 - i_3) = 0$$
$$6(i_3 - i_1) + 3i_3 - 36 + 2(i_3 + i_5) = 0 \qquad\text{(2.4-10)}$$
$$36 - 2(i_3 + i_5) - 2i_5 = 0$$

or

$$9i_1 - 6i_3 = -18$$
$$-6i_1 + 11i_3 + 2i_5 = 36 \qquad\qquad\text{(2.4-11)}$$
$$2i_3 + 4i_5 = 36$$

which yield

$$i_1 = -\tfrac{4}{3}, \qquad i_3 = 1, \qquad i_5 = \tfrac{17}{2}$$

From Eqs. 2.4-8 and 2.4-1,

$$i_2 = \tfrac{7}{3}, \qquad i_4 = -\tfrac{19}{2}$$
$$e_1 = 14, \qquad e_2 = 14, \qquad e_3 = 3, \qquad e_4 = 17, \qquad e_5 = 17$$

which is the same as the previous solution. Section 2.6 shows how Eqs. 2.4-10 or 2.4-11, which are called *loop equations*, can be written down directly from the circuit.

The above solutions for the voltages and currents in Fig. 2.4-2 may seem unnecessarily elaborate for such a simple problem, but they illustrate the detailed mathematical steps behind the node and loop equations, which are discussed in the next two sections. In order to summarize these steps in sufficiently general and precise terms, it is necessary to examine certain concepts from network topology. After doing this, we shall relate these concepts to the above solutions.

The characteristics of any circuit depend upon the elements used and upon how these elements are connected together. *Network topology* (or *linear graph theory*) is concerned only with the manner in which the elements are interconnected and not in the nature of the elements themselves. A large number of useful properties can be deduced from such a study, and these properties can be applied not only in circuit theory, but in many other disciplines as well. Although we shall state the basic properties in general terms, we shall also illustrate them by frequent reference

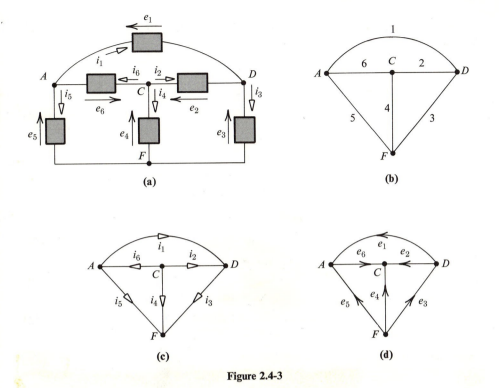

Figure 2.4-3

to the typical configuration in Fig. 2.4-3a. Several of the proofs rely on intuitive concepts, but can be made completely rigorous in more advanced courses.

Consider a circuit that has B branches and N nodes. For the circuit in Fig. 2.4-2b, $B = 5$ and $N = 3$, while, for the circuit in Fig. 2.4-3a, $B = 6$ and $N = 4$. A linear graph (or more simply a graph) is formed from the circuit by disregarding the nature of the branches, replacing them with ordinary lines, and representing the nodes by dots. The graph for the circuit in Fig. 2.4-3a is shown in Fig. 2.4-3b. Remember that the lines joining the nodes really represent branches and are not short circuits.[†]

A *directed* graph has an orientation arrow assigned to each branch. It is customary to use either the current *or* the voltage arrows in the original circuit as the orientation arrows, as in Figs. 2.4-3c and 2.4-3d. If the branch voltage arrows are consistently drawn opposite to the current arrows, either one of the directed graphs is sufficient to reconstruct the original voltage and current references.

A *connected* graph is one in which there is a path between any two nodes.

[†] Nonelectrical systems can also be represented by linear graphs. See, for example, S. Seshu and M. B. Reed, *Linear Graphs and Electrical Networks* (Reading, Mass.: Addison-Wesley, 1961), Section 2.1, for an interesting problem solved by Euler in 1736, where the nodes of a graph represent land masses and the branches represent bridges over a river.

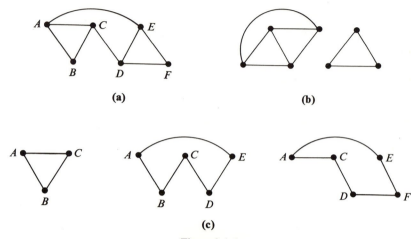

(a)

(b)

(c)

Figure 2.4-4

Figures 2.4-4a and 2.4-4b are examples of connected and unconnected graphs, respectively. An unconnected graph consists of two or more isolated parts, corresponding to isolated circuits. Since an unconnected graph can always be analyzed as two or more separate problems, it is not discussed further.

A *loop* of a graph is a set of branches which forms a single closed path. Every node that is in a loop must have exactly two branches of the loop connected to it. Several of the loops contained in the graph of Fig. 2.4-4a are shown in Fig. 2.4-4c.

A *tree* of a graph is a connected subgraph that connects all the nodes but contains no loops. Among the many trees for Fig. 2.4-3b are those in Figs. 2.4-5a and 2.4-5b. The branch-by-branch construction of the tree in part (a) is shown in part (c). The first branch selected connects two nodes, while the second branch joins these two nodes to a third. Each additional tree branch adds one new node to the previous ones, so that when all N nodes have been connected, the tree contains exactly

$$N - 1 \quad \text{tree branches} \qquad (2.4\text{-}12)$$

For a particular choice of tree, the branches in a graph can be divided into two mutually exclusive categories: tree branches and *links* (which are branches not belonging to the tree and which are also called *chords*). If the number of links is denoted by L, $B = (N - 1) + L$, or

$$L = B - N + 1 \qquad (2.4\text{-}13)$$

The links for the tree in Fig. 2.4-5a are shown in Fig. 2.4-5d. A set of links may or may not include one or more loops.

In order to relate this discussion to Kirchhoff's laws, recall that a circuit containing B branches will have $2B$ unknown voltages and currents and will require $2B$ independent equations for its solution. One-half of these are furnished by the

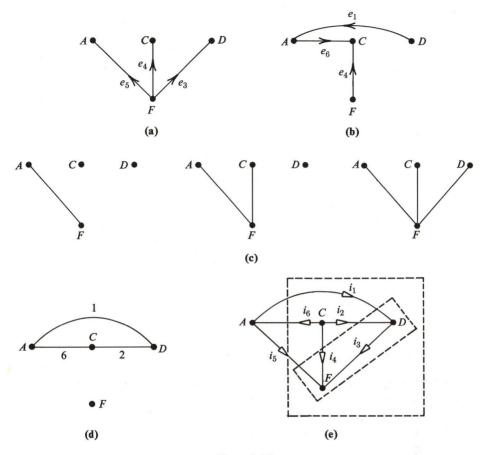

Figure 2.4-5

voltage-current relationships for the individual branches, and the other B equations must come from Kirchhoff's laws.

Theorem 2.4-1. There are exactly $N - 1$ independent Kirchhoff current-law equations, which can be obtained by setting the algebraic sum of the currents leaving each of $N - 1$ nodes equal to zero.

To see that there must be at least $N - 1$ independent equations, examine the branch-by-branch construction of the tree in Fig. 2.4-5c, where the first branch joined two nodes and each subsequent branch connected a new node to the previous ones. If the nodes are considered in *reverse order*, each node except the last one is connected to a tree branch that is not joined to any of the previous nodes. If in the complete graph a current-law equation is written at each node in this reverse order, each equation except the last one will contain a tree branch current not appearing in any previous equation. Corresponding to the tree construction in Fig. 2.4-5c, the nodes are considered in the order D, C, F, and A. The entire directed graph is repeated in Fig. 2.4-5e. At node D,

$$-i_1 - i_2 + i_3 = 0$$

The current-law equation at C,

$$i_2 + i_4 + i_6 = 0$$

contains the new tree branch current i_4, while the one at F,

$$-i_3 - i_4 - i_5 = 0$$

includes the new tree branch current i_5. Since each of the $N - 1$ equations contains a new variable, they must be independent.

Consider the sum of the current-law equations at nodes D and F. The current i_3, which flows from D to F, contributes a positive term at D and a negative term at F, and so does not contribute to the desired sum. The summation includes, therefore, only those branches with one terminal at D or F and the other at A or C, i.e., the branches that pass through a surface surrounding nodes D and F, as indicated by the smaller dashed rectangle in Fig. 2.4-5e. The algebraic sum of the currents flowing out of this surface is

$$-i_1 - i_2 - i_4 - i_5 = 0$$

which is indeed the sum of the current-law equations at D and F. By a similar argument, the algebraic sum of the currents leaving the large closed surface surrounding nodes C, D, and F,

$$-i_1 - i_5 + i_6 = 0$$

is the sum of the current-law equations at the nodes C, D, and F.

The set of branches passing through a closed surface is called a *cut set*, because cutting these branches would divide the graph into two isolated parts. Any cut-set equation (summing the currents leaving the surface) is the sum of the current-law equations at the enclosed nodes; hence, it is not an additional independent equation. Finally, notice that all the currents leaving the surface enclosing C, D, and F must enter node A. Therefore, the corresponding cut-set equation must be the negative of the current-law equation at node A.

$$i_1 + i_5 - i_6 = 0$$

Hence, the equation at node A is not independent of those at the other nodes.

In summary, if Kirchhoff's current law is applied at each of $N - 1$ nodes, no additional *independent* current-law equation can be written. Since we previously saw that the number of independent current-law equations was *at least* $N - 1$, we now know that it must be *exactly* $N - 1$.

Theorem 2.4-2. All branch voltages can be expressed in terms of $N - 1$ independent voltages by Kirchhoff's voltage law.

Since a tree connects all the nodes, there is a path consisting of only tree branches between any two nodes. Thus the voltage between the two nodes can be expressed as the algebraic sum of the tree voltages in that path.

Reconsider the directed graph in Fig. 2.4-3d. Since branches 3, 4, and 5 constitute a tree, all voltages can be expressed in terms of e_3, e_4, and e_5.

$$e_1 = -e_3 + e_5, \qquad e_4 = e_4$$

$$e_2 = -e_3 + e_4, \qquad e_5 = e_5$$

$$e_3 = e_3, \qquad e_6 = -e_5 + e_4$$

For the tree consisting of branches 1, 4, and 6,

$$e_1 = e_1, \qquad e_4 = e_4$$

$$e_2 = e_1 + e_6, \qquad e_5 = -e_6 + e_4$$

$$e_3 = e_4 - e_6 - e_1, \qquad e_6 = e_6$$

There are $N - 1$ branches in a tree, and the voltage of any branch in the graph can be expressed in terms of the $N - 1$ tree voltages.

The tree with branches 3, 4, and 5 is particularly significant, because it connects all the nodes directly to a common node F, which is called the *reference* or *datum node*.[†] The tree voltages are then the voltages of nodes A, C, and D with respect to the reference node, and if the potential of the reference node is arbitrarily defined to be zero, the tree voltages represent the potentials of the other nodes. Such a set of $N - 1$ independent voltages is known as *node-to-datum* (or simply *node*) *voltages*.

Even if the given circuit does not possess a tree having the characteristics described in the preceding paragraph, the potential of any one node may be assigned the value zero, and the potentials of the remaining $N - 1$ nodes may be taken as the $N - 1$ independent voltages in Theorem 2.4-2. Every branch voltage is the difference in the potentials of the two nodes to which the branch is connected, so it certainly can be expressed in terms of this set of $N - 1$ voltages. This approach is the one used in Section 2.5.

Theorem 2.4-3. There are $L = B - N + 1$ independent loops, around which a voltage-law equation may be written.

One way to obtain $B - N + 1$ independent loops is first to select a tree of the graph. The addition of a single link to the tree will form a loop consisting of this link and one or more tree branches. In the same way, each of the $B - N + 1$ links, together with some tree branches but no other links, defines a unique loop. These loops are independent, because each one includes a link that does not belong to any of the other loops.

[†] Such a tree is often called a *starlike tree*.

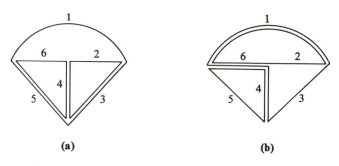

(a)   (b)

Figure 2.4-6

For the tree with branches 3, 4, and 5 in Fig. 2.4-3d, the independent loops are defined by the links 1, 2, and 6 and are shown in Fig. 2.4-6a. Figure 2.4-6b gives the independent loops corresponding to the tree branches 1, 4, and 6.

Although the links of any tree may be always used to construct a set of L independent loops, this is not the only satisfactory method. A set of L loops is certainly independent if each loop that is selected contains a branch that was not part of any previous loop. In order to indicate one important application of this fact, we first introduce two more definitions.

A graph is *planar* if it can be drawn on a plane surface so that its branches intersect only at the nodes. The graph in Fig. 2.4-7a is planar even though branches 2 and 4 intersect, because it can be redrawn as in Fig. 2.4-7b. Figure 2.4-7c is an example of a nonplanar graph.

In a graph without intersecting branches, a *mesh* is defined as a loop that does not enclose any other branches.[†] If the branches are visualized as the wooden partitions in a window, then each window pane defines a mesh. The four meshes in Fig. 2.4-8a are the loops labelled I through IV in Fig. 2.4-8b.

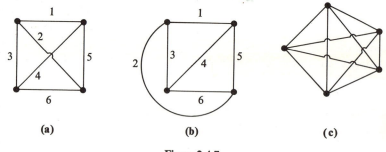

(a)   (b)   (c)

Figure 2.4-7

[†] Instead of defining a mesh as a special kind of a loop, some authors use the two terms interchangeably.

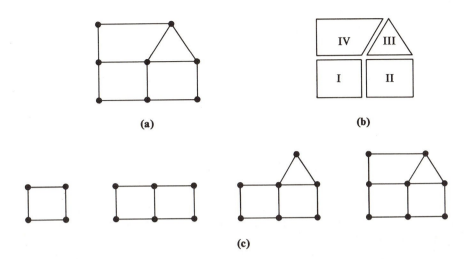

(a) (b)

(c)

Figure 2.4-8

If the branches in the first mesh are inserted in sequence, the first branch joins two nodes and each additional branch connects a new node to the previous ones, except for the last branch, which is joined to the very first node and not to a new node. Thus the mesh has the same number of branches and nodes. If the graph is reconstructed mesh by mesh, as in Fig. 2.4-8c, the number of new branches added by each additional mesh is one more than the number of new nodes. Thus, for a graph containing M meshes, B branches, and N nodes, $B = N + (M-1)$ or $M = B - N + 1$, which is exactly equal to the number of links and to the number of independent loops. Then the number of independent loops can be found by counting the number of meshes. Furthermore, the meshes are an independent set of L loops, since each new one contains at least one branch that is not part of the previous meshes.

Once a set of $B - N + 1$ independent loops have been chosen, Kirchhoff's voltage law may be applied to each of them, giving $B - N + 1$ independent equations relating the voltages. These equations can be used to express $B - N + 1$ of the B voltages in terms of the remaining $N - 1$ ones, which agrees with Theorem 2.4-2.

Theorem 2.4-4. All branch currents can be expressed in terms of $L = B - N + 1$ independent currents by Kirchhoff's current law.

Suppose that all the $B - N + 1$ link currents, corresponding to a particular choice of a tree, are made zero by replacing the links by open circuits. Since the tree itself contains no closed paths, *all* the currents must be zero. If, however, any tree branch current were independent of the link currents, it could not be forced to vanish by specifying the link currents. All the tree branch currents must therefore

Figure 2.4-9

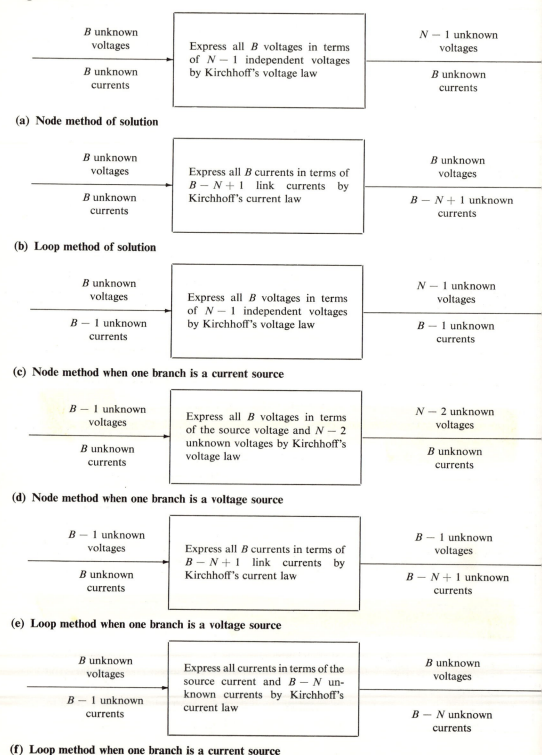

B unknown
voltages

B unknown
currents

Express all B voltages in terms
of $N-1$ independent voltages
by Kirchhoff's voltage law

$N-1$ unknown
voltages

B unknown
currents

(a) Node method of solution

B unknown
voltages

B unknown
currents

Express all B currents in terms of
$B-N+1$ link currents by
Kirchhoff's current law

B unknown
voltages

$B-N+1$ unknown
currents

(b) Loop method of solution

B unknown
voltages

$B-1$ unknown
currents

Express all B voltages in terms
of $N-1$ independent voltages
by Kirchhoff's voltage law

$N-1$ unknown
voltages

$B-1$ unknown
currents

(c) Node method when one branch is a current source

$B-1$ unknown
voltages

B unknown
currents

Express all B voltages in terms
of the source voltage and $N-2$
unknown voltages by Kirchhoff's
voltage law

$N-2$ unknown
voltages

B unknown
currents

(d) Node method when one branch is a voltage source

$B-1$ unknown
voltages

B unknown
currents

Express all B currents in terms of
$B-N+1$ link currents by
Kirchhoff's current law

$B-1$ unknown
voltages

$B-N+1$ unknown
currents

(e) Loop method when one branch is a voltage source

B unknown
voltages

$B-1$ unknown
currents

Express all currents in terms of the
source current and $B-N$ unknown currents by Kirchhoff's
current law

B unknown
voltages

$B-N$ unknown
currents

(f) Loop method when one branch is a current source

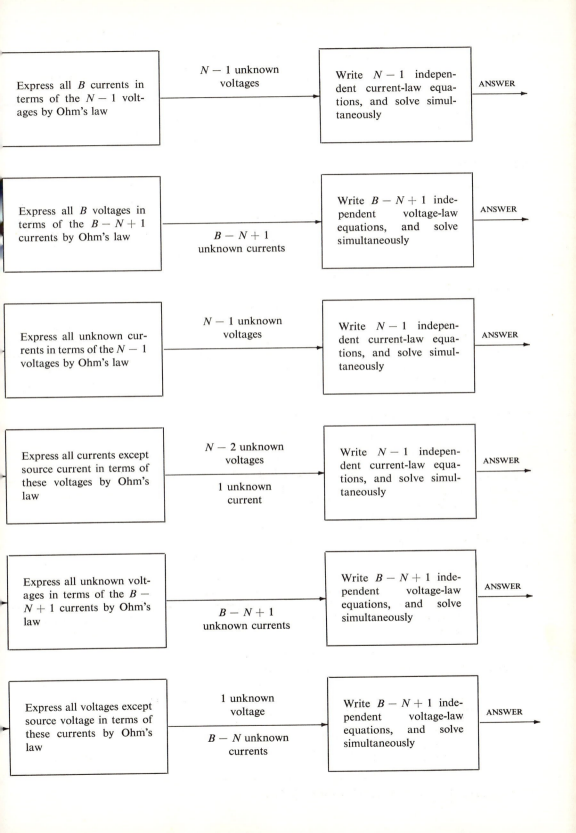

| Express all B currents in terms of the $N - 1$ voltages by Ohm's law | $N - 1$ unknown voltages | Write $N - 1$ independent current-law equations, and solve simultaneously | ANSWER |

| Express all B voltages in terms of the $B - N + 1$ currents by Ohm's law | $B - N + 1$ unknown currents | Write $B - N + 1$ independent voltage-law equations, and solve simultaneously | ANSWER |

| Express all unknown currents in terms of the $N - 1$ voltages by Ohm's law | $N - 1$ unknown voltages | Write $N - 1$ independent current-law equations, and solve simultaneously | ANSWER |

| Express all currents except source current in terms of these voltages by Ohm's law | $N - 2$ unknown voltages / 1 unknown current | Write $N - 1$ independent current-law equations, and solve simultaneously | ANSWER |

| Express all unknown voltages in terms of the $B - N + 1$ currents by Ohm's law | $B - N + 1$ unknown currents | Write $B - N + 1$ independent voltage-law equations, and solve simultaneously | ANSWER |

| Express all voltages except source voltage in terms of these currents by Ohm's law | 1 unknown voltage / $B - N$ unknown currents | Write $B - N + 1$ independent voltage-law equations, and solve simultaneously | ANSWER |

depend upon the $B - N + 1$ link currents, i.e., can be expressed in terms of them. This conclusion is also consistent with Theorem 2.4-1, which states that there are $N - 1$ independent equations relating the currents. Using these equations, we may express $N - 1$ of the B currents in terms of a set of $B - (N - 1)$ independent ones.

Theorems 2.4-1 and 2.4-3 indicate that there are $N - 1$ independent current-law equations and $B - N + 1$ independent voltage-law equations. Together with the B voltage-current relationships for the individual branches, these equations are sufficient to solve for the $2B$ unknowns.

To further illustrate the above theorems, we reconsider Fig. 2.4-2 and examine the significance of the steps leading to the node equations 2.4-6 and the loop equations 2.4-10. The node method of solution is based upon Theorems 2.4-1 and 2.4-2, and it can be summarized by the chart in Fig. 2.4-9a. Initially, there are B unknown voltages and B unknown currents. According to Theorem 2.4-2, however, all the voltages can be expressed in terms of $N - 1$ tree voltages or $N - 1$ node-to-datum voltages. If the tree selected consists of branches 2 and 4, an inspection of the circuit shows that

$$e_1 = e_2, \qquad e_3 = e_4 - e_2, \qquad e_5 = e_4$$

as in Eqs. 2.4-4. This particular tree is a starlike tree; so the above equations are identical with those obtained by choosing node D as a reference node with potential of zero and letting e_2 and e_4 denote the potentials of nodes A and C, respectively. Kirchhoff's voltage law was used implicitly in expressing $B - N + 1$ of the voltages in terms of the remaining $N - 1$ ones by an inspection of the circuit.

Now the unknown voltages have been reduced to $N - 1$ voltages, namely, e_2 and e_4, and all currents can be expressed in terms of these voltages by Ohm's law, $i = e/R$. In Fig. 2.4-2, the voltages across R_1 and R_4 are $e_2 - e_x$ and $e_4 - e_y$, respectively. Thus,

$$i_1 = \tfrac{1}{3}(e_2 - e_x) = \tfrac{1}{3}(e_2 - 18), \qquad i_2 = \tfrac{1}{6}e_2, \qquad i_3 = \tfrac{1}{3}e_3 = \tfrac{1}{3}(e_4 - e_2)$$

$$i_4 = \tfrac{1}{2}(e_4 - e_y) = \tfrac{1}{2}(e_4 - 36), \qquad i_5 = \tfrac{1}{2}e_4$$

which follow directly from the figure, but which are the same as Eqs. 2.4-5.

The last step is to write a current-law equation at $N - 1$ nodes, with the currents expressed in terms of $N - 1$ independent voltages. At nodes A and C,

$$\tfrac{1}{3}(e_2 - 18) + \tfrac{1}{6}e_2 + \tfrac{1}{3}(e_2 - e_4) = 0$$

$$\tfrac{1}{3}(e_4 - e_2) + \tfrac{1}{2}(e_4 - 36) + \tfrac{1}{2}e_4 = 0$$

which are identical with Eqs. 2.4-6, and which may be solved simultaneously for e_2 and e_4. In Section 2.5 we discuss how such a set of equations may be easily written down by an inspection of any given circuit.

The loop method of solution, which is based on Theorems 2.4-3 and 2.4-4, is summarized in Fig. 2.4-9b. For a tree consisting of branches 2 and 4, the link currents are i_1, i_3, and i_5. By Theorem 2.4-4, all currents can be expressed in terms of these link currents, leaving only $B - N + 1$ unknown currents. From Fig. 2.4-2, and as a result of Kirchhoff's current law,

$$i_2 = i_3 - i_1, \qquad i_4 = -i_3 - i_5$$

as in Eqs. 2.4-8.

Now all the voltages can be expressed in terms of these $B - N + 1$ currents by the relationship $e = Ri$. From the circuit diagram,

$$e_1 = e_x + 3i_1 = 18 + 3i_1, \qquad e_2 = 6(i_3 - i_1)$$

$$e_3 = 3i_3, \qquad e_4 = e_y + 2i_4 = 36 - 2(i_3 + i_5), \qquad e_5 = 2i_5$$

Next we must write $B - N + 1$ voltage-law equations around a set of independent loops. The loops defined by the links 1, 3, and 5 happen to be the same as the three meshes.

$$(18 + 3i_1) - 6(i_3 - i_1) = 0$$

$$6(i_3 - i_1) + 3i_3 - 36 + 2(i_3 + i_5) = 0$$

$$36 - 2(i_3 + i_5) - 2i_5 = 0$$

which are Eqs. 2.4-10, and which may be solved simultaneously for the $B - N + 1$ link currents.

Up to this point, we have assumed that each of the B branches is characterized by a voltage-current relationship that depends upon the nature of that branch. It is often convenient, however, to regard a source as a separate branch (instead of including within the branch a resistance as well as the source). In this event, either the branch voltage or current is specified, while the value of the other variable depends upon the rest of the circuit. Now we shall investigate how the charts of Figs. 2.4-9a and 2.4-9b should be modified to allow for such an approach.

Suppose that we wish to form a set of node equations when one branch is a current source. As indicated in Fig. 2.4-9c, there will be B unknown voltages but only $B - 1$ unknown currents. All voltages, including the unknown voltage of the source, still may be expressed in terms of $N - 1$ independent voltages, and then the unknown currents can be expressed in terms of these voltages by Ohm's law. The node procedure still will yield $N - 1$ equations with $N - 1$ unknown voltages.

If one branch of the circuit is a voltage source, there are $B - 1$ unknown voltages and B unknown currents including the current of the source, as indicated in Fig. 2.4-9d. Since one branch voltage is known, all the voltages can be expressed in terms of this voltage and $N - 2$ unknown voltages. All currents except the source current can be expressed in terms of these voltages, so the $N - 1$ independent current-law equations contain sufficient information for us to solve for the $N - 1$ unknowns. It should be pointed out that if both of the nodes to which the voltage source is connected are avoided, the current-law equations at the other $N - 2$ nodes will not involve the source current and can be solved for the $N - 2$ unknown voltages. If the current of the voltage source is sought, it can be found by an additional step.

The formation of a set of loop equations, when one branch is a voltage source, is summarized in Fig. 2.4-9e. All currents, including the unknown current of the source, may be expressed in terms of $B - N + 1$ link currents; then the unknown voltages may be expressed in terms of these same currents. The procedure results in $B - N + 1$ equations with the same number of unknown currents.

If one branch is a current source, as is assumed in Fig. 2.4-9f, all currents can be expressed in terms of the source current and $B - N$ unknown currents. Then all voltages except the source voltage can be expressed in terms of these same currents, but the voltage of the current source will be an additional unknown. The $B - N + 1$ independent voltage-law equations are sufficient for us to solve for the $B - N$ unknown currents and for the voltage of the current source.

2.5

Node Equations

As discussed in the previous section, node-to-datum (or node) equations result when one particular systematic method is used for the simultaneous solution of the voltage-current relationships for the branches and the necessary Kirchhoff-law equations. Since many of the steps should be carried out mentally without writing them down, the branches should be kept relatively simple. In general, it is best to treat each circuit element as a separate branch, except for resistances in series, which should be combined into a single branch.

Assume that a number of branches (not shown) are connected between the five nodes in Fig. 2.5-1a. The four voltages shown by the solid arrows in the figure represent the voltages of four of the nodes with respect to node F, which is called the reference or datum node. If the potential of the reference node is arbitrarily assigned a value of zero, the four voltages represent the potentials of nodes A through D. In Fig. 2.5-1b, the ground symbol attached to node F indicates that its potential is zero, while the potentials of the other nodes are denoted by the symbols alongside them instead of by arrows drawn from the reference node. Although the representations in parts (a) and (b) of the figure are equivalent, it is customary to use the notation of part (b) when writing node equations. The voltages e_A through e_D are called node-to-datum (or node) voltages. The voltage of any branch in the circuit is the difference of the potentials of the two nodes to which it is connected and can be expressed in terms of the node voltages by inspection. The voltage associated with the dashed reference arrow, for example, is $e_C - e_D$. Since the voltages of the individual branches are so easily expressed, they usually are not shown explicitly on the circuit diagram.

The current through any purely resistive branch is given by $i = e/R$, where e is the difference in potential between the two ends of the branch, as discussed above. If the only sources are independent current sources, all the unknown currents can be easily expressed in terms of the $N - 1$ node voltages, where N denotes the number of nodes. Applying Kirchhoff's current law at each of $N - 1$ nodes

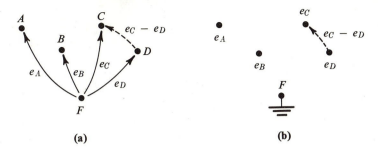

Figure 2.5-1

(usually choosing the reference node as the one to be omitted), we obtain $N - 1$ independent equations, which can be solved for the node voltages.

Example 2.5-1. Find the voltage across and the power supplied by each of the current sources in Fig. 2.5-2a.

Solution. The bottom line in the figure has been chosen as the reference node, and the voltages of the other two nodes are denoted by e_A and e_B. Figure 2.5-2b shows the voltage and current of every element connected to node A. If the algebraic sum of the currents leaving this node is set equal to zero,

$$-5 + \tfrac{1}{4}e_A + \tfrac{1}{2}(e_A - e_B) = 0$$

The current-law equation at node B follows from Fig. 2.5-2c:

$$\tfrac{1}{2}(e_B - e_A) + \tfrac{1}{3}e_B + 2 + \tfrac{1}{6}e_B = 0$$

Collecting like terms, we get

$$(\tfrac{1}{4} + \tfrac{1}{2})e_A - \tfrac{1}{2}e_B = 5$$

$$-\tfrac{1}{2}e_A + (\tfrac{1}{2} + \tfrac{1}{3} + \tfrac{1}{6})e_B = -2$$

$$(2.5\text{-}1)$$

Solving these equations simultaneously, we obtain $e_A = 8$ volts and $e_B = 2$ volts. The power supplied is $p_A = (8)(5) = 40$ watts for the left-hand source, and $p_B = -(2)(2) = -4$ watts for the right-hand source.

Several comments about the above example should be helpful. Notice that the expressions for the currents in Fig. 2.5-2 indicate that $i_3 = -i_4$, which is consistent with the fact that reversing a current's reference arrow reverses its sign. The current i_3 leaving node A is equivalent to a current $i_4 = -i_3$ entering the node, so the first node equation could have been written

$$-5 + \tfrac{1}{4}e_A - \tfrac{1}{2}(e_B - e_A) = 0$$

We prefer, however, to express any current through a resistance as a current leaving the node, so that all such terms in the original node equation have the same sign. Notice that a current leaving node A through a resistive branch always has the form $G(e_A - e_x)$, where e_x is the potential of the other node to which the branch is

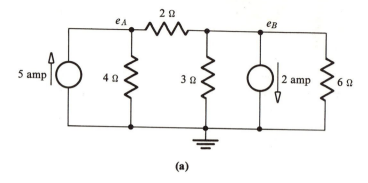

(a)

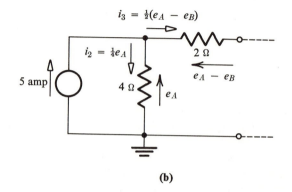

(b)

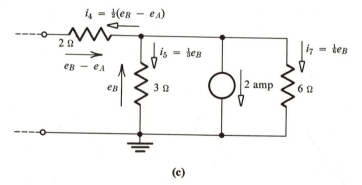

(c)

Figure 2.5-2

joined and G is the conductance of the branch. If the resistance is connected between node A and the reference node, e_x is of course zero. An analogous statement
can be made concerning the currents leaving node B through resistive branches. In
practice, therefore, the original node equations are written down without drawing
Figs. 2.5-2b and 2.5-2c.

When the equation for node A is simplified by collecting like terms, as in Eqs.

2.5-1, the coefficient of e_A is the sum of the conductances of all the resistive branches connected to the node, while the coefficient of e_B is the negative sum of the conductances of the resistive branches connected between nodes A and B. If the terms representing the current sources attached to A are put on the *right-hand* side of the equation, they represent the algebraic sum of the source currents *entering* the node. The simplified equation at node B can be interpreted in an analogous way.

In general, for a circuit of N nodes that contains only independent current sources, the simplified node equations have the following form:

$$G_{11}e_1 - G_{12}e_2 - \cdots - G_{1,N-1}e_{N-1} = i_1$$

$$-G_{21}e_1 + G_{22}e_2 - \cdots - G_{2,N-1}e_{N-1} = i_2 \qquad (2.5\text{-}2)$$

$$\cdots\cdots\cdots\cdots\cdots\cdots\cdots\cdots\cdots\cdots\cdots\cdots$$

$$-G_{N-1,1}e_1 - G_{N-1,2}e_2 - \cdots + G_{N-1,N-1}e_{N-1} = i_{N-1}$$

where e_1 through e_{N-1} denote the voltages of $N-1$ nodes with respect to the reference node and where

G_{jj} = the self-conductance at node j
   = the sum of the conductances of all resistive branches that have one terminal at node j
$G_{jk} = G_{kj}$ = the mutual conductance between nodes j and k
   = the sum of the conductances of the resistive branches that are connected directly between nodes j and k ($j \neq k$)
i_j = the algebraic sum of the currents entering node j through any current sources connected to that node

It is assumed that a current-law equation is written at every node except the reference node and that the voltages in each equation are in the same order in which the $N-1$ nodes are examined. In Fig. 2.5-2a, $G_{AA} = \frac{1}{4} + \frac{1}{2}$, $G_{BB} = \frac{1}{2} + \frac{1}{3} + \frac{1}{6}$, $G_{AB} = G_{BA} = \frac{1}{2}$, $i_A = 5$, and $i_B = -2$, so that the simplified node equations 2.5-1 may be written down directly from the circuit.

In writing Eqs. 2.5-2 it is assumed that the only algebraic manipulation performed on the original node equations is collecting and rearranging the terms. The coefficients on the left side are sometimes written in a rectangular array, which is called a *matrix*:

$$\begin{bmatrix} G_{11} & -G_{12} & \cdots & -G_{1,N-1} \\ -G_{21} & G_{22} & \cdots & -G_{2,N-1} \\ \cdots\cdots\cdots & \cdots\cdots\cdots & & \cdots\cdots \\ -G_{N-1,1} & -G_{N-1,2} & \cdots & G_{N-1,N-1} \end{bmatrix}$$

The terms on the main diagonal (a line drawn from the upper left-hand corner to the lower right-hand corner) are positive. Since $G_{jk} = G_{kj}$, the off-diagonal terms are not only negative, but are symmetrical about the main diagonal. Even if we do

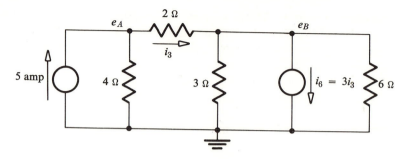

Figure 2.5-3

not use Eqs. 2.5-2 when writing node equations, this symmetry property serves as a check on the work.

In the above discussion, including Eqs. 2.5-2, it is assumed that the only sources present are independent current sources. If a controlled source is present, the procedure is the same except that an extra step is required. After writing the node equations, the value of the controlled source must be expressed in terms of the node voltages, in order to limit the number of unknowns to $N - 1$.

Example 2.5-2. Find the voltage across each of the current sources in Fig. 2.5-3.

Solution. This figure is identical with Fig. 2.5-2a, except that one of the current sources has been changed from 2 amp to $3i_3$. The node equations are therefore the same as Eqs. 2.5-1, except for the term representing this current source.

$$\tfrac{3}{4}e_A - \tfrac{1}{2}e_B = 5$$
$$-\tfrac{1}{2}e_A + e_B = -3i_3$$

Since $i_3 = \tfrac{1}{2}(e_A - e_B)$,

$$\tfrac{3}{4}e_A - \tfrac{1}{2}e_B = 5$$
$$e_A - \tfrac{1}{2}e_B = 0$$

from which $e_A = -20$ volts and $e_B = -40$ volts.

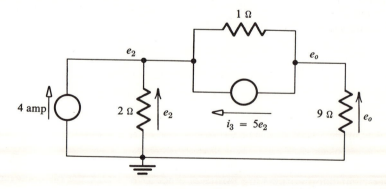

Figure 2.5-4

In Example 2.5-2, notice that the coefficients on the left side of the final set of node equations are no longer symmetrical. No general statement can be made about the signs or symmetry of the coefficients when controlled sources are present.

Example 2.5-3. Find the voltage e_o in Fig. 2.5-4.

Solution. The choice of a reference node and the node voltages e_2 and e_o are indicated on the figure. The two node equations are

$$(\tfrac{1}{2} + 1)e_2 - e_o = 4 + i_3$$

$$-e_o + (1 + \tfrac{1}{9})e_o = -i_3$$

Since $i_3 = 5e_2$,

$$-\tfrac{7}{2}e_2 - e_o = 4$$

$$4e_2 + \tfrac{10}{9}e_o = 0$$

which yield $e_o = -144$ volts. Once again, expressing the controlled source in terms of the node voltages has destroyed the symmetry of the coefficients in the simplified node equations.

If one of the branches in a circuit is an independent voltage source, its current is an unknown that cannot be easily expressed in terms of the node voltages. Since the source voltage is known, however, there are only $N - 2$ instead of $N - 1$ unknown node voltages, so $N - 1$ current-law equations will still provide a complete solution. Furthermore, if the two nodes to which the voltage source is connected are avoided when the current-law equations are written, the unknown current of the source does not appear, and the equations at the remaining $N - 2$ nodes are sufficient to solve for the $N - 2$ unknown voltages. If the current of the voltage source is sought, it can then be found by an additional step.

Example 2.5-4. Find e_4 and the total resistance seen by the voltage source in Fig. 2.3-6a, which is repeated in Fig. 2.5-5a.

Solution. The voltages of the three nodes with respect to the indicated reference node are denoted by e_1, e_2, and e_3. First, we write a current-law equation at each of the nodes whose voltage is unknown:

$$\frac{e_2 - 6}{1} + \frac{e_2}{2} + \frac{e_2 - e_3}{1} = 0$$

$$\frac{e_3 - e_2}{1} + \frac{e_3 - 6}{4} + \frac{e_3}{2} = 0$$

Collecting like terms, we obtain

$$\tfrac{5}{2}e_2 - e_3 = 6$$

$$-e_2 + \tfrac{7}{4}e_3 = \tfrac{3}{2}$$

(2.5-3)

the solution of which is $e_2 = \tfrac{32}{9}$ volts and $e_3 = \tfrac{26}{9}$ volts. Then $e_4 = e_2 - e_3 = \tfrac{2}{3}$ volt,

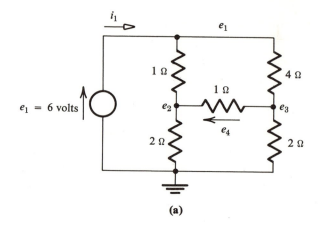

(a)

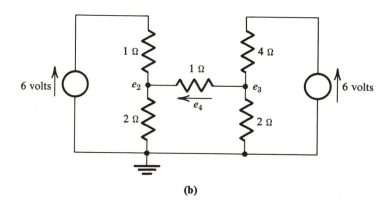

(b)

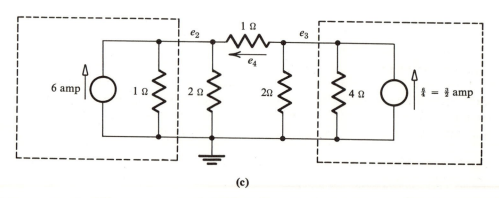

(c)

Figure 2.5-5

which is consistent with the answer to Example 2.3-4. Since the current i_1 is the sum of the currents through the two top resistances,

$$i_1 = \frac{6 - e_2}{1} + \frac{6 - e_3}{4} = \frac{22}{9} + \frac{7}{9} = \frac{29}{9} \text{ amp}$$

The total resistance seen by the source is $e_1/i_1 = \frac{54}{29} \,\Omega$.

As long as the only operation performed on the original equations is collecting like terms, the comments following Eqs. 2.5-2 on the sign and symmetry of the co-efficients are valid for any circuit with no controlled sources. The fact that Eqs. 2.5-3 obey these rules provides a partial check on the work in the last example.

We have not developed a general method of writing the simplified node equations directly from the circuit when voltage sources are present. In many problems, a series resistance can be associated with the voltage source. Such a combination may be converted to a current source in parallel with the resistance, as in Fig. 2.3-15 and Eq. 2.3-1. Even if there is no resistance in series with the voltage source, it is still possible to redraw the circuit with only current sources.

Excessive manipulations to replace voltage by current sources are not necessary, since writing a satisfactory set of node equations for a circuit with voltage sources is not difficult. For completeness, however, we shall indicate how the voltage source in Fig. 2.5-5a may be eliminated. The source forces the potential at the top of the 1 and 4 Ω resistances to be 6 volts with respect to the reference node. The *two* voltage sources in Fig. 2.5-5b perform this same function, so the currents through the resistances are unchanged. (If there were three other branches connected to one terminal of the voltage source, the original source could be replaced by a separate source in each of the three branches.) Now, each series combination of a voltage source in series with a resistance may be converted to a current source in parallel with the resistance, as in Fig. 2.5-5c. By Eqs. 2.5-2,

$$(1 + \tfrac{1}{2} + 1)e_2 - e_3 = 6$$
$$-e_2 + (1 + \tfrac{1}{2} + \tfrac{1}{4})e_3 = \tfrac{3}{2}$$

which is consistent with Eqs. 2.5-3.

The node analysis of a circuit with controlled voltage sources again involves the extra step of expressing the values of the controlled sources in terms of the node voltages. When like terms are collected, the comments following Eqs. 2.5-2 about the sign and symmetry of the coefficients will not be valid.

Example 2.5-5.   Find e_2/e_1 and e_o/e_1 in Fig. 2.5-6a, which is a model for two identical vacuum-tube amplifiers that have a common, unbypassed cathode resistance.

Solution.   In Fig. 2.5-6b, the choice of the reference node has been indicated by the heavy line, the series resistances R_o and r_p have been replaced by the single resistance $R_o + r_p$, and the potential of every node has been expressed in terms of the node voltage e_2 and the voltage sources. When writing a node equation at e_2, we notice that

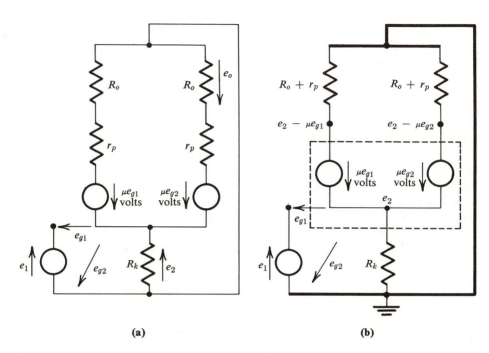

Figure 2.5-6

the current through each controlled source is also the current through $R_o + r_p$. The resulting equation is equivalent to setting the algebraic sum of the currents leaving the dashed rectangle equal to zero or to regarding a voltage source in series with a resistance as a single branch.

$$\frac{e_2}{R_k} + \frac{e_2 - \mu e_{g1}}{R_o + r_p} + \frac{e_2 - \mu e_{g2}}{R_o + r_p} = 0$$

Since $e_{g1} = -e_2 + e_1$ and $e_{g2} = -e_2$,

$$\left[\frac{1}{R_k} + \frac{2(1 + \mu)}{R_o + r_p}\right] e_2 = \frac{\mu}{R_o + r_p} e_1$$

or

$$\frac{e_2}{e_1} = \frac{\mu/(R_o + r_p)}{(1/R_k) + 2(1 + \mu)/(R_o + r_p)} = \frac{\mu R_k}{R_o + r_p + 2(1 + \mu)R_k}$$

By the voltage divider rule,

$$e_o = \frac{R_o}{R_o + r_p}(e_2 - \mu e_{g2}) = \frac{R_o(1 + \mu)e_2}{R_o + r_p}$$

$$\frac{e_o}{e_1} = \frac{\mu(1 + \mu)R_o R_k}{(R_o + r_p)[R_o + r_p + 2(1 + \mu)R_k]}$$

2.6

Loop Equations

Loop equations result when another systematic method, based upon the discussion in Section 2.4, is used for the analysis of circuits. In this method, it is usually best to treat each element as a separate branch, except for resistances in parallel, which should be combined into a single branch. In Section 2.4 it was demonstrated that, in a circuit containing B branches and N nodes, all the currents can be expressed in terms of $L = B - N + 1$ independent currents by Kirchhoff's current law. This is the first and crucial step in the loop method. The voltage of each purely resistive branch is given by $e = Ri$, so, if all the sources are independent voltage sources, all the unknown voltages can be expressed in terms of the L independent currents. Kirchhoff's voltage law will then yield L independent equations, which can be solved simultaneously for these unknown currents. The method is developed in the following example, which should be carefully studied.

Example 2.6-1.   Find all the currents in Fig. 2.6-1a.

Solution.   The first task is to express all the currents in terms of $L = B - N + 1$ $= 6 - 5 + 1 = 2$ independent currents. A general discussion of how to choose the necessary number of independent currents follows this example.

One procedure, which is suggested and justified by the comments immediately following Theorem 2.4-4, is summarized here. A tree, which is a collection of branches joining all the nodes without forming any closed paths, is chosen first. The currents of the links, which are those branches not belonging to the tree, always constitute a satisfactory set of independent currents. For the tree indicated by the heavy lines in Fig. 2.6-1b, the links are the 3 and 4 Ω resistances. As indicated in the figure, the current of every element is easily expressed in terms of the two link currents i_1 and i_2.

Instead of formally showing the current of every branch, the concept of *loop currents* may be used. Each of the two links, together with part of the tree but no other links, forms a unique closed path or loop. Loop currents are formed by letting the link currents circulate around these loops, as indicated in Fig. 2.6-1c. The actual current through any branch is understood to be the algebraic sum of the loop currents passing through it. Thus the current flowing down through the 6 Ω resistance is $i_1 - i_2$, which is consistent with Fig. 2.6-1b. Equivalently, the current up through this resistance is $i_2 - i_1$.

The act of drawing the loop currents automatically satisfied Kirchhoff's current law at every node. Since each loop current circulates around a closed path, the loop currents that enter a node always leave by another branch.

Now, by Ohm's law, the voltage across each resistance can be expressed in terms of the loop currents. The discussion following Theorem 2.4-3 showed that the loops defined by the links of a tree are independent (because each loop contains a different link), and that Kirchhoff's voltage law applied to each of the loops yields a set of L independent equations, which may be solved for the L loop currents.

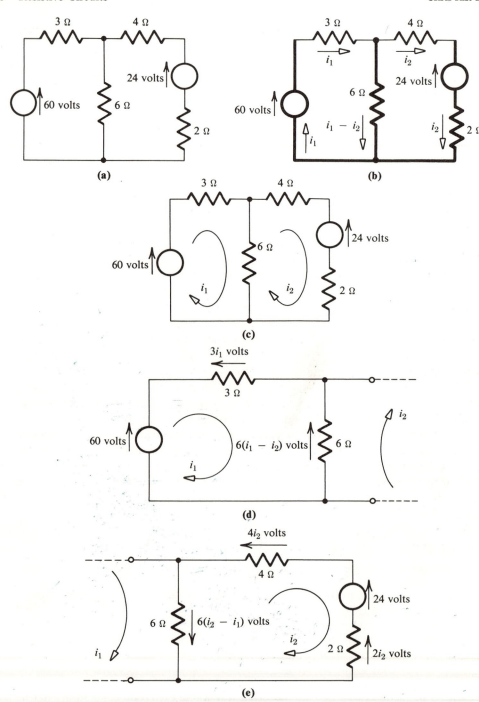

Figure 2.6-1

The voltages in the left-hand loop are shown in Fig. 2.6-1d. By Kirchhoff's voltage law,

$$6(i_1 - i_2) + 3i_1 - 60 = 0$$

For the right-hand loop, which has the voltages shown in Fig. 2.6-1e,

$$2i_2 + 24 + 4i_2 + 6(i_2 - i_1) = 0$$

Collecting like terms, we obtain

$$(6 + 3)i_1 - 6i_2 = 60$$
$$-6i_1 + (2 + 4 + 6)i_2 = -24$$

(2.6-1)

When these equations are solved simultaneously, $i_1 = 8$ amp and $i_2 = 2$ amp. The current flowing down through the 6 Ω resistance is $i_1 - i_2 = 6$ amp.

After Kirchhoff's voltage law has been applied to each of the L independent loops, the resulting equations are usually simplified by collecting like terms. In order to see whether the simplified equations can be written down directly from the circuit, recall Example 2.6-1. In forming the original loop equations, the first loop was traversed in a direction opposite to the loop-current arrow, and the algebraic sum of the voltages equated to zero. If only i_1 passes through a resistance, the voltage is Ri_1; but if another loop current i_x also passes through the resistance, the voltage becomes $R(i_1 \pm i_x)$. The sign is plus or minus, respectively, according to whether i_x flows through the resistance in the same or opposite direction as i_1.

When the voltage-law equation for the first loop is simplified, as in Eqs. 2.6-1, the coefficient of i_1 is the sum of the resistances contained in the loop. The coefficient of i_2 is plus or minus the sum of the resistances that are common to loops 1 and 2. Its sign is plus if the two loop currents flow through the common resistances in the same direction. The equation for the first loop also contains a term representing the voltage source. If any terms representing sources are put on the right-hand side of the equation, they represent the algebraic sum of the voltage sources in the loop and are positive if the sources tend to make the loop current i_1 flow in the assumed positive direction.

For a circuit that has L loop currents and only sources that are independent voltage sources, the simplified loop equations have the following form:

$$R_{11}i_1 \pm R_{12}i_2 \pm \cdots \pm R_{1L}i_L = e_1$$
$$\pm R_{21}i_1 + R_{22}i_2 \pm \cdots \pm R_{2L}i_L = e_2$$
$$\cdots\cdots\cdots\cdots\cdots\cdots\cdots\cdots$$
$$\pm R_{L1}i_1 \pm R_{L2}i_2 \pm \cdots + R_{LL}i_L = e_L$$

(2.6-2)

where i_1 through i_L denote the loop currents and where

R_{jj} = the sum of the resistances in loop j

$R_{jk} = R_{kj}$ = the sum of the resistances that are common to loops j and k $(j \neq k)$

e_j = the algebraic sum of the source voltages in loop j, where a positive term represents a source that tends to produce a current in the direction of i_j

It is assumed that a voltage-law equation is written around each of the loops defined by these currents and that the currents in each equation appear in the order in which the loops are examined.[†]

The coefficients on the left-hand side of Eqs. 2.6-2 may be arranged in a matrix:

$$\begin{bmatrix} R_{11} & \pm R_{12} & \cdots & \pm R_{1L} \\ \pm R_{21} & R_{22} & \cdots & \pm R_{2L} \\ \cdots\cdots\cdots\cdots\cdots\cdots\cdots\cdots \\ \pm R_{L1} & \pm R_{L2} & \cdots & R_{LL} \end{bmatrix}$$

Since $R_{jk} = R_{kj}$, the off-diagonal coefficients are symmetrical about the main diagonal. The sign before R_{jk} and R_{kj} is positive if and only if the loop currents i_j and i_k flow in the same direction through the resistances that are common to these two loops. With these rules, Eqs. 2.6-1 can be written down directly from Fig. 2.6-1c. Even if the loop equations are not immediately written down in simplified form, the symmetry property provides a useful check on the work.

It is important to be able to draw a set of $L = B - N + 1$ independent loop currents for a given circuit rapidly. Figure 2.6-1 is an example of a planar circuit, i.e., one that can be drawn on a plane surface with no intersecting branches (except where there is an electrical connection). There is a particularly simple way of selecting a satisfactory set of loop currents for a circuit that is drawn with no crossing wires. The discussion following Theorem 2.4-3 defined a mesh of such a circuit as a loop that does not enclose any other branches. If the branches are visualized as the wooden partitions in a window, then each window pane defines a mesh. In Section 2.4 it was shown that there are L meshes and that the meshes constitute a set of independent loops. Thus, the number of loop currents that are required may be found by counting the meshes. Furthermore, the currents circulating around the meshes constitute a satisfactory set of loop currents. In Fig. 2.6-1c, the currents i_1 and i_2 flow around the two meshes.

The simplified set of loop equations is particularly easy to obtain when the loop currents coincide with the meshes. In such a case, if all the loop currents are drawn clockwise (or all counterclockwise), the signs of the off-diagonal terms are negative. Some of the later examples show that the set of loop currents defined by the meshes is not always the most convenient one. Furthermore, the use of meshes is restricted to planar circuits.

Example 2.6-2.   Find the voltage e_4 in Fig. 2.6-2a.

[†] It is possible, although very unusual, to write the voltage-law equations around a different set of L independent loops than those defined by the loop currents. In such a case, the following comments about the sign and symmetry of the coefficients do not apply.

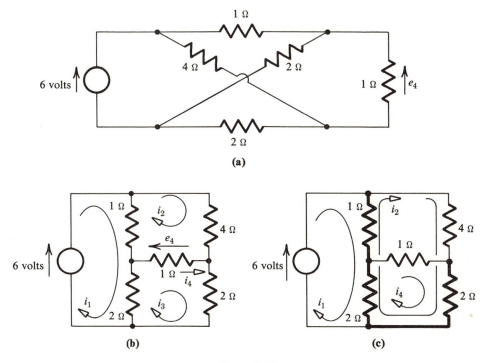

Figure 2.6-2

Solution. The circuit is planar, because it can be redrawn without crossing wires, as in Fig. 2.6-2b. From the discussion associated with Eqs. 2.6-2, the simplified loop equations are

$$3i_1 - i_2 - 2i_3 = 6$$

$$-i_1 + 6i_2 - i_3 = 0$$

$$-2i_1 - i_2 + 5i_3 = 0$$

When the equations are solved simultaneously, $i_2 = \frac{7}{9}$ amp and $i_3 = \frac{13}{9}$ amp, so $i_4 = i_3 - i_2 = \frac{2}{3}$ amp and $e_4 = \frac{2}{3}$ volt, which is consistent with Example 2.5-4.

In this solution, it was necessary to solve for two of the unknown loop currents in order to find the desired voltage. The computational work is reduced if different loop currents are chosen, so that only a single loop current passes through the resistance whose voltage is sought. For the choice of loops in Fig. 2.6-2c,[†] the three simplified loop equations are

$$3i_1 - 3i_2 - 2i_4 = 6$$

$$-3i_1 + 9i_2 + 4i_4 = 0$$

$$-2i_1 + 4i_2 + 5i_4 = 0$$

[†] The heavy lines in the figure constitute the tree whose links correspond to the loop currents.

Notice that the coefficient matrix for the left side of the equations is symmetrical about the main diagonal, but that the off-diagonal terms are not all negative. The solution of these equations gives $i_4 = \frac{2}{3}$ amp, so $e_4 = \frac{2}{3}$ volt.

After determining the number of independent loop currents (by counting meshes or by using the relationship $L = B - N + 1$), it is not necessary to make the loop currents coincide with either the meshes or with the links of some tree. One way to select a set of independent loop currents is to let each loop include a branch that was not part of any previous loop. It is not, in general, sufficient to merely make sure that at least one loop current passes through each element, and it is never permissible to choose fewer than L loop currents.

As a flagrant example of an incorrect choice of loop currents, consider Fig. 2.6-3a, in which at least one current does pass through each element. If a voltage-law equation is written around each of the two loops,

$$8i_1 + 3i_2 = -18$$
$$3i_1 + 11i_2 = 36$$

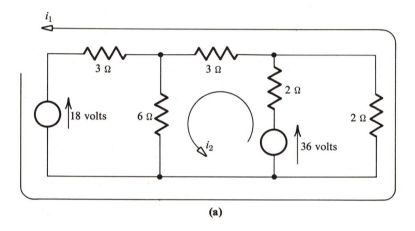

(a)

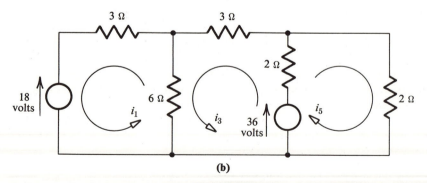

(b)

Figure 2.6-3

from which $i_1 = -3.87$ amp and $i_2 = 4.33$ amp, which is not consistent with the answers found in Section 2.4 for the same circuit. A correct answer should not be expected, since the loops in Fig. 2.6-3a infer that the current through the 6 Ω resistance and the 36 volt source (and also through the resistances on the far left and far right) are the same, which is not necessarily true. From another point of view, the solution satisfies only two of the three independent voltage-law equations that may be written. For the left-hand mesh, $18 + 3i_1 - 6i_2 = 0$, which is not satisfied by the above values. One correct set of loop currents, which corresponds to the solution in Section 2.4, is shown in Fig. 2.6-3b.

If one of the branches of a circuit is a current source, its voltage is an additional unknown, which cannot be easily expressed in terms of the loop currents. If, however, exactly one loop current is drawn through the source, it will have the value of the current source, leaving only $L - 1$ unknown currents. To solve for these unknown currents, a voltage-law equation is written around every loop except the one containing the current source, which should be avoided, since its voltage is unknown. The voltage across the current source can be found by an additional step at the conclusion of the problem.

Example 2.6-3.   Find the current through the 2 Ω resistance in Fig. 2.6-4a.

Solution.   Only one loop current should be drawn through each of the current sources, and, if possible, only one loop current should be drawn through the 2 Ω resistance.[†] The choice of loops shown in the figure does have these characteristics. Next, a voltage-law equation is written around each of the loops for which the current is unknown:

$$12i_1 - 6i_2 \qquad - 4i_4 = 0$$

$$-6i_1 + 18i_2 + 3i_3 - 8i_4 = 0$$

Since $i_3 = 10$ and $i_4 = 12$,

$$12i_1 - 6i_2 = 48$$

$$-6i_1 + 18i_2 = 66$$

$$(2.6\text{-}3)$$

Solved simultaneously, these equations yield $i_1 = 7$ amp.

In the development of the simplified loop equations 2.6-2, it was assumed that the only sources are independent voltage sources. Although, as in Example 2.6-3, it is not difficult to write loop equations when current sources are present, it is always possible to convert all sources to voltage sources. A current source in parallel with a resistance may be changed to a voltage source in series with the resistance, as in Fig. 2.3-15 and Eq. 2.3-1. There is, however, no resistance in parallel with the 12 amp source in Fig. 2.6-4a. This source causes 12 amp to leave node C and enter

---

[†] For any circuit that has a solution, it is possible to find some tree such that all the current sources are links. One way to ensure that only one loop current passes through each current source is to use these links to define the loops, as explained in the discussion following Theorem 2.4-3.

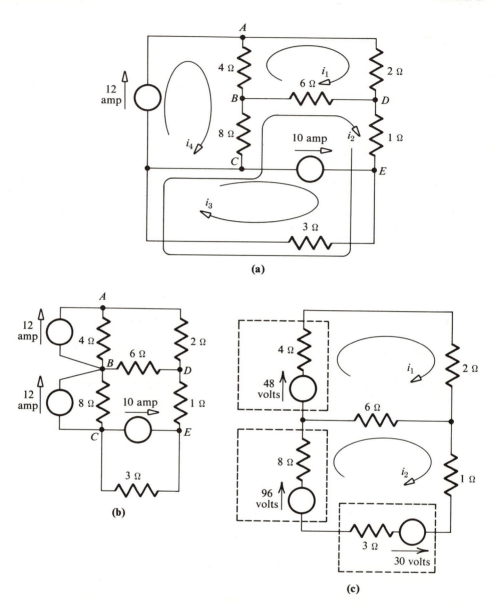

Figure 2.6-4

node A, a function that can be performed equally well by the *two* 12 amp sources in Fig. 2.6-4b. Note that these two sources supply no net current to node B. Now all the current sources can be converted to voltage sources, as in Fig. 2.6-4c, from which Eqs. 2.6-3 may be written directly.

The only sources in the examples considered thus far have been independent sources. If controlled sources are present, an extra step is required. After a set of loop equations is written, the values of the controlled sources must be expressed in terms of the loop currents. This extra step means that Eqs. 2.6-2 and the previous comments about the signs and symmetry of the coefficients are not valid for the final form of the loop equations for circuits with controlled sources.

Example 2.6-4. Find the voltage e_o in Fig. 2.5-4, which is repeated in Fig. 2.6-5, by writing loop equations.

Solution. The loop currents i_1, i_2, and i_3 are selected so that only a single loop current passes through a current source and only one loop current passes through the resistance whose voltage is sought. The voltage-law equation around the loop not containing a current source is

$$-2i_1 + 12i_2 + i_3 = 0$$

Since $i_1 = 4$ amp and $i_3 = 5e_2 = 5(2)(4 - i_2)$,

$$-8 + 40 - 10i_2 + 12i_2 = 0$$

so $i_2 = -16$ amp, and $e_o = -144$ volts, which is consistent with Example 2.5-3.

Example 2.6-5. Find expressions for e_2 and e_o in Fig. 2.6-6.

Solution. The loop currents i_1 and i_2 are shown in the figure. The two voltage-law equations are

$$2(R_o + r_p)i_1 - (R_o + r_p)i_2 = \mu(e_{g1} - e_{g2})$$

$$-(R_o + r_p)i_1 + (R_o + r_p + R_k)i_2 = \mu e_{g2}$$

Since $e_{g1} - e_{g2} = e_1$ and $e_{g2} = -R_k i_2$,

$$2(R_o + r_p)i_1 - (R_o + r_p)i_2 = \mu e_1$$

$$-(R_o + r_p)i_1 + [R_o + r_p + (1 + \mu)R_k] i_2 = 0$$

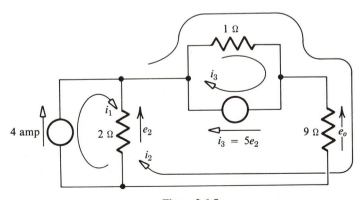

Figure 2.6-5

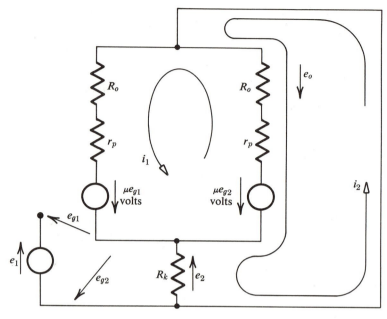

Figure 2.6-6

Solving these equations simultaneously, we obtain

$$i_1 = \frac{\mu[R_o + r_p + (1 + \mu)R_k]e_1}{(R_o + r_p)[R_o + r_p + 2(1 + \mu)R_k]}$$

$$i_2 = \frac{\mu(R_o + r_p)e_1}{(R_o + r_p)[R_o + r_p + 2(1 + \mu)R_k]}$$

Then

$$e_2 = R_k i_2 = \frac{\mu R_k e_1}{R_o + r_p + 2(1 + \mu)R_k}$$

$$e_o = R_o(i_1 - i_2) = \frac{\mu(1 + \mu)R_o R_k e_1}{(R_o + r_p)[R_o + r_p + 2(1 + \mu)R_k]}$$

which are the same as the answers to Example 2.5-5.

2.7

Duals

Suppose that the sets of equations describing two different circuits are identical, except for a change in symbols. As soon as one of the circuits is solved, the solution of the other can be immediately written down. This situation is examined briefly in this section and is discussed more thoroughly in Chapter 11.

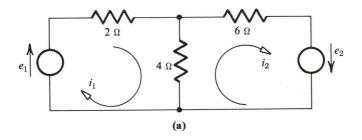

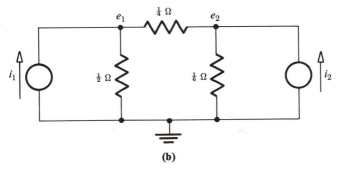

Figure 2.7-1

The loop equations for Fig. 2.7-1a are

$$4(i_1 - i_2) + 2i_1 = e_1$$
$$6i_2 + 4(i_2 - i_1) = e_2$$

(2.7-1)

while the node equations for Fig. 2.7-1b are

$$4(e_1 - e_2) + 2e_1 = i_1$$
$$6e_2 + 4(e_2 - e_1) = i_2$$

(2.7-2)

Except for the symbols, the two sets of equations are identical. If the currents in Fig. 2.7-1a are found, the voltages in Fig. 2.7-1b are also known. Whenever the loop equations of one circuit are identical with the node equations of another, except for interchanging the roles of voltage and current, the circuits are called *duals*.

To construct the dual of a given circuit, we can replace i by e, and e by i in the original equations, and then try to draw a circuit corresponding to the new equations. If, for example, Fig. 2.7-1a were given, Eqs. 2.7-2 could be obtained by this procedure. When trying to draw the dual circuit, notice that the terms on the right side of the new equations are currents, so that the coefficients on the left side must have units of mhos. The presence of e_1 and e_2 indicates that there must be two nodes in addition to the reference node. The first equation requires a $\frac{1}{4}$ Ω resistance between nodes 1 and 2 and a $\frac{1}{2}$ Ω resistance between node 1 and the reference node. For the second equation, there must be a $\frac{1}{4}$ Ω resistance between nodes 1 and 2 and

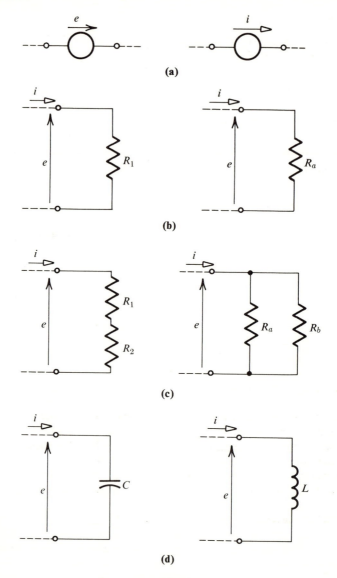

Figure 2.7-2

a $\frac{1}{6}$ Ω resistance between node 2 and the reference node, which results in Fig. 2.7-1b. A similar process may be used to construct Fig. 2.7-1a from Fig. 2.7-1b.

All planar circuits have a unique dual, which may be found by the method used in the previous paragraph.[†] It is instructive, however, to see when individual parts

---

[†] A simple but general graphical method, which eliminates the need for writing equations, is described in M. G. Gardner and J. L. Barnes, *Transients in Linear Systems* (New York: Wiley, 1942), pp. 47–49.

of two circuits are duals. The two sources in Fig. 2.7-2a are duals when their values are numerically equal. In Fig. 2.7-2b, $e = R_1 i$ and $i = G_a e$, so the components are duals if R_1 and $G_a = 1/R_a$ have the same values. One special case is $R_1 = 0$ and $G_a = 0$, so that short and open circuits are duals. In Fig. 2.7-2c, $e = (R_1 + R_2)i$ and $i = (G_a + G_b)e$, so the series and parallel combinations are duals provided that corresponding components are duals.

The discussion can be extended to circuits with capacitance and inductance. Since $i = C(de/dt)$ and $e = L(di/dt)$, a capacitance and an inductance with the same numerical value are duals. A partial list of dual quantities is given in Table 2.7-1. The use of this table to construct a dual circuit is illustrated by the following examples.

TABLE 2.7-1.   Dual Quantities

Voltage	Current
Resistance	Conductance
Capacitance	Inductance
Short circuit	Open circuit
Series connection	Parallel connection
Loop	Node

Example 2.7-1.   Find the dual of Fig. 2.7-3a.

Solution.   The parallel combination of the 3 and 6 Ω resistances is replaced by a series combination of $\frac{1}{3}$ and $\frac{1}{6}$ Ω. Since the original combination was in series with a 4 Ω resistance and a voltage source, in the dual of Fig. 2.7-3b it is in parallel with a $\frac{1}{4}$ Ω resistance and a current source. As a check, the student may show that the loop equations for Fig. 2.7-3a are identical, except for a change in symbols, with the node equations for Fig. 2.7-3b. If e_1 and i_1 are numerically equal, then so are e_o and i_o.

Example 2.7-2.   Construct the dual of Fig. 2.7-4a.

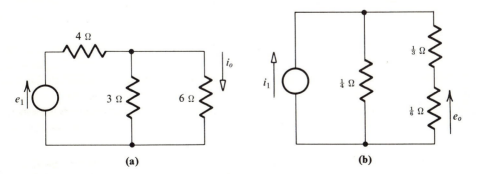

(a)   (b)

Figure 2.7-3

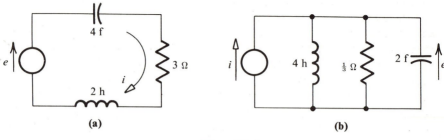

(a)   **(b)**

Figure 2.7-4

Solution. The dual follows immediately from Table 2.7-1 and is given in Fig. 2.7-4b. The loop equation for (a) and the node equation for (b) are, respectively,

$$2\frac{di}{dt} + 3i + \frac{1}{4}\int i\,dt = e$$

and

$$2\frac{de}{dt} + 3e + \frac{1}{4}\int e\,dt = i$$

The properties and theorems of circuit analysis remain valid if all words are replaced by their dual quantities. The dual of Thévenin's theorem (stating that a two-terminal circuit may be replaced by its open-circuit voltage in series with the dead network) is Norton's theorem (stating that a two-terminal circuit may be replaced by its short-circuit current in parallel with the dead network).

2.8

Summary

Any circuit may be solved by the direct application of the element laws and Kirchhoff's laws. However, it is often helpful to replace a combination of elements by an equivalent circuit when calculating external voltages and currents. If the combination has only two external terminals and consists only of resistances and possibly controlled sources, it may be replaced by a single resistance R_{eq}. A two-terminal network that contains independent sources may be replaced by the Thévenin or Norton equivalent circuits in Figs. 2.3-1b or 2.3-1c. The dead network in this figure is formed by replacing independent (but not controlled) voltage and current sources by short and open circuits, respectively. If the dead network does not contain capacitances or inductances, it may be replaced by a single resistance. Also useful is the superposition theorem, which states that the response to several independent sources is the sum of the responses to each independent source with the remaining independent sources dead.

Loop and node equations are two systematic methods for the simultaneous solution of the element laws and Kirchhoff's laws, and they are especially valuable in complicated problems. In the node method, the voltage of each node with respect to a reference node (whose potential may be assigned a value of zero) is indicated on the circuit diagram. Then the current through each resistive branch may be mentally expressed in terms of these $N - 1$ voltages, where N is the number of nodes. A current-law equation may be written at each node whose voltage is unknown in order to obtain the necessary number of independent equations. An equation is not needed at the reference node, and equations involving the unknown current of a voltage source should be avoided.

In the second method, loop currents are drawn around each of $L = B - N + 1$ independent loops, and a voltage-law equation is written around each of these loops (except those that contain a current source). It is advisable to draw only one loop current through any current source and to avoid an equation involving the unknown voltage of such a source. The actual current of any branch is the algebraic sum of the loop currents passing through it.

The concept of duality is introduced in Section 2.7, because the solution of a circuit can be immediately written down if its dual has been already solved. The student should now be able to systematically analyze any circuit containing only sources and resistances. The next chapter begins a study of circuits that include capacitances and inductances.

PROBLEMS

2.1 Find the power supplied to the 6 Ω resistance in Fig. P2.1.

2.2 Find the equivalent resistance for each of the two-terminal networks in Fig. P2.2.

2.3 Choose R_1 and R_2 in Fig. P2.3 so that $i_m = 2$ milliamp (ma) when $R_x = 0$, and $i_m = 1$ ma when $R_x = 2000$ Ω. If R_m represents the resistance of an ammeter (a

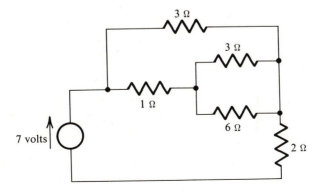

Figure P2.1

current-measuring device), the circuit within the dashed lines is the model of an ohmmeter, which can be used to measure the value of an unknown resistance R_x. (HINT: Find an expression for i_m when $R_x = 0$. Then notice that all currents will be halved when R_x equals the equivalent resistance for R_1, R_2, and R_m.)

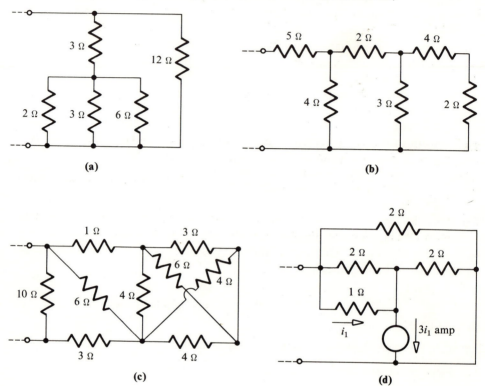

(a)   **(b)**

(c)   **(d)**

Figure P2.2

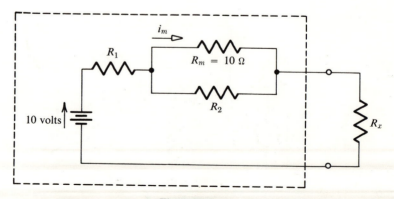

Figure P2.3

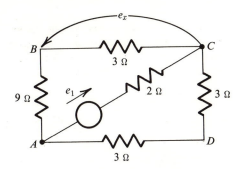

Figure P2.4

2.4 A voltmeter indicates that point B is 4 volts positive with respect to point D in Fig. P2.4. Find the voltage e_x from point C to point B.

2.5 Verify the results given in Fig. 2.1-4. The elements are assumed to contain no stored energy at $t = 0$.

2.6 Find the current i_4 in Fig. P2.6 by temporarily assuming that $e_3 = 3$ volts and by using the procedure and arguments in Examples 2.2-1 and 2.2-2.

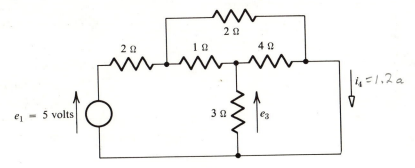

Figure P2.6

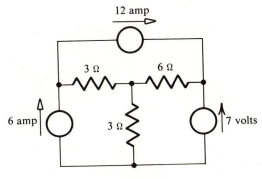

Figure P2.7

2.7 Find the power supplied by each source in Fig. P2.7 by using the superposition theorem. Verify that the net power supplied by the sources equals the power dissipated in the resistances.

2.8 Find $e_o(t)$ in Fig. P2.8.

2.9 Find the voltage between points A and B in Fig. P2.9.

2.10 Under what conditions will $e_o = 0$ in Fig. P2.10?

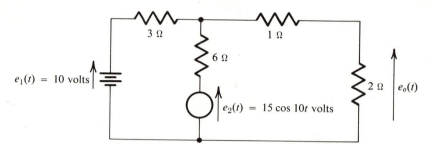

Figure P2.8

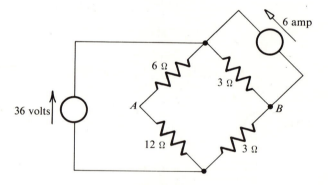

Figure P2.9

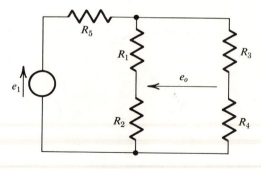

Figure P2.10

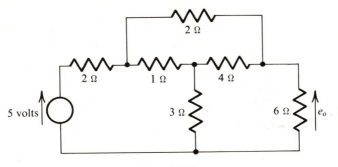

Figure P2.11

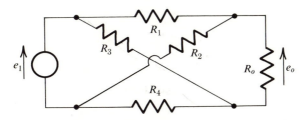

Figure P2.12

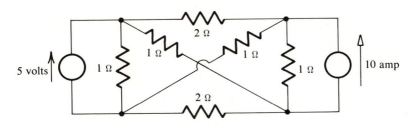

Figure P2.13

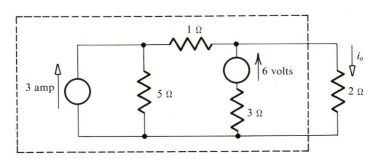

Figure P2.14

2.11 Use the results of the previous problem to find e_o for the bridged-T network in Fig. P2.11.

2.12 Under what conditions will $e_o = 0$ in Fig. P2.12? If $R_4 = R_1$ and $R_3 = R_2$, derive (by Thévenin's theorem) a general expression for e_o.

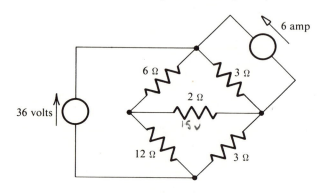

6 amp

6 Ω   3 Ω

2 Ω

36 volts

15∨

12 Ω   3 Ω

Figure P2.15

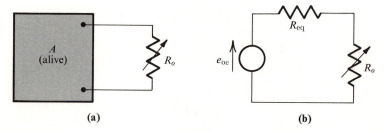

A
(alive)

R_o

e_{oc} R_{eq} R_o

(a)   (b)

Figure P2.16

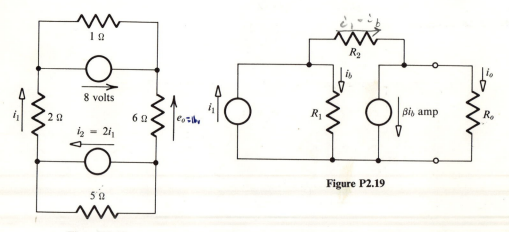

1 Ω

8 volts

i_1 $2\,\Omega$ $6\,\Omega$ e_o

$i_2 = 2i_1$

$5\,\Omega$

Figure P2.18

$\dot{c}_1 \cdot \dot{c}_b$

R_2

i_b i_o

i_1

R_1 βi_b amp R_o

Figure P2.19

2.13 Use the results of Problem 2.12 to find the voltage across the current source in Fig. P2.13.

2.14 Replace the network within the dashed lines in Fig. P2.14 by a Thévenin equivalent circuit, and then find the current i_o.

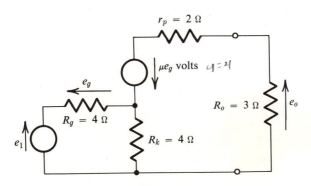

Figure P2.20

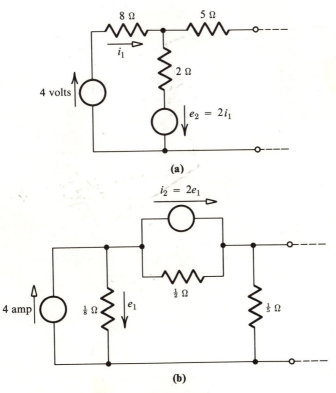

(a)

(b)

Figure P2.21

2.15 By the use of Thévenin's theorem and the answer to Problem 2.9, find the power dissipated in the 2 Ω resistance in Fig. P2.15.

2.16 Figure P2.16a shows a variable "load resistance" R_o connected to the terminals of network A. In Fig. P2.16b, the network A is represented by its Thévenin equivalent

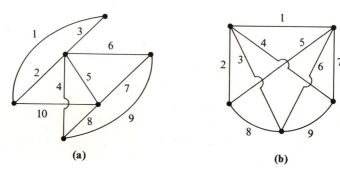

(a)   (b)

Figure P2.22

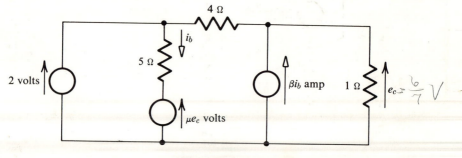

Figure P2.23

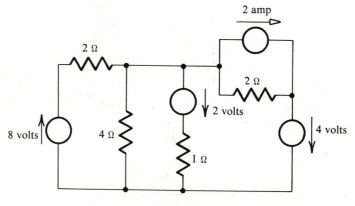

Figure P2.24

circuit. Find the value of $R_o(R_o \geq 0)$ that will result in the maximum power delivered to R_o.

2.17 Solve Problem 2.11, *without* using the results of Problem 2.10, by replacing everything except the 6 Ω resistance by a Thévenin equivalent circuit. (HINT: Calculate the equivalent resistance of the dead network by Eq. 2.3-1 and the results of Problem 2.6.)

2.18 Replace everything except the 6 Ω resistance in Fig. P2.18 by a Thévenin equivalent circuit, and then find e_o.

2.19 Use Thévenin's or Norton's theorem to find the current gain i_o/i_1 in Fig. P2.19, which is a simplified model of a transistor feedback amplifier.

2.20 Use Thévenin's or Norton's theorem to find the voltage gain e_o/e_1 in the vacuum-tube circuit model of Fig. P2.20 when $\mu = 4$.

2.21 Find the Thévenin and Norton equivalent circuits for the networks in Fig. P2.21.

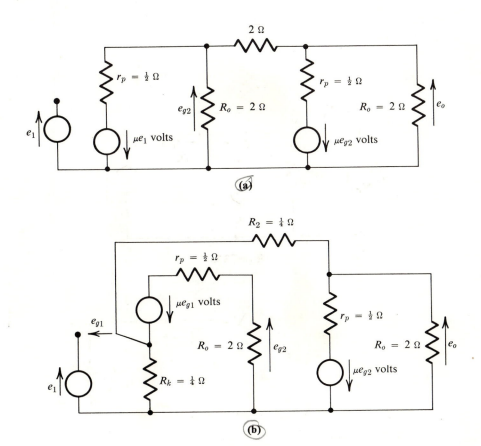

Figure P2.25

2.22 How many independent loops does each of the graphs in Fig. P2.22 contain? In each case, list a maximum set of independent loops.

LOOP AND NODE EQUATIONS. In addition to the following problems, many of the previous circuits can be conveniently solved by these methods.

2.23 Find the voltage across the current source in Fig. P2.23 (a) by writing a node equation, and (b) by writing three loop equations.

2.24 Find e_c in Fig. P2.24 by the node- and loop-equation methods when $\beta = 10$ and $\mu = 2$.

2.25 Use node equations to find the voltage gain e_o/e_1 for each of the circuits in Fig. P2.25 when $\mu = 10$.   Fig (a),(b)

2.26 Find the current gain i_o/i_1 in Fig. P2.26 (a) when $R_2 = 0$ and (b) when $R_2 = 1\ \Omega$.

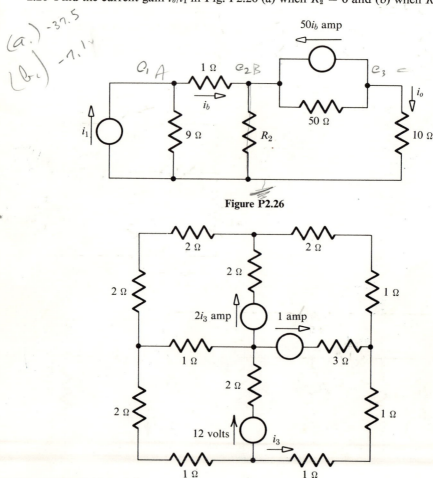

Figure P2.26

Figure P2.27

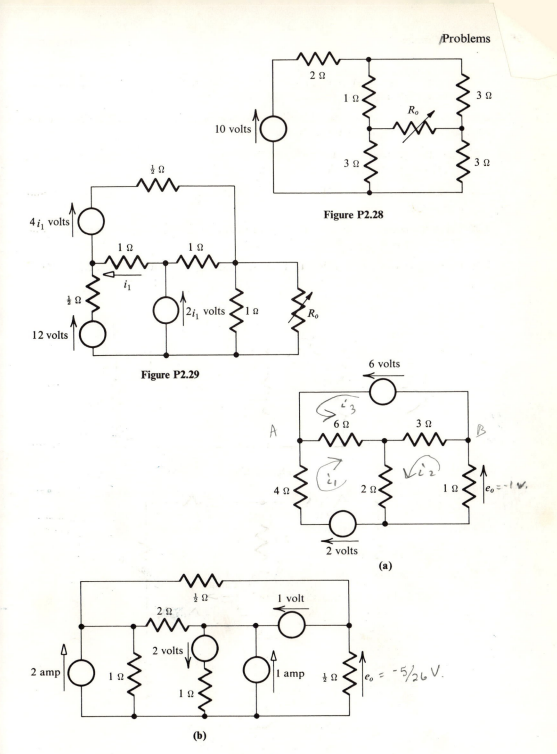

Figure P2.28

Figure P2.29

(a)

(b)

Figure P2.30

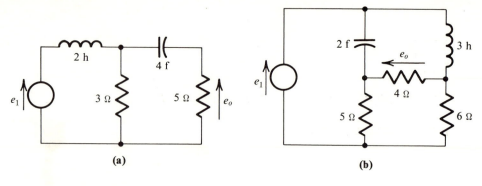

(a)

(b)

K closes at t = 0

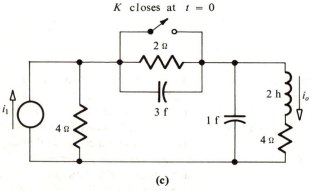

(c)

Figure P 2.31

2.27 Find i_3 in Fig. P2.27 by writing loop equations.

2.28 What value of R_o in Fig. P2.28 will result in the maximum power delivered to R_o? (The results of Problem 2.16 are useful in the solution.)

2.29 Repeat Problem 2.28 for the circuit shown in Fig. P2.29.

2.30 Find e_o for the circuits in Fig. P2.30 by writing node equations.

2.31 Construct the dual circuit for each of the networks in Fig. P2.31.

2.32 Construct and solve the duals of Problems 1.11, 1.12, and 1.13.

3

Circuits with Capacitance

or Inductance

The resistive circuits of the previous chapter were described by ordinary algebraic equations. This is a consequence of the fact that the voltage and current of a resistance are directly proportional, and that the value of any controlled source is proportional to the quantity that controls it. If such a circuit has a single input $x(t)$ and a single output $y(t)$, then the ratio $y(t)/x(t)$ is a constant that depends only upon the circuit, a fact that was used in Examples 2.2-1 and 2.2-2. The output and input waveshapes are proportional to each other, and the output at any time t_0 depends upon the input at that instant, but not upon previous values of the input.

Since capacitance and inductance are defined in terms of derivatives and integrals, circuits containing these elements are described by integral-differential equations. In general, the input and output waveshapes are not similar, and the response at time t_0 depends upon the values of the input for all $t \leq t_0$.

This chapter is restricted to circuits that may be described by first-order, or by very simple second-order differential equations. Among the inputs that are considered are the functions shown in Fig. 3.1-1: (a) an input of constant value suddenly applied at $t = 0$, (b) a linearly increasing input, and (c) a rectangular pulse. These and certain other inputs are closely related to a class of functions known as *singularity functions* and are discussed in Section 3.2. The response to the singularity functions often can be determined more easily than the response to some arbitrary input. In Sections 3.4 through 3.6, which may be omitted without any

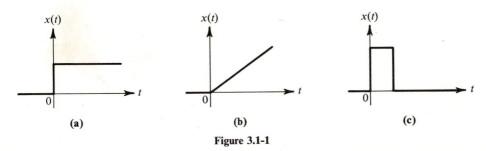

Figure 3.1-1

125

serious lack of continuity, arbitrary inputs are decomposed into the sum of singularity functions. Once the responses to the singularity functions are known, the response to the arbitrary input can be found by the superposition theorem.

In Chapter 4 the classical solution of differential equations is reviewed and the response of more complicated circuits is considered.

3.1

The Free and the Step Response

The circuits in this section can be described by a single differential equation for all values of t greater than some reference time, which is usually taken to be $t = 0$. The input is given for all $t > 0$, so the differential equation may be solved for the desired output. The result will be an expression that is valid for all $t > 0$, but that contains a constant of integration. This constant can be evaluated, however, if the value of the voltage across the capacitance or the current through the inductance is known at $t = 0+$. The "$0+$" notation refers to an instant immediately after $t = 0$.

It is necessary to know that, if all voltages and currents remain finite, the voltage across a capacitance and the current through an inductance cannot change instantaneously. This means that the values of these quantities are the same at $t = 0+$ and at $t = 0-$ (just before $t = 0$). In order to justify this statement, recall from Section 1.3 that the energy stored in the field of a capacitance is

$$w(t) = \tfrac{1}{2} C e^2(t)$$

while for an inductance

$$w(t) = \tfrac{1}{2} L i^2(t)$$

The power supplied to any component is

$$p(t) = e(t)i(t) = \frac{dw}{dt} = \lim_{\Delta t \to 0} \frac{\Delta w}{\Delta t}$$

If $e(t)$ and $i(t)$ are finite, $p(t)$ remains finite and Δw must approach zero as Δt approaches zero. In such a case, $w(t)$ and hence $e_C(t)$ and $i_L(t)$ cannot change instantaneously.

> **Example 3.1-1.** Suppose that at some time in the past a charge has been placed upon the capacitance in the circuit in Fig. 3.1-2a (by some circuit that is not shown), so that $e(0-) = A$ volts. If the switch K closes at $t = 0$, find $e(t)$ for all $t > 0$.

> **Solution.** For $t > 0$, $e(t)$ is the voltage across R and C, as in Fig. 3.1-2b. We set the algebraic sum of the currents leaving the upper node equal to zero to obtain

$$\frac{1}{R} e + C \frac{de}{dt} = 0$$

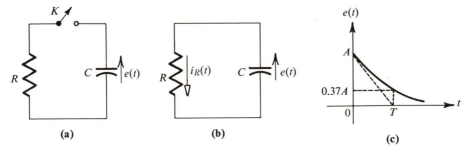

Figure 3.1-2

and separate the variables $e(t)$ and t.

$$\frac{de}{e} = -\frac{dt}{RC}$$

Integration of both sides of this equation gives

$$\ln e(t) = -\frac{t}{RC} + \ln K$$

where $\ln K$ is a constant of integration. Thus $\ln [e(t)/K] = -t/RC$, and

$$e(t) = K\epsilon^{-t/RC}$$

for $t > 0$. Because the voltage across the capacitance does not change instantaneously, $e(0+) = e(0-) = A$. The expression for $e(t)$ for $t > 0$ must be valid at $t = 0+$, so we may replace $e(t)$ and t by A and zero, respectively. Thus, $K = A$, and for all $t > 0$

$$e(t) = A\epsilon^{-t/RC}$$

which is sketched in Fig. 3.1-2c. The shape of the curve can be anticipated by noting that $i_R(t) = e(t)/R$ in Fig. 3.1-2b. The initial voltage $e(0+) = A$ produces a current through the resistance, which begins to discharge the capacitance. The reduced voltage in turn reduces the size of the current, so the capacitance is discharged more slowly, and the slope of the curve of $e(t)$ becomes less negative. Both the voltage and the current decay toward zero.

Since the circuit in Fig. 3.1-2b is excited by some energy initially stored in the capacitance, and not by an external source, the response is called the *free response* or the *natural behavior*. For some simple circuits, the free response has the form $K\epsilon^{-t/T}$, where T is called the *time constant* of the circuit. It is the time required for the response to decay to $K\epsilon^{-1}$, which is 37% of its initial value, as indicated in Fig. 3.1-2c. Another interpretation of the time constant is based upon the fact that the initial slope of the response is

$$\left[\frac{d}{dt} K\epsilon^{-t/T}\right]_{t=0+} = -\frac{K}{T}$$

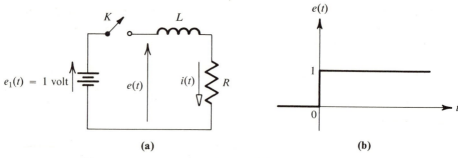

(a)   **(b)**

Figure 3.1-3

If this initial rate of decay were maintained, it would take $K/(K/T) = T$ seconds for the response to decay to zero.

Figure 3.1-3a shows an example of a circuit with an external source. For $t < 0$, the switch K is open, and $i(t) = 0$. The switch closes at $t = 0$, and the desired response is $i(t)$ for $t > 0$. Since the RL branch is unexcited for $t < 0$, $e(t) = 0$ for $t < 0$ and 1 volt for $t > 0$, as in Fig. 3.1-3b. In Example 1.4-2, the current was shown to be

$$i(t) = \frac{1}{R}\left(1 - \epsilon^{-(R/L)t}\right)$$

for $t > 0$. This response is often called the *step response*, because there is a step, or discontinuity in the plot of $e(t)$.

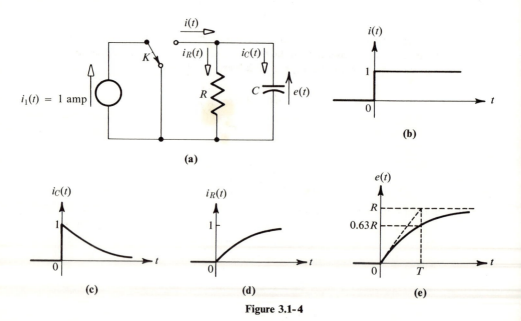

(a)

(c)   **(d)**   **(e)**

Figure 3.1-4

Example 3.1-2. For $t < 0$, the switch K in Fig. 3.1-4a is in the lower position and $e(t) = 0$. At $t = 0$, the switch is thrown to the upper position, so $i(t)$ has the waveshape shown in Fig. 3.1-4b. Find and sketch $e(t)$.

Solution. Let us try to anticipate the output waveshape before writing and solving the differential equation describing the circuit. The voltage $e(t) = e_R(t) = e_C(t)$, which is zero for $t < 0$ and which is across the capacitance, does not change instantaneously, since the source is finite. At $t = 0+$, no part of the source current can pass through the resistance, since the voltage $e(t) = Ri_R(t)$ would then jump to a nonzero value. Thus, the entire source current initially flows into the capacitance, which begins to accumulate charge. As the charge and voltage increase, the current $i_R(t) = e(t)/R$ increases, so $i_C(t)$ decreases, which in turn reduces the rate at which the voltage rises. When $i_R(t)$ becomes nearly equal to the 1 amp source current, $e(t) = R(1)$, $i_C(t)$ approaches zero, and there is no further increase in voltage. Curves of $i_C(t)$, $i_R(t)$, and $e(t)$ are sketched in Fig. 3.1-4. In Example 3.1-1, the circuit's speed of response was described by the time constant $T = RC$ in the exponential factor $\epsilon^{-t/T}$. Assuming that the same time constant applies, we should expect that the curves in Fig. 3.1-4 would be described by the following equations for $t > 0$:

$$i_C(t) = \epsilon^{-t/RC}$$

$$i_R(t) = 1 - \epsilon^{-t/RC}$$

$$e(t) = R(1 - \epsilon^{-t/RC})$$

To verify these results, we first write a node equation for $t > 0$ at the upper right-hand node.

$$C\frac{de}{dt} + \frac{1}{R}e = 1$$

The shortest procedure is to express the solution as the sum of a complementary and a particular component, as in the next chapter. However, the differential equation can also be solved by separating the variables, as follows:

$$C\,de = \left(1 - \frac{e}{R}\right)dt$$

$$dt = \frac{C\,de}{1 - e/R}$$

$$\frac{-dt}{RC} = \frac{-de/R}{1 - e/R}$$

Integrating both sides, we obtain

$$\frac{-t}{RC} = \ln\left[1 - \frac{e(t)}{R}\right] - \ln K$$

$$\epsilon^{-t/RC} = \frac{1}{K}\left[1 - \frac{e(t)}{R}\right]$$

$$e(t) = R(1 - K\epsilon^{-t/RC})$$

Since the voltage across the capacitance cannot change instantaneously, $e(t)$ must be zero when t is replaced by zero, so $K = 1$ and

$$e(t) = R(1 - \epsilon^{-t/RC})$$

for $t > 0$, which agrees with the expected answer.

When t equals the time constant $T = RC$, the response $R(1 - \epsilon^{-t/T})$ becomes $R(1 - \epsilon^{-1}) = 0.63R$, so that the output voltage is within 37% of its final value. Since the initial slope of the curve in Fig. 3.1-4e is

$$\left[\frac{R}{T} \epsilon^{-t/T} \right]_{t=0} = \frac{R}{T}$$

T is the time it would take for the response to reach its final value if the initial build-up rate were maintained.

Figures 3.1-3a and 3.1-4a are dual circuits. Replacing all quantities in the expression for $e(t)$ in Fig. 3.1-4a by their dual quantities, as given in Table 2.7-1, we obtain

$$i(t) = \frac{1}{R} (1 - \epsilon^{-(R/L)t})$$

for $t > 0$. As expected, this is the solution for Fig. 3.1-3a.

Before extending the discussion to circuits with other types of inputs, we turn to a systematic consideration of singularity functions.

3.2

The Singularity Functions

The first singularity function considered is the *unit step function*, which is denoted by $U_{-1}(t)$, and which is shown in Fig. 3.2-1a. The function is zero for negative t, unity for positive t, and has a discontinuity at $t = 0$. The symbol U is suggested by the adjective *unit*, but the reasons for the subscript -1 will not become entirely clear until Chapter 10. If a function $f(t)$ is plotted versus t, replacing t by $(t - a)$ to form $f(t - a)$ shifts the curve a units to the right. Thus, a unit step function having its discontinuity at $t = a$, as in Fig. 3.2-1b, is denoted by

$$U_{-1}(t - a) = \begin{cases} 0 & \text{for} \quad t < a \\ 1 & \text{for} \quad t > a \end{cases} \tag{3.2-1}$$

The unit step function has the value zero or unity when the quantity in parentheses is negative or positive, respectively. $AU_{-1}(t - a)$ is called a step function of value A, and it is 0 for $t < a$ and A for $t > a$, as in Fig. 3.2-1c.

We have already seen in Section 3.1 how a source of constant value together with a switch may produce a step function. If, for example, circuit B in Fig. 3.2-1d

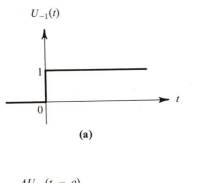

(a)

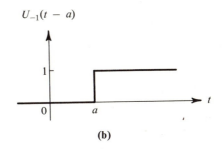

(b)

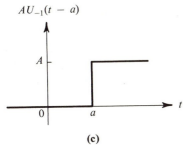

(c)

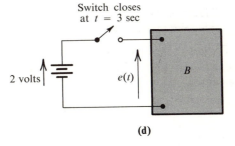

(d)

Figure 3.2-1

contains no independent sources of any kind, then $e(t) = 0$ for $t < 3$ and $e(t) = 2$ volts for $t > 3$, so $e(t) = 2U_{-1}(t - 3)$. Although step functions frequently do arise when switching operations are involved, this is by no means the only reason for their importance.

An infinite number of other singularity functions may be obtained from the unit step function by successive integrations or differentiations. The *unit ramp function*, denoted by $U_{-2}(t)$, is defined by

$$U_{-2}(t) = \int_{-\infty}^{t} U_{-1}(\lambda)\, d\lambda$$

$$\frac{d}{dt} U_{-2}(t) = U_{-1}(t)$$

(3.2-2)

and is shown in Fig. 3.2-2a. It is the total area underneath the curve in Fig. 3.2-1a up to the time t and equals zero for $t < 0$, and t for $t > 0$. Still another singularity function, $U_{-3}(t) = \int_{-\infty}^{t} U_{-2}(\lambda)\, d\lambda$, is sketched in Fig. 3.2-2b.

The pattern of the subscripts for the singularity functions is suggested by the common use of negative superscripts to indicate integration:

$$f^{-1}(t) = \int f(t)\, dt, \qquad f(t) = \frac{d}{dt} f^{-1}(t)$$

$$f^{-2}(t) = \int f^{-1}(t)\, dt, \qquad f^{-1}(t) = \frac{d}{dt} f^{-2}(t)$$

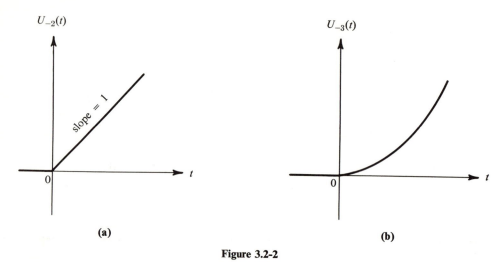

Figure 3.2-2

All the singularity functions are related by

$$U_{n-1}(t) = \int_{-\infty}^{t} U_n(\lambda)\, d\lambda$$

$$U_{n+1}(t) = \frac{d}{dt}\, U_n(t) \tag{3.2-3}$$

Obtaining other singularity functions by successive differentiation of the unit step function is not as straightforward as by integration, because the derivative of $U_{-1}(t)$ is zero for $t \neq 0$ and does not exist for $t = 0$. First, consider Figs. 3.2-3a and 3.2-3b, where the function $f_{-1}(t)$ approximates $U_{-1}(t)$ when Δ is small, and where

$$f_0(t) = \frac{d}{dt} f_{-1}(t)$$

$$f_{-1}(t) = \int_{-\infty}^{t} f_0(\lambda)\, d\lambda \tag{3.2-4}$$

Note that the area underneath $f_0(t)$ is unity regardless of the value of Δ. As Δ is decreased, $f_0(t)$ becomes a narrower, higher pulse whose area remains unchanged. In the limit as Δ approaches zero, $f_{-1}(t)$ becomes $U_{-1}(t)$, and $f_0(t)$ becomes the *unit impulse* $U_0(t)$, which may be crudely visualized as a pulse of infinitesimal width, and of infinite height and unit area. It is represented symbolically in Fig. 3.2-3c, where the number 1 alongside the arrow pointing toward infinity stands for the area underneath the impulse. Its properties may be summarized as follows:

$$U_0(t) = 0 \quad \text{for} \quad t \neq 0$$

$$\int_{-a}^{b} U_0(t)\, dt = 1 \quad \text{for} \quad a > 0, b > 0 \tag{3.2-5}$$

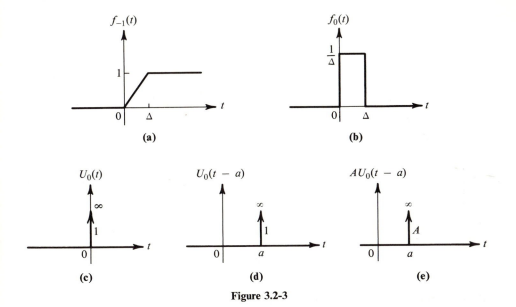

Figure 3.2-3

The unit step and impulse are related by the equations

$$U_0(t) = \frac{d}{dt} U_{-1}(t)$$

$$U_{-1}(t) = \int_{-\infty}^{t} U_0(\lambda) \, d\lambda$$

(3.2-6)

A unit impulse that occurs at $t = a$, as shown in Fig. 3.2-3d, is described by

$$U_0(t - a) = 0 \quad \text{for} \quad t \neq a$$

$$\int_{b}^{c} U_0(t - a) \, dt = 1 \quad \text{for} \quad b < a < c$$

(3.2-7)

An impulse of value A at $t = a$ is denoted by $AU_0(t - a)$ and is the derivative of the step function $AU_{-1}(t - a)$. In Fig. 3.2-3e, the A alongside the arrow pointing toward infinity represents the area underneath the impulse. The phrase "of value A" does not refer to the height of the impulse, which is infinite, but to its area and to the size of the step whose derivative it represents.

Finally, consider

$$\int_{b}^{c} f(t) \, U_0(t - a) \, dt$$

where a is some instant of time in the range $b < t < c$, and where $f(t)$ is any function that is continuous at $t = a$. Since the integrand is zero except at $t = a$, the value of $f(t)$ is immaterial except at this instant, and $f(t)$ can be replaced by the constant $f(a)$. Then, by Eqs. 3.2-7,

$$\int_{b}^{c} f(t) \, U_0(t - a) \, dt = f(a) \quad \text{for} \quad b < a < c$$

(3.2-8)

The effect of the integration is to sample $f(t)$ at the instant $t = a$, so Eq. 3.2-8 is often called the *sampling property* of impulses.

The unit impulse proves to be very useful in many later parts of this book. For example, in Section 3.4 we discuss how under certain circumstances the response to a narrow pulse, such as $f_0(t)$ in Fig. 3.2-3, may be approximated by the response to an impulse of equal area. There are, however, some difficulties in rigorously applying conventional mathematics to the impulse. It should be emphasized that these difficulties, which are outlined below, do not prevent the student from confidently using impulses in the solution of circuit problems.

The unit impulse, which is also called the *Dirac delta function*, is not a function at all in the strict mathematical sense. In contrast to Eqs. 3.2-5 and 3.2-7, the definite integral of any function that is zero at all points but one must have a value of zero. From another point of view, $f(t)$ is a function of t if and only if it can be completely described by a point-by-point relationship, i.e., by assigning a unique value of f to every value of t within the range of interest. A statement that $f(t)$ is zero for $t \neq 0$, and that it does not exist at $t = 0$, defines a satisfactory function at all points. Although equations, including integrals, may be used to indirectly define a function, they cannot contain information that could not also be deduced from the direct point-by-point description. The additional statement that $U_0(t)$ has unit area is therefore inadmissible.

Notice also that we assumed that Eqs. 3.2-6 follow from Eqs. 3.2-4, because the functions $f_{-1}(t)$ and $f_0(t)$ in Fig. 3.2-3 become $U_{-1}(t)$ and $U_0(t)$, respectively, as Δ approaches zero. The assumption is correct, however, only if

$$\lim_{\Delta \to 0} \left[\frac{d}{dt} f_{-1}(t) \right] = \frac{d}{dt} \left[\lim_{\Delta \to 0} f_{-1}(t) \right]$$

$$\lim_{\Delta \to 0} \left[\int_{-\infty}^{t} f_0(\lambda) \, d\lambda \right] = \int_{-\infty}^{t} \left[\lim_{\Delta \to 0} f_0(\lambda) \right] d\lambda$$

Since the basic definitions of differentiation and integration involve a limiting process, we have in effect interchanged the order of two limiting processes, which is not always justifiable.

The difficulties with the unit impulse may be resolved in one of two ways. The unit step and impulse could be defined by the curves of $f_{-1}(t)$ and $f_0(t)$ for an arbitrarily small but nonzero value of Δ. This would correspond more closely to the actual voltages and currents in a physical circuit. Waveshapes *exactly* like $U_{-1}(t)$ and $U_0(t)$ are never encountered, because it takes a finite time (even though it may be only a nanosecond or so) for their values to change, and they must, of course, remain finite. Unfortunately, this method would also make the mathematical analysis of circuit problems more cumbersome.

The preferable approach is to try to justify by more advanced mathematics the results obtained from the definition of $U_0(t)$. This can be done by considering the unit impulse to be a "generalized function" or "distribution," which includes the ordinary mathematical functions as a special case. It is not appropriate to pursue

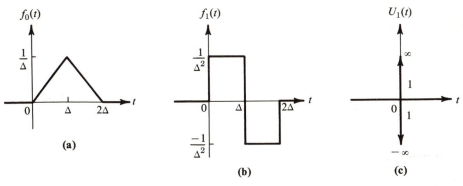

Figure 3.2-4

this matter in an undergraduate textbook, but the student should be aware that the results of circuit analysis involving impulses may be verified by rigorous mathematics. In summary, Eqs. 3.2-5 through 3.2-8 *do* constitute the principal properties of impulses, even though not all of them have been rigorously justified in our discussion.

The function in Fig. 3.2-3b is not the only one that leads to the unit impulse in the limit. For example, the function $f_0(t)$ in Fig. 3.2-4a, whose derivative $f_1(t)$ is shown in Fig. 3.2-4b, satisfies Eqs. 3.2-5 as Δ approaches zero, and hence becomes a unit impulse. The limit of $f_1(t)$ as Δ approaches zero is known as the *unit doublet* $U_1(t)$, and is shown symbolically in part (c) of the figure. The number 1 alongside the arrows does not refer to an area, but indicates that $U_1(t)$ is derived from a *unit* impulse. Although the same mathematical problems exist for the doublet as for the impulse, we write

$$U_1(t) = \frac{d}{dt}\, U_0(t)$$

$$U_0(t) = \int_{-\infty}^{t} U_1(\lambda)\, d\lambda$$

$$(3.2\text{-}9)$$

Although still other singularity functions may be developed, the ones of greatest importance in this book are the ramp, step, and impulse functions—especially the latter two. Even though the step and impulse functions do not appear as signals in any physical device, they may approximate some physical signals. A more important reason for considering them, however, is that the response of any linear circuit to an arbitrary input may be found, once its response to a unit step or impulse is known.

3.3

The Response to the Singularity Functions

The response of some circuits to a unit step function has already been considered in Section 3.1, but in this section we amplify and extend that discussion. First, we present three theorems concerning the voltage across a capacitance and the current through an inductance. Recall that, for a capacitance,

$$i = C \frac{de}{dt}, \qquad w(t) = \frac{1}{2} Ce^2 \tag{3.3-1}$$

where $w(t)$ is the stored energy. For an inductance,

$$e = L \frac{di}{dt}, \qquad w(t) = \frac{1}{2} Li^2 \tag{3.3-2}$$

Theorem 3.3-1. If all voltages and currents remain finite, the voltage across a capacitance and the current through an inductance cannot change instantaneously.

Assume, for example, that a capacitance's voltage does change instantaneously, so that there is a step in the curve of $e(t)$ versus t. Then de/dt, and hence the current $i(t)$, must contain an impulse and does not remain finite. The same conclusion was reached in Section 3.1 by considering the second of Eqs. 3.3-1. An instantaneous change in $e(t)$ corresponds to an instantaneous change in the stored energy $w = dp/dt$, which requires an infinite power flow and hence an infinite current. By examining Eqs. 3.3-2 in the same manner, we can conclude that an infinite voltage is required to instantaneously change the current through an inductance.

Theorem 3.3-2. A unit impulse of current flowing into a capacitance instantaneously changes its voltage by $1/C$ volts, while a unit impulse of voltage across an inductance instantaneously changes the current by $1/L$ amperes.

An impulse of current or voltage is capable of instantaneously inserting some energy into a circuit. When a unit impulse of current flows into an initially uncharged capacitance,

$$e(t) = \frac{1}{C} \int_{-\infty}^{t} U_0(\lambda) \, d\lambda = \frac{1}{C} U_{-1}(t)$$

from Eqs. 1.3-3 and 3.2-6. The current impulse instantaneously places a charge of 1 coulomb on the capacitance and $1/2C$ joules of energy in its electric field, changing the voltage from zero to $1/C$ volts. A unit impulse of voltage across an inductance that had no current for $t < 0$ produces the current

$$i(t) = \frac{1}{L} \int_{-\infty}^{t} U_0(\lambda) \, d\lambda = \frac{1}{L} U_{-1}(t)$$

An energy of $1/2L$ joules is instantaneously inserted into the magnetic field, and the current changed from zero to $1/L$ amperes.

Theorem 3.3-3. The voltage across a capacitance and the current through an inductance must always remain finite.

Theorem 3.3-3 is equivalent to postulating that the stored energy of any element must remain finite, or equivalently that no voltage or current can be the unit doublet $U_1(t)$. If the current to an initially uncharged capacitance were allowed to be $i(t) = U_1(t)$, then

$$e(t) = \frac{1}{C} \int_{-\infty}^{t} U_1(\lambda) \, d\lambda = \frac{1}{C} U_0(t)$$

Any circuit that must violate Theorem 3.3-3 in order to satisfy the element laws and Kirchhoff's laws is regarded as unsolvable. In practice, such a circuit would be an extremely poor model for the device it purports to represent and should be revised.

Notice that the three theorems above place no restrictions upon the voltage and current of a resistance, the current of a capacitance, and the voltage of an inductance. For example, these quantities may change instantaneously even if all voltages and currents in the circuit remain finite.

The response of a circuit containing no stored energy at $t = 0-$ to a unit impulse or to a unit step function is sufficiently important to be given the special symbol $h(t)$ or $r(t)$, respectively. Thus, if the input and output are denoted by $x(t)$ and $y(t)$,

$$h(t) = y(t) \quad \text{when} \quad x(t) = U_0(t)$$
$$r(t) = y(t) \quad \text{when} \quad x(t) = U_{-1}(t)$$

$$(3.3\text{-}3)$$

The terms *impulse response* and *step response* always refer to the response to a *unit* singularity function. The use of these terms also assumes that the circuit contains no stored energy before the singularity function is applied.

In Fig. 3.3-1 are tabulated the step and impulse responses of several simple circuits. Some of the results given in the figure are derived in the following examples. In each case, the input is the independent source, and the desired output is the voltage or current that is followed by a question mark.

Example 3.3-1.   Find the step and the impulse response for Fig. 3.3-1a.

Solution.   In Example 3.1-2 we found that, if $i(t) = U_{-1}(t)$ and if $e(t) = 0$ for $t < 0$, then

$$e(t) = R(1 - \epsilon^{-t/RC}) \quad \text{for} \quad t > 0$$

This is the expression for $r(t)$ in Fig. 3.3-1a, if the phrase "for $t > 0$" is replaced by the factor $U_{-1}(t)$. The presence of the $U_{-1}(t)$ factor does not change the expression for $t > 0$, since it equals unity for positive t; however, it does emphasize that the response is zero for negative values of t, since $U_{-1}(t)$ is zero for $t < 0$. Frequently, both the $U_{-1}(t)$ factor and the phrase "for $t > 0$" are omitted, with the tacit understanding that the response of a realizable circuit cannot start before the input is applied.

When $i(t) = U_0(t)$, the source current is zero except at $t = 0$. For $t > 0$, therefore,

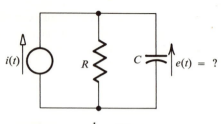

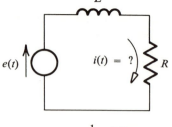

(a) $h(t) = \dfrac{1}{C}\, \epsilon^{-t/RC} U_{-1}(t)$

$r(t) = R(1 - \epsilon^{-t/RC}) U_{-1}(t)$

(b) $h(t) = \dfrac{1}{L}\, \epsilon^{-(R/L)t} U_{-1}(t)$

$r(t) = \dfrac{1}{R}\, (1 - \epsilon^{-(R/L)t}) U_{-1}(t)$

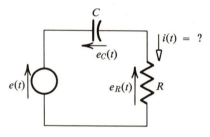

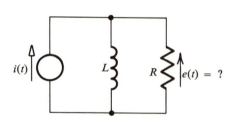

(c) $h(t) = \dfrac{1}{R}\, U_0(t) - \dfrac{1}{R^2 C}\, \epsilon^{-t/RC} U_{-1}(t)$

$r(t) = \dfrac{1}{R}\, \epsilon^{-t/RC} U_{-1}(t)$

(d) $h(t) = R U_0(t) - \dfrac{R^2}{L}\, \epsilon^{-(R/L)t} U_{-1}(t)$

$r(t) = R \epsilon^{-(R/L)t} U_{-1}(t)$

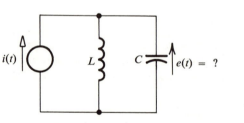

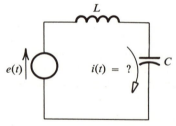

(e) $h(t) = \dfrac{1}{C}\, \cos \dfrac{t}{\sqrt{LC}}\, U_{-1}(t)$

$r(t) = \sqrt{\dfrac{L}{C}}\, \sin \dfrac{t}{\sqrt{LC}}\, U_{-1}(t)$

(f) $h(t) = \dfrac{1}{L}\, \cos \dfrac{t}{\sqrt{LC}}\, U_{-1}(t)$

$r(t) = \sqrt{\dfrac{C}{L}}\, \sin \dfrac{t}{\sqrt{LC}}\, U_{-1}(t)$

Figure 3.3-1

the source is dead and can be replaced by an open circuit. The node equation for $t > 0$ is

$$\frac{1}{R} e + C \frac{de}{dt} = 0$$

whose solution, as in Example 3.1-1, is

$$e(t) = K \epsilon^{-t/RC}$$

At $t = 0$, the impulse of current from the source must flow through the capacitance and not through the resistance. Since $e(t) = e_C(t) = e_R(t) = Ri_R(t)$, a current impulse through the resistance would create an infinite voltage across the capacitance, in violation of Theorem 3.3-3. By Theorem 3.3-2, a unit impulse of current through the capacitance instantaneously changes its voltage from zero to $1/C$ volts, so $e(0+) = 1/C$. Thus, $K = 1/C$ and

$$e(t) = \frac{1}{C} \epsilon^{-t/RC} U_{-1}(t)$$

Notice that the current impulse merely inserts some energy into the circuit at $t = 0$, and that the source is dead for $t \neq 0$. Thus, the impulse response has the form of the free response and describes the natural behavior of the circuit. The expression for $h(t)$ is the same, for all $t > 0$, as the response in Example 3.1-1, if the initial voltage on the capacitance is $A = 1/C$ volts.

Example 3.3-2. Verify the expressions for $h(t)$ and $r(t)$ in Fig. 3.3-1c.

Solution.   The loop equation describing the circuit is

$$Ri + \frac{1}{C} \int i \, dt = e$$

For a unit impulse of voltage, $e(t) = 0$ for $t > 0$, while for a unit step function $e(t) = 1$ for $t > 0$. Note that in both of these two cases $de/dt = 0$. Therefore, when the equation is differentiated term by term to remove the integral sign,

$$R\frac{di}{dt} + \frac{1}{C}i = 0$$

which has the solution $i(t) = K\epsilon^{-t/RC}$ for $t > 0$.

Although the form of the solution for $t > 0$ is the same whether the source is an impulse or step, the initial conditions and hence the constant of integration are different. When $e(t) = e_R(t) + e_C(t) = U_{-1}(t)$, there are no impulses, so the voltage across the capacitance cannot change instantaneously and $e_R(0+) = 1$. Then $i(0+) = 1/R$, so $K = 1/R$ and

$$r(t) = \left(\frac{1}{R} \epsilon^{-t/RC}\right) U_{-1}(t)$$

When $e(t) = U_0(t)$, the voltage impulse at $t = 0$ must appear across the resistance, since $e_C(t)$ must always remain finite. Thus, the current $i(t) = e_R(t)/R$ contains the impulse $(1/R)U_0(t)$, which in turn causes $e_C(t)$ to jump from zero to $1/RC$. Since $e_C(0+) = 1/RC$, $i(0+) = -1/R^2C$ and $K = -1/R^2C$. The complete impulse response is

$$h(t) = \frac{1}{R} U_0(t) - \left(\frac{1}{R^2C} \epsilon^{-t/RC}\right) U_{-1}(t)$$

The expressions for $r(t)$ and $h(t)$ are sketched in Fig. 3.3-2.

The task of verifying the other results in Fig. 3.3-1 is left to the student. The circuits in parts (e) and (f) are described by second-order instead of first-order differential equations, so it might be preferable to defer their analysis until after reading Section 4.1. Notice that parts (b), (d), and (f) of the figure are the duals of parts (a), (c), and (e), respectively. The effort required to derive the six sets of equations may be cut in half by taking advantage of this fact.

If $x(t)$ and $y(t)$ are the input and output of a linear circuit with no initial stored energy, it can be shown that the response to the new input dx/dt is dy/dt, and that the response to $\int_{-\infty}^{t} x(\lambda)\, d\lambda$ is $\int_{-\infty}^{t} y(\lambda)\, d\lambda$. In summary, if the input is differentiated or integrated, so is the response. Since

$$U_0(t) = \frac{d}{dt}\, U_{-1}(t) \quad \text{and} \quad U_{-1}(t) = \int_{-\infty}^{t} U_0(\lambda)\, d\lambda$$

it follows that

$$h(t) = \frac{d}{dt}\, r(t)$$

$$r(t) = \int_{-\infty}^{t} h(\lambda)\, d\lambda$$

(3.3-4)

where dr/dt is the slope of the step response, and where $\int_{-\infty}^{t} h(\lambda)\, d\lambda$ is the net area underneath the impulse response up to the time t. We shall not prove these statements at this time, but Eqs. 3.3-4 are specifically justified in Section 3.5.

Example 3.3-3.   Show that the expressions in Figs. 3.3-1a and 3.3-1c satisfy Eqs. 3.3-4.

Solution.   For $t > 0$ in Fig. 3.3-1a,

$$\frac{d}{dt}\, r(t) = \frac{d}{dt}\, (R - R\epsilon^{-t/RC}) = \frac{1}{C}\, \epsilon^{-t/RC}$$

which is $h(t)$.

$$\int_{-\infty}^{t} h(\lambda)\, d\lambda = \int_{-\infty}^{0} 0\, d\lambda + \int_{0}^{t} \frac{1}{C}\, \epsilon^{-\lambda/RC}\, d\lambda$$

$$= 0 + \left[-R\epsilon^{-\lambda/RC} \right]_{0}^{t} = R(1 - \epsilon^{-t/RC})$$

which is $r(t)$.

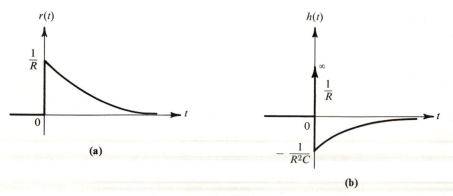

(a)

(b)

Figure 3.3-2

In Fig. 3.3-1c,

$$\frac{d}{dt} r(t) = \frac{d}{dt}\left(\frac{1}{R}\, \epsilon^{-t/RC}\right) = -\frac{1}{R^2 C}\, \epsilon^{-t/RC}$$

for $t > 0$. The impulse in $h(t)$ may be accounted for by considering the complete sketch of $r(t)$ in Fig. 3.3-2a. Since the derivative of the unit step is the unit impulse, differentiating the discontinuity of height $1/R$ gives an impulse of value $1/R$. Thus, the complete derivative is

$$\frac{d}{dt} r(t) = \left(-\frac{1}{R^2 C}\, \epsilon^{-t/RC}\right) U_{-1}(t) + \frac{1}{R}\, U_0(t)$$

which is the impulse response $h(t)$ shown in Fig. 3.3-2b.[†] Again for $t > 0$ in Fig. 3.3-1c,

$$\int_{-\infty}^{t} h(\lambda)\, d\lambda = \frac{1}{R} \int_{-\infty}^{t} U_0(\lambda)\, d\lambda - \frac{1}{R^2 C} \int_{0}^{t} \epsilon^{-\lambda/RC}\, d\lambda$$

$$= \frac{1}{R} + \frac{1}{R}\left[\epsilon^{-\lambda/RC}\right]_{0}^{t} = \frac{1}{R}\, \epsilon^{-t/RC}$$

which is $r(t)$.

Example 3.3-4.   The theorem that leads to Eqs. 3.3-4 also can be used to find the response to the unit ramp function $U_{-2}(t)$. Find $e(t)$ in Fig. 3.3-1a when $i(t) = U_{-2}(t)$.

Solution.   Since $U_{-2}(t) = \int_{-\infty}^{t} U_{-1}(\lambda)\, d\lambda$, the unit ramp response is $\int_{-\infty}^{t} r(\lambda)\, d\lambda$, which for Fig. 3.3-1a gives

$$\int_{0}^{t} R(1 - \epsilon^{-\lambda/RC})\, d\lambda = \left[R\lambda + R^2 C \epsilon^{-\lambda/RC}\right]_{0}^{t}$$

$$= Rt + R^2 C(\epsilon^{-t/RC} - 1) \quad \text{for} \quad t > 0$$

Any circuit containing one capacitance or one inductance, but having several resistances, can be described by a first-order differential equation. Its step and impulse responses contain an exponential with a time constant having the form $T = RC$ or L/R.

[†] This same result may be obtained by applying the rule for the derivative of a product to the complete expression $r(t) = (1/R)\epsilon^{-t/RC}\, U_{-1}(t)$, giving

$$\frac{d}{dt} r(t) = \frac{-1}{R^2 C}\, \epsilon^{-t/RC}\, U_{-1}(t) + \frac{1}{R}\, \epsilon^{-t/RC}\, U_0(t)$$

The exponential factor $\epsilon^{-t/RC}$ in the second term is important only at $t = 0$ and can be replaced by $\epsilon^{-0} = 1$, since the entire term is zero except at $t = 0$. If this is done, the expression for dr/dt is the same as $h(t)$. This procedure is subject to the usual criticism about the mathematical rigor of operations involving impulses, but it is always justifiable. The method of looking for discontinuities in the sketch of $r(t)$ is, however, usually preferred, since it gives more insight into the reason for impulses in $h(t)$.

Example 3.3-5.   Find and sketch the step and impulse response of Fig. 3.3-3a.

Solution.   As soon as one of these two responses is known, the other can be found by Eqs. 3.3-4. The differential equation is often slightly easier to solve when the source is an impulse, although the evaluation of the constant of integration may be slightly more difficult. The node equation describing the circuit is

$$C \frac{de_o}{dt} + G_2 e_o + G_1(e_o - e_1) = 0$$

where $G_1 = 1/R_1$ and $G_2 = 1/R_2$. If the source is the unit impulse, $e_1(t) = 0$ for $t > 0$ and

$$C \frac{de_o}{dt} + \frac{1}{R} e_o = 0$$

where $R = 1/(G_1 + G_2) = R_1 R_2/(R_1 + R_2)$. Then $e_o(t) = K\epsilon^{-t/RC}$ for $t > 0$. Since the voltage across the capacitance must remain finite, the voltage impulse at $t = 0$ appears across R_1, which produces a current impulse of value $1/R_1$. This current impulse flows into the capacitance, so $e_o(0+) = 1/R_1 C$. Thus,

$$h(t) = \frac{1}{R_1 C} \epsilon^{-t/RC} \quad \text{for} \quad t > 0$$

$$r(t) = \int_0^t \frac{1}{R_1 C} \epsilon^{-\lambda/RC} \, d\lambda$$

$$= \left[\frac{-R}{R_1} \epsilon^{-\lambda/RC} \right]_0^t = \frac{R_2}{R_1 + R_2} (1 - \epsilon^{-t/RC}) \quad \text{for} \quad t > 0$$

which are sketched in Figs. 3.3-3b and 3.3-3c.

The impulse response can be sketched directly from the circuit by noting that the capacitance is instantaneously charged to $1/R_1 C$ volts at $t = 0$, and that for $t > 0$ the dead voltage source acts like a short circuit, as in Fig. 3.3-3d. The capacitance then discharges through the resistances R_1 and R_2 in parallel, and the voltage decays to zero with a time constant $T = [R_1 R_2/(R_1 + R_2)]C$.

For a unit step input, the voltage across the capacitance cannot change instantaneously, so $e_o(0+) = 0$ and $i_1(0+) = i_C(0+) = 1/R_1$. As the voltage $e_o(t)$ builds up, a larger proportion of the current from the source passes through R_2 instead of the capacitance. Eventually, the currents through R_1 and R_2 are the same, so that the voltage divider rule becomes applicable, and $e_2(t) = [R_2/(R_1 + R_2)]e_1(t)$. In summary, the initial and final voltages are zero and $[R_2/(R_1 + R_2)]e_1(t)$, respectively, and the time for the response to reach 63% of its final value equals the time constant $T = [R_1 R_2/(R_1 + R_2)]C$.

For any circuit containing resistances and a single capacitance or inductance, the time constant can be determined by inspecting the circuit when the source is dead. The initial value of the output can be found from Theorems 3.3-1 through 3.3-3. The final value of the impulse response is zero, since the current through the resistances dissipates the energy that was initially supplied by the impulse. The

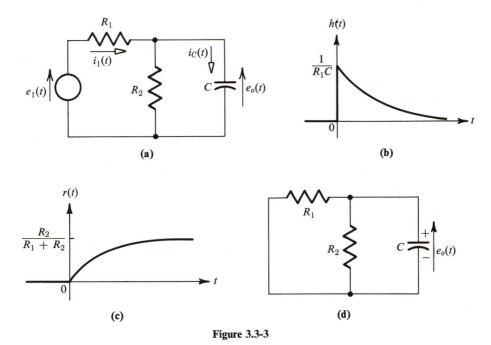

Figure 3.3-3

final value of the step response is not necessarily zero, since the source can continue to supply energy, but it can be easily calculated by replacing the capacitance by an open circuit, or the inductance by a short circuit. One reason for this, which is discussed more fully in the next chapter, is that when the source has a constant value for all positive values of t, all other voltages and currents eventually become constants also. The expressions $i = C(de/dt)$ and $e = L(di/dt)$ then become zero. Based upon the time constant and the initial and final response, the complete response can be sketched without formally writing and solving a differential equation.

The Linearity Theorems. The superposition, Thévenin, and Norton theorems were derived and then applied to resistive circuits in Sections 2.2 and 2.3. The superposition theorem states that the response to several independent sources is the sum of the response to each of them with the remaining independent sources dead. Thévenin's and Norton's theorems state that, when calculating external voltages and currents, any two-terminal subnetwork may be replaced by the open-circuit voltage in series with the dead network (Fig. 2.3-1b) or by the short-circuit current in parallel with the dead network (Fig. 2.3-1c). Sometimes the results shown in Fig. 3.3-1 can be extended to more complicated circuits by the use of these theorems.

Example 3.3-6. Find the unit step response of the circuits in Figs. 3.3-4a and 3.3-5a by the use of Norton's theorem.

Solution. Replacing the part of the circuit within the dashed lines in Fig. 3.3-4a by its Norton equivalent circuit, we obtain the circuit of part (b). Next we combine the two resistances in parallel to form the single resistance $R = R_1 R_2/(R_1 + R_2)$, and use the expression for $r(t)$ in Fig. 3.3-1a.

$$e_o(t) = \frac{R_2}{R_1 + R_2} (1 - \epsilon^{-t/RC}) U_{-1}(t)$$

which agrees with the answer to Example 3.3-5.

When replacing the circuit enclosed by the dashed lines in Fig. 3.3-5a by its Norton equivalent, note that

$$i_{sc}(t) = C_2 \frac{de_1}{dt} + \frac{1}{R_1} e_1 = 2U_0(t) + U_{-1}(t)$$

Combining like elements in the equivalent circuit of Fig. 3.3-5b, we obtain part (c). From the results in Fig. 3.3-1a,

$$e_o(t) = [2(\tfrac{1}{8})\epsilon^{-2t/9} + \tfrac{3}{4}(1 - \epsilon^{-2t/9})] U_{-1}(t)$$

$$= (\tfrac{3}{4} - \tfrac{5}{12}\epsilon^{-2t/9}) U_{-1}(t)$$

Notice that the voltage $e_o(t)$ across C_4 in Fig. 3.3-5a does change instantaneously, even though the original source voltage remains finite. Since $e_1(t) = e_o(t) + e_2(t)$,

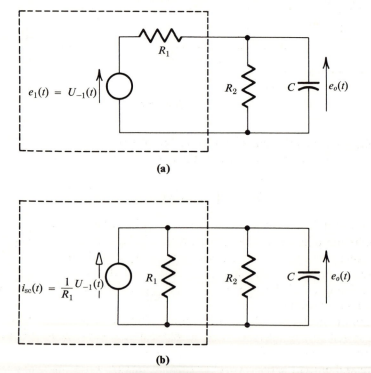

(a)

(b)

Figure 3.3-4

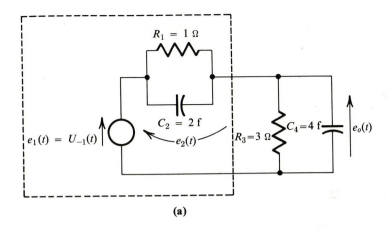

(a)

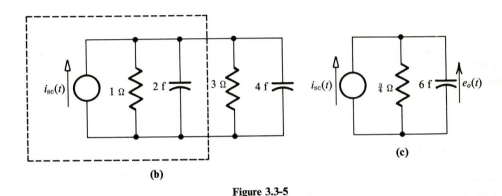

(b)

(c)

Figure 3.3-5

the voltages across the capacitances are forced to change when the source voltage jumps from 0 to 1 volt. Notice that the current through C_4 is

$$4\frac{de_o}{dt} = 4\left[\frac{1}{3}U_0(t) + \frac{5}{54}\epsilon^{-2t/9}U_{-1}(t)\right]$$

and does contain an impulse of value $\frac{4}{3}$. Impulses of current can flow in response to a finite jump in the source voltage, if the circuit contains a loop composed only of the voltage source and capacitances [consisting of $e_1(t)$, C_2, and C_4 in Fig. 3.3-5a]. Similarly, a finite discontinuity in the waveshape of a current source can produce voltage impulses, if the only branches connected to some node are the current source and inductances.

In the application of the superposition, Thévenin, and Norton theorems, any initial stored energy is treated like any other independent source of energy. By the superposition theorem, the output is the sum of the response to the independent voltage and current sources (with any initial energy in the inductances and capacitances made equal to zero) and the response to the initial stored energy (with the other independent sources dead). Any initial energy in a capacitance or inductance is retained when calculating $e_{oc}(t)$ or $i_{sc}(t)$ for the Thévenin or Norton equivalent circuit, but it should be removed when forming the dead network.

Example 3.3-7.   In Fig. 3.3-6, $e_o(0-) = 2$ volts and the switch K closes at $t = 0$. Find $e_o(t)$ for all $t > 0$.

Solution.   With the initial energy stored in the capacitance removed, $i(t) = 3\epsilon^{-2t}$ from Fig. 3.3-1c. Then $e_o(t) = 6 - e_R(t) = 6 - 6\epsilon^{-2t}$ for $t > 0$. With the voltage source dead, the response to the initial stored energy is $e_o(t) = 2\epsilon^{-2t}$ for $t > 0$. The complete output for $t > 0$ is

$$e_o(t) = 6 - 4\epsilon^{-2t}$$

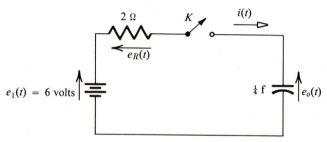

Figure 3.3-6

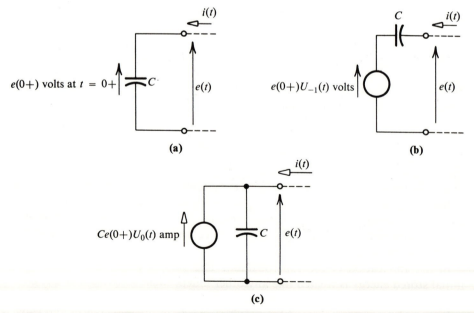

Figure 3.3-7

This answer may also be obtained without using the superposition theorem by noting that the initial and final values of $e_o(t)$ must be 2 and 6 volts, respectively, and that the time constant is $RC = \frac{1}{2}$ sec. Alternatively, a current-law equation at the upper right-hand node may be written and solved.

Figure 3.3-7a shows a capacitance with an initial voltage $e(0+)$. The open-circuit voltage is $e(0+)$ for all positive values of time, while the dead network consists of an initially uncharged capacitance. This suggests the Thévenin equivalent circuit in Fig. 3.3-7b. As a further demonstration that these two circuits have the same external behavior when they are part of a larger network, notice that in both cases

$$e(t) = e(0+) + \frac{1}{C}\int_{0+}^{t} i(\lambda)\, d\lambda$$

for all $t > 0$. In Section 3.1 it was shown that a unit impulse of current flowing into a capacitance instantaneously changes its voltage by $1/C$ volts. The open-circuit voltage in Fig. 3.3-7c is $e(0+)$ for all positive values of time, so this figure is another equivalent circuit for Fig. 3.3-7a. (It is also the Norton equivalent of Fig. 3.3-7b.)

Next, consider an inductance with an initial current $i(0+)$ shown in Fig. 3.3-8a. If a short circuit is placed across the terminals, the current remains equal to $i(0+)$ for all $t > 0$, because, if $e(t) = L(di/dt) = 0$, then $i(t)$ cannot change. Because the short-circuit current and the dead networks are the same in Figs. 3.3-8a and

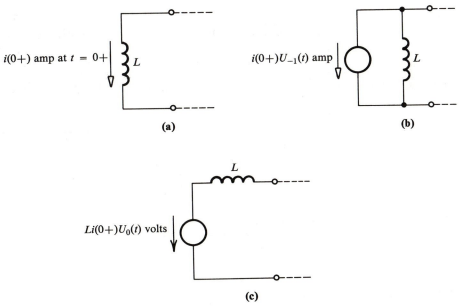

(a)   (b)

(c)

Figure 3.3-8

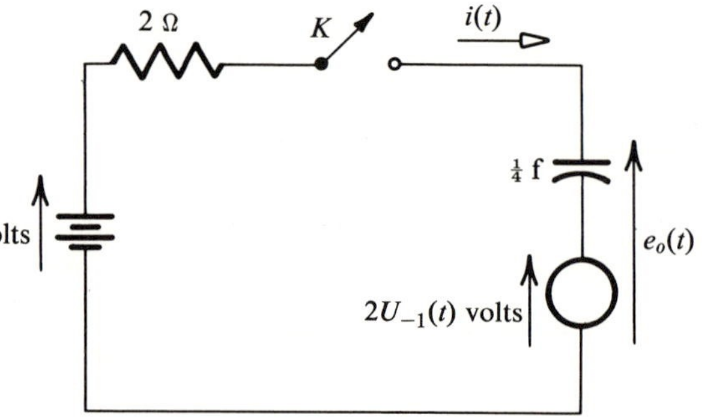

Figure 3.3-9

3.3-8b, the circuits are equivalent. Since a unit impulse of voltage across an inductance instantaneously changes its current by $1/L$ amperes, Fig. 3.3-8c is still another equivalent circuit.

The equivalent circuits in Figs. 3.3-7 and 3.3-8 show how any initial stored energy in a capacitance or inductance may be represented by an added independent source. Although the figures will be used rarely, until Chapter 10, one application is the following alternative solution of Fig. 3.3-6.

The initial capacitance voltage of 2 volts leads to the additional source in Fig. 3.3-9. When the switch closes at $t = 0$, a net voltage of 4 volts is applied to the series RC combination, so for $t > 0$

$$i(t) = \tfrac{4}{2}\epsilon^{-2t}$$

$$e_o(t) = e_1(t) - 2i(t) = 6 - 4\epsilon^{-2t}$$

which agrees with Example 3.3-7.

3.4

Representing Signals as the Sum of Singularity Functions

The circuits in this and the following section are understood to contain no stored energy before the input is applied, but the inputs are allowed to have arbitrary waveshapes. If the input waveshape consists of straight lines, it can be decomposed into a finite number of step and ramp functions. The response to each such component can be determined by the methods of the previous section, and the entire response can be written down by superposition.

Example 3.4-1.   Find the response of the circuit in Fig. 3.4-1a to the rectangular pulse shown in Fig. 3.4-1b.

Solution.   The given input is the sum of the two step functions shown in Fig. 3.4-1c.

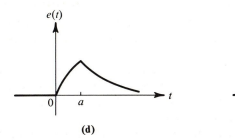

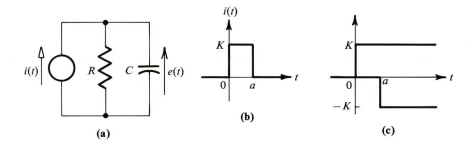

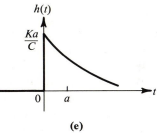

(a) **(b)** **(c)**

(d) **(e)**

Figure 3.4-1

The response to the first component is $Kr(t)$, where

$$r(t) = R(1 - \epsilon^{-t/RC})U_{-1}(t)$$

is the unit step response given in Fig. 3.3-1a. The second component in the input is the same as the first, except for a minus sign and a time delay of a, so the corresponding response is $-Kr(t - a)$. The complete response is

$$e(t) = KR(1 - \epsilon^{-t/RC})U_{-1}(t) - KR(1 - \epsilon^{-(t-a)/RC})U_{-1}(t - a)$$

The first term is zero for $t < 0$, and the second term is zero for $t < a$, so the answer may be written in two parts.

$$e(t) = \begin{cases} KR(1 - \epsilon^{-t/RC}) & \text{for } 0 < t < a \\ KR(1 - \epsilon^{-t/RC} - 1 + \epsilon^{-(t-a)/RC}) = KR(\epsilon^{a/RC} - 1)\epsilon^{-t/RC} & \text{for } t > a \end{cases}$$

which is sketched in Fig. 3.4-1d.

It is important to later discussions to compare the pulse response in Example 3.4-1 with the response to an impulse of equal area. From Fig. 3.3-1a, the response to $KaU_0(t)$ is

$$Kah(t) = \frac{Ka}{C} \epsilon^{-t/RC} U_{-1}(t)$$

which is shown in Fig. 3.4-1e. Notice that while the nature of the two responses is quite different for $0 < t < a$, they seem to be similar for $t > a$. With the infinite series expansion

$$\epsilon^x = 1 + x + \frac{1}{2!} x^2 + \cdots \quad \text{for} \quad |x| < 1$$

the response to the finite pulse for $t > a$ can be rewritten as

$$e(t) = KR \left[1 - \left(\frac{a}{RC} \right) + \frac{1}{2} \left(\frac{a}{RC} \right)^2 - \cdots - 1 \right] \epsilon^{-t/RC}$$

$$\doteq KR \left(\frac{a}{RC} \right) \epsilon^{-t/RC} = Kah(t)$$

provided that $a \ll RC$, i.e., provided that the pulse width is small compared with the time constant of the circuit. The finite pulse supplies power to the circuit continuously for $0 < t < a$, while the impulse instantaneously inserts all the energy at $t = 0$. If the response to the impulse does not have time to change appreciably for $0 < t < a$, the response to each of the two inputs for $t > a$ will be approximately the same. The symbol $\doteq$ is read "is approximately equal to."

Example 3.4-2.   Express the "staircase" input of Fig. 3.4-2a as the sum of step functions.

Solution.   The six step functions comprising the given input are shown in Fig. 3.4-2b. Notice that there is a step function corresponding to each discontinuity in the original input. Thus,

$$x(t) = 2U_{-1}(t) + U_{-1}(t - 1) - 2U_{-1}(t - 2) - 2U_{-1}(t - 3)$$
$$- \tfrac{1}{2}U_{-1}(t - 4) + \tfrac{3}{2}U_{-1}(t - 5)$$

Example 3.4-3.   Find the response of the circuit in Fig. 3.4-1a when the current source has the waveshape in Fig. 3.4-3a and when $R = C = 1$.

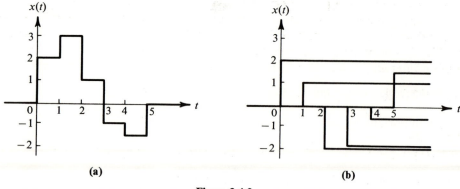

(a)

(b)

Figure 3.4-2

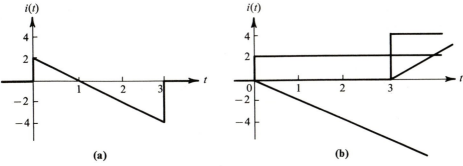

Figure 3.4-3

Solution. The given input can be represented as the sum of two step functions and two ramp functions, as shown in Fig. 3.4-3b. A step function is required by each discontinuity, as in the previous examples. The first ramp function is needed because the slope of $i(t)$ is -2 for $0 < t < 3$, while the ramp starting at $t = 3$ makes the total slope zero for $t > 3$. By Fig. 3.3-1a and Example 3.3-4, the circuit's responses to $U_{-1}(t)$ and $U_{-2}(t)$ are $(1 - \epsilon^{-t})U_{-1}(t)$ and $(t + \epsilon^{-t} - 1)U_{-1}(t)$, respectively. The complete response to $i(t) = 2U_{-1}(t) - 2U_{-2}(t) + 4U_{-1}(t - 3) + 2U_{-2}(t - 3)$ is, therefore,

$$e(t) = (2 - 2\epsilon^{-t} - 2t - 2\epsilon^{-t} + 2)U_{-1}(t)$$
$$+ [4 - 4\epsilon^{-(t-3)} + 2(t - 3) + 2\epsilon^{-(t-3)} - 2]U_{-1}(t - 3)$$

$$= \begin{cases} 4 - 2t - 4\epsilon^{-t} & \text{for } 0 < t < 3 \\ -2(\epsilon^3 + 2)\epsilon^{-t} & \text{for } t > 3 \end{cases}$$

Since $e(t)$ is the voltage across the capacitance and does not change instantaneously, the last two expressions are identical when t is replaced by 3.

The method used in the above examples is applicable to any input waveshape that consists only of straight lines, but most inputs cannot be exactly represented by a finite number of singularity functions. Two methods of determining the approximate output are illustrated in Fig. 3.4-4. If the given input is the solid curve in Fig. 3.4-4a, it may be approximated by the staircase function shown by the dashed lines and discussed in Example 3.4-2. Thus,

$$x(t) \doteq 2U_{-1}(t) + U_{-1}(t - 1) - 2U_{-1}(t - 2) - 2U_{-1}(t - 3)$$
$$- \tfrac{1}{2}U_{-1}(t - 4) + \tfrac{3}{2}U_{-1}(t - 5)$$

and

$$y(t) \doteq 2r(t) + r(t - 1) - 2r(t - 2) - 2r(t - 3) - \tfrac{1}{2}r(t - 4) + \tfrac{3}{2}r(t - 5)$$

The responses to the individual components may be easily determined, and their sum will approximate the complete output if the circuit's time constant is large compared with the 1 sec increments used in constructing the staircase function.

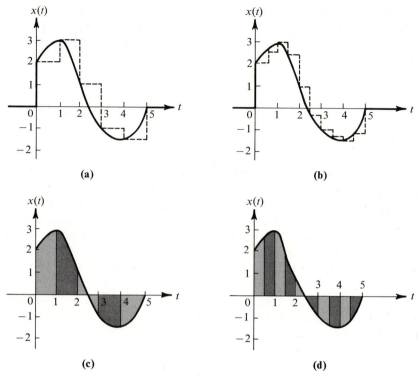

Figure 3.4-4

The staircase approximation to the input and the calculation of the complete response will be more accurate if the size of the time increments is cut in half, as in Fig. 3.4-4b. This, of course, doubles the number of terms in the approximate expression for $y(t)$. The approximations should become exact if the size of the time increments is allowed to approach zero, and the number of terms used allowed to approach infinity. From his knowledge of the calculus, the student should realize that in the limit the summations become integrations. This matter is investigated further in the next section, where it is shown that the exact response to an arbitrary input can be found by an appropriate integration.

Another method of finding the response to the solid curve in Fig. 3.4-4a is based upon the fact that the response to a pulse whose width is small compared with the time constant of the circuit can be approximated by the response to an impulse of equal area. This approach is satisfactory whether the finite pulse is rectangular or has some other shape. In Fig. 3.4-4c, the original input has been broken up into five parts, having areas of about 2.6, 2.1, −0.1, −1.3, and −1.0, respectively. If each of these components is replaced by an impulse of equal area occurring when the corresponding component begins,

$$y(t) \doteq 2.6\,h(t) + 2.1\,h(t-1) - 0.1\,h(t-2) - 1.3\,h(t-3) - 1.0\,h(t-4)$$

Once again, the approximation is improved if the time increment is cut in half, as in Fig. 3.4-4d, causing the number of terms in the approximate expression for $y(t)$ to be doubled. The approximation becomes exact as the size of the time increments approaches zero.

3.5
The Convolution Theorem

In this section we examine more closely the approximations suggested at the end of the previous section. Consider the arbitrary input $x(\lambda)$ shown by the solid curve in Fig. 3.5-1a. The dummy variable λ has been introduced to represent the general distance along the abscissa from the origin (and will eventually become a dummy variable of integration), while t represents a particular time at which we wish to know the response. The abscissa has been divided into increments of equal length $\Delta\lambda$, and the staircase approximation to the input is shown by the dashed line. The symbol $q(\lambda)$ denotes the height of a typical discontinuity in the staircase function, and depends upon the value of λ at which the discontinuity takes place. If $\Delta\lambda$ is sufficiently small so that, in any one interval, the input waveshape has a nearly constant slope of $dx(\lambda)/d\lambda$,

$$q(\lambda) \doteq \frac{dx(\lambda)}{d\lambda}\,\Delta\lambda$$

As discussed in the previous section, the staircase approximation to the input is the sum of step functions, where $q(\lambda)$ is the height of the step function occurring at time λ. If $r(t)$ denotes the circuit's response at time t to a unit step starting at time zero, the response at time t to a unit step starting at time λ and having a height $q(\lambda)$ is $q(\lambda)\, r(t - \lambda)$. The quantity $t - \lambda$ is simply the elapsed time from the application of the step to the instant at which the response is sought. The complete response

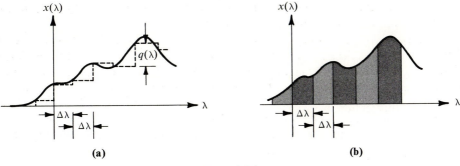

(a)   (b)

Figure 3.5-1

at time t is found by summing the individual responses to all the steps in the staircase function up to the time t.

$$y(t) \doteq \sum \left[q(\lambda) \, r(t - \lambda) \right] = \sum \left[\frac{dx(\lambda)}{d\lambda} \, r(t - \lambda) \, \Delta\lambda \right]$$

The summation is understood to be over all values of λ up to the time t, in increments of $\Delta\lambda$. If the input starts at time zero, the quantity within the brackets is to be evaluated at $\lambda = 0, \Delta\lambda, 2\,\Delta\lambda, \ldots$.

In the limit, as the increment $\Delta\lambda$ approaches zero, the above expression for the output becomes exact, $\Delta\lambda$ can be replaced by the differential $d\lambda$, and the summation can be replaced by an integration. Thus,

$$y(t) = \int_{-\infty}^{t} \frac{dx(\lambda)}{d\lambda} \, r(t - \lambda) \, d\lambda \qquad (3.5\text{-}1)$$

The lower limit of integration is really the time at which the input starts, as the above expression implies, since $dx(\lambda)/d\lambda = 0$ prior to the start of the input. In most problems, the input is applied at time zero, and the lower limit may be changed to zero.

In Fig. 3.5-1b, the given input is divided into sections, where each section is to be replaced by an impulse of equal area. The area of the section starting at time λ is approximately equal to $x(\lambda)\,\Delta\lambda$. If the circuit's response at time t to a unit impulse occurring at time zero is $h(t)$, the response at time t to an impulse of value $x(\lambda)\,\Delta\lambda$ at time λ is $x(\lambda)\,\Delta\lambda\,h(t - \lambda)$. The complete response at time t to all the impulses is

$$y(t) \doteq \sum x(\lambda)\,\Delta\lambda\,h(t - \lambda)$$

where the summation is in increments of $\Delta\lambda$ up to the time t. In the limit as $\Delta\lambda$ approaches zero, the equation becomes exact, and

$$y(t) = \int_{-\infty}^{t} x(\lambda)\,h(t - \lambda)\,d\lambda \qquad (3.5\text{-}2)$$

Once again, the lower limit of integration may be changed to zero, if the input is not applied until time zero.

As stated at the beginning of Section 3.4, in these sections we assume that the circuit contains no stored energy before the input is applied. Any initial stored energy must be the result of the previous history of the circuit, so its effect could be accounted for by including all previous inputs when using the above equations. A simpler procedure is to calculate the response to any initial energy separately (with the external source dead), and to add the result to the above equations. In the examples in this section, we assume that there is no initial stored energy and that the input is zero for $t < 0$.

Equations 3.5-1 and 3.5-2 were derived by decomposing the input into the sum of either step functions or impulses, and finding the total response by superposition. The expressions, which are valid for any linear system, are sometimes called *superposition integrals.* Other names that are used are *Carson, DuHamel,* and *convolution integrals.*

The *convolution* of two functions $f_1(t)$ and $f_2(t)$ is a mathematical operation denoted by $f_1(t) * f_2(t)$, and is defined by[†]

$$f_1(t) * f_2(t) = \int_{-\infty}^{\infty} f_1(\lambda) f_2(t - \lambda)\, d\lambda$$

$$= \int_{-\infty}^{\infty} f_1(t - \lambda) f_2(\lambda)\, d\lambda \tag{3.5-3}$$

The two integrals in this equation always yield identical results, so that $f_1(t) * f_2(t) = f_2(t) * f_1(t)$. A geometrical interpretation of the convolution operation is given in Section 3.6.

The upper limit of integration in both Eqs. 3.5-1 and 3.5-2 may be changed to infinity without affecting the result. The quantity $r(t - \lambda)$ is the response at time t to a unit step function applied at time λ and is therefore zero for $\lambda > t$. Similarly, the unit impulse response $h(t - \lambda)$ is zero for $\lambda > t$. Thus,

$$y(t) = x(t) * h(t) = \frac{dx}{dt} * r(t) \tag{3.5-4}$$

i.e.,

$$y(t) = \int_{-\infty}^{\infty} x(\lambda)\, h(t - \lambda)\, d\lambda \tag{3.5-5}$$

$$= \int_{-\infty}^{\infty} \frac{dx(\lambda)}{d\lambda}\, r(t - \lambda)\, d\lambda \tag{3.5-6}$$

$$= \int_{-\infty}^{\infty} x(t - \lambda)\, h(\lambda)\, d\lambda \tag{3.5-7}$$

$$= \int_{-\infty}^{\infty} \frac{dx(t - \lambda)}{d(t - \lambda)}\, r(\lambda)\, d\lambda \tag{3.5-8}$$

For a realizable system whose input is not applied until time zero, $x(\lambda)$, $dx(\lambda)/d\lambda$, $h(\lambda)$, and $r(\lambda)$ are zero for $\lambda < 0$. Also $h(t - \lambda)$, $r(t - \lambda)$, $x(t - \lambda)$, and $dx(t - \lambda)/d(t - \lambda)$ are zero for $\lambda > t$. Then the lower and upper limits of integration may be changed to zero and t in all four equations, which will be done in all of the examples. In the event that the integrand contains an impulse at $\lambda = 0$ or $\lambda = t$, it is understood that the impulse is included within the range of integration. Equations 3.5-7 and 3.5-8 are sometimes more difficult to apply and are less frequently used.

The above equations indicate that the response of a linear circuit containing no initial stored energy may be found for *any* input, if its step or impulse response is known. The quantity $x(\lambda)$ is formed from $x(t)$ by replacing every t by the dummy variable λ, while $h(t - \lambda)$ is $h(t)$ with t replaced by $t - \lambda$. In carrying out the integration with respect to λ, t is treated as a constant.

---

† In many textbooks, the limits of integration in the definition of convolution are zero to t. The more general definition given here is necessary in certain applications, especially when the variable t does not refer to time.

Example 3.5-1.   Find the response of the circuit in Fig. 3.5-2a when $i(t) = \epsilon^{-t}U_{-1}(t)$, when $i(t)$ is the half-cycle of a sine wave, as shown in Fig. 3.5-2b, and when $i(t)$ has the waveshape in Fig. 3.5-2c.

Solution.   The unit impulse response is $h(t) = \epsilon^{-t}$ for $t > 0$, so $h(t - \lambda) = \epsilon^{-(t-\lambda)}$. When $x(t) = i(t) = \epsilon^{-t}$ for $t > 0$, Eq. 3.5-5 gives

$$y(t) = e(t) = \int_0^t \epsilon^{-\lambda}\, \epsilon^{-(t-\lambda)}\, d\lambda = \epsilon^{-t}\int_0^t d\lambda = t\epsilon^{-t}$$

for all $t > 0$.

For the input in Fig. 3.5-2b, and for $0 < t < \pi$, $i(t) = \sin t$ and

$$e(t) = \int_0^t (\sin \lambda)\epsilon^{-(t-\lambda)}\, d\lambda = \epsilon^{-t}\int_0^t \epsilon^{\lambda} \sin \lambda\, d\lambda$$

$$= \tfrac{1}{2}\epsilon^{-t}[\epsilon^{\lambda}(\sin \lambda - \cos \lambda)]_0^t = \tfrac{1}{2}(\sin t - \cos t + \epsilon^{-t})$$

For values of t greater than π, the integration from zero to t should be broken up into two parts, because the form of the input is different for $t < \pi$ and for $t > \pi$.

$$e(t) = \int_0^{\pi} i(\lambda)\, h(t - \lambda)\, d\lambda + \int_{\pi}^t i(\lambda)\, h(t - \lambda)\, d\lambda$$

Since $i(\lambda) = \sin \lambda$ for $0 < \lambda < \pi$ and $i(\lambda) = 0$ for $\lambda > \pi$,

$$e(t) = \tfrac{1}{2}\epsilon^{-t}\left[\epsilon^{\lambda}(\sin \lambda - \cos \lambda)\right]_0^{\pi} = \tfrac{1}{2}(1 + \epsilon^{\pi})\epsilon^{-t}$$

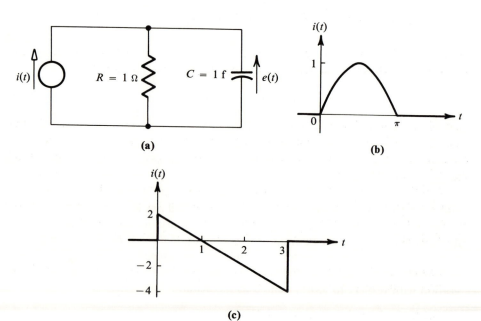

(a)

(b)

(c)

Figure 3.5-2

The response to Fig. 3.5-2b also may be found from Eq. 3.5-6. The unit step response is $r(t) = 1 - \epsilon^{-t}$ for $t > 0$, and di/dt equals $\cos t$ for $0 < t < \pi$ and zero for $t > \pi$. Thus, for $0 < t < \pi$,

$$e(t) = \int_0^t \cos \lambda \, (1 - \epsilon^{-(t-\lambda)}) \, d\lambda = \tfrac{1}{2}(\sin t - \cos t + \epsilon^{-t})$$

where the details of the integration have been omitted. For $t > \pi$,

$$e(t) = \int_0^\pi \cos \lambda \, (1 - \epsilon^{-(t-\lambda)}) \, d\lambda = \tfrac{1}{2}(1 + \epsilon^\pi)\epsilon^{-t}$$

Finally, consider the input in Fig. 3.5-2c. For $0 < t < 3$, $i(t) = 2 - 2t$ and

$$e(t) = \int_0^t (2 - 2\lambda)\epsilon^{-(t-\lambda)} \, d\lambda = 4 - 2t - 4\epsilon^{-t}$$

where the details of the integration again have been omitted. For $t > 3$,

$$e(t) = \int_0^3 (2 - 2\lambda)\epsilon^{-(t-\lambda)} \, d\lambda = -2(\epsilon^3 + 2)\epsilon^{-t}$$

The last two equations agree with the answer to Example 3.4-3.

One situation that requires special attention is a circuit for which $h(t)$ contains an impulse, as in Figs. 3.3-1c and 3.3-1d. Suppose that the unit impulse response has the form

$$h(t) = KU_0(t) + h_1(t)$$

where $h_1(t)$ does not contain an impulse. From Eq. 3.5-5,

$$y(t) = K \int_{-\infty}^\infty x(\lambda) \, U_0(t - \lambda) \, d\lambda + \int_{-\infty}^\infty x(\lambda) \, h_1(t - \lambda) \, d\lambda$$

In the first of the two integrands, $U_0(t - \lambda)$ is zero except as $\lambda = t$, so $x(\lambda)$ can be replaced by $x(t)$. Since the area underneath the unit impulse is unity,

$$y(t) = Kx(t) + \int_{-\infty}^\infty x(\lambda) \, h_1(t - \lambda) \, d\lambda \qquad (3.5\text{-}9)$$

as could have been immediately written down from Eq. 3.2-8. The result indicates, as might have been anticipated, that in this special case the input signal appears as part of the output. When there is no input for $t < 0$, the limits of integration again become zero to t.

An alternative proof of the last equation is based upon the fact that $x(t) * h(t) = h(t) * x(t)$; i.e., the roles of the impulse response and the input may be interchanged. If a circuit having an impulse response $x(t)$ is subjected to the input signal $KU_0(t) + h_1(t)$, Eq. 3.5-9 follows immediately by superposition.

Example 3.5-2. Find an expression for the current in the circuit in Fig. 3.3-1c, which is repeated in Fig. 3.5-3, when $e(t) = \epsilon^{-t}$ for $t > 0$ and when $R = C = 1$.

Solution. Since $h(t) = U_0(t) - \epsilon^{-t}U_{-1}(t)$,

$$i(t) = \epsilon^{-t} - \int_0^t \epsilon^{-\lambda}\epsilon^{-(t-\lambda)} \, d\lambda = (1 - t)\epsilon^{-t}$$

for all $t > 0$.

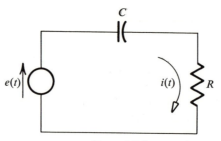

Figure 3.5-3

The use of Eq. 3.5-6 when there are discontinuities in the input waveshape also requires special consideration. As discussed in Section 3.2, differentiating a function having a discontinuity of height A produces an impulse of area A. The impulse in $dx(t)/dt$ and $dx(\lambda)/d\lambda$ complicates the evaluation of the integral. Two methods of handling such problems are illustrated in the following example.

Example 3.5-3.   Find the response of the circuit in Fig. 3.5-2a to $i(t) = \epsilon^{-t} U_{-1}(t)$.

Solution.   The input waveshape $i(t)$, shown in Fig. 3.5-4a, has a discontinuity at the time origin. Such discontinuities are easy to overlook, especially if the input is written as $i(t) = \epsilon^{-t}$ for $t > 0$, but they do affect the solution. The derivative of the input is shown in Fig. 3.5-4b and is

$$\frac{di}{dt} = U_0(t) - \epsilon^{-t} U_{-1}(t)$$

Since $r(t) = 1 - \epsilon^{-t}$ for all $t > 0$, Eq. 3.5-6 gives

$$e(t) = \int_0^t U_0(\lambda)(1 - \epsilon^{-(t-\lambda)})\, d\lambda - \int_0^t \epsilon^{-\lambda}(1 - \epsilon^{-(t-\lambda)})\, d\lambda$$

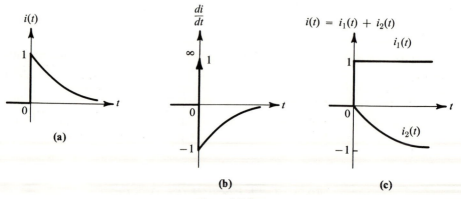

Figure 3.5-4

The first integrand is zero except at $\lambda = 0$, so

$$e(t) = 1 - \epsilon^{-t} - \int_0^t (\epsilon^{-\lambda} - \epsilon^{-t})\, d\lambda = t\epsilon^{-t}$$

for all $t > 0$, which agrees with Example 3.5-1.

Another approach is to decompose the input into the two components $i_1(t)$ and $i_2(t)$ shown in Fig. 3.5-4c and to find the response to each component separately. The response to the unit step function $i_1(t)$ is

$$1 - \epsilon^{-t} \quad \text{for} \quad t > 0$$

Since $i_2(t)$ does not contain any discontinuities, there are no impulses in the convolution integral. The slope of $i_2(t)$ is the same as the slope of $i(t)$ for all $t > 0$, so the circuit's response to $i_2(t)$ is

$$- \int_0^t \epsilon^{-\lambda}(1 - \epsilon^{-(t-\lambda)})\, d\lambda$$

The sum of the two responses is the same as the previous answer.

Consider the more general input waveshape in Fig. 3.5-5a, which jumps from zero to $x(0+)$ at the time origin, and which also has discontinuities at t_2 and t_3. The output is the sum of the responses to the four components in Fig. 3.5-5b.

$$y_1(t) = x(0+)\, r(t)$$
$$y_2(t) = -Ar(t - t_2)$$
$$y_3(t) = Br(t - t_3)$$
$$y_4(t) = \int_0^t \frac{dx_4(\lambda)}{d\lambda}\, r(t - \lambda)\, d\lambda$$

Notice that $dx_4(\lambda)/d\lambda = dx(\lambda)/d\lambda$ except at the discontinuities. In summary, the response contains a term for each discontinuity in the input, plus the result of Eq. 3.5-6 with the discontinuities ignored. The components $y_2(t)$ and $y_3(t)$ do not contribute to the output until $t = t_2$ and t_3, respectively, since $r(t - t_0)$ is zero for $t < t_0$.

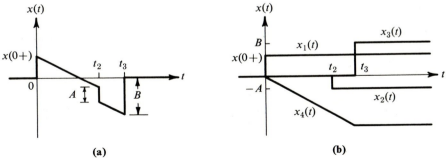

(a)   (b)

Figure 3.5-5

Example 3.5-4.   Find the response of the circuit in Fig. 3.5-2a to the input of Fig. 3.5-2c.

Solution.   The input has discontinuities at $t = 0$ and $t = 3$, and its slope is -2 for $0 < t < 3$ and zero for $t > 3$. For $0 < t < 3$,

$$e(t) = 2(1 - \epsilon^{-t}) + \int_0^t (-2)(1 - \epsilon^{-(t-\lambda)}) \, d\lambda = 4 - 2t - 4\epsilon^{-t}$$

For $t > 3$,

$$e(t) = 2(1 - \epsilon^{-t}) + 4[1 - \epsilon^{-(t-3)}] + \int_0^3 (-2)(1 - \epsilon^{-(t-\lambda)}) \, d\lambda = -2(\epsilon^3 + 2)\epsilon^{-t}$$

which agrees with Example 3.5-1.

Frequently, the output of one circuit is used as the input to another one. In Fig. 3.5-6, $z(t)$ is the output signal of network N_1 and the input signal of N_2. The two networks are said to be connected *in cascade*. They are said to be *isolated* if the presence of N_2 does not affect the output of N_1. Let $h_1(t)$ and $h_2(t)$ denote the unit impulse responses of two isolated networks that are connected in cascade.

$$h_1(t) = z(t) \quad \text{when} \quad x(t) = U_0(t)$$

$$h_2(t) = y(t) \quad \text{when} \quad z(t) = U_0(t)$$

We now apply the convolution theorem to the second network when $x(t) = U_0(t)$. Then $y(t) = z(t) * h_2(t) = h_1(t) * h_2(t)$, so the unit impulse response of the cascaded combination is

$$h(t) = h_1(t) * h_2(t) \tag{3.5-10}$$

If $r_1(t)$ and $r_2(t)$ denote the unit step responses of the individual networks, the output of N_2 when $x(t) = U_{-1}(t)$ is

$$y(t) = \frac{dz(t)}{dt} * r_2(t) = \frac{dr_1(t)}{dt} * r_2(t)$$

Thus the unit step response of the combination is

$$r(t) = \frac{dr_1(t)}{dt} * r_2(t) \tag{3.5-11}$$

Example 3.5-5.   Find the unit impulse response of the circuit in Fig. 3.5-7a.

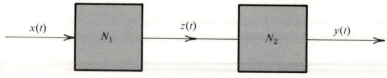

Figure 3.5-6

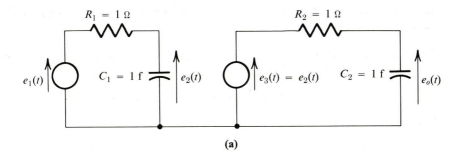

(a)

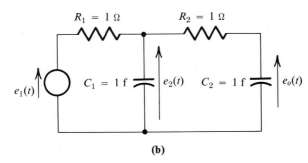

(b)

Figure 3.5-7

Solution.   The unit impulse response of each of the two RC networks is ϵ^{-t} for $t > 0$. By Eq. 3.5-10, the unit impulse response of the entire circuit is

$$h(t) = \int_0^t \epsilon^{-\lambda} \epsilon^{-(t-\lambda)} \, d\lambda = t\epsilon^{-t}$$

for $t > 0$.

The unit impulse response of Fig. 3.5-7b is *not* given by the answer to Example 3.5-5, because the two RC networks are not isolated. In part (a) of the figure, all of the current through R_1 must flow through C_1. In part (b), however, the $R_2 C_2$ branch provides another path for the current through R_1, and $e_2(t)$ no longer equals ϵ^{-t} when $e_1(t) = U_0(t)$. This effect is often described by saying that the $R_2 C_2$ branch *loads down* the previous network. Because of this loading effect, a circuit like Fig. 3.5-7b cannot be conveniently analyzed as a cascaded combination. By the techniques discussed in the next chapter, the unit impulse response of this circuit can be shown to be

$$h(t) = 0.45 \left(\epsilon^{-0.38t} - \epsilon^{-2.62t}\right)$$

for $t > 0$.

Equations 3.3-4 stated without proof that $h(t) = dr/dt$ and $r(t) = \int_{-\infty}^t h(\lambda) \, d\lambda$. For a realizable circuit, $h(\lambda) = 0$ for $\lambda < 0$ and the lower limit of integration may be changed to zero. To obtain these results, let $x(t) = U_{-1}(t)$ and $y(t) = r(t)$ in the equation

$$y(t) = \int_0^t x(t - \lambda)\, h(\lambda)\, d\lambda$$

Then,

$$r(t) = \int_0^t h(\lambda)\, d\lambda$$

When both sides of this equation are differentiated with respect to t,

$$\frac{dr}{dt} = h(t)$$

One of the properties of the convolution process is that

$$\frac{d}{dt}\left[f_1(t) * f_2(t) \right] = \frac{df_1(t)}{dt} * f_2(t) = f_1(t) * \frac{df_2(t)}{dt}$$

Since the response of a circuit to the input $x(t)$ is $y(t) = x(t) * h(t)$, the response to dx/dt is

$$\frac{dx}{dt} * h(t) = \frac{d}{dt}\left[x(t) * h(t) \right] = \frac{dy}{dt}$$

This is consistent with the statement in Section 3.3 that if the input is differentiated, so is the response.

3.6
Graphical Interpretation of Convolution

If a circuit's response to a unit impulse or unit step function is known, its response to an arbitrary input can be found by the convolution theorem of Eq. 3.5-4. For some inputs, however, the analytical evaluation of the integral may be difficult, particularly if the input waveshape is given graphically and cannot be accurately described by a simple expression. In other cases, perhaps an approximate curve for $h(t)$ or $r(t)$ might have been determined experimentally by applying a very short pulse (to approximate an impulse) or by suddenly applying a constant input signal.

A graphical approach might be used for any of the above-mentioned reasons. Furthermore, graphical methods sometimes give additional insight into the significance of the results. Any general properties that do not depend upon knowing the exact shape of the input waveshape are often more readily recognized and understood.

In the graphical interpretation of

$$f(t) = f_1(t) * f_2(t) = \int_{-\infty}^{\infty} f_1(\lambda)\, f_2(t - \lambda)\, d\lambda$$

the quantities $f_1(\lambda)$ and $f_2(t - \lambda)$ are plotted versus λ for a specific value of t. Then the two curves are multiplied to form the integrand, and the value of $f(t)$ equals the total area underneath the product curve.

Example 3.6-1. In the first part of Example 3.5-1, $x(t) = h(t) = \epsilon^{-t}$ for $t > 0$. Interpret graphically the convolution operation

$$y(t) = x(t) * h(t) = \int_{-\infty}^{\infty} x(\lambda)\, h(t - \lambda)\, d\lambda$$

Solution. The curves $x(\lambda)$ versus λ and $h(t - \lambda)$ versus $t - \lambda$ are identical with the curves of $x(t)$ and $h(t)$ versus t, and are shown in parts (a) and (b) of Fig. 3.6-1. Replacing the variable $t - \lambda$ on the abscissa of part (b) by $\lambda - t$ folds the curve of $h(t - \lambda)$ about the vertical axis, since $h(t - \lambda) = h[-(\lambda - t)]$. Adding t to the abscissa variable shifts the waveshape t units in the horizontal direction, as in part (d). The product of parts (a) and (d) is the integrand of the convolution integral and is shown in part (e). Certainly, the product curve is zero for $\lambda < 0$ and for $\lambda > t$ (which means that the limits of integration could have been written as zero to t). It is also easy to see that the product must equal ϵ^{-t} at $\lambda = 0+$ and at $\lambda = t-$, although it might not be immediately obvious that the curve is a straight line. The value of $y(t)$ equals the area underneath the product curve, and is shaded in part (e) of the figure.

The graphical procedure consists of four steps. **Fold** one of the two functions of time about the vertical axis, **slide** it forward a distance t into the other curve,

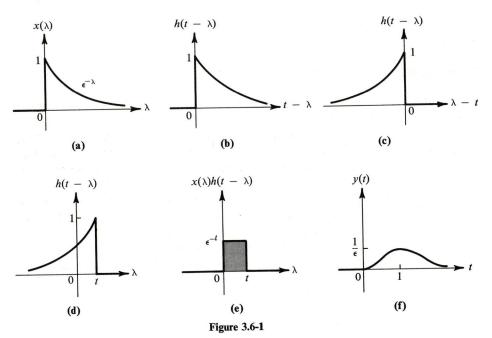

Figure 3.6-1

multiply the two curves, and **integrate** by calculating the area underneath the product curve. In Example 3.6-1, the product curve is a horizontal line, so $y(t)$ is seen to be te^{-t} for $t > 0$. In general, the area underneath the product curve must be obtained graphically for several different values of t. Plotting the resulting values versus t and drawing a smooth curve through the plotted points, we shall obtain a curve of $y(t)$ versus t. In the above example, $y(t)$ is small when t is very small because the curves of $x(\lambda)$ and $h(t - \lambda)$ versus λ overlap for only a small range of λ. As t increases, $h(t - \lambda)$ is moved forward into $x(\lambda)$, and $y(t)$ increases. For very

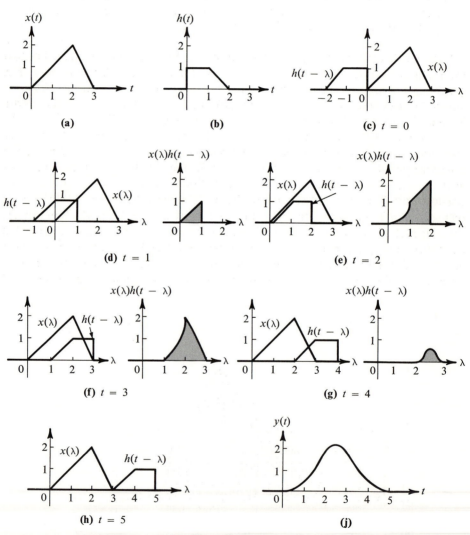

Figure 3.6-2

large values of t, the value of at least one of the two component curves is very small for any value of λ, and $y(t)$ decreases.

> **Example 3.6-2.**   The convolution of two functions, each of which is nonzero for only a finite range of t, is important in some applications of the Fourier transform (which is discussed in Chapter 9). Graphically, find $y(t) = x(t) * h(t)$ when $x(t)$ and $h(t)$ are given in parts (a) and (b) of Fig. 3.6-2.
>
> **Solution.**   In Fig. 3.6-2c through 3.6-2h, $x(\lambda)$ and $h(t - \lambda)$ are sketched versus λ for several different values of t. The area underneath the product curve is shaded and is zero for $t \leq 0$, $\frac{1}{2}$ for $t = 1$, $\frac{11}{6}$ for $t = 2$, $\frac{11}{6}$ for $t = 3$, $\frac{1}{3}$ for $t = 4$, and zero for $t \geq 5$. A sketch of $y(t)$ versus t is given in part (j).

Suppose that the waveshapes of $x(t)$ and $h(t)$ in Fig. 3.6-2 are changed, but that the functions are still zero except for $0 < t < 3$ and $0 < t < 2$, respectively. The graphical construction in the figure demonstrates that the convolution $x(t) * h(t)$ still will be zero except for $0 < t < 5$. In general, if $x(t)$ and $h(t)$ are pulses of width T_1 and T_2, respectively, their convolution is a pulse of width $T_1 + T_2$.

The graphical approach suggests another physical interpretation for the impulse response. Consider the input and impulse response in Figs. 3.6-2a and 3.6-2b, and the corresponding curves of $x(\lambda)$ and $h(t - \lambda)$ in Fig. 3.6-3. The value of the input at time t (which may be defined as the present time) is given by the solid dot on the curve of $x(\lambda)$. That part of the curve to the left of the dot represents the input prior to the time t. In constructing the product curve of $x(\lambda) h(t - \lambda)$, the input at the time t is multiplied by $h(0)$, the initial value of the impulse response. The input that existed t_1 seconds before the present time is multiplied by $h(t_1)$. As should have been anticipated, the product curve and hence $y(t)$ depend not only upon the input at time t, but also upon the past values of the input. The relative importance of

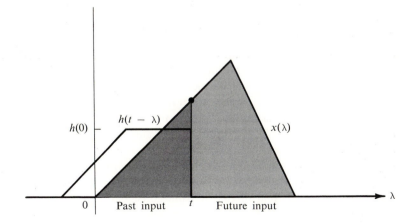

Figure 3.6-3

different parts of the past input to the present output $y(t)$, which is the area underneath the product curve, is determined by the impulse response.

If $h(t)$ is given by Fig. 3.6-2b, all values of the input from the present to 1 sec in the past are equally important in determining the output. The importance of inputs that occurred more than 1 sec in the past steadily diminishes, while any input signals that are more than 2 sec old have no effect at all on the output. The circuits in this book have an impulse response $h(t)$ that is nonzero for all $t > 0$, in which case the output is theoretically affected by all previous input signals. Except for circuits with no resistance and some circuits with controlled sources, however, $h(t)$ approaches zero as t becomes very large, so that the input in the distant past is less important than recent values of the input. Since the contribution of past values of the input is weighted by the impulse response, $h(t)$ is sometimes called the *weighting function* of the circuit.

3.7

Summary

A circuit composed of resistances and either one capacitance or one inductance may be described by a first-order differential equation. The three theorems at the start of Section 3.3 are needed to evaluate the constant of integration that appears in the solution of the differential equation.

The symbols $r(t)$ and $h(t)$ denote the response of a circuit that contains no initial stored energy to the unit step and to the unit impulse, respectively. A constant input that is suddenly applied may be represented by a step function. An input equal to the unit impulse $U_0(t)$ instantaneously inserts some energy into the circuit at $t = 0$, but is dead for all $t > 0$. The form of the resulting response is a characteristic of the circuit and describes its natural behavior. The response to a pulse whose width is small compared to the time constants of the circuit may be approximated by the response to an impulse of equal area.

The step and impulse responses for circuits with only one capacitance or one inductance may be written down by noting the circuit's time constant and the initial and final values of the response. The method is described following Example 3.3-5. The two responses are related by the equation $h(t) = dr/dt$.

The application of the superposition, Thévenin, and Norton theorems is discussed at the end of Section 3.3. Any initial stored energy is to be treated like any other independent source of energy.

In Section 3.4 input waveshapes that consist of straight lines are expressed as the sum of a finite number of step and ramp functions. The output is equal to the sum of the responses to the individual singularity functions. This concept is extended to other inputs in Sections 3.5 and 3.6, where the summation is replaced by an integration. If $h(t)$ or $r(t)$ is known, the response to an arbitrary input can be found by the convolution formulas.

PROBLEMS

3.1 If in the circuit in Fig. P3.1 the switch K_1 closes and K_2 remains open, $i_1(t)$ will eventually approach $1/R_1 = \frac{1}{2}$ amp, according to the equation associated with Fig. 3.1-3. Assume that K_1 has already been closed long enough so that $i_1(t)$ is approximately equal to $\frac{1}{2}$, when K_2 closes at $t = 0$. Find an expression for $i_1(t)$ for all $t > 0$.

3.2 For $t < 0$ in Fig. 3.1-4a, the switch K is in the lower position and $e(t) = 0$. The switch is thrown to the upper position at $t = 0$ and is returned to the lower position at $t = 1$ sec. Find and sketch $e(t)$ for all $t > 0$, if $R = 2\,\Omega$ and $C = 1$ f.

3.3 For the circuit in Fig. P3.3, find $e_o(t)$ for all $t > 0$ if

(a) $e_1(t) = 1$ volt for $t > 0$ and $e_C(0+) = 0$.

(b) $e_1(t) = 1$ volt for $t > 0$ and $e_C(0+) = 2$ volts.

3.4 The model in Fig. 1.4-8a representing the horizontal deflection coil in an oscilloscope is modified in Fig. P3.4a. The added capacitance C represents the stray (or parasitic) capacitance. If $L = 5$ h, $R = 1\,\Omega$, and $C = 1$ f, and if the current $i_L(t)$ should have the waveshape in Fig. P3.4b, what must be the waveshape of the source current?

3.5 The switch K in Fig. P3.5 closes at $t = 0$, and the circuit contains no stored energy

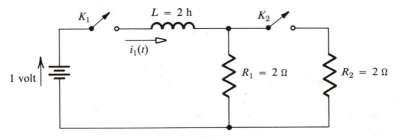

Figure P3.1

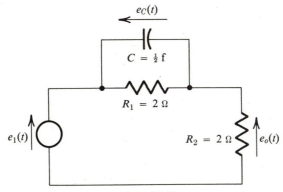

Figure P3.3

for $t < 0$. Sketch to scale $i_1(t)$ and $e_o(t)$ when

(a) $L = 1$ h and $R_1 = 1 \ \Omega$.

(b) $L = 2$ h and $R_1 = 2 \ \Omega$.

(c) $L = 2$ h and $R_1 = 1 \ \Omega$.

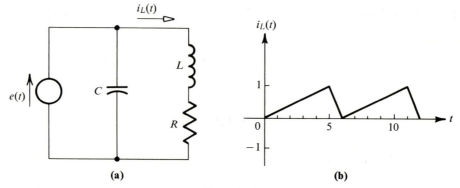

(a)   (b)

Figure P3.4

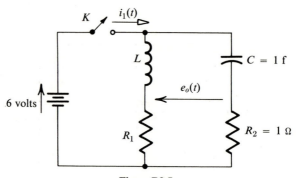

Figure P3.5

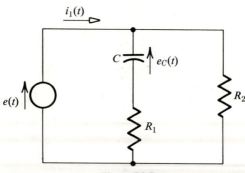

Figure P3.7

3.6 Verify the expressions for $r(t)$ and $h(t)$ in Figs. 3.3-1e and 3.3-1f.

3.7 Find $e_C(t)$ and $i_1(t)$ in Fig. P3.7 when

(a) $e(t) = U_{-1}(t)$.

(b) $e(t) = U_0(t)$.

3.8 Find the Norton equivalent circuit for the two-terminal network in Fig. P3.8.

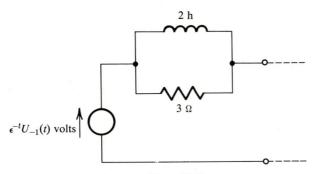

Figure P3.8

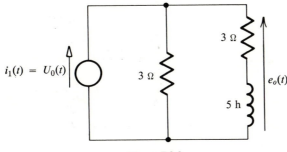

Figure P3.9

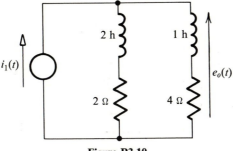

Figure P3.10

3.9   If the circuit in Fig. P3.9 contains no stored energy for $t < 0$, find and sketch $e_o(t)$. Solve the problem without altering the circuit, and then by using Thévenin's theorem and Fig. 3.3-1.

3.10   Find the unit step response for the circuit in Fig. P3.10.

3.11   Find $i_o(t)$ if the circuit in Fig. P3.11 contains no stored energy for $t < 0$, and if $e_1(t) = U_o(t)$.

3.12   Find the unit step and the unit impulse responses for each of the circuits in Fig. P3.12. Do this by the procedure that is described following Example 3.3-5 instead of by explicitly solving a differential equation. Check the answers by the equation $h(t) = dr(t)/dt$.

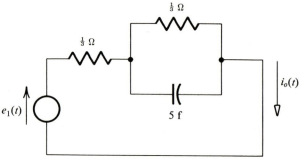

Figure P3.11

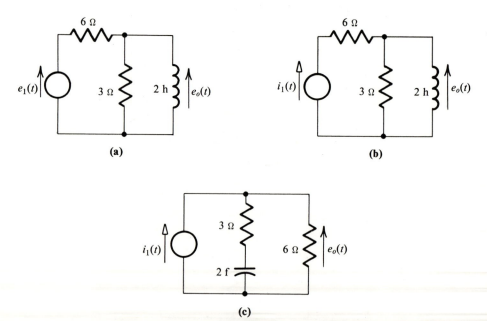

(a)

(b)

(c)

Figure P3.12

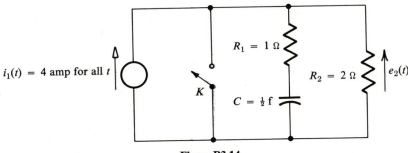

Figure P3.14

3.13 The response of a circuit with no initial stored energy to the unit step function $U_{-1}(t)$ is $(t + 2)\epsilon^{-t}$ for $t > 0$. Find its response to a unit impulse and to a unit ramp function.

3.14 The switch K in Fig. P3.14 has been open long enough so that the voltage $e_2(t)$ has reached a constant value. If K closes at $t = 0$, find the current flowing through the switch for all $t > 0$.

3.15 If the circuit in Fig. P3.15 contains no stored energy for $t < 0$, find $e_o(t)$ when $i_1(t) = U_o(t)$.

3.16 Find the unit step response and the unit ramp response for the circuit in Fig. P3.16.

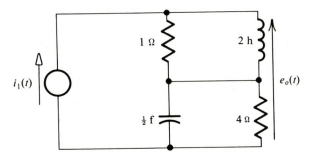

Figure P3.15

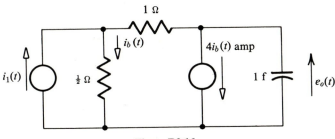

Figure P3.16

The solution may be carried out by writing two node equations, by writing one loop equation, or by using Thévenin's or Norton's theorem.

3.17 Find the energy supplied by the source in Fig. 3.4-1a, if $i(t)$ has the waveshape in Fig. 3.4-1b and if $a \ll RC$. Compare the answer with the energy stored in the capacitance at $t = a$.

3.18 Find the response of Fig. 3.4-1a for $t > a$ to the triangular pulse in Fig. P3.18. Show that if $a \ll RC$, then the response for $t > a$ is approximately equal to $Kah(t)$. This problem further substantiates the statement that the response to a pulse of any shape may be approximated by the response to an impulse of equal area, if the pulse width is small compared with the time constant of the circuit. Examples 3.3-4 and 3.4-1 are helpful in the solution.

3.19 A circuit's unit step response is known to be $10\epsilon^{-5t}$ for $t > 0$. Find its response to each of the inputs in Fig. P3.19 at $t = 2$ by the methods of Section 3.4. Assume that the circuit contains no stored energy for $t < 0$.

3.20 Solve the previous problem by using the convolution formulas.

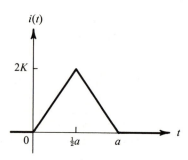

Figure P3.18

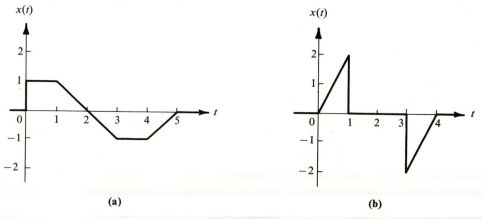

(a)

(b)

Figure P3.19

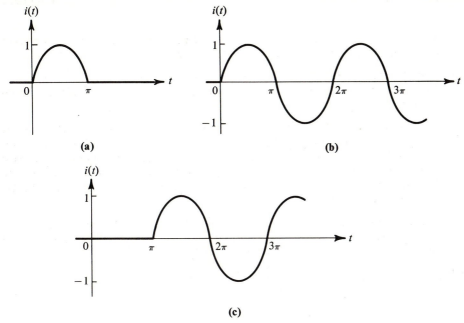

(a)

(b)

(c)

Figure P3.21

3.21 In Example 3.5-1 we found that the response of the circuit in Fig. 3.5-2a to the input in Fig. 3.5-2b, which is repeated in Fig. P3.21a, is $\frac{1}{2}(\sin t - \cos t + \epsilon^{-t})$ for $0 < t < \pi$. Write down the response to the input in Fig. P3.21b for all $t > 0$. Use the fact that the waveshape in Fig. P3.21a is the sum of the waveshapes in parts (b) and (c) of the figure in order to obtain the response to Fig. P3.21a for $t > \pi$.

3.22 Use the results of the previous problem and the discussion associated with Eq. 3.3-4 to find the response of the circuit in Fig. 3.5-2a to each of the inputs in Fig. P3.22.

3.23 Solve Problem 3.22 by the direct application of the convolution formulas.

$i(t) = (\cos t)U_{-1}(t)$

$i(t) = \cos t$ for $0 < t < \dfrac{\pi}{2}$

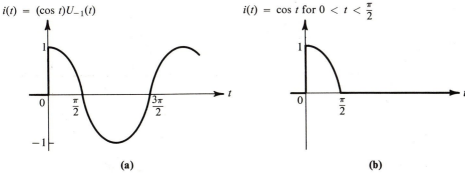

(a)

(b)

Figure P3.22

3.24 Show that $f(t) * U_0(t - a) = f(t - a)$.

3.25 A circuit's response to the unit ramp function $U_{-2}(t)$ is $1 - \epsilon^{-2t}$ for $t > 0$. If the circuit contains no stored energy for $t < 0$, find its response to $\epsilon^{-t}U_{-1}(t)$.

3.26 If N_1 and N_2 in Fig. 3.5-6 represent two isolated networks in cascade, will the relationship between $x(t)$ and $y(t)$ be changed if the two networks are interchanged? Repeat the problem if N_1 and N_2 are two nonisolated networks. Justify your answers.

3.27 Let N_1 and N_2 in Fig. 3.5-6 represent two isolated networks in cascade. If the unit impulse response of each network is $t\epsilon^{-2t}$, find the unit step response of the combination.

3.28 If the circuit in Fig. P3.28 contains no stored energy for $t < 0$, find $i_o(t)$ when $i_1(t) = U_{-1}(t)$.

3.29 The unit impulse response of a circuit is approximated by the curve in Fig. P3.29a. If the circuit contains no initial stored energy, sketch the response to the inputs in Figs. P3.29b and P3.29c. A graphical solution is preferred.

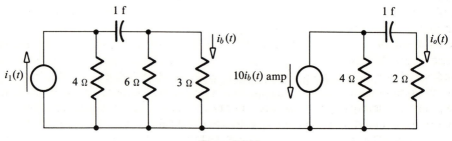

Figure P3.28

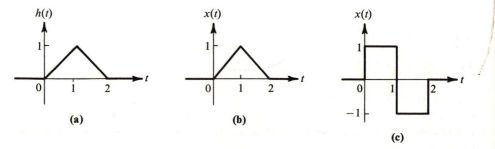

(a)

(b)

(c)

Figure P3.29

4

The Classical Solution
of Circuits

In Chapter 3 we considered the response of some simple circuits to the singularity functions. In this chapter we discuss the solution of more complicated circuits, which may or may not contain initial stored energy, and which are subjected to an arbitrary input. There are three basic steps in the classical solution of circuits: finding a differential equation relating the input and the output, obtaining the general solution of this equation, and evaluating all the arbitrary constants of integration. The general solution of differential equations is usually included in undergraduate mathematics courses, but it is reviewed in Section 4.1. Obtaining the initial conditions needed for the evaluation of the constants of integration is the step that frequently requires the most thought. It is discussed in detail in Section 4.2, before we proceed to the complete solution of circuit problems in Section 4.3. In later sections the emphasis is placed on the physical significance of the steps in the solution and an examination of the response to certain inputs that are frequently encountered in practice.

4.1

The General Solution of Differential Equations

Several simple examples of a differential equation relating the input and the output of a circuit can be found in Sections 1.4, 3.1, 3.3, and 4.3. Any linear, time-invariant circuit can be described by a differential equation with constant coefficients, having the general form

$$a_n \frac{d^n y}{dt^n} + a_{n-1} \frac{d^{n-1} y}{dt^{n-1}} + \cdots + a_1 \frac{dy}{dt} + a_0 y$$

$$= b_m \frac{d^m x}{dt^m} + \cdots + b_1 \frac{dx}{dt} + b_0 x \qquad (4.1\text{-}1)$$

175

where $x(t)$ and $y(t)$ are the input and output, and where the a's and b's are constants.[†] Since $x(t)$ is a known input, the right side of Eq. 4.1-1 is a known function of time, called the *forcing function*, and denoted by $F(t)$.

$$a_n \frac{d^n y}{dt^n} + a_{n-1} \frac{d^{n-1} y}{dt^{n-1}} + \cdots + a_1 \frac{dy}{dt} + a_0 y = F(t) \tag{4.1-2}$$

If $F(t)$ is identically zero, this equation reduces to the *homogeneous* differential equation

$$a_n \frac{d^n y}{dt^n} + a_{n-1} \frac{d^{n-1} y}{dt^{n-1}} + \cdots + a_1 \frac{dy}{dt} + a_0 y = 0 \tag{4.1-3}$$

For a nonzero forcing function, Eq. 4.1-2 is called a *nonhomogeneous* differential equation.

Some of the important properties of differential equations may be summarized as follows. The homogeneous equation 4.1-3 has n different, linearly independent solutions, which may be denoted by $y_1(t)$, $y_2(t)$, $\ldots$, $y_n(t)$. The most general solution is

$$y_H(t) = K_1 y_1(t) + K_2 y_2(t) + \cdots + K_n y_n(t) \tag{4.1-4}$$

where the K's are arbitrary constants, and where the subscript H refers to the solution of the homogeneous equation. The most general, or "complete," solution of the nonhomogeneous equation 4.1-2 is

$$y(t) = y_H(t) + y_P(t) \tag{4.1-5}$$

The term $y_H(t)$ is the solution of the related homogeneous equation, as given in Eq. 4.1-4, and is called the *complementary* solution. The term $y_P(t)$ is any one solution, no matter how achieved, which satisfies Eq. 4.1-2, and is called the *particular* solution. The particular solution contains no arbitrary constants.

Solution of the Homogeneous Differential Equation. While separation of variables and other special techniques can be used in simple cases, including the first-order differential equations in Chapter 3, we seek a systematic procedure that will always yield the general solution. Assume that the solutions of Eq. 4.1-3 have the form

$$y(t) = \epsilon^{rt}$$

where r is a constant to be determined. Substituting the assumed solution into Eq. 4.1-3, we obtain

$$(a_n r^n + a_{n-1} r^{n-1} + \cdots + a_1 r + a_0) \epsilon^{rt} = 0$$

In order for this equation to be satisfied for all values of t,

$$a_n r^n + a_{n-1} r^{n-1} + \cdots + a_1 r + a_0 = 0 \tag{4.1-6}$$

[†] All the circuits in this book are linear and time invariant. If the coefficients in Eq. 4.1-1 are functions of y or x, the circuit is nonlinear. If they are explicit functions of t, the circuit is time varying. These statements are consistent with Examples 1.1-1 through 1.1-3 and with the results of Problem 1.2.

Equation 4.1-6 is called the *characteristic* or *auxiliary equation* and is usually written down directly from Eq. 4.1-3. It is an algebraic equation that has n solutions (given by the zeros of the nth-order *characteristic polynomial* on the left-hand side), denoted by $r_1, r_2, \ldots, r_n$. The solutions of the homogeneous differential equation are then

$$y_1(t) = \epsilon^{r_1 t}, \qquad y_2(t) = \epsilon^{r_2 t}, \qquad \ldots, \qquad y_n(t) = \epsilon^{r_n t}$$

If all the zeros of the characteristic polynomial are distinct, the above n solutions are independent, and the general solution is

$$y_H(t) = K_1 \epsilon^{r_1 t} + K_2 \epsilon^{r_2 t} + \cdots + K_n \epsilon^{r_n t} \qquad \text{Transient term} \qquad (4.1\text{-}7)$$

If some of the roots of the characteristic equation are complex, $y_H(t)$ should be written in a different form. The coefficients in the differential equation describing any circuit are real, and any complex roots of the characteristic equation must occur in complex conjugate pairs.[†] If, for example, one root is $r_1 = -\alpha + j\beta$, where α and β are real, then another root must be $r_2 = -\alpha - j\beta$. By Euler's identity $\epsilon^{\pm j\theta} = \cos\theta \pm j\sin\theta$, the terms $K_1 \epsilon^{r_1 t} + K_2 \epsilon^{r_2 t}$ may be combined as follows:

$$\epsilon^{-\alpha t}(K_1 \epsilon^{j\beta t} + K_2 \epsilon^{-j\beta t}) = \epsilon^{-\alpha t}[(K_1 + K_2)\cos\beta t + j(K_1 - K_2)\sin\beta t]$$

$$= \epsilon^{-\alpha t}(K_3 \cos\beta t + K_4 \sin\beta t)$$

$$= K_5 \epsilon^{-\alpha t}\cos(\beta t + \phi) \qquad (4.1\text{-}8)$$

The last expression must be a real function of time, so K_1 and K_2 will be complex conjugates, while K_3, K_4, K_5, and ϕ will be real constants. Whenever the characteristic equation has complex roots, the corresponding terms in $y_H(t)$ should be expressed as one of the last two forms in Eq. 4.1-8.

If there are repeated roots, the terms in Eq. 4.1-7 are not independent, and the equation does not represent the most general solution. If $r_1 = r_2$, for example, $K_1 \epsilon^{r_1 t} + K_2 \epsilon^{r_2 t} = (K_1 + K_2)\epsilon^{r_1 t} = K\epsilon^{r_1 t}$. In this event, $y_1(t) = \epsilon^{r_1 t}$ and $y_2(t) = t\epsilon^{r_1 t}$ can be shown to be independent solutions, so that the general solution is

$$y_H(t) = (K_1 + K_2 t)\epsilon^{r_1 t} + K_3 \epsilon^{r_3 t} + \cdots + K_n \epsilon^{r_n t} \qquad (4.1\text{-}9)$$

If the root r_1 is repeated k times, so that $r_1 = r_2 = \cdots = r_k$, then the most general solution is

$$y_H(t) = (K_1 + K_2 t + \cdots + K_k t^{k-1})\epsilon^{r_1 t} + \cdots + K_n \epsilon^{r_n t} \qquad (4.1\text{-}10)$$

Example 4.1-1. Find the general solution to the following differential equation, which will be frequently encountered in later sections:

$$\frac{d^2 y}{dt^2} + 2\alpha \frac{dy}{dt} + \omega_0^2 y = 0$$

[†] It can be shown that this statement is one consequence of the requirement that the response to any real input must be another real function of time.

Solution. The characteristic equation $r^2 + 2\alpha r + \omega_0{}^2 = 0$ has roots at $-\alpha \pm \sqrt{\alpha^2 - \omega_0{}^2}$. If $\alpha^2 > \omega_0{}^2$, the two roots are real and distinct, and

$$y_H(t) = K_1 \epsilon^{(-\alpha + \sqrt{\alpha^2 - \omega_0{}^2})t} + K_2 \epsilon^{(-\alpha - \sqrt{\alpha^2 - \omega_0{}^2})t}$$

If $\alpha = \omega_0$, there is a repeated root at $-\alpha$, and

$$y_H(t) = (K_1 + K_2 t)\,\epsilon^{-\alpha t}$$

Finally, if $\alpha^2 < \omega_0{}^2$, the two roots are complex. If $\omega_d = j\sqrt{\omega_0{}^2 - \alpha^2}$, the roots are given by $r = -\alpha \pm j\omega_d$, and

$$y_H(t) = K\epsilon^{-\alpha t} \cos(\omega_d t + \phi)$$

Solution of the Nonhomogeneous Differential Equation. The solution of the nonhomogeneous equation 4.1-2 is made up of two parts: the complementary solution $y_H(t)$, which is found by replacing $F(t)$ by zero and by using Eqs. 4.1-6 through 4.1-10, and the particular solution $y_P(t)$. The two standard methods of finding $y_P(t)$ are the *variation of parameters* and the *method of undetermined coefficients*. This discussion is restricted to the latter method, because in most circuit problems it is the simpler. It does, however, have two disadvantages. The forcing function $F(t)$ must possess only a finite number of linearly independent derivatives [e.g., $F(t)$ can be t^3 or $\cos t$, but not $\sqrt{t}$], and the method cannot be extended to the solution of time-varying circuits.

The method of undetermined coefficients consists in examining the forcing function $F(t)$ and guessing the form of $y_P(t)$. The sizes of the terms in the assumed solution are chosen so that $y_P(t)$ satisfies the original differential equation for all values of t. In the event that the assumed $y_P(t)$ cannot be made to satisfy the differential equation, that form of $y_P(t)$ is not suitable.

Such a procedure is acceptable, since the most general solution is the sum of $y_H(t)$ and *any* one solution to the original differential equation. The form of $y_P(t)$ that should be chosen is determined by the forcing function and is a linear combination of the terms in $F(t)$ and its derivatives, with each term multiplied by a constant to be determined. The procedure is illustrated by the following examples, some of which are used in succeeding sections.

Example 4.1-2. Solve the differential equation

$$C\frac{de_o}{dt} + \frac{1}{R}e_o = i_1$$

for $e_o(t)$ when $i_1(t) = 1$ amp, and when $i_1(t) = \cos \omega t$.

Solution. Since the solution of the characteristic equation $Cr + 1/R = 0$ is $r = -1/RC$, the complementary solution is

$$(e_o)_H = K\epsilon^{-t/RC}$$

When $i_1(t) = 1$ amp, the particular solution has the form $(e_o)_P = A$. To satisfy the original differential equation, $A = R$, so the general solution is

$$e_o(t) = K\epsilon^{-t/RC} + R$$

In the case $i_1(t) = \cos \omega t$, the assumed particular solution is $(e_o)_P = A \cos \omega t + B \sin \omega t$. When these two expressions are substituted into the differential equation,

$$C(-A\omega \sin \omega t + B\omega \cos \omega t) + \frac{1}{R}(A \cos \omega t + B \sin \omega t) = \cos \omega t$$

or

$$(A + \omega RCB) \cos \omega t + (B - \omega RCA) \sin \omega t = R \cos \omega t$$

We equate corresponding coefficients, so that this result is satisfied for all values of t.

$$A + \omega RCB = R \quad \text{and} \quad B - \omega RCA = 0$$

Solving these two equations simultaneously, we find that

$$A = \frac{R}{1 + (\omega RC)^2} \quad \text{and} \quad B = \frac{\omega R^2 C}{1 + (\omega RC)^2}$$

The general solution is

$$e_o(t) = K\epsilon^{-t/RC} + \frac{R}{1 + (\omega RC)^2}(\cos \omega t + \omega RC \sin \omega t)$$

Example 4.1-3. Find the general solution to

$$\frac{de_o}{dt} + 3e_o = \epsilon^{-t} \cos t$$

Solution. The complementary solution is $(e_o)_H = K\epsilon^{-3t}$. The particular solution has the form $(e_o)_P = A\epsilon^{-t} \cos t + B\epsilon^{-t} \sin t$, which is substituted into the differential equation.

$$[(B - A)\epsilon^{-t} \cos t - (A + B)\epsilon^{-t} \sin t] + 3(A\epsilon^{-t} \cos t + B\epsilon^{-t} \sin t) = \epsilon^{-t} \cos t$$

$$(2A + B)\epsilon^{-t} \cos t + (-A + 2B)\epsilon^{-t} \sin t = \epsilon^{-t} \cos t$$

Equating corresponding coefficients, we obtain $2A + B = 1$ and $-A + 2B = 0$, from which $A = \frac{2}{5}$ and $B = \frac{1}{5}$. The complete solution is

$$e_o(t) = K\epsilon^{-3t} + \tfrac{1}{5}\epsilon^{-t}(2 \cos t + \sin t)$$

Example 4.1-4. Solve the differential equation

$$L\frac{d^2i}{dt^2} + 2R\frac{di}{dt} + \frac{1}{C}i = \frac{1}{C}$$

when $R^2 = L/C$.

Solution. Since $1/C = R^2/L$, the characteristic equation $Lr^2 + 2Rr + 1/C = 0$ may be rewritten as $r^2 + (2R/L)r + (R/L)^2 = (r + R/L)^2 = 0$. Then

$$i_H(t) = (K_1 + K_2 t)\epsilon^{-Rt/L}$$

The particular solution is $i_P(t) = 1$, so the general solution is

$$i(t) = (K_1 + K_2 t)\, \epsilon^{-Rt/L} + 1$$

Example 4.1-5.   Find the solution to

$$\frac{d^2 y}{dt^2} + 2\frac{dy}{dt} + 5y = t$$

Solution.   The characteristic equation $r^2 + 2r + 5 = 0$ has roots at $r = -1 \pm j2$, so

$$y_H(t) = \epsilon^{-t}(K_1 \cos 2t + K_2 \sin 2t)$$

Substitution of the assumed particular solution $y_P(t) = At + B$ into the differential equation yields

$$2A + 5(At + B) = t$$

which requires $5A = 1$ and $2A + 5B = 0$; i.e., $A = \frac{1}{5}$ and $B = -\frac{2}{25}$.

$$y(t) = \epsilon^{-t}\,(K_1 \cos 2t + K_2 \sin 2t) + \tfrac{1}{25}\,(5t - 2)$$

A modification is necessary in the assumed form of the particular solution when a term in $F(t)$ has exactly the same form as a term in the complementary solution. Consider the differential equation $(dy/dt) + 2y = \epsilon^{-2t}$, which has the complementary solution $y_H(t) = K\epsilon^{-2t}$. If an assumed particular solution $y_P(t) = A\epsilon^{-2t}$ is substituted into the differential equation, the resulting equation

$$-2A\epsilon^{-2t} + 2A\epsilon^{-2t} = \epsilon^{-2t}$$

cannot be identically satisfied by any value of the constant A. Whenever a term in $F(t)$ duplicates a term in $y_H(t)$, as does ϵ^{-2t} in this example, the corresponding terms in $y_P(t)$ should be multiplied by t. In the example under consideration, the assumed particular solution should be $y_P(t) = At\epsilon^{-2t}$. When this is substituted into the differential equation,

$$(A\epsilon^{-2t} - 2At\epsilon^{-2t}) + 2At\epsilon^{-2t} = \epsilon^{-2t}$$

which is satisfied for all values of t when $A = 1$. Thus the general solution is $y(t) = K\epsilon^{-2t} + t\epsilon^{-2t}$.

Whenever a term in $F(t)$ corresponds to a *repeated* root of the characteristic equation (say an mth-order root), the corresponding terms in the assumed particular solution should be multiplied by t^m.

Example 4.1-6.   Solve the differential equation

$$\frac{d^2 y}{dt^2} + 4\frac{dy}{dt} + 4y = \epsilon^{-2t}$$

Solution.   The characteristic equation $r^2 + 4r + 4 = (r + 2)^2 = 0$ has a double root at $r = -2$, and

$$y_H(t) = (K_1 + K_2 t)\, \epsilon^{-2t}$$

Therefore, the normal particular solution of $A\epsilon^{-2t}$ is replaced by $y_P(t) = At^2\epsilon^{-2t}$. Substitution of this expression into the differential equation gives

$$(2 - 8t + 4t^2)\, A\epsilon^{-2t} + (8t - 8t^2)\, A\epsilon^{-2t} + 4At^2\epsilon^{-2t} = \epsilon^{-2t}$$

from which $A = \frac{1}{2}$. The complete solution is $y(t) = (K_1 + K_2 t)\,\epsilon^{-2t} + \frac{1}{2}t^2\epsilon^{-2t}$.

The general solution of an nth-order differential equation contains n arbitrary constants, $K_1, K_2, \ldots, K_n$. These constants must be evaluated from a knowledge of the *initial conditions*, which cannot be obtained from the differential equation itself, but which must be found from the circuit.[†] Usually, the differential equation describing the circuit applies for all $t > 0$, and the initial conditions needed are the values of the response and its first $n - 1$ derivatives at $t = 0+$ (an infinitesimal interval after the reference time $t = 0$). The techniques for finding these conditions are discussed in the next section.

4.2

Initial Conditions

In general, the order of the differential equation relating the input and output of a circuit is the sum of the number of capacitances and the number of inductances, although in some cases the order may be less than this. The number of arbitrary constants in the solution, and hence the number of initial conditions needed, is equal to the order of the differential equation.

Most of the examples in this chapter involve a second-order differential equation and require a knowledge of $y(0+)$ and $[dy/dt](0+)$. The initial value of the output can be found by the application of Theorems 3.3-1 through 3.3-3. If there are no impulses present, the voltage across a capacitance and the current through an inductance cannot change instantaneously and must be the same at $t = 0-$ and $t = 0+$. Once these quantities have been determined, all voltages and currents at $t = 0+$ may be found by Kirchhoff's laws.

Frequently, the given circuit contains no stored energy for $t < 0$; i.e., $e_C(0-) = i_L(0-) = 0$. If, in addition, the applied source does not cause any impulses, then $e_C(0+) = i_L(0+) = 0$.[††] For the purpose of calculating the initial voltages and currents in a circuit with no initial stored energy, therefore, the capacitances may be replaced by short circuits and the inductances by open circuits, and the entire network may be reduced to a resistive circuit. It should be emphasized that the

---

[†] If it is known that there is no energy stored within the circuit for $t < 0$, it is possible to find the initial conditions from the differential equation, but the technique is not obvious and will not be discussed here.

[††] The discussion following Example 3.3-6 describes the only situations in which a finite source voltage or current can cause an impulse.

quantities de_C/dt and di_L/dt are not necessarily zero at $t = 0+$, even if $e_C(0+)$ and $i_L(0+)$ are zero.

The quantity $[dy/dt](0+)$ is understood to be $[dy(t)/dt]_{t=0+}$ and not $d[y(0+)]/dt$, which would always be zero. When the arbitrary constants are evaluated, $y(t)$ is not yet known over a continuous range of time that includes $t = 0+$, and the initial value of the derivative must be found by indirect and sometimes ingenious methods. If the desired response is a capacitance's voltage or an inductance's current, the initial value of the derivative may be found from the element law $i_C = C(de/dt)$ or $e_L = L(di/dt)$.

Example 4.2-1.   If the circuit in Fig. 4.2-1 contains no stored energy for $t < 0$, find the values of $e_o(t)$ and de_o/dt at $t = 0+$.

Solution.   Since the voltage across C and the current through L cannot change instantaneously, $e_o(0+) = i_o(0+) = 0$, and therefore $i_1(0+) = e_o(0+)/R_1 = 0$. Then $i_C(0+) = i(0+) = 1$ amp, and

$$\frac{de_o}{dt}(0+) = \frac{1}{C} i_C(0+) = \frac{1}{C} \text{ volts per second}$$

In many problems, determining the derivative of the output at $t = 0+$ requires more than the relationships $i_C = C(de/dt)$ and $e_L = L(di/dt)$. Usually, an appropriate equation that is valid for $t > 0$ must be written and differentiated term by term. If possible, one of the terms in the equation should be the desired response, and the derivative of the other terms at $t = 0+$ should be known. The student should keep in mind that any equation that is to be differentiated must be valid over a range of t including $t = 0+$, and not valid *only* at $t = 0+$.

Example 4.2-2.   Find $e_o(t)$ and de_o/dt at $t = 0+$ in Fig. 4.2-2, if there is no initial stored energy.

Solution.   Since $e_2(0+) = e_4(0+) = 0$, the source voltage must appear initially across the 2 and 3 Ω resistances; i.e., $e_o(0+) = e_3(0+) = e_1(0+) = 6$ volts. Then $i_o(0+) = \frac{6}{3} = 2$ amp, $i_3(0+) = \frac{6}{2} = 3$ amp, and $i_4(0+) = 2 + 3 = 5$ amp. The derivative of the capacitances' voltages can be found from the initial currents as in the previous example, but this does not give the derivative of the output directly. For all values of t, however, $e_1(t) = e_o(t) + e_2(t) + e_4(t)$, so

$$\frac{de_1}{dt} = \frac{de_o}{dt} + \frac{de_2}{dt} + \frac{de_4}{dt}$$

The value of all of the derivatives at $t = 0+$, except de_o/dt, may be easily obtained from the circuit. For $t > 0$, $e_1(t) = 6 \cos 2t$ and $de_1/dt = -12 \sin 2t$, so

$$\frac{de_1}{dt}(0+) = 0$$

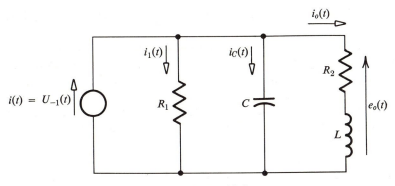

Figure 4.2-1

Furthermore,

$$\frac{de_2}{dt}(0+) = \frac{1}{C_2} i_o(0+) = 1 \quad \text{and} \quad \frac{de_4}{dt}(0+) = \frac{1}{C_4} i_4(0+) = \frac{5}{4}$$

so $\dfrac{de_o}{dt}(0+) = -\dfrac{9}{4}$ volts per sec.

Example 4.2-3. Find the values of $e_4(t)$, de_4/dt, $e_o(t)$, and de_o/dt at $t = 0+$, if the circuit in Fig. 4.2-3 contains no initial stored energy.

Solution. The quantities that are immediately known at $t = 0+$ are $i(0+) = 1$ amp, $e_C(0+) = 0$, and $i_o(0+) = 0$. Then $i_1(0+) = i(0+) - i_o(0+) = 1$ amp, $e_1(0+) = 2i_1(0+) = 2$ volts, and $i_2(0+) = i_3(0+) = 3e_1(0+) = 6$ amp. Also, $e_4(0+) = 0$ and $e_L(0+) = -e_4(0+) + e_1(0+) + e_2(0+) + e_C(0+) = 8$ volts. Then

$$\frac{di_o}{dt}(0+) = \frac{1}{L} e_L(0+) = 4$$

Since $e_4(t) = 2i_o(t)$, $de_4/dt = 2di_o/dt$, and

$$\frac{de_4}{dt}(0+) = 8 \text{ volts per sec}$$

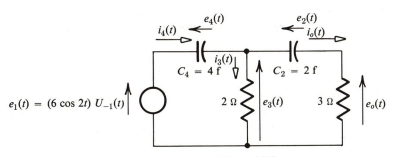

Figure 4.2-2

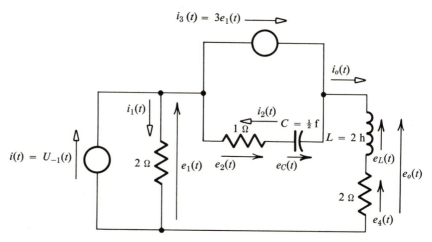

Figure 4.2-3

If the voltage $e_o(t)$ were required for all $t > 0$, the simplest procedure would be to find $i_o(t)$ for $t > 0$, using the initial conditions already found. Then $e_o(t) = 2i_o(t) + 2di_o/dt$. For practice, however, we shall find the initial values of $e_o(t)$ and de_o/dt, perhaps because the response for all $t > 0$ is not desired. By Kirchhoff's voltage law, $e_o(t) = e_1(t) + e_2(t) + ec(t)$ and

$$\frac{de_o}{dt} = \frac{de_1}{dt} + \frac{de_2}{dt} + \frac{dec}{dt}$$

The initial value of $e_o(t)$ is $e_o(0+) = 8$ volts, and the initial value of the derivative can be obtained if the three terms on the right-hand side of the above equation can be evaluated at $t = 0+$. First,

$$\frac{dec}{dt}(0+) = \frac{1}{C} i_2(0+) = 12$$

Since $i(t) = i_1(t) + i_o(t)$, $di/dt = di_1/dt + di_o/dt$ and

$$\frac{de_1}{dt}(0+) = 2\frac{di_1}{dt}(0+) = 2\left[\frac{di}{dt}(0+) - \frac{di_o}{dt}(0+)\right] = 2(0 - 4) = -8$$

Next, $i_2(t) = i_3(t) - i_o(t) = 3e_1(t) - i_o(t)$, so

$$\frac{de_2}{dt}(0+) = \frac{di_2}{dt}(0+) = 3\frac{de_1}{dt}(0+) - \frac{di_o}{dt}(0+) = -24 - 4 = -28$$

Finally,

$$\frac{de_o}{dt}(0+) = -8 - 28 + 12 = -24 \text{ volts per sec}$$

Example 4.2-4. The circuit in Fig. 4.2-4 contains a capacitance and two inductances and is described by a third-order differential equation. Assuming that the circuit has no initial stored energy, find the initial values of $e_o(t)$, de_o/dt, and d^2e_o/dt^2, which would

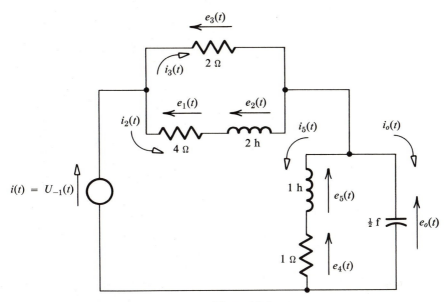

Figure 4.2-4

be needed to evaluate the three arbitrary constants in the general solution. (Since the rest of this chapter does not require the solution of third-order differential equations, the student may prefer to omit this example.)

Solution.   The quantities that are immediately known are $i(0+) = 1$ amp, $i_2(0+) = i_5(0+) = 0$, and $e_o(0+) = 0$. Then $i_3(0+) = 1$ amp, $e_3(0+) = 2$ volts, $e_1(0+) = 0$, $e_2(0+) = 2$ volts, $i_o(0+) = 1$ amp, $e_4(0+) = 0$, and $e_5(0+) = 0$. Next,

$$\frac{de_o}{dt}(0+) = \frac{1}{C}i_o(0+) = 2 \text{ volts per sec}$$

$$\frac{di_2}{dt}(0+) = \frac{1}{L_2}e_2(0+) = 1 \quad \text{and} \quad \frac{di_5}{dt}(0+) = \frac{1}{L_1}e_5(0+) = 0$$

Since $di_o/dt = C\,d^2e_o/dt^2$, the solution will be completed if we can find the initial value of di_o/dt. Because $i_o(t) = i(t) - i_5(t)$,

$$\frac{di_o}{dt}(0+) = \frac{di}{dt}(0+) - \frac{di_5}{dt}(0+) = 0 \quad \text{and} \quad \frac{d^2e_o}{dt^2}(0+) = 0$$

In the examples thus far we have assumed that no energy was stored in the circuit at $t = 0-$. In many problems, however, the circuit has been excited for $t < 0$ and does contain stored energy at $t = 0-$. Because we are not yet in a position to fully discuss the means by which the initial energy was supplied, we shall present only one such example at this time. Other examples are found in Section 4.5 and in later chapters.

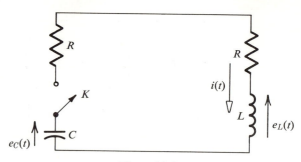

Figure 4.2-5

Example 4.2-5. The capacitance in Fig. 4.2-5 has been previously charged, so that $e_C(0-) = 1$ volt. The switch K suddenly closes at $t = 0$, causing a current $i(t)$ to flow. Find $i(t)$ and di/dt at $t = 0+$.

Solution. Since the voltage across C and the current through L cannot change instantaneously, $e_C(0+) = 1$ volt and $i(0+) = 0$. There is no initial voltage across the two resistances, so $e_L(0+) = 1$ volt. Then

$$\frac{di}{dt}(0+) = \frac{1}{L} e_L(0+) = \frac{1}{L}$$

4.3

The Complete Solution of Circuits

In the previous two sections we discussed two of the steps encountered in the classical solution of a circuit: solving a given differential equation, and finding the initial conditions required for an evaluation of the arbitrary constants of integration. Because the initial conditions apply to the entire response $y(t)$, the arbitrary constants cannot be evaluated until both the complementary solution $y_H(t)$ and the particular solution $y_P(t)$ have been found. The examples in this section incorporate the techniques of the previous sections into the complete solution of circuits. The first example involves only a first-order differential equation, similar to those in Chapter 3, but the next two circuits are described by a loop or a node equation that contains both a derivative and an integral. Differentiation of the equation term by term to remove the integral sign yields a second-order differential equation.

Example 4.3-1. If the capacitance in Fig. 4.3-1 is uncharged for $t < 0$, find $e_o(t)$ for $t > 0$ when $i_1(t) = U_{-1}(t)$ and when $i_1(t) = (\cos \omega t) U_{-1}(t)$.

Solution. The node equation

$$C \frac{de_o}{dt} + \frac{1}{R} e_o = i_1$$

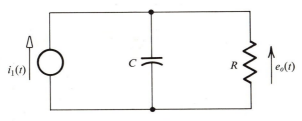

Figure 4.3-1

is the same differential equation considered in Example 4.1-2. When the source current equals the unit step function, $i_1(t) = 1$ amp for $t > 0$, and

$$e_o(t) = K\epsilon^{-t/RC} + R$$

Since the voltage of the capacitance cannot change instantaneously, $e_o(0+) = 0$, so $K = -R$ and

$$e_o(t) = R(1 - \epsilon^{-t/RC})$$

for $t > 0$, agreeing with the expression for $r(t)$ in Fig. 3.3-1a.

When $i_1(t) = \cos \omega t$ for $t > 0$,

$$e_o(t) = K\epsilon^{-t/RC} + \frac{R}{1 + (\omega RC)^2} (\cos \omega t + \omega RC \sin \omega t)$$

Since $e_o(0+)$ is again zero, $K = -R/[1 + (\omega RC)^2]$.

Example 4.3-2. The circuit in Fig. 4.3-2a contains no stored energy for $t < 0$. If $R^2 = L/C$ and $i_1(t) = U_{-1}(t)$, find $i_o(t)$, $i_C(t)$, and $e_1(t)$ for $t > 0$.

Solution. The only unknown loop current in Fig. 4.3-2b is $i_o(t)$. The voltage-law equation around this loop is

$$L\frac{di_o}{dt} + Ri_o + R(i_o - i_1) + \frac{1}{C}\int (i_o - i_1)\, dt = 0$$

In Chapter 1 it was suggested that $e_C(t)$ be written as a definite integral in order to avoid a constant of integration, but this is not necessary when the equation is to be immediately differentiated.

$$L\frac{d^2i_o}{dt^2} + 2R\frac{di_o}{dt} + \frac{1}{C}i_o = R\frac{di_1}{dt} + \frac{1}{C}i_1$$

When $i_1(t) = U_{-1}(t)$, the right-hand side equals $1/C$ for $t > 0$. From the results of Example 4.1-4, the general solution is

$$i_o(t) = (K_1 + K_2 t)\, \epsilon^{-Rt/L} + 1$$

$$\frac{di_o}{dt} = -\frac{R}{L}(K_1 + K_2 t)\, \epsilon^{-Rt/L} + K_2\epsilon^{-Rt/L}$$

for $t > 0$. The current through the inductance cannot change instantaneously, so

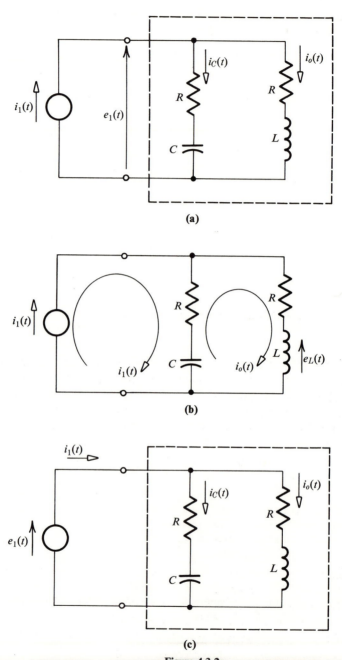

(a)

(b)

(c)

Figure 4.3-2

$i_o(0+) = 0$ and $i_1(0+) = 1$ amp. Since the voltage across the capacitance must be zero at $t = 0+$, $e_L(0+) = Ri_1(0+)$ and

$$\frac{di_o}{dt}(0+) = \frac{1}{L} e_L(0+) = \frac{R}{L}$$

From these initial values of $i_o(t)$ and its derivative, $K_1 = -1$ and $K_2 = 0$.

$$i_o(t) = 1 - \epsilon^{-Rt/L}$$

$$i_C(t) = i_1(t) - i_o(t) = \epsilon^{-Rt/L}$$

The voltage $e_1(t)$ may be found from either of the following two equations:

$$e_1 = L\frac{di_o}{dt} + Ri_o$$

$$e_1 = \frac{1}{C}\int_0^t i_C(\lambda)\, d\lambda + Ri_C$$

from which we obtain

$$e_1(t) = R$$

for all $t > 0$. Notice that it is important to use the proper limits in the last integral, as in Eq. 1.3-2.

It is quite surprising that the input voltage in the last example should have a constant value for all $t > 0$, even though the currents $i_o(t)$ and $i_C(t)$ vary with time. Since $e_1(t)/i_1(t) = R$ for all values of time, the circuit enclosed by the dashed lines in Fig. 4.3-2a behaves like a single resistance. In general, a combination of elements of different kinds is not equivalent to a single element and neither is the circuit under consideration if $R^2 \neq L/C$. This is more clearly seen when the circuit is excited by a voltage source $e_1(t) = U_{-1}(t)$, as in Fig. 4.3-2c. Then, as in Fig. 3.3-1,

$$i_C(t) = \frac{1}{R} \epsilon^{-t/RC}$$

$$i_o(t) = \frac{1}{R}(1 - \epsilon^{-Rt/L})$$

for $t > 0$. The total current $i_1(t)$ equals $1/R$ for $t > 0$ if and only if the two time constants RC and L/R are equal, i.e., if $R^2 = L/C$.

One unanswered question is whether or not the fact that a circuit behaves like a resistance for a step function input ensures that it will be equivalent to a resistance regardless of the external connections. In Section 3.5 we demonstrated that the response of a circuit having no initial stored energy to an arbitrary input may be found as soon as its response to a unit step function is known. Therefore, two two-terminal networks with no initial stored energy are always equivalent if they have the same unit step response.

Example 4.3-3.   Assuming that there is no initial stored energy in either the inductance or the capacitance in Fig. 4.3-3, find $e_o(t)$ for $t > 0$.

Solution.   If the bottom line is taken as the reference node, the current-law equation at node o is

$$6 \int e_o \, dt + \frac{de_o}{dt} + e_o = 4e_g = 4(e_1 - e_o)$$

Differentiating and rearranging, we obtain

$$\frac{d^2 e_o}{dt^2} + 5 \frac{de_o}{dt} + 6e_o = 4 \frac{de_1}{dt}$$

Since the right side of the last equation is zero for $t > 0$, the general solution is the complementary solution

$$e_o(t) = K_1 \epsilon^{-2t} + K_2 \epsilon^{-3t}$$

for $t > 0$. At $t = 0+$, the current through the inductance and the voltage across the capacitance remain zero. The current $i_2(0+) = 4$ amp must flow into the capacitance, since there can be no current through a resistance when there is no voltage across it. Thus,

$$e_o(0+) = 0, \qquad \frac{de_o}{dt}(0+) = \frac{1}{C} i_C(0+) = 4$$

from which $K_1 = 4$ and $K_2 = -4$, and

$$e_o(t) = 4(\epsilon^{-2t} - \epsilon^{-3t}) \quad \text{for} \quad t > 0$$

When several loop or node equations are needed to describe a circuit, they should be solved simultaneously to yield a single differential equation relating the output and the input. This is usually done by systematically eliminating the undesired unknowns, as in the following example. Once such a differential equation has been found, the rest of the solution is similar to the previous examples.

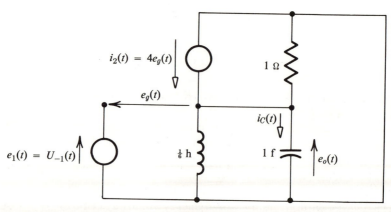

Figure 4.3-3

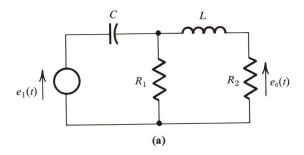

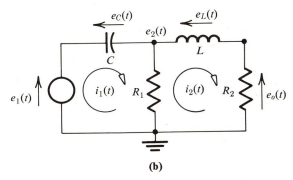

Figure 4.3-4

Example 4.3-4. Find a differential equation relating $e_o(t)$ and $e_1(t)$ in Fig. 4.3-4a. Then find the unit step response when $R_1 = R_2 = C = L = 1$.

Solution. The circuit may be described by either two loop or two node equations. When the loop currents in Fig. 4.3-4b are used, the voltage-law equations around these loops are

$$\left(R_1 i_1 + \frac{1}{C} \int i_1 \, dt \right) - R_1 i_2 = e_1$$

$$- R_1 i_1 + \left[(R_1 + R_2) i_2 + L \frac{di_2}{dt} \right] = 0$$

The first equation is differentiated to remove the integral sign:

$$\left(R_1 \frac{di_1}{dt} + \frac{i_1}{C} \right) - R_1 \frac{di_2}{dt} = \frac{de_1}{dt}$$

The second equation may be solved for i_1, and the result may be substituted into the first equation:

$$\left[(R_1 + R_2) \frac{di_2}{dt} + L \frac{d^2 i_2}{dt^2} \right] + \frac{1}{R_1 C} \left[(R_1 + R_2) i_2 + L \frac{di_2}{dt} \right] - R_1 \frac{di_2}{dt} = \frac{de_1}{dt}$$

Collecting terms, we obtain

$$L \frac{d^2 i_2}{dt^2} + \left(R_2 + \frac{L}{R_1 C} \right) \frac{di_2}{dt} + \frac{R_1 + R_2}{R_1 C} i_2 = \frac{de_1}{dt}$$

Since $e_o(t) = R_2 i_2(t)$,

$$\frac{d^2 e_o}{dt^2} + \left(\frac{R_1 R_2 C + L}{R_1 LC} \right) \frac{de_o}{dt} + \frac{R_1 + R_2}{R_1 LC} e_o = \frac{R_2}{L} \frac{de_1}{dt}$$

An alternative solution is based upon the choice of a reference node shown in Fig. 4.3-4b. The current-law equations at nodes 2 and o are

$$C \frac{d}{dt} (e_2 - e_1) + \frac{e_2}{R_1} + \frac{1}{L} \int (e_2 - e_o) \, dt = 0$$

$$\frac{1}{L} \int (e_o - e_2) \, dt + \frac{e_o}{R_2} = 0$$

or

$$\left(C \frac{d^2 e_2}{dt^2} + \frac{1}{R_1} \frac{de_2}{dt} + \frac{e_2}{L} \right) - \frac{e_o}{L} = C \frac{d^2 e_1}{dt^2}$$

$$- \frac{e_2}{L} + \left(\frac{1}{R_2} \frac{de_o}{dt} + \frac{e_o}{L} \right) = 0$$

Solving the second equation for $e_2(t)$ and substituting the result into the first equation, we get

$$\left(\frac{LC}{R_2} \frac{d^3 e_o}{dt^3} + C \frac{d^2 e_o}{dt^2} \right) + \left(\frac{L}{R_1 R_2} \frac{d^2 e_o}{dt^2} + \frac{1}{R_1} \frac{de_o}{dt} \right) + \left(\frac{1}{R_2} \frac{de_o}{dt} + \frac{e_o}{L} \right) - \frac{e_o}{L} = C \frac{d^2 e_1}{dt^2}$$

or

$$\frac{LC}{R_2} \frac{d^3 e_o}{dt^3} + \left(C + \frac{L}{R_1 R_2} \right) \frac{d^2 e_o}{dt^2} + \left(\frac{1}{R_1} + \frac{1}{R_2} \right) \frac{de_o}{dt} = C \frac{d^2 e_1}{dt^2}$$

or

$$\frac{d^2 e_o}{dt^2} + \left(\frac{R_1 R_2 C + L}{R_1 LC} \right) \frac{de_o}{dt} + \frac{R_1 + R_2}{R_1 LC} e_o = \frac{R_2}{L} \frac{de_1}{dt}$$

which agrees with the previous answer.

If each of the passive elements has a numerical value of unity and if $e_1(t) = U_{-1}(t)$, the differential equation for $t > 0$ is

$$\frac{d^2 e_o}{dt^2} + 2 \frac{de_o}{dt} + 2e_o = 0$$

The particular solution is zero, and the characteristic equation $r^2 + 2r + 2 = 0$ has roots at $r = -1 \pm j1$. By Eq. 4.1-8,

$$e_o(t) = \epsilon^{-t}(K_1 \cos t + K_2 \sin t)$$

for $t > 0$. If the circuit contains no initial stored energy, $e_C(0+) = 0$ and $i_2(0+) = 0$. Thus $e_o(0+) = 0$, $e_L(0+) = e_1(0+) = 1$ volt, and

$$\frac{de_o}{dt}(0+) = R\frac{di_2}{dt}(0+) = \frac{R}{L}e_L(0+) = 1$$

By the use of these initial conditions, it is found that $K_1 = 0$ and $K_2 = 1$, and

$$e_o(t) = \epsilon^{-t}\sin t \quad \text{for} \quad t > 0$$

In some circuits, the unwanted variables cannot be eliminated from a set of loop or node equations as easily as in Example 4.3-4. In Fig. 4.3-5, for example, the node equations are

$$\frac{de_2}{dt} + 2e_2 - \frac{de_o}{dt} - e_o = e_1$$

$$-\frac{de_2}{dt} - e_2 + 2\frac{de_o}{dt} + e_o = 0$$

If the desired output is $e_o(t)$, one method would be to add the two equations in order to obtain an expression for $e_2(t)$ in terms of $e_o(t)$ and $e_1(t)$ and their derivatives. This expression could then be substituted into either of the above equations, yielding a differential equation in $e_o(t)$ and $e_1(t)$.

When solving simultaneous differential equations, some students prefer to introduce the differentiation operator p, which is formally defined by

$$py(t) = \frac{d}{dt}y(t) \tag{4.3-1}$$

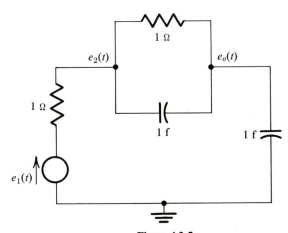

Figure 4.3-5

In most respects, the operator p may be treated as an algebraic quantity, as in the identities

$$p(py) = p^2 y = \frac{d^2 y}{dt^2}$$

$$p^m(p^n y) = p^{m+n} y$$

$$(k_1 p^2 + k_2 p)y = k_1 p^2 y + k_2 py$$

$$p(k_1 y_1 + k_2 y_2) = k_1 py_1 + k_2 py_2$$

$$(p + k_1)(p + k_2)y = [p^2 + (k_1 + k_2)p + k_1 k_2]y$$

where k_1 and k_2 are constants and m and n are positive integers. These properties may be easily verified by the definition of differentiation. It is important to know that the operator p does *not* commute with functions, i.e., $p(ty) \neq t(py)$ and $p(y_1 y_2) \neq y_1(py_2)$, and that the position of the parentheses is important, e.g., $p(y_1 y_2) \neq (py_1)y_2$.

If the operator p is used, the node equations for Fig. 4.3-5 become

$$(p + 2)e_2(t) - (p + 1)e_o(t) = e_1(t)$$

$$-(p + 1)e_2(t) + (2p + 1)e_o(t) = 0$$

If every term in the first equation is premultiplied by the operator $(p + 1)$ and every term in the second equation is premultiplied by $(p + 2)$,

$$(p + 1)(p + 2)e_2(t) - (p + 1)(p + 1)e_o(t) = (p + 1)e_1(t)$$

$$-(p + 2)(p + 1)e_2(t) + (p + 2)(2p + 1)e_o(t) = 0$$

This operation is equivalent to replacing the first equation by the sum of itself and its derivative and to replacing the second equation by the sum of its derivative and twice itself. When the equations are added, the $e_2(t)$ terms cancel, leaving

$$(2p^2 + 5p + 2 - p^2 - 2p - 1)e_o(t) = (p + 1)e_1(t)$$

or

$$(p^2 + 3p + 1)e_o(t) = (p + 1)e_1(t)$$

or

$$\frac{d^2 e_o}{dt^2} + 3\frac{de_o}{dt} + e_o = \frac{de_1}{dt} + e_1$$

which is the differential equation relating the input and the output.

Sometimes, an input cannot be easily described by a single equation for all $t > 0$, but is given by several expressions that are valid for different ranges of time. For the waveshape in Fig. 4.3-6, $i_1(t) = 2 - 2t$ for $0 < t < 3$ and $i_1(t) = 0$ for $t > 3$. In the classical solution for the response to such an input, the circuit's differential equation should be solved separately for each range of time, with each solution containing a different set of arbitrary constants.

Example 4.3-5. Find $e_o(t)$ for $t > 0$ in Fig. 4.3-1, if the capacitance is initially uncharged, if $R = C = 1$, and if $i_1(t)$ has the waveshape shown in Fig. 4.3-6.

Solution.   For $0 < t < 3$,

$$\frac{de_o}{dt} + e_o = 2 - 2t$$

The complementary solution is $(e_o)_H = K_1 \epsilon^{-t}$. Substituting the assumed particular solution $(e_o)_P = At + B$ into the differential equation, we obtain

$$A + At + B = 2 - 2t$$

from which $A = -2$ and $B = 4$. The general solution is

$$e_o(t) = 4 - 2t + K_1 \epsilon^{-t}$$

Since $e_o(0+) = 0$, $K_1 = -4$ and

$$e_o(t) = 4 - 2t - 4e^{-t} \quad \text{for} \quad 0 < t < 3$$

For $t > 3$, $i_1(t) = 0$, and the differential equation

$$\frac{de_o}{dt} + e_o = 0$$

has only the complementary solution

$$e_o(t) = K_2 \epsilon^{-t}$$

Since the voltage across the capacitance cannot change instantaneously, $e_o(t)$ is the same at $t = 3-$ and at $t = 3+$. From the solution for $0 < t < 3$,

$$e_o(3-) = -2 - 4\epsilon^{-3}$$

which must equal $K_2 \epsilon^{-3}$. Thus $K_2 = -2\epsilon^3 - 4$ and

$$e_o(t) = -2(\epsilon^3 + 2)\epsilon^{-t} \quad \text{for} \quad t > 3$$

which agrees with the answer to Example 3.4-3.

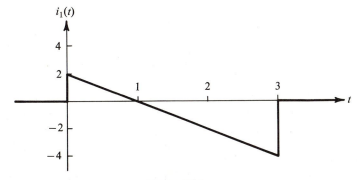

Figure 4.3-6

Occasionally, the superposition, Thévenin, and Norton theorems may be useful in the classical solution of a circuit. These theorems were first presented in Sections 2.2 and 2.3, and their application in circuits containing capacitances and inductances was discussed at the end of Section 3.3.

Example 4.3-6.   The capacitance C_1 in Fig. 4.3-7a has been previously charged, so that $e_1(0-) = 1$ volt. The other elements contain no initial stored energy. If the switch K closes at $t = 0$, find $e_o(t)$ for all $t > 0$. Replace the part of the circuit within the dashed lines by a Thévenin equivalent circuit.

Solution.   The network for which a Thévenin equivalent is desired is shown in Fig. 4.3-7b. Under open-circuit conditions,

$$L_1 \frac{di_1}{dt} + 2R_1 i_1 + \frac{1}{C_1} \int i_1 \, dt = 0$$

for $t > 0$. When the equation is differentiated and numerical values inserted,

$$\frac{d^2 i_1}{dt^2} + 2 \frac{di_1}{dt} + i_1 = 0$$

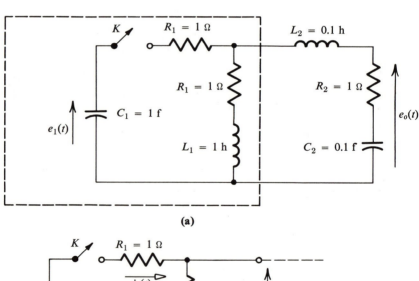

(a)

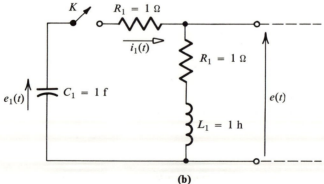

(b)

Since the characteristic equation $r^2 + 2r + 1 = 0$ has a double root at -1,

$$i_1(t) = (K_1 + K_2 t)\epsilon^{-t}$$

$$\frac{di_1}{dt} = -(K_1 + K_2 t)\epsilon^{-t} + K_2 \epsilon^{-t}$$

The student should show (by arguments similar to those in Example 4.2-5) that

$$i_1(0+) = 0, \qquad \frac{di_1}{dt}(0+) = 1$$

Thus $K_1 = 0$, $K_2 = 1$, and $i(t) = t\epsilon^{-t}$. The open-circuit voltage is

$$e_{oc} = L_1 \frac{di_1}{dt} + R_1 i_1 = \epsilon^{-t}$$

The dead network for $t > 0$ is formed by closing the switch and by removing the initial stored energy from C_1 in Fig. 4.3-7b. From the results of Example 4.3-2, the resulting circuit is equivalent to a 1 Ω resistance. Thévenin's theorem then allows

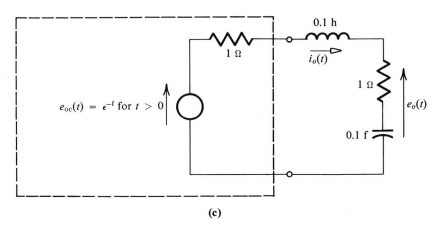

(c)

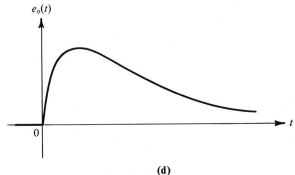

(d)

Figure 4.3-7

Fig. 4.3-7a to be replaced by Fig. 4.3-7c. The theorem is particularly helpful in this example because the dead network happens to reduce to a single element. For $t > 0$,

$$0.1 \frac{di_o}{dt} + 2i_o + 10 \int i_o \, dt = \epsilon^{-t}$$

or

$$\frac{d^2 i_o}{dt^2} + 20 \frac{di_o}{dt} + 100 i_o = -10 \epsilon^{-t}$$

The complementary solution is $(K_3 + K_4 t)\epsilon^{-10t}$. When the assumed particular solution $(i_o)_P = A\epsilon^{-t}$ is substituted into the differential equation,

$$A^2 - 20A + 100A = -10$$

Thus $A = -\frac{10}{81}$ and

$$i_o(t) = (K_3 + K_4 t)\epsilon^{-10t} - \frac{10}{81} \epsilon^{-t}$$

$$e_o(t) = e_{oc}(t) - i_o(t) - 0.1 \frac{di_o}{dt}$$

$$= \frac{10}{9}\epsilon^{-t} - 0.1 \, K_4 \epsilon^{-10t}$$

Since $e_o(0+) = 0$, the output for $t > 0$ is

$$e_o(t) = \tfrac{10}{9}(\epsilon^{-t} - \epsilon^{-10t})$$

which is sketched in Fig. 4.3-7d.

4.4

Physical Significance of the Complementary and Particular Solutions

The complementary solution $y_H(t)$ is the solution of the related homogeneous differential equation, with the input and hence the forcing function $F(t)$ made equal to zero. The form of $y_H(t)$ is completely determined by the roots of the characteristic equation, which in turn depends only upon the circuit. If there is no external source for $t > 0$, as occurs when the circuit is excited only by some initial stored energy (or, equivalently, subjected to an impulse function at $t = 0$), then $y_P(t) = 0$, and $y_H(t)$ is the complete solution. Since the complementary solution represents the behavior of the circuit when the source is dead, it is also called the *free response*. The particular solution $y_P(t)$ is called the *forced response*.

The nature of the free response can be seen by examining the roots of the characteristic equation. Since the roots may be either real or complex numbers, they may be plotted as points in a complex plane, as in Fig. 4.4-1. Let the real, nonnegative constants α and β denote the distance of a root from the vertical and horizontal axes, respectively. If there is a real, negative root $r_1 = -\alpha_1$, $y_H(t)$ will contain the term $K_1 \epsilon^{-\alpha_1 t}$, which is a decaying exponential, with a time constant of $1/\alpha_1$. A root

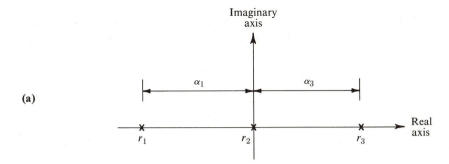

(a)

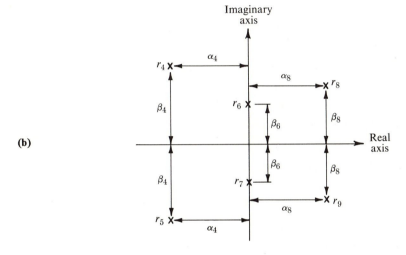

(b)

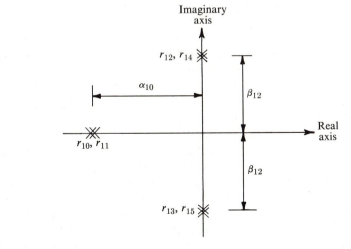

(c)

Figure 4.4-1

r_2 at the origin yields the constant term $K_2 \epsilon^{0t} = K_2$, while a real, positive root $r_3 = \alpha_3$ gives the growing exponential $K_3 \epsilon^{\alpha_3 t}$.

Suppose that the characteristic equation has a pair of complex roots $r_4 = -\alpha_4 + j\beta_4$ and $r_5 = -\alpha_4 - j\beta_4$, which are plotted as two points in the left half of the complex plane in Fig. 4.4-1b. As in Eq. 4.1-8, the corresponding terms in the free response can be combined as $K_4 \epsilon^{-\alpha_4 t} \cos(\beta_4 t + \phi_4)$. This is a decaying oscillation, with an angular frequency β_4. The pair of purely imaginary roots $r_6 = j\beta_6$ and $r_7 = -j\beta_6$ yield the constant-amplitude oscillation $K_6 \cos(\beta_6 t + \phi_6)$, while the complex roots r_8 and r_9 in the right half plane yield the growing oscillation $K_8 \epsilon^{\alpha_8 t} \cos(\beta_8 t + \phi_8)$.

Figure 4.4-1c shows some typical *repeated* roots of the characteristic equation. The double root $r_{10} = r_{11} = -\alpha_{10}$ on the negative real axis leads to the terms $K_{10} \epsilon^{-\alpha_{10} t} + K_{11} t \epsilon^{-\alpha_{10} t}$ in $y_H(t)$. Although the second of these two terms is an indeterminant form as t approaches infinity, it can be shown (e.g., by l'Hôpital's rule) that it approaches zero. The pair of repeated roots on the imaginary axis produce the terms $K_{12} \cos(\beta_{12} t + \phi_{12}) + K_{14} t \cos(\beta_{12} t + \phi_{14})$, which is a growing oscillation.

The term *open* left half plane is used to denote that part of the complex plane to the left of, but *not including*, the imaginary axis. We have seen that roots of the characteristic equation that lie in the open left half plane give rise to terms in $y_H(t)$ that decay to zero as t approaches infinity. The reciprocal of the distance of the roots from the imaginary axis is a measure of how fast the terms decay, while the distance from the real axis is the angular frequency of oscillation (if any). Roots in the open right half plane, or repeated roots on the imaginary axis (including the origin), give terms that increase without limit as t approaches infinity.

The above discussion shows how the nature of the free response is immediately evident from a plot of the roots of the characteristic equation. In certain applications in communication and control systems, it is necessary to know how these roots move in the complex plane when one of the circuit elements is varied.

Example 4.4-1. Sketch the locus traced out by the roots of the characteristic equation as the resistance R in Fig. 4.4-2a is varied.

Solution. Differentiation of the loop equation

$$L \frac{di}{dt} + Ri + \frac{1}{C} \int i \, dt = e$$

gives

$$\frac{d^2 i}{dt^2} + \frac{R}{L} \frac{di}{dt} + \frac{1}{LC} i = \frac{1}{L} \frac{de}{dt}$$

In order to make use of Example 4.1-1 and to be able to apply the results of this example to other circuits, the characteristic equation $r^2 + (R/L)r + 1/LC = 0$ is rewritten as

$$r^2 + 2\alpha r + \omega_0^2 = 0 \qquad (4.4\text{-}1)$$

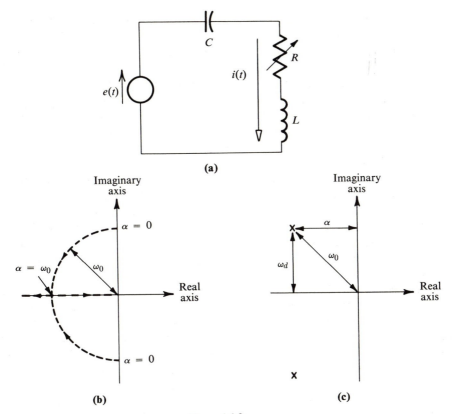

Figure 4.4-2

where $\alpha = R/2L$ and $\omega_0^2 = 1/LC$. Varying the value of R is equivalent to varying α and keeping ω_0^2 fixed.

When R (and hence α) is zero, the roots are at $\pm j\omega_0$ and lie on the imaginary axis of the complex plane, as shown in Fig. 4.4-2b. The free response is then a sinusoidal oscillation of constant amplitude, as should be expected. The free response is the circuit's behavior when the voltage source is dead, i.e., replaced by a short circuit. The lack of any resistance means that no power can be dissipated as heat, so the initial stored energy will not be lost, but will be continually interchanged between the inductance and capacitance.

When α is positive but less than ω_0, the roots are at

$$-\alpha \pm j\omega_d = -\alpha \pm j\sqrt{\omega_0^2 - \alpha^2}$$

as shown in Fig. 4.4-2c. The distance of the roots from the origin of the complex plane is $\sqrt{\alpha^2 + \omega_d^2} = \omega_0$. As α is varied, therefore, the locus traced out by the roots is a circle of radius ω_0, as indicated by the dashed lines in Fig. 4.4-2b. The arrows indicate the direction of increasing α. The free response is now a decaying oscillation, with energy being lost to the circuit because of the resistance. As R is increased, the complex roots move farther away from the imaginary axis, so that the oscillations are damped out more rapidly. The symbol ω_d is called the *damped natural angular frequency*, since it is the angular frequency of the free response when R is present, while ω_0 is the *undamped natural angular frequency*.

When $\alpha = \omega_0$, i.e., when $R = 2\sqrt{L/C}$, the characteristic equation has a double root at $-\alpha$, and the free response contains no oscillations. When $\alpha > \omega_0$, there are two distinct roots on the negative real axis, which correspond to two exponential terms. From a physical point of view, the resistance is now large enough to prevent the continual interchange of energy between the inductance and capacitance.

The *damping ratio* ζ is a dimensionless quantity that is defined by $\alpha = \zeta\omega_0$, so the characteristic equation may be rewritten as

$$r^2 + 2\zeta\omega_0 r + \omega_0{}^2 = 0 \tag{4.4-2}$$

From the above discussion, the free response is a constant-amplitude sinusoid when $\zeta = 0$, is a damped oscillation when $0 < \zeta < 1$, and contains no oscillations when $\zeta > 1$. The conditions where $0 \leq \zeta < 1$, $\zeta = 1$, and $\zeta > 1$ are often called the underdamped, critically damped, and overdamped cases, respectively.

If the value of R, and hence α and ζ, were able to be negative, the roots of the characteristic equation would be in the right half plane, and the free response would increase without limit. A negative resistance is a source of power and can be represented by a controlled source plus a positive resistance, as in Fig. 1.4-10. Whenever the dead circuit contains a source of energy, then it is indeed possible for $y_H(t)$ to be a growing function of time.

As in the previous example, circuits that contain only inductances and capacitances are characterized by roots on the imaginary axis of the complex plane. The presence of resistances normally indicates that power will be dissipated and that the roots will lie in the open left half plane. Since controlled sources can supply energy, the roots for such circuits may lie anywhere in the complex plane.

The student may show that the locus in Fig. 4.4-2b also describes the free response of the circuit in Fig. 4.4-3, if $\alpha = 1/2RC$ and $\omega_0{}^2 = 1/LC$. Since Figs. 4.4-2a and 4.4-3 are duals, this result should be expected.

It should be pointed out that the roots of the characteristic equation and hence the form of the free response do depend upon whether the input is a voltage source or a current source, although not upon the waveshape of the source. The differential equation for the circuit in Fig. 4.4-4a, for example, is

$$C\frac{de_o}{dt} + (G_1 + G_2)e_o = G_2 e_1$$

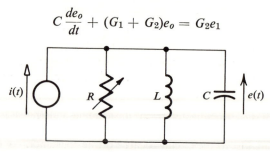

Figure 4.4-3

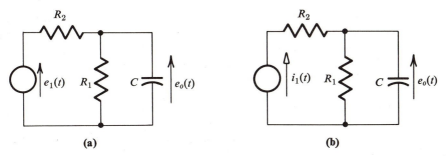

Figure 4.4-4

The characteristic equation has a root at $-(G_1 + G_2)/C$, so the free response has the form

$$(e_o)_H = K\epsilon^{-t/R_{12}C}$$

where $R_{12} = 1/(G_1 + G_2)$ is the resistance of R_1 and R_2 in parallel. For the circuit in Fig. 4.4-4b,

$$C\frac{de_o}{dt} + \frac{1}{R_1} e_o = i_1$$

and the free response is

$$(e_o)_H = K\epsilon^{-t/R_1C}$$

The reason for the difference in the free response of the two circuits is that a dead voltage source is a short circuit and a dead current source an open circuit. Thus the time constant should be $R_{12}C$ in Fig. 4.4-4a and R_1C in Fig. 4.4-4b.

A circuit is said to be *unstable* if its free response increases without limit as t approaches infinity. An unstable circuit has at least one root of the characteristic equation in the open right half plane or repeated roots on the imaginary axis. Physically, such a circuit must contain an energy source inside it. If the terms in $y_H(t)$ decay to zero or have a constant amplitude, the circuit is *stable*. (A few authors classify a circuit whose free response contains a constant or a constant-amplitude oscillation, corresponding to distinct roots on the imaginary axis, as unstable.)

Unstable circuits are not desirable, since the growing free response masks the forced response to a given input. Vacuum-tube and transistor circuits, whose models contain controlled sources, may or may not be stable. The form of the free response can be changed by varying the value of the controlled source in the model.

An *oscillator* is an exception to the normal requirement that the free response should die out as t approaches infinity. This device produces a constant-amplitude oscillation when there is no external signal. One simple model for an oscillator is considered in the following example; others may be found in Sections 7.2 and 10.6.

Example 4.4-2. Find a differential equation relating $e_2(t)$ and $e_1(t)$ in Fig. 4.4-5a. For what values of the real, positive constant A will the circuit be unstable? For what value of A will the free response consist of a constant-amplitude oscillation, and what will be the angular frequency of this oscillation?

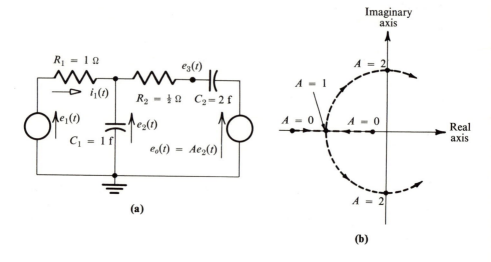

(a)

(b)

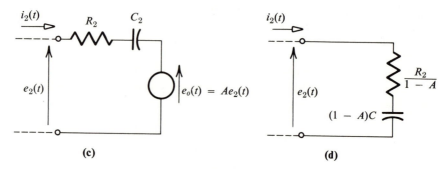

(c)

(d)

Figure 4.4-5

Solution. The current-law equations at nodes 2 and 3 are

$$\frac{1}{R_1}(e_2 - e_1) + C_1 \frac{de_2}{dt} + \frac{1}{R_2}(e_2 - e_3) = 0$$

$$\frac{1}{R_2}(e_3 - e_2) + C_2 \frac{d}{dt}(e_3 - e_0) = 0$$

When e_0 is replaced by Ae_2 and when the given element values are used, these equations become

$$\frac{de_2}{dt} + 3e_2 - 2e_3 = e_1$$

$$-2A \frac{de_2}{dt} - 2e_2 + 2 \frac{de_3}{dt} + 2e_3 = 0$$

When the first equation is solved for $2e_3$ and the result is inserted into the second equation, the result is

$$- 2A \frac{de_2}{dt} - 2e_2 + \left(\frac{d^2e_2}{dt^2} + 3 \frac{de_2}{dt} - \frac{de_1}{dt}\right) + \left(\frac{de_2}{dt} + 3e_2 - e_1\right) = 0$$

or

$$\frac{d^2e_2}{dt^2} + (4 - 2A) \frac{de_2}{dt} + e_2 = \frac{de_1}{dt} + e_1$$

The characteristic equation $r^2 + (4 - 2A)\, r + 1 = 0$ has the form of Eq. 4.4-1,

$$r^2 + 2\alpha r + \omega_0{}^2 = 0$$

with $\alpha = 2 - A$ and $\omega_0 = 1$. Varying the value of A is equivalent to varying the value of α, so the discussion associated with Fig. 4.4-2b is pertinent to this example. The locus traced out by the roots of the characteristic equation as A increases from zero is shown in Fig. 4.4-5b. When $A > 2$, the roots are in the right half plane, and the circuit is unstable. When $A = 2$, the roots are at $\pm j1$, and the free response contains a constant-amplitude, sinusoidal oscillation at an angular frequency of 1 rad per sec. In practice, if the circuit were to be used to produce a sinusoidal oscillation when $e_1(t) = 0$, A would be chosen slightly larger than 2. This would result in a growing oscillation, but the size of the output would eventually be limited by nonlinearities that are not included in the circuit model.

Part of the circuit in Fig. 4.4-5a is redrawn in Fig. 4.4-5c. If C_2 contains no initial stored energy,

$$e_2 = Ae_2 + \frac{1}{C_2} \int_0^t i_2 \, dt + R_2 i_2$$

or

$$e_2 = \frac{1}{(1 - A)C_2} \int_0^t i_2 \, dt + \frac{R_2}{1 - A} i_2$$

which describes Fig. 4.4-5d. For $A > 1$, the resistance and capacitance have negative values, and a negative resistance is a source of energy. If the power supplied by the negative resistance exceeds the power consumed in R_1, the free response will increase without limit.

A problem involving the negative resistance that appears in the model of a tunnel diode can be found at the end of the chapter.

The form of the particular solution $y_P(t)$ is determined by the form of the input, as was seen in the method of undetermined coefficients discussed in Section 4.1. The magnitudes of the terms in the forced response depend upon both the input and the values of the circuit elements. It is, of course, possible that the magnitude of some of the terms that are assumed to be in $y_P(t)$ may turn out to be zero. In this case, the circuit refuses to respond to a particular component of the input. The only other situation where the circuit influences the form of the forced response is when a term in the forcing function duplicates a term in $y_H(t)$. The corresponding terms in $y_P(t)$ are then multiplied by t. Physically, this means that the circuit is being excited in one of its natural modes; i.e., it is being asked to respond as it would if no external source were present. In such a case, the circuit responds more enthusiastically than usual, as is indicated mathematically by the multiplying factor of t. This situation will be investigated further in Chapter 6.

The form and magnitude of the forced response and the form of the free response are independent of any initial stored energy in the circuit. The stored energy at $t = 0$, which reflects the history of the circuit for $t < 0$, does influence the arbitrary constants, i.e., the magnitudes of the terms in the free response. The arbitrary constants also depend upon the source and upon the values of the circuit elements. Because the arbitrary constants depend upon all these factors, they cannot be evaluated until both $y_H(t)$ and $y_P(t)$ have been determined.

It is possible to think of the forced response as being immediately established by the application of the input. The values of the arbitrary constants in the free response are chosen so as to provide a proper transition from an unexcited circuit to a circuit under the dominance of the input. The magnitudes of the arbitrary constants depend in part upon how greatly the character of the input differs from the natural behavior of the circuit.

Instead of dividing the response of a circuit into a free and a forced component, it may be divided into a transient and steady-state component:

$$y(t) = y_T(t) + y_{ss}(t) \tag{4.4-3}$$

The *transient* component $y_T(t)$ is the part of the solution that decays to zero as t becomes large. The *steady-state* component $y_{ss}(t)$ consists of the terms that are left after the transient terms have disappeared.

If all the roots of the characteristic equation are in the open left half plane, and if the forced response does not vanish as t approaches infinity, then $y_T(t) = y_H(t)$ and $y_{ss}(t) = y_P(t)$. In the following section and in Chapter 5 it is assumed that $y_{ss}(t) = y_P(t)$.

4.5
The D-C Steady State

A source that is equal to a constant for all values of t is known as a *d-c* (*direct current*) *source*. Thus, a step function input is equivalent to a d-c source suddenly applied to the circuit.

When considering the forced response of a circuit, the form of all the voltages and currents is determined by the form of the input. In this section, we assume that the inputs are d-c sources and that the free response $y_H(t)$ decays to zero, so that $y_P(t) = y_{ss}(t)$. Then all the voltages and currents will have constant values in the steady state. Since the derivative of a constant is zero, $i_C(t) = e_L(t) = 0$, as indicated in Table 4.5-1, and a capacitance or inductance may be replaced by an open or short circuit, respectively.

In the d-c steady state, all circuits reduce to purely resistive circuits. For this

TABLE 4.5-1

Element	Defining equation	D-C steady-state behavior
Resistance	$e = R\,i$	$\dfrac{e}{i} = R$
Capacitance	$i = C\dfrac{de}{dt}$	$i = 0$ (open circuit)
Inductance	$e = L\dfrac{di}{dt}$	$e = 0$ (short circuit)

reason, resistive circuits are often called d-c circuits, but this does not imply that only d-c sources may be used. No matter what the input to a resistive circuit may be, the output and the input waveshapes have the same shape, and their relative sizes are uniquely determined by the values of the resistances and by any controlled sources. Thus, the results of Chapter 2 are not restricted to d-c inputs.

In many problems involving switching operations, the necessary initial conditions may be found from d-c steady-state analysis, as illustrated in the following examples. If a switch has been in one position long enough for any previous transient to have died out, but is suddenly moved to a new position at $t = 0$, the voltages across any capacitances and the currents through any inductances at $t = 0-$ may be found by using Table 4.5-1. Since these quantities cannot change instantaneously unless impulses are present, their values at $t = 0+$ will be known, and they will provide the basis for evaluating $y(0+)$ and $[dy/dt](0+)$.

Example 4.5-1. The switch K in Fig. 4.5-1a has been closed long enough for any previous transient to have died out. If K reopens at $t = 0$, find the voltage across the switch for $t > 0$.

Solution. In the d-c steady state, the inductance acts like a short circuit, so $i_2(0-) = i_L(0-) = \frac{10}{2} = 5$ amp. Although $i_L(0+) = 5$ amp, the other currents are forced to

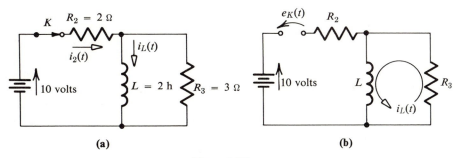

(a)   (b)

Figure 4.5-1

change when K opens. Since no current can flow through the open switch, the current $i_L(t)$ flows around the right-hand loop for $t > 0$, as shown in Fig. 4.5-1b. From the discussion in Section 3.3, or by solving a first-order differential equation,

$$i_L(t) = 5\epsilon^{-R_3 t/L} = 5\epsilon^{-3t/2}$$

$$e_K(t) = 10 + R_3 i_L(t) = 10 + 15\epsilon^{-3t/2}$$

for $t > 0$.

At $t = 0+$, the voltage across the open switch is $10 + R_3 i_L(0+) = 25$ volts, which is considerably larger than the d-c source voltage. The initial value of $e_K(t)$ would be increased still further if R_3 were increased, and would contain an impulse if R_3 were removed. A sufficiently large voltage across the switch could cause the air between the terminals to break down, with a resulting spark or arc. Arcing should not be unexpected when the current to an inductive circuit is interrupted.

Example 4.5-2.   The switch K in Fig. 4.5-2 has been closed for a long time, before it suddenly opens at $t = 0$. Find an expression for the voltage across the capacitance for all $t > 0$.

Solution.   In the d-c steady state with the switch closed, the inductance and the capacitance may be replaced by short and open circuits, respectively. Thus, $i_1(0-) = i_2(0-) = \frac{16}{4} = 4$ amp, $i_L(0-) = 4 + 2 = 6$ amp, and $e_C(0-) = 8$ volts. Immediately after the switch opens, $i_2(0+) = i_L(0+) = 6$ amp and $i_1(0+) = \frac{1}{4}[16 - e_C(0+)] = 4$ amp. Thus, $i_C(0+) = -2$ amp and the capacitance starts to discharge. After the switching transients have died out, however, the voltage across the capacitance will return to 8 volts.

For $t > 0$, the node equations at nodes C and B are

$$\frac{1}{2}(e_C - 16) + \frac{1}{8}\frac{de_C}{dt} + \frac{1}{2}(e_C - e_B) = 0$$

$$\frac{1}{2}(e_B - e_C) + 2\int e_B \, dt = 0$$

or

$$\frac{1}{8}\frac{de_C}{dt} + e_C - \frac{1}{2}e_B = 8$$

$$-\frac{1}{2}\frac{de_C}{dt} + \frac{1}{2}\frac{de_B}{dt} + 2e_B = 0$$

Solving the first equation for e_B and inserting the result into the second equation, we obtain

$$\frac{1}{8}\frac{d^2e_C}{dt^2} + \frac{de_C}{dt} + 4e_C = 32$$

The characteristic equation $\frac{1}{8}r^2 + r + 4 = 0$ or $r^2 + 8r + 32 = 0$ has roots at $-4 \pm j4$, and therefore the complementary solution is, as in Eq. 4.1-8, $\epsilon^{-4t}(K_1 \sin 4t + K_2 \cos 4t)$. The complete response for $t > 0$ is

$$e_C(t) = 8 + \epsilon^{-4t}(K_1 \sin 4t + K_2 \cos 4t)$$

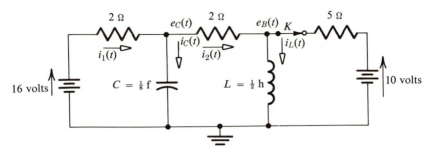

Figure 4.5-2

Since $e_C(0+) = 8$ volts, $K_2 = 0$ and

$$e_C(t) = 8 + K_1\epsilon^{-4t} \sin 4t$$

$$\frac{de_C}{dt} = K_1\epsilon^{-4t}(4\cos 4t - 4\sin 4t)$$

We previously saw that $i_C(0+) = -2$ amp, so

$$\frac{de_C}{dt}(0+) = \frac{1}{C}e_C(0+) = -16$$

Thus, $K_1 = -4$ and $e_C(t) = 8 - 4\epsilon^{-4t} \sin 4t$ for $t > 0$.

This section is concluded with two additional examples of finding the initial conditions that are needed for an evaluation of the arbitrary constants of integration.

Example 4.5-3.   Any transient associated with the connection of the voltage source to the circuit in Fig. 4.5-3a is assumed to have disappeared before the reference time $t = 0$. At $t = 0$, the lower right-hand, $1\ \Omega$ resistance suddenly burns out, becoming an open circuit. Find the values of $e_o(t)$ and de_o/dt at $t = 0+$.

Solution.   From Table 4.5-1, the voltages and currents at $t = 0-$ are as shown in Fig. 4.5-3b. The required initial conditions can be found from Fig. 4.5-3c, which shows the circuit at $t = 0+$. Notice that the voltages across the capacitances do not change instantaneously. Then $e_o(0+) = 8$ volts, $e_D(0+) = 2$ volts, $i_1(0+) = 2$ amp, and $i_3(0+) = \frac{8}{5}$ amp, so $i_2(0+) = \frac{4}{5}$ amp. From the defining equation $i = C(de/dt)$ for a capacitance,

$$\frac{de_A}{dt}(0+) = 1, \qquad \frac{de_B}{dt}(0+) = \frac{4}{5}$$

Since $e_o(t) = e_A(t) + e_B(t)$,

$$\frac{de_o}{dt}(0+) = \frac{de_A}{dt}(0+) + \frac{de_B}{dt}(0+) = \frac{9}{5} \text{ volts per sec}$$

Example 4.5-4.   The switch K in Fig. 4.5-4, after having been closed for a long time, opens at $t = 0$. Find the values of $i_1(t)$ and di_1/dt at $t = 0+$.

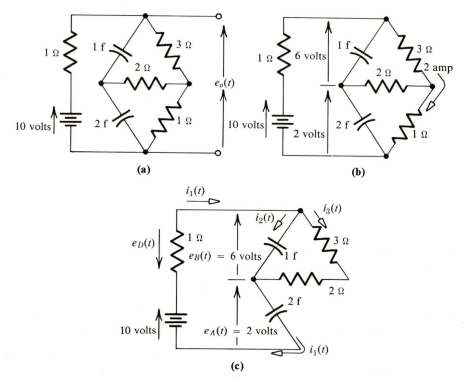

(a)

(b)

(c)

Figure 4.5-3

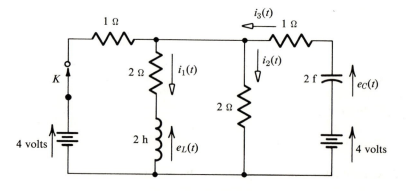

Figure 4.5-4

Solution. At $t = 0-$, the right-hand voltage source does not contribute to $i_1(t)$, since the capacitance acts like an open circuit in the d-c steady state. Thus, $i_1(0-) = \frac{1}{2}[4/(1 + 1)] = 1$ amp, and $e_C(0-) = -4 + 2i_1(0-) = -2$ volts.

For $t > 0$, the switch is open and $i_2(t) = i_3(t) - i_1(t)$. By Kirchhoff's voltage law,

$$2[i_3(t) - i_1(t)] + i_3(t) - e_C(t) - 4 = 0$$

Since $i_1(0+) = 1$ amp and $e_C(0+) = -2$ volts, $i_3(0+) = \frac{1}{3}$ amp. Next,

$$e_L(t) = 2[i_3(t) - i_1(t)] - 2i_1(t)$$

which equals $-\frac{4}{3}$ volts at $t = 0+$. Finally,

$$\frac{di_1}{dt}(0+) = \frac{1}{L} e_L(0+) = -\frac{2}{3} \text{ amp per sec}$$

4.6

The Forced Response to ϵ^{st}

The response to an input having the form ϵ^{st}, where s is an independent parameter that is a constant with respect to t, is important to the material in Chapters 5, 6, 9, and 10. The particular solution has the form $y_P(t) = H(s) \epsilon^{st}$, where $H(s)$ is a multiplying factor whose value depends upon the parameter s but not upon t.

$$y_P(t) = H(s) \epsilon^{st} \quad \text{when} \quad x(t) = \epsilon^{st} \tag{4.6-1}$$

The general form of the differential equation relating a circuit's output and input, as given in Eq. 4.4-1, is

$$a_n \frac{d^n y}{dt^n} + \cdots + a_1 \frac{dy}{dt} + a_0 y = b_m \frac{d^m x}{dt^m} + \cdots + b_1 \frac{dx}{dt} + b_0 x \tag{4.6-2}$$

Substituting Eqs. 4.6-1 into 4.6-2, we obtain

$$(a_n s^n + \cdots + a_1 s + a_0) H(s) \epsilon^{st} = (b_m s^m + \cdots + b_1 s + b_0) \epsilon^{st}$$

so

$$H(s) = \frac{b_m s^m + \cdots + b_1 s + b_0}{a_n s^n + \cdots + a_1 s + a_0} \tag{4.6-3}$$

Note that $H(s)$ is the quotient of two polynomials in s and that the numerator and denominator coefficients are identical with the coefficients on the right and left sides, respectively, of Eq. 4.6-2. Thus, $H(s)$, which is often called the *network function*, can be written down by inspection of the circuit's differential equation. (Conversely, if the network function is known, the differential equation can be reconstructed.)

Example 4.6-1. Find the forced response of Fig. 4.3-1 to $i_1(t) = \epsilon^{-t}$.

Solution. From the differential equation

$$C \frac{de_o}{dt} + \frac{1}{R} e_o = i_1$$

the network function is

$$H(s) = \frac{1}{sC + 1/R} = \frac{R}{1 + sCR}$$

so the forced response to $i_1(t) = \epsilon^{st}$ is $[R/(1 + sCR)]\epsilon^{st}$. When $s = -1$, $i_1(t) = \epsilon^{-t}$ and

$$(e_o)_P = \frac{R}{1 - CR} \epsilon^{-t}$$

In order to fully exploit the fact that the forced response to ϵ^{st} can be found so easily, we allow the parameter s to be either a real or a complex quantity. Although the input to a physical circuit is always a real function of time, it may nevertheless be useful to determine the forced response to a complex input. The reason for doing this is based upon the following theorem.

Theorem 4.6-1. If $y_P(t)$ is the forced response to the complex input $x(t)$, the forced response to the real part of $x(t)$ is the real part of $y_P(t)$. (The theorem remains valid if the word "real" is replaced by the word "imaginary.")

To prove the theorem, let the response to

$$x(t) = x_1(t) + jx_2(t)$$

be denoted by

$$y_P(t) = y_1(t) + jy_2(t)$$

where x_1, x_2, y_1, and y_2 are real functions of time. When these expressions are substituted into Eq. 4.6-2,

$$a_n \left(\frac{d^n y_1}{dt^n} + j\frac{d^n y_2}{dt^n}\right) + \cdots + a_1 \left(\frac{dy_1}{dt} + j\frac{dy_2}{dt}\right) + a_0(y_1 + jy_2)$$

$$= b_m \left(\frac{d^m x_1}{dt^m} + j\frac{d^m x_2}{dt^m}\right) + \cdots + b_0(x_1 + jx_2)$$

Two complex quantities are equal only if their real parts are equal and their imaginary parts are equal, so

$$a_n \frac{d^n y_1}{dt^n} + \cdots + a_1 \frac{dy_1}{dt} + a_0 y_1 = b_m \frac{d^m x_1}{dt^m} + \cdots + b_0 x_1$$

$$a_n \frac{d^n y_2}{dt^n} + \cdots + a_1 \frac{dy_2}{dt} + a_0 y_2 = b_m \frac{d^m x_2}{dt^m} + \cdots + b_0 x_2$$

From the first of these two equations, $y_1(t)$ must be the forced response to $x_1(t)$. From the second equation, $y_2(t)$ is the forced response to $x_2(t)$.

Example 4.6-2. Find the forced response of the circuit in Fig. 4.3-1 to $i_1(t) = \cos \omega t$.

Solution. From Euler's identity

$$\epsilon^{j\omega t} = \cos \omega t + j \sin \omega t$$

cos ωt is the real part of $\epsilon^{j\omega t}$. From Example 4.6-1 with $s = j\omega$, the forced response to
$i_1(t) = \epsilon^{j\omega t}$ is

$$(e_o)_P = \frac{R}{1 + j\omega CR} \epsilon^{j\omega t}$$

We multiply and divide this expression by $1 - j\omega CR$ and again use Euler's identity:

$$(e_o)_P = \frac{R(1 - j\omega CR)}{1 + (\omega CR)^2} \epsilon^{j\omega t} = \frac{R}{1 + (\omega CR)^2} (1 - j\omega CR)(\cos \omega t + j \sin \omega t)$$

The forced response to $i_1(t) = \cos \omega t$ is the real part of this expression, namely,

$$(e_o)_P = \frac{R}{1 + (\omega CR)^2} (\cos \omega t + \omega CR \sin \omega t)$$

which agrees with Example 4.3-1.

Although this indirect method of finding the forced response to sinusoidal inputs may seem more difficult than the direct approach, it is considerably easier in complicated problems. The method will be considered in greater detail in Chapters 5 and 6.

4.7

Summary

If a circuit has a single input $x(t)$ and a single output $y(t)$, a differential equation relating these two quantities always can be found. The output is the sum of the complementary solution $y_H(t)$ and the particular solution $y_P(t)$. The final (and sometimes difficult) step in the analysis is to evaluate the arbitrary constants of integration that appear in the complementary solution.

The complementary solution (or free response) is the complete solution when the circuit is excited only by some initial stored energy instead of by an external input. The nature of the free response is immediately evident if the roots of the characteristic equation (which can be written down by an inspection of the differential equation) are plotted in a complex plane. If there are any roots in the open right half plane or any repeated roots on the imaginary axis, the circuit is unstable and the free response increases without limit as t approaches infinity. Circuits with controlled sources or with negative resistances may or may not be stable.

The form of the particular solution (or forced response) is determined by the input signal. The forced responses to a constant input and to $x(t) = \epsilon^{st}$ are discussed in Sections 4.5 and 4.6. In Chapters 5 and 6 we consider the forced responses to a constant-amplitude sinusoid and to an exponentially growing or decaying input.

The transient response is the part of the output that decays to zero as t becomes large, while the remaining part is the steady-state response. If all the roots of the characteristic equation are in the open left half plane, and if the forced response does not disappear as t approaches infinity, the transient and steady-state components are the complementary and particular solutions, respectively.

PROBLEMS

4.1 Find a differential equation

Mon (a) that relates $e_o(t)$ and $i_1(t)$ for the circuit in Fig. P4.1.

(b) that relates $i_o(t)$ and $i(t)$ for the circuit in Fig. 4.2-3.

4.2 Find the general solution to each of the following differential equations. Evaluate the arbitrary constants of integration by using the given initial conditions.

Mon (a) $\dfrac{d^2y}{dt^2} + 2\dfrac{dy}{dt} = t;\qquad y(0+) = 1;\qquad \dfrac{dy}{dt}(0+) = 0$

wed (b) $\dfrac{dy}{dt} + 3y = t\cos 2t;\qquad y(0+) = 1$ $y = \frac{174}{169}\varepsilon^{-3t} + \frac{1}{13}\left[(3t-\frac{5}{13})\cos 2t + (2t-\frac{12}{13})\sin\right]$

Mon (c) $\dfrac{d^2y}{dt^2} + 3\dfrac{dy}{dt} + 2y = \epsilon^{-2t};\qquad y(0+) = \dfrac{dy}{dt}(0+) = 0$

wed (d) $\dfrac{d^2y}{dt^2} + 4y = \cos 2t;\qquad y(0+) = \dfrac{dy}{dt}(0+) = 0$ $y = (\frac{1}{4}t)\sin(2t)$

(e) $\dfrac{d^2y}{dt^2} + 2\dfrac{dy}{dt} + 10y = t\epsilon^{-t};\qquad y(0+) = \dfrac{dy}{dt}(0+) = 0$

(f) $\dfrac{d^2y}{dt^2} + 2\dfrac{dy}{dt} + y = t\epsilon^{-t};\qquad y(0+) = \dfrac{dy}{dt}(0+) = 0$

fri (g) $\dfrac{d^2y}{dt^2} + 2\dfrac{dy}{dt} + 2y = t^2;\qquad y(0+) = 0;\qquad \dfrac{dy}{dt}(0+) = 1$

(h) $\dfrac{d^2y}{dt^2} + 4\dfrac{dy}{dt} + 4y = \sin 2t;\qquad y(0+) = \dfrac{dy}{dt}(0+) = 0$

(i) $\dfrac{d^2y}{dt^2} + 4\dfrac{dy}{dt} + 4y = \sin\left(2t - \dfrac{\pi}{4}\right);\qquad y(0+) = \dfrac{dy}{dt}(0+) = 0$

4.3 For the circuit in Fig. P4.3, find the values of $i_L(t)$, di_L/dt, $e_C(t)$, and de_C/dt at $t = 0+$, if $e_1(t) = U_{-1}(t)$ and if there is no stored energy for $t < 0$. Repeat the problem when $e_1(t) = U_o(t)$.

4.4 If the circuit in Fig. P4.4 contains no stored energy for $t < 0$, find the energy stored in the capacitance and in the inductance at $t = 0+$.

4.5 If the circuit in Fig. P4.5 contains no stored energy for $t < 0$, find the values of $e_o(t)$ and de_o/dt at $t = 0+$.

4.6 The circuit in Fig. P4.6 contains no stored energy for $t < 0$. Find $e_o(t)$ and de_o/dt at $t = 0+$ if $e_1(t) = U_{-1}(t)$. Repeat, when $e_1(t) = U_o(t)$.

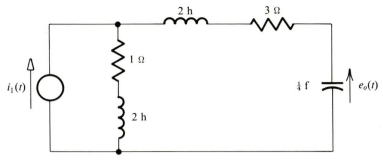

Figure P4.1

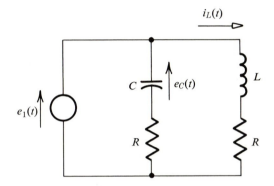

Figure P4.3

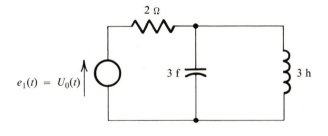

Figure P4.4

4.7 Find the values of $e_o(t)$, $e_1(t)$, de_o/dt, and de_1/dt at $t = 0+$, if the circuit in Fig. P4.7 contains no stored energy at $t = 0-$ and if $i_1(t) = U_{-1}(t)$. Repeat, when $i_1(t) = U_0(t)$.

4.8 If the circuit in Fig. P4.8 contains no initial stored energy, find $e_o(t)$ and its first three derivatives at $t = 0+$.

4.9 If the circuit in Fig. P4.9 contains no initial stored energy, and if $e_1(t) = te^{-2t}$ for $t > 0$, find $e_o(t)$ for all $t > 0$.

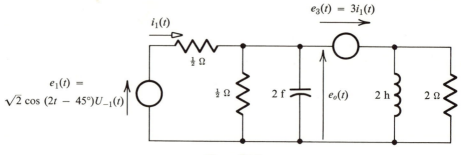

Figure P4.5

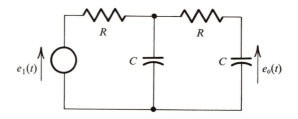

Figure P4.6

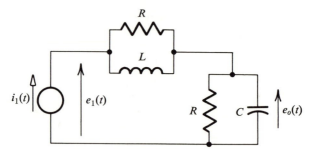

Figure P4.7

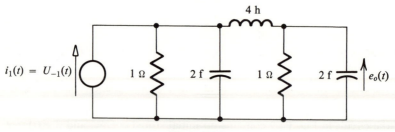

Figure P4.8

4.10 Find the unit impulse response for the circuit in Fig. 3.5-7b, and compare the result with the answer to Example 3.5-5.

4.11 Find the unit step response and the unit impulse response for the circuit in Fig. P4.11.

4.12 If the circuit in Fig. P4.12a contains no initial stored energy, find $e_o(t)$ for all $t > 0$ when $e_1(t)$ has the waveshape in Fig. P4.12b.

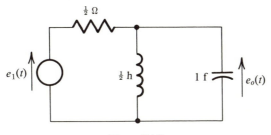

Figure P4.9

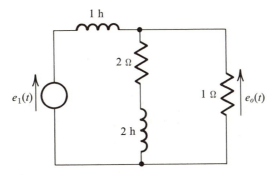

Figure P4.11

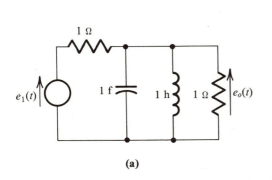

(a)

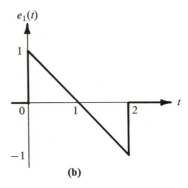

(b)

Figure P4.12

4.13 Find $i_o(t)$ in Fig. P4.13a for all $t > 0$ when the input waveshape is given in Fig. P4.13b. Assume that there is no stored energy for $t < 0$.

4.14 For the circuit in Fig. P4.14, find the largest value of R that will prevent oscillations in the waveshape of $e_o(t)$.

4.15 Can the circuit in Fig. P4.15 be unstable for any positive value of the real constant β? Can it be unstable for any negative value of β (corresponding to reversing the polarity of the controlled source)? Can the free response ever contain a constant-amplitude, sinusoidal oscillation?

4.16 The portion of the circuit within the dashed lines in Fig. P4.16 is the model for a tunnel diode operating in its linear region. Notice that the model contains a negative

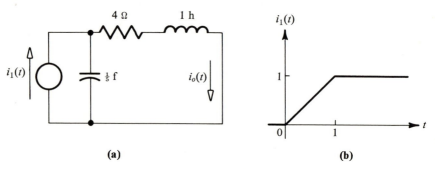

(a)   (b)

Figure P4.13

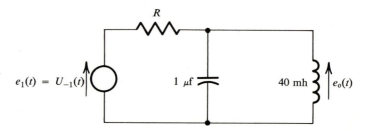

Figure P4.14

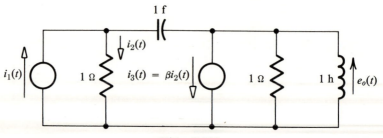

Figure P4.15

resistance. Find, by writing two loop equations, a differential equation relating $e_o(t)$ and $e_1(t)$. What is the smallest value of R and the largest value of L for which the circuit is stable?

4.17 At $t = 0-$ the circuit in Fig. P4.17 is operating in the steady state with the switch K closed. If the switch opens at $t = 0$, find $e_o(t)$ and de_o/dt at $t = 0+$.

4.18 After having been closed for a long time, the switch K in Fig. P4.18 opens at $t = 0$. Find $i_2(t)$ for all $t > 0$.

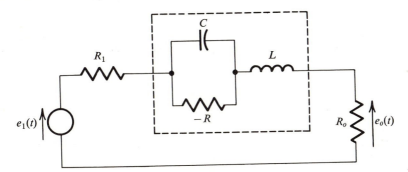

Figure P4.16

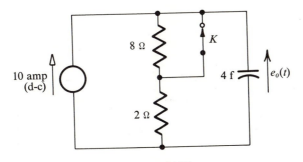

Figure P4.17

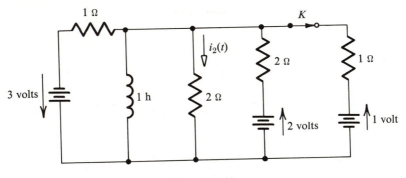

Figure P4.18

4.19 In Fig. P4.19 the switch K has been closed long enough for any previous transient to have disappeared. If the switch opens at $t = 0$, find and sketch $i(t)$ for all $t > 0$. Determine the numerical value of $i(t)$ at the first maximum and at the first minimum.

4.20 The circuit in Fig. P4.20 has been operating in the steady state with the switch K open for $t < 0$. If there is no initial charge on the capacitance and if the switch closes at $t = 0$, find $e_o(t)$ for all $t > 0$.

4.21 After steady-state conditions have been reached, the switch K in Fig. P4.21 closes at $t = 0$. Find $i_o(t)$ and di_o/dt at $t = 0+$.

4.22 The switch K in Fig. P4.22 opens after steady state conditions have been reached. Find the voltage across the switch immediately after it opens.

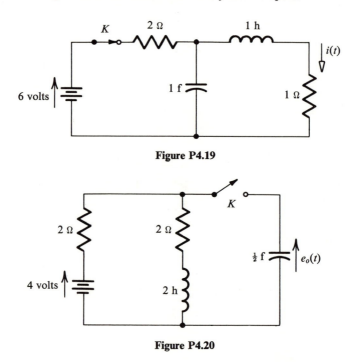

Figure P4.19

Figure P4.20

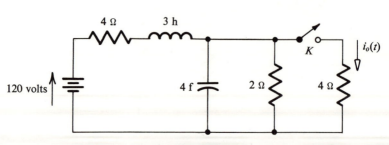

Figure P4.21

4.23 The circuit in Fig. P4.23 has been operating in the steady state with the switch K open for $t < 0$. If the switch closes at $t = 0$, find $e_o(t)$ and de_o/dt at $t = 0+$.

4.24 Find the particular solution in Problems 4.2h and 4.2i by the method used in Example 4.6-2. Can this method also be used in Problem 4.2d?

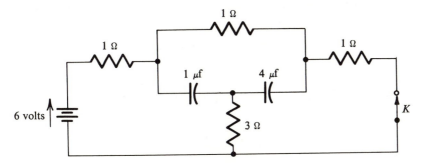

Figure P4.22

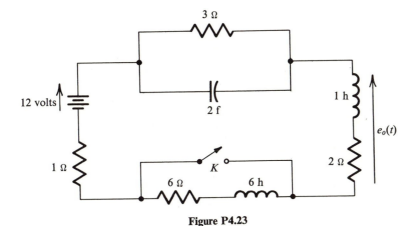

Figure P4.23

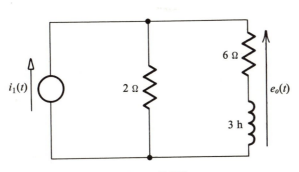

Figure P4.25

4.25 Find the network function $H(s)$ for the circuit in Fig. P4.25. Then find the forced component of the output voltage $e_o(t)$ when $i_1(t) = \epsilon^{-2t}$ and when $i_1(t) = \sin 2t$.

4.26 All the sources in the circuit in Fig. P4.26 are d-c sources. After having been closed for $t < 0$, the switch K opens at $t = 0$. Find the voltage across the 1 amp source for all $t > 0$ by the loop-equation method.

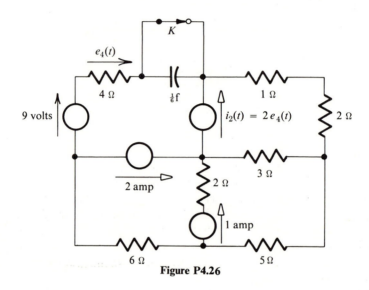

Figure P4.26

5

A-C Steady-State
Circuit Theory

A sinusoidal source of constant amplitude is known as an *a-c* (*alternating current*) *source*. Sinusoidal sources are common in both power and electronic systems, and thus represent a very important special case. Furthermore, as discussed in Chapter 9, the steady-state response to any periodic input can be found by the Fourier series, once the steady-state response to sinusoidal inputs is known. Although sinusoidal inputs may be handled by the methods of Chapter 4, it is worthwhile to develop techniques that will reduce computational work and eventually provide some further insight into this special case.

In this chapter, we assume that the free response $y_H(t)$ decays to zero as t becomes large, so that the forced response $y_P(t)$ is the steady-state response $y_{ss}(t)$. As discussed in Section 4.4, this requires that none of the roots of the characteristic equation lie on the imaginary axis or in the right half of the complex plane. We shall assume that we are interested only in the steady-state and not in the transient response. With this understanding, the subscripts P and ss are frequently omitted in this chapter. Examples in which both the steady-state and transient behavior are considered can be found in the problems at the end of the chapter.

The steady-state response to a sinusoid can be found by writing and solving the circuit's differential equation in a straightforward manner, as in Examples 4.1-2 and 5.2-1. As suggested by Example 4.6-2, however, this method tends to become laborious, and it is usually easier to first find the response to $\epsilon^{j\omega t}$. Because this input is a complex rather than a real function of time, it is important that the student thoroughly master the handling of complex numbers. Although the student may already be acquainted with most of the material in Section 5.1, he is urged to review it and to notice the nomenclature used before proceeding further.

5.1

Complex Algebra

A complex quantity z is represented in *rectangular form* as

$$z = x + jy \qquad (5.1\text{-}1)$$

where the real quantities x and y are called the real and imaginary parts of z, respectively.[†] If the notations Re z and Im z are used for the phrases "the real part of z" and "the imaginary part of z," respectively,

$$x = \text{Re } z \quad \text{and} \quad y = \text{Im } z$$

Two complex quantities $z_1 = x_1 + jy_1$ and $z_2 = x_2 + jy_2$ are *equal* if and only if $x_1 = x_2$ and $y_1 = y_2$. The common algebraic operations are similar to those for real numbers, with $j^2 = -1$. *Addition* and *subtraction* are defined by

$$z_1 + z_2 = (x_1 + x_2) + j(y_1 + y_2)$$
$$z_1 - z_2 = (x_1 - x_2) + j(y_1 - y_2) \qquad (5.1\text{-}2)$$

The *multiplication* of complex quantities is given by

$$z_1 z_2 = (x_1 + jy_1)(x_2 + jy_2) = (x_1 x_2 + j^2 y_1 y_2) + j(x_1 y_2 + x_2 y_1)$$
$$= (x_1 x_2 - y_1 y_2) + j(x_1 y_2 + x_2 y_1) \qquad (5.1\text{-}3)$$

The *division* of two complex quantities, $z_3 = z_1/z_2$, is defined by $z_1 = z_2 z_3$ if $z_2 \neq 0$. The division process can be carried out by *rationalization* as follows.

$$\frac{z_1}{z_2} = \frac{x_1 + jy_1}{x_2 + jy_2} = \left(\frac{x_1 + jy_1}{x_2 + jy_2}\right)\left(\frac{x_2 - jy_2}{x_2 - jy_2}\right)$$

$$= \left(\frac{x_1 x_2 + y_1 y_2}{x_2{}^2 + y_2{}^2}\right) + j\left(\frac{x_2 y_1 - x_1 y_2}{x_2{}^2 + y_2{}^2}\right) \qquad (5.1\text{-}4)$$

The *complex conjugate* of $z = x + jy$ is denoted and defined by

$$z^* = x - jy \qquad (5.1\text{-}5)$$

and can be formed from z by replacing j by $-j$. Thus, the rationalization process used in Eq. 5.1-4 consists in multiplying the numerator and denominator of the given fraction by the complex conjugate of the denominator.

[†] Mathematicians usually use i in place of the symbol j, but we wish to reserve i for current. For the reasons stated in the preface, we do not, in general, use boldface type or other special symbols to denote complex quantities. The context should make clear whether or not the letters in a given equation represent complex quantities. Boldface type is used only for a phasor, which is a particular kind of complex quantity that is discussed on page 232.

It can be easily shown that the above definitions result in the usual commutative, associative, and distributive laws.

$$z_1 z_2 = z_2 z_1$$

$$z_1 + z_2 = z_2 + z_1$$

$$z_1(z_2 z_3) = (z_1 z_2)z_3$$

$$z_1 + (z_2 + z_3) = (z_1 + z_2) + z_3 \tag{5.1-6}$$

$$z_1(z_2 + z_3) = z_1 z_2 + z_1 z_3$$

$$(z_1 + z_2)z_3 = z_1 z_3 + z_2 z_3$$

A complex number z can be represented geometrically by a point in a plane, as shown in Fig. 5.1-1a. Every combination of x and y determines a unique point in the plane. In the equivalent geometrical interpretation shown in Fig. 5.1-1b, z is represented by a directed line drawn from the origin to the point $x + jy$.[†] Then the addition or subtraction of two complex quantities can be carried out graphically, as in parts (c) and (d) of the figure, using the usual rules for vector addition and subtraction. Figure 5.1-1e shows the graphical addition of z and its complex conjugate z^*. Notice that $z + z^* = 2x = 2\,\mathrm{Re}\,z$.

The symbol $|z|$ denotes the length of the directed line representing z and is called the *magnitude* (or *absolute value* or *modulus*) of z. The angle of the directed line (measured counterclockwise from the positive real axis) is denoted by θ or $\measuredangle z$ and is called the *angle* or *argument* of z. From Fig. 5.1-1f,

$$x = |z| \cos \theta$$

$$y = |z| \sin \theta \tag{5.1-7}$$

$$|z| = \sqrt{x^2 + y^2}$$

$$\theta = \tan^{-1} \frac{y}{x}$$

When the last equation is used, the quadrant of θ must be determined by inspection.

Euler's formulas

$$\epsilon^{j\theta} = \cos \theta + j \sin \theta$$

$$\epsilon^{-j\theta} = \cos \theta - j \sin \theta \tag{5.1-8}$$

and the two corollaries

$$\cos \theta = \frac{\epsilon^{j\theta} + \epsilon^{-j\theta}}{2}$$

$$\sin \theta = \frac{\epsilon^{j\theta} - \epsilon^{-j\theta}}{j2} \tag{5.1-9}$$

[†] The directed line is called a *vector* or a *phasor*, depending upon the application. The distinction between the two terms is discussed later in this chapter.

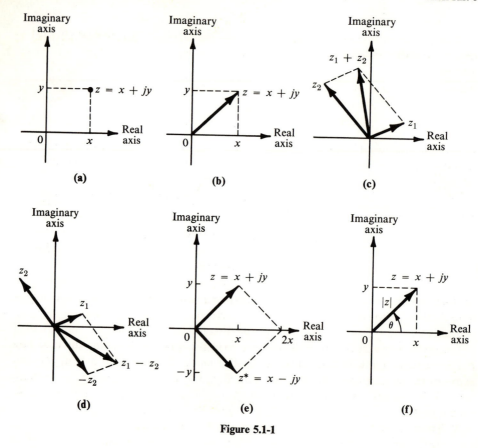

Figure 5.1-1

are frequently needed when handling complex quantities. From Eqs. 5.1-8 and 5.1-7,

Thus,

$$|z|\epsilon^{j\theta} = |z|(\cos\theta + j\sin\theta) = x + jy = z$$

$$z = |z|\epsilon^{j\theta} = |z|\underline{/\theta} \qquad (5.1\text{-}10)$$

which is called the *polar form* of z. The symbol $\underline{/\theta}$ is just a shorthand notation used by engineers for $\epsilon^{j\theta}$.

The proper interpretation of the factor $\epsilon^{j\theta}$ requires that the angle θ be expressed in radians. In practice, the angle is frequently given in degrees (1 degree = $\pi/180$ radians), but whenever there is apt to be any confusion in the application of Eqs. 5.1-8 through 5.1-10, θ should be converted to radians. Writing $\epsilon^{j\theta}$ with θ expressed in degrees is usually considered bad practice, but writing $\underline{/\theta}$ with θ in degrees is quite common.

Equations 5.1-8 and 5.1-10 lead to another interpretation of j, which was previously defined by $j^2 = -1$. If $\theta = \pi/2$ in these equations, $\epsilon^{j\pi/2} = j$, so j is a complex number with a magnitude of unity and an angle of 90°. Notice also that $1/j = j/j^2 = -j$.

It is frequently necessary to convert a complex number from the rectangular to polar, or polar to rectangular form, which can be done using Eqs. 5.1-7. For example,

$$4 + j3 = \sqrt{4^2 + 3^2} \, \underline{/\tan^{-1}\left(\tfrac{3}{4}\right)} = 5\underline{/36.9°}$$

or

$$5\underline{/36.9°} = 5 \cos 36.9° + j5 \sin 36.9° = 4 + j3$$

It is important to be able to change complex numbers from one form to another rapidly. Many slide-rule manuals develop techniques that are more efficient than Eqs. 5.1-7.[†]

The multiplication and division of complex quantities is easily carried out in polar form. If $z_1 = |z_1|\epsilon^{j\theta_1}$ and $z_2 = |z_2|\epsilon^{j\theta_2}$,

$$z_1 z_2 = |z_1||z_2|\epsilon^{j(\theta_1+\theta_2)}$$

$$\frac{z_1}{z_2} = \frac{|z_1|}{|z_2|}\,\epsilon^{j(\theta_1-\theta_2)}$$

(5.1-11)

In forming a product, therefore, the magnitudes are multiplied, and the angles added. In forming a quotient, the magnitudes are divided, and the angles subtracted.

It should be noticed that multiplying a complex quantity by $j = 1\underline{/90°}$ does not change the magnitude, but adds 90° to the angle. If the original quantity was represented by a directed line in the complex plane, the multiplication by j rotates the line 90° in the counterclockwise direction. Notice also that the complex conjugate of $z = x + jy = |z|\underline{/\theta}$ is $z^* = x - jy = |z|\underline{/-\theta}$, so $zz^* = |z|^2$.

Raising a complex number to a power is also easily handled with the number in polar form. If $z = |z|\epsilon^{j\theta}$,

$$z^n = |z|^n \epsilon^{jn\theta}$$

(5.1-12)

Occasionally, the square or cube roots of a number may be required. Since

$$z = z\epsilon^{j\theta} = |z|\epsilon^{j(2\pi+\theta)}$$

[†] If $z = x + jy$ is a complex number in the first quadrant of the complex plane, it can be converted to polar form by a Keuffel and Esser Decitrig slide rule as follows. Set the index over the larger of the two numbers x and y on the D scale. Move the hairline over the smaller number, also on the D scale.

If $\frac{1}{10}x < y < 10x$, the angle θ is underneath the hairline on the T scale. If $x > y$, the angle in black is used; if $x < y$, the angle is in red. Next move the slide until the angle θ appears underneath the hairline on the S scale. The magnitude $|z|$ is then under the index on the D scale.

If the larger of the two numbers x and y is from ten to 100 times the smaller, the index and hairline are initially set as before. The angle θ appears underneath the hairline on the SRT scale if $x > y$. Otherwise, θ is the complement of the angle on the SRT scale. In most cases, it is sufficiently accurate to let $|z|$ be equal to the larger of the two numbers x and y, since the error in doing so is always less than $\frac{1}{2}\%$.

The student should carry out these steps for such typical conversions as $5 + j12 = 13\underline{/67.4°}$ and $50 + j2 = 50\underline{/2.3°}$. For complex numbers in quadrants other than the first, the above procedure still yields the proper value of $|z|$, but care must be exercised when determining θ. A rough sketch in the complex plane is very helpful in finding θ.

the two square roots of z are

$$\sqrt{|z|}\,\epsilon^{j\theta/2}, \qquad \sqrt{|z|}\,\epsilon^{j\,(\pi+\theta/2)}$$

Similarly, the nth roots of z are

$$\sqrt[n]{|z|}\,\epsilon^{j\theta/n}, \qquad \sqrt[n]{|z|}\,\epsilon^{j\left(\frac{2\pi}{n}+\frac{\theta}{n}\right)}, \qquad \dots$$

for any positive integer n.

Example 5.1-1.   Find $z_1 = \dfrac{j100(1 - j2)^2}{(-3 + j4)(-1 - j)}$ in rectangular form.

Solution.   If we express each of the factors in polar form,

$$z_1 = \frac{(100\underline{/90°})(\sqrt{5}\underline{/-63.4°})^2}{(5\underline{/126.8°})(\sqrt{2}\underline{/-135°})} = 70.7\underline{/-28.6°} = 62 - j34$$

Alternatively, if the factors are left in rectangular form, and the resulting fraction is rationalized,

$$z_1 = \left[\frac{100(4 - j3)}{7 - j1}\right]\left(\frac{7 + j1}{7 + j1}\right) = \frac{100(31 - j17)}{49 + 1} = 62 - j34$$

5.2

Representing Sinusoidal Functions of Time

The sinusoidal function

$$f(t) = F_m \cos (\omega t + \phi) \tag{5.2-1}$$

is shown in Fig. 5.2-1a. The period T is related to the frequency f (cycles per second) and to the angular frequency ω (radians per second, rad per sec) by

$$\omega = 2\pi f = \frac{2\pi}{T} \tag{5.2-2}$$

The real number F_m is the maximum value (or amplitude) of the function, while ϕ is the phase angle with respect to a pure cosine wave. Figure 5.2-1b shows two sinusoidal functions with the same angular frequency but with different phase angles. The function $f_2(t)$ is said to "lag" $f_1(t)$ by the angle α. Equivalently, $f_1(t)$ "leads" $f_2(t)$ by this angle.

The straightforward determination of the steady-state response to an input having the form of Eq. 5.2-1 is illustrated in Examples 4.3-1 and 4.1-2, and in the following example. This direct method is, however, quite laborious even for a simple circuit, and the principal aim of this section is to develop a procedure that

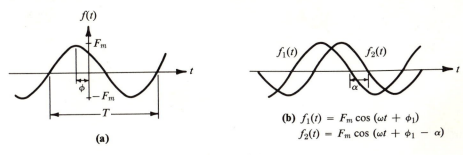

Figure 5.2-1

(b) $f_1(t) = F_m \cos(\omega t + \phi_1)$
$f_2(t) = F_m \cos(\omega t + \phi_1 - \alpha)$

(a)

requires less computational work. Since the direct method will not be used in subsequent sections, it is not important for the student to verify the algebraic details in the following example, but only to understand the approach used and to realize that it is apt to become cumbersome.

Example 5.2-1. Find an expression for the steady-state current in Fig. 5.2-2 if $e(t) = E_m \cos(\omega t + \phi_1)$.

Solution. The loop equation

$$L\frac{di}{dt} + Ri + \frac{1}{C}\int i\,dt = e$$

may be differentiated to give

$$L\frac{d^2i}{dt^2} + R\frac{di}{dt} + \frac{1}{C}i = \frac{de}{dt}$$

The form that should be assumed for the particular (or steady-state) solution is $i(t) = A\cos(\omega t + \phi_1) + B\sin(\omega t + \phi_1)$, where A and B are constants to be determined. When the expressions for $e(t)$ and $i(t)$ are substituted into the differential equation,

$$L[-\omega^2 A\cos(\omega t + \phi_1) - \omega^2 B\sin(\omega t + \phi_1)]$$
$$+ R[-\omega A\sin(\omega t + \phi_1) + \omega B\cos(\omega t + \phi_1)]$$
$$+ \frac{1}{C}\left[A\cos(\omega t + \phi_1) + B\sin(\omega t + \phi_1)\right] = -\omega E_m\sin(\omega t + \phi_1)$$

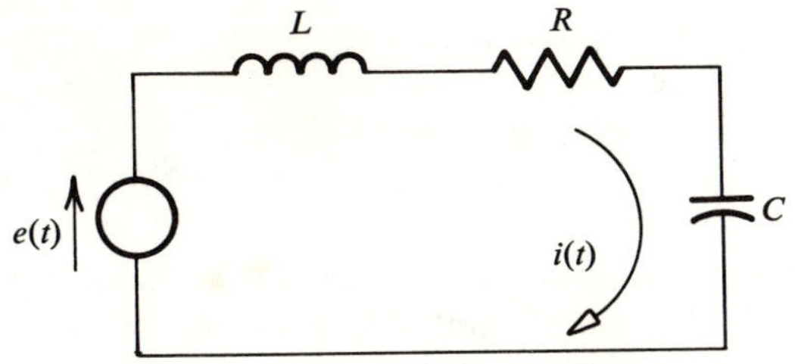

Figure 5.2-2

[Handwritten annotations:]
See physics C page 319
$m\frac{d^2x}{dt^2} + r\frac{dx}{dt} + kx = F_m \cos \omega t$
Electrical
Equivalents —
$L \Rightarrow m$
$R \Rightarrow r = $ damping constant
$C \Rightarrow K = $ force constant

Equating corresponding coefficients, we obtain

$$-\omega^2 LA + \omega RB + \frac{1}{C}A = 0$$

$$-\omega^2 LB - \omega RA + \frac{1}{C}B = -\omega E_m$$

Solving these equations simultaneously, we find that

$$A = \frac{RE_m}{R^2 + (\omega L - 1/\omega C)^2}, \qquad B = \frac{(\omega L - 1/\omega C)E_m}{R^2 + (\omega L - 1/\omega C)^2}$$

so

$$i(t) = \frac{E_m}{R^2 + (\omega L - 1/\omega C)^2} [R \cos(\omega t + \phi_1) + (\omega L - 1/\omega C) \sin(\omega t + \phi_1)]$$

The steady-state response can be better appreciated and more easily plotted if it is expressed in the form of Eq. 5.2-1. The trigonometric identity

$$A \cos \alpha + B \sin \alpha = \sqrt{A^2 + B^2} \cos\left(\alpha - \tan^{-1}\frac{B}{A}\right) \qquad (5.2\text{-}3)$$

may be verified by expanding the right side as $\cos(\alpha - \beta) = \cos \alpha \cos \beta + \sin \alpha \sin \beta$. The two trigonometric terms in $i(t)$ can be combined as follows by using Eq. 5.2-3:

$$i(t) = \frac{E_m}{\sqrt{R^2 + (\omega L - 1/\omega C)^2}} \cos\left(\omega t + \phi_1 - \tan^{-1}\frac{\omega L - 1/\omega C}{R}\right)$$

which has the form $i(t) = I_m \cos(\omega t + \phi_2)$.

The method in Example 5.2-1 soon becomes unwieldly when more complicated circuits are considered. Recall that the forced response to the input $x(t) = \epsilon^{st}$ can be easily found, as in Section 4.6, and that sinusoidal functions can be expressed as the sum of two exponential terms, as in Eqs. 5.1-8 and 5.1-9. By the use of Eqs. 5.1-9, Eq. 5.2-1 may be rewritten as

$$f(t) = F_m \cos(\omega t + \phi) = \tfrac{1}{2}(F_m \epsilon^{j(\omega t + \phi)} + F_m \epsilon^{-j(\omega t + \phi)})$$

$$= \tfrac{1}{2}(F_m \epsilon^{j\phi} \epsilon^{j\omega t} + F_m \epsilon^{-j\phi} \epsilon^{-j\omega t})$$

$$= \tfrac{1}{2}(\mathbf{F}\epsilon^{j\omega t} + \mathbf{F}^* \epsilon^{-j\omega t}) \qquad (5.2\text{-}4)$$

where $\mathbf{F} = F_m \epsilon^{j\phi}$ is a complex constant, and where $\mathbf{F}^*$ is its complex conjugate.

For a given value of t, the two complex functions $\mathbf{F}\epsilon^{j\omega t}$ and $\mathbf{F}^* \epsilon^{-j\omega t}$ can be represented by directed lines in a complex plane, as in Fig. 5.2-3a. In Fig. 5.2-3b, the two functions are represented graphically at three different instants of time. As t increases, the two directed lines rotate in opposite directions, tracing out the locus

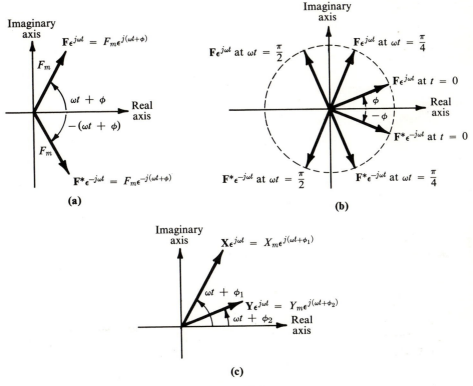

Figure 5.2-3

of a circle; for this reason they are called *counter-rotating phasors*.[†] At $\omega t = 2\pi$, $4\pi, \ldots$ (i.e., at $t = 2\pi/\omega$, $4\pi/\omega$, $\ldots$), they are in the same position as at $t = 0$. They thus make $\omega/2\pi$ revolutions per second, and have an angular frequency of ω radians per second.

For all values of time, one-half the sum of the counter-rotating phasors is the real function $f(t)$ in Eq. 5.2-4. Notice that $f(t)$ is also equal to the projection of either one of them upon the real axis, e.g.,

$$f(t) = \text{Re } (\mathbf{F}\epsilon^{j\omega t}) \tag{5.2-5}$$

There is no need, therefore, to explicitly show both of the rotating phasors in order to be able to reconstruct $f(t)$. By convention, the counterclockwise (CCW) one is retained.

[†] Although directed lines are commonly called vectors, this terminology is avoided in this application. In field theory, quantities that have an orientation in three-dimensional space and that are sinusoidal functions of time are encountered. The term "vector" then refers to the spacial orientation, and the term "rotating phasor" to the exponential terms in Eq. 5.2-4. Since both phasors and vectors are complex quantities, phasors are added and subtracted in the same way as vectors.

The forced response to the CCW rotating phasor $\mathbf{X}\epsilon^{j\omega t}$ has the form $\mathbf{Y}\epsilon^{j\omega t}$, which is another CCW rotating phasor with the same angular frequency of rotation, as indicated in Fig. 5.2-3c. If $\mathbf{X} = X_m\epsilon^{j\phi_1}$ and $\mathbf{Y} = Y_m\epsilon^{j\phi_2}$, the projection of the rotating phasors on the real axis give the sinusoidal functions $x(t) = X_m \cos(\omega t + \phi_1)$ and $y(t) = Y_m \cos(\omega t + \phi_2)$, respectively.

The CCW phasors for the input and output rotate in the same direction and with the same angular frequency, so their magnitudes and the angle *between* them remain fixed for all values of time. Since the magnitudes and relative angle are the usual quantities of interest, it is sufficient to show the CCW phasors at a single instant of time, which by convention is $t = 0$.

The term *phasor*, unless there is a modifying adjective, is understood to be the value of the CCW rotating phasor at the instant $t = 0$. Thus, the phasor corresponding to the function $f(t)$ in Eq. 5.2-4 is

$$\mathbf{F} = F_m\epsilon^{j\phi} = F_m\underline{/\phi}$$

In later sections, there will be little confusion between phasors and rotating phasors, since the latter will be rarely used.

Capital boldface letters denote the phasors for the corresponding sinusoidal functions of time. Notice that the magnitude and angle of a phasor are the amplitude and phase angle (with respect to a pure cosine wave) of the sinusoid that it represents. The only information that is lost by using the phasor $\mathbf{F}$ instead of the rotating phasor $\mathbf{F}\epsilon^{j\omega t}$ is the angular frequency. In an a-c steady-state problem, however, the angular frequency of all voltages and currents is known to be the same as that of the source. If the angular frequency is specified separately, there is a one-to-one correspondence between the phasor $\mathbf{F} = F_m\epsilon^{j\phi}$ and the sinusoidal function $f(t) = F_m \cos(\omega t + \phi)$.

The forced response to an input $x(t) = \mathbf{X}\epsilon^{j\omega t}$ may be easily found, since the input has the form $\mathbf{X}\epsilon^{st}$ with $s = j\omega$. If we use Eq. 4.6-1, $y_P(t) = \mathbf{Y}\epsilon^{j\omega t}$, where

$$\mathbf{Y} = H(j\omega)\mathbf{X} \tag{5.2-6}$$

and where the network function $H(j\omega)$ may be written down directly from the circuit's differential equation, as in Section 4.6. Since all three factors in Eq. 5.2-6 are complex quantities,

$$|\mathbf{Y}| = |H(j\omega)|\,|\mathbf{X}|$$
$$\underline{*}\,\mathbf{Y} = \underline{*}\,H(j\omega) + \underline{*}\,\mathbf{X} \tag{5.2-7}$$

If $\mathbf{X} = X_m\epsilon^{j\phi_1}$, $H(j\omega) = |H(j\omega)|\epsilon^{j\theta}$, and $\mathbf{Y} = Y_m\epsilon^{j\phi_2}$, then

$$Y_m = |H(j\omega)|X_m, \qquad \phi_2 = \theta + \phi_1 \tag{5.2-8}$$

In a similar manner, the forced response to $x(t) = \mathbf{X}^*\epsilon^{-j\omega t}$ is $y_P(t) = \mathbf{Y}^*\epsilon^{-j\omega t}$. If the input is

$$x(t) = X_m \cos(\omega t + \phi_1) = \tfrac{1}{2}(\mathbf{X}\epsilon^{j\omega t} + \mathbf{X}^*\epsilon^{-j\omega t}) \tag{5.2-9}$$

then the forced response may be found by superposition, and is

$$y_P(t) = \tfrac{1}{2}(\mathbf{Y}\epsilon^{j\omega t} + \mathbf{Y}^*\epsilon^{-j\omega t}) = Y_m \cos(\omega t + \phi_2) \qquad (5.2\text{-}10)$$

where the last expression follows from Eq. 5.2-4, and where Y_m and ϕ_2 are given in Eq. 5.2-8. Thus if $H(j\omega)$, which is called the *a-c steady-state network function*, is known, the forced response to a sinusoidal input may be written down immediately.

Another viewpoint which leads to the same conclusion is based upon Theorem 4.6-1 and Eqs. 5.1-8. The forced response to

$$x(t) = X_m \epsilon^{j\phi_1}\epsilon^{j\omega t} = X_m[\cos(\omega t + \phi_1) + j \sin(\omega t + \phi_1)]$$

is

$$y_P(t) = Y_m \epsilon^{j\phi_2}\epsilon^{j\omega t} = Y_m[\cos(\omega t + \phi_2) + j \sin(\omega t + \phi_2)]$$

where Y_m and ϕ_2 are again given by Eqs. 5.2-8. If the input is the real part of the expression for $x(t)$, then the forced response is the real part of the above $y_P(t)$, which gives Eqs. 5.2-9 and 5.2-10.

From the first viewpoint, the responses to the two counter-rotating phasors $\mathbf{X}\epsilon^{j\omega t}$ and $\mathbf{X}^*\epsilon^{-j\omega t}$ are found, and the desired answer is one-half their sum. From the second point of view, the sinusoidal functions of time are regarded as the real-axis projections of the CCW rotating phasors.†

Example 5.2-2. Solve Example 5.2-1 by finding the a-c steady-state network function for the circuit in Fig. 5.2-2.

Solution. By an inspection of the differential equation

$$L\frac{d^2i}{dt^2} + R\frac{di}{dt} + \frac{1}{C}i = \frac{de}{dt}$$

the network function is $H(s) = \dfrac{s}{Ls^2 + Rs + 1/C} = \dfrac{1}{Ls + R + 1/sC}$, and

$$H(j\omega) = \frac{1}{j\omega L + R + 1/j\omega C} = \frac{1}{R + j(\omega L - 1/\omega C)}$$

---

† In Chapter 8, it is convenient to let the phasor corresponding to $f(t) = F_m \cos(\omega t + \phi)$ be $(F_m/\sqrt{2})\,\epsilon^{j\phi}$ instead of $F_m\epsilon^{j\phi}$. As long as the student is consistent, either of these two conventions may be used when finding the steady-state output signal. Suppose that the a-c steady-state network function $H(j\omega)$ and the input $x(t) = X_m \cos(\omega t + \phi_1)$ are given. The steady-state response will have the form $y(t) = Y_m \cos(\omega t + \phi_2)$. If $\mathbf{X} = (X_m/\sqrt{2})\underline{/\phi_1}$, $H(j\omega) = |H(j\omega)|\,\underline{/\theta}$, and $\mathbf{Y} = (Y_m/\sqrt{2})\,\underline{/\phi_2}$, then Eq. 5.2-7 gives

$$\frac{Y_m}{\sqrt{2}}\underline{/\phi_2} = \left[|H(j\omega)|\,\underline{/\theta}\right]\left(\frac{X_m}{\sqrt{2}}\underline{/\phi_1}\right)$$

so $Y_m = |H(j\omega)|X_m$ and $\phi_2 = \theta + \phi_1$, as in Eq. 5.2-9. Since the arbitrarily introduced factor $\sqrt{2}$ cancels out when the maximum value and the angle of the steady-state response are calculated, it is not used in this chapter.

When the denominator is converted from rectangular to polar form,

$$H(j\omega) = \frac{1}{\sqrt{R^2 + (\omega L - 1/\omega C)^2}} \bigg/ -\tan^{-1}\frac{\omega L - 1/\omega C}{R}$$

so the steady-state response to $e(t) = E_m \cos(\omega t + \phi_1)$ is

$$i(t) = \frac{E_m}{\sqrt{R^2 + (\omega L - 1/\omega C)^2}} \cos\left(\omega t + \phi_1 - \tan^{-1}\frac{\omega L - 1/\omega C}{R}\right)$$

which agrees with the answer to Example 5.2-1.

5.3

Impedance and Admittance

The discussion in Section 5.2 applies to the a-c steady-state solution of any fixed, linear system whose differential equation is known. The student will find this same approach useful in electromagnetic field theory and in the solution of nonelectrical systems. We next consider the steady-state behavior of individual circuit elements in more detail. This study leads to a method of determining $H(j\omega)$ and the steady-state response without having to first find a differential equation relating $y(t)$ and $x(t)$.

For a sinusoidal input of angular frequency ω, all the steady-state currents and voltages can be found as the projection on the real axis of their CCW rotating phasors. We consider, therefore, the input $x(t) = \mathbf{X}\epsilon^{j\omega t}$. The form of the forced response is determined by the input, so in the steady state the voltage and the current of the typical two-terminal component in Fig. 5.3-1 have the form

$$e(t) = \mathbf{E}\epsilon^{j\omega t}, \qquad i(t) = \mathbf{I}\epsilon^{j\omega t} \tag{5.3-1}$$

where the phasors $\mathbf{X}$, $\mathbf{E}$, and $\mathbf{I}$ are complex numbers. If the component is a resistance, the defining equation $e(t) = Ri(t)$ becomes

$$\mathbf{E}\epsilon^{j\omega t} = R\mathbf{I}\epsilon^{j\omega t}$$

so that

$$\frac{\mathbf{E}}{\mathbf{I}} = R \tag{5.3-2}$$

For a capacitance, the defining equation $i = C(de/dt)$ becomes

$$\mathbf{I}\epsilon^{j\omega t} = Cj\omega\mathbf{E}\epsilon^{j\omega t}$$

so

$$\frac{\mathbf{E}}{\mathbf{I}} = \frac{1}{j\omega C} = -j\frac{1}{\omega C} = \frac{1}{\omega C}\big/-90° \tag{5.3-3}$$

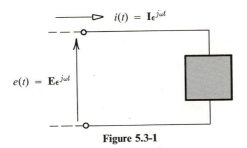

Figure 5.3-1

Finally, for an inductance, $e = L(di/dt)$ and

$$\mathbf{E}\epsilon^{j\omega t} = Lj\omega\mathbf{I}\epsilon^{j\omega t}$$

so

$$\frac{\mathbf{E}}{\mathbf{I}} = j\omega L = \omega L\underline{/90°} \qquad\qquad (5.3\text{-}4)$$

Notice that for each of the passive elements the ratio of the steady-state voltage and current in Eqs. 5.3-1 is $\mathbf{E}\epsilon^{j\omega t}/\mathbf{I}\epsilon^{j\omega t} = \mathbf{E}/\mathbf{I}$, which is a complex constant, and not a function of time. By analogy to the definition of resistance, the *impedance Z* is defined by

$$Z = \frac{\mathbf{E}}{\mathbf{I}} \qquad\qquad (5.3\text{-}5)$$

which is similar to Ohm's law, except for the fact that complex rather than real numbers are involved. The *admittance Y* is defined as the reciprocal of the impedance.

$$Y = \frac{1}{Z} = \frac{\mathbf{I}}{\mathbf{E}} \qquad\qquad (5.3\text{-}6)$$

The impedance and admittance for each of the passive elements are summarized in Table 5.3-1. The units for Z and Y are ohms and mhos, respectively.

If a circuit's input is $x(t) = X_m \cos(\omega t + \phi_1)$, the steady-state voltage and current for a typical passive component have the form $e(t) = E_m \cos(\omega t + \alpha)$ and

TABLE 5.3-1

Element	Impedance	Admittance
Resistance	$Z = R$	$Y = \dfrac{1}{R} = G$
Capacitance	$Z = \dfrac{1}{j\omega C} = \dfrac{1}{\omega C}\underline{/-90°}$	$Y = j\omega C = \omega C\underline{/90°}$
Inductance	$Z = j\omega L = \omega L\underline{/90°}$	$Y = \dfrac{1}{j\omega L} = \dfrac{1}{\omega L}\underline{/-90°}$

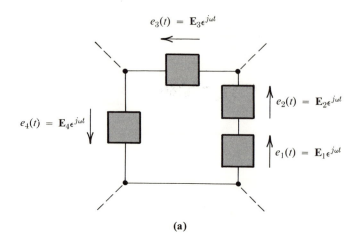

$e_3(t) = \mathbf{E}_3 \epsilon^{j\omega t}$

$e_2(t) = \mathbf{E}_2 \epsilon^{j\omega t}$

$e_4(t) = \mathbf{E}_4 \epsilon^{j\omega t}$

$e_1(t) = \mathbf{E}_1 \epsilon^{j\omega t}$

(a)

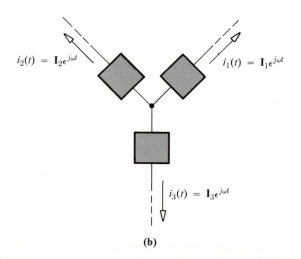

$i_2(t) = \mathbf{I}_2 \epsilon^{j\omega t}$

$i_1(t) = \mathbf{I}_1 \epsilon^{j\omega t}$

$i_3(t) = \mathbf{I}_3 \epsilon^{j\omega t}$

(b)

$i(t) = I_m \cos(\omega t + \beta)$. Although the quotient of the corresponding phasors $\mathbf{E} = E_m \epsilon^{j\alpha}$ and $\mathbf{I} = I_m \epsilon^{j\beta}$ is the impedance Z, the quotient

$$\frac{E_m \cos(\omega t + \alpha)}{I_m \cos(\omega t + \beta)}$$

would not be a constant, but instead would be a function of time. This is another reason for introducing phasors and for considering sinusoidal functions as the real-axis projection of the CCW rotating phasors.

Recall that, if the angular frequency ω is specified separately, there is a one-to-one correspondence between the phasor $\mathbf{F} = F_m \underline{/\phi}$ and the sinusoidal function $f(t) = F_m \cos(\omega t + \phi)$. Thus the a-c steady-state response can be written down as soon

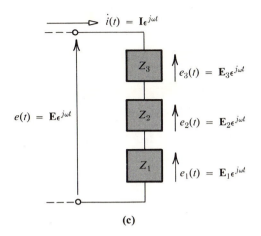

(c)

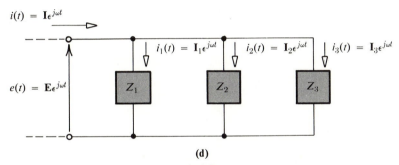

(d)

Figure 5.3-2

as the output phasor has been determined. According to Eq. 5.3-5, the voltage and current phasors obey a relationship similar to Ohm's law, so we might next inquire whether or not the phasors obey *all* the rules for resistive circuits. If they do, then the output phasor can be found by the simple methods used in Chapter 2, and the writing of differential equations can be completely avoided.

Consider again the input $x(t) = \mathbf{X}\epsilon^{j\omega t}$, so that the steady-state voltages and currents have the form of Eqs. 5.3-1. Then in Fig. 5.3-2a, $\mathbf{E}_1\epsilon^{j\omega t} + \mathbf{E}_2\epsilon^{j\omega t} + \mathbf{E}_3\epsilon^{j\omega t} + \mathbf{E}_4\epsilon^{j\omega t} = 0$, so

$$\mathbf{E}_1 + \mathbf{E}_2 + \mathbf{E}_3 + \mathbf{E}_4 = 0$$

In Fig. 5.3-2b, $\mathbf{I}_1\epsilon^{j\omega t} + \mathbf{I}_2\epsilon^{j\omega t} + \mathbf{I}_3\epsilon^{j\omega t} = 0$, so

$$\mathbf{I}_1 + \mathbf{I}_2 + \mathbf{I}_3 = 0$$

so that the voltage and current phasors satisfy Kirchhoff's laws.

The block in Fig. 5.3-1 may contain several passive elements instead of only one, in which case the total impedance $Z = \mathbf{E}/\mathbf{I}$ is the ratio of the external voltage and current phasors. For the series connection in Fig. 5.3-2c, where $Z_1 = E_1/I$, $Z_2 = E_2/I$, and $Z_3 = E_3/I$,

$$\mathbf{E}\epsilon^{j\omega t} = \mathbf{E}_1\epsilon^{j\omega t} + \mathbf{E}_2\epsilon^{j\omega t} + \mathbf{E}_3\epsilon^{j\omega t} = (Z_1 + Z_2 + Z_3)\mathbf{I}\epsilon^{j\omega t}$$

so the total impedance is

$$Z = Z_1 + Z_2 + Z_3 \tag{5.3-7}$$

For the parallel connection in Fig. 5.3-2d, $Y_1 = I_1/E$, $Y_2 = I_2/E$, $Y_3 = I_3/E$, and

$$\mathbf{I}\epsilon^{j\omega t} = \mathbf{I}_1\epsilon^{j\omega t} + \mathbf{I}_2\epsilon^{j\omega t} + \mathbf{I}_3\epsilon^{j\omega t} = (Y_1 + Y_2 + Y_3)\mathbf{E}\epsilon^{j\omega t}$$

so the total admittance is

$$Y = Y_1 + Y_2 + Y_3 \tag{5.3-8}$$

or

$$\frac{1}{Z} = \frac{1}{Z_1} + \frac{1}{Z_2} + \frac{1}{Z_3} \tag{5.3-9}$$

Equations 5.3-7 through 5.3-9 are analogous to Eqs. 2.1-1 through 2.1-3, so the impedances of series and parallel elements may be combined as in Chapter 2, *even if the elements are of different kinds.* In summary, if sinusoidal voltages and currents are characterized by their phasors, and the passive elements by their impedances, all the procedures for resistive circuits can be carried over to the calculation of the unknown phasors. The only modification is that, in general, the numbers involved will be complex.

Example 5.3-1.   Find the steady-state current in Fig. 5.2-2 if $e(t) = E_m \cos(\omega t + \phi_1)$.

Solution.   The phasor representing the source voltage is $\mathbf{E} = E_m\underline{/\phi_1}$. The total imped-ance of the series combination of L, R, and C is

$$Z = j\omega L + R + \frac{1}{j\omega C} = R + j(\omega L - 1/\omega C) = |Z|\underline{/\theta}$$

where $|Z| = \sqrt{R^2 + (\omega L - 1/\omega C)^2}$ and $\theta = \tan^{-1}[(\omega L - 1/\omega C)/R]$. The current phasor is $\mathbf{I} = \mathbf{E}/Z = (E_m/|Z|)\underline{/\phi_1 - \theta}$, and so the sinusoidal steady-state current is

$$i(t) = \frac{E_m}{|Z|}\cos(\omega t + \phi_1 - \theta)$$

which agrees with the answer to Examples 5.2-1 and 5.2-2.

The techniques developed in Sections 2.2 through 2.6, including the linearity theorems and loop and node equations, can be applied to the determination of the unknown phasors. Several of these methods are illustrated in the following example.

Example 5.3-2.   For the circuit in Fig. 5.3-3a, find the output voltage $e_o(t)$ in the steady state.

Solution.   In Fig. 5.3-3b, the voltages and currents are characterized by their phasors (complex numbers whose magnitudes and angles are equal to the amplitudes and phase angles of the sinusoidal functions of time). <mark>Passive elements are characterized by their impedances,</mark> using Table 5.3-1 and $\omega = 10$ (the angular frequency of the source). Now the unknown phasors can be determined by the methods of Chapter 2.

Method I.   By combining impedances in series and in parallel, we obtain

$$Z_3 = \frac{1}{2} + j2 + \frac{1}{2} = 1 + j2$$

$$Z_2 = \frac{(1 - j)(1 + j2)}{(1 - j) + (1 + j2)} = 1.414/\!-8° = 1.40 - j0.20$$

$$Z_1 = (j1) + (1.40 - j0.20) = 1.40 + j0.80 = 1.61/29.7°$$

where Z_1, Z_2, and Z_3 are the impedances seen looking to the right at the dashed lines. Then,

$$I_1 = \frac{E_1}{Z_1} = \frac{10/20°}{1.61/29.7°} = 6.19/\!-9.7°$$

$$E_2 = I_1 Z_2 = (6.19/\!-9.7°)(1.414/\!-8°) = 8.75/\!-17.7°$$

By the voltage divider rule,

$$E_o = \frac{(8.75/\!-17.7°)(0.5)}{1 + j2} = 1.96/\!-81.3°$$

Since the output phasor has been found, the steady-state output voltage immediately follows.

$$e_o(t) = 1.96 \cos(10t - 81.3°)$$

Method II.   The previous method is probably the least attractive of any. Temporarily assume $E_o = 0.5/0°$, so that $I_o = 1/0°$ and $E_2 = 1 + j2$. Then

$$I_1 = I_o + I_2 = 1 + \frac{1 + j2}{1 - j} = 0.5 + j1.5$$

$$E_1 = E_2 + (j1)I_1 = 1 + j2 - 1.5 + j0.5 = 2.55/101.3°$$

Since the given source voltage is $E_1 = 10/20°$,

$$E_o = (0.5/0°)\left(\frac{10/20°}{2.55/101.3°}\right) = 1.96/\!-81.3°$$

$$e_o(t) = 1.96 \cos(10t - 81.3°)$$

Method III.   Write two loop equations, using the loop currents shown in Fig. 5.3-3c.

$$I_1 - (1 - j)I_o = 10/20°$$

$$-(1 - j)I_1 + (2 + j)I_o = 0$$

When these equations are solved simultaneously,

$$I_o = 3.92/\!-81.3°, \qquad E_o = \tfrac{1}{2}I_o = 1.96/\!-81.3°$$

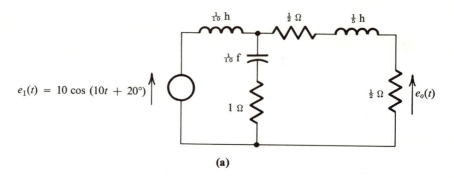

(a)

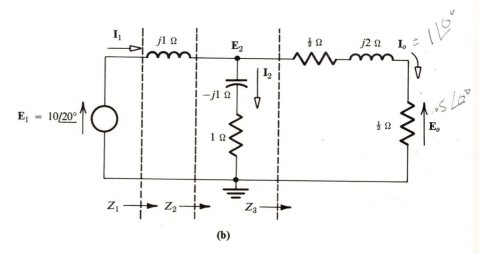

(b)

Method IV. The node equation at node 2 is

$$\frac{E_2 - 10\underline{/20^\circ}}{j} + \frac{E_2}{1 - j} + \frac{E_2}{1 + j2} = 0$$

which yields

$$E_2 = 8.75\underline{/-17.7^\circ}, \qquad E_o = \left(\frac{0.5}{1 + j2}\right) E_2 = 1.96\underline{/-81.3^\circ}$$

Method V. The part of the network within the dashed lines in Fig. 5.3-3c may be replaced by its Thévenin equivalent circuit. The open-circuit voltage is, by the voltage divider rule,

$$E_{oc} = (10\underline{/20^\circ}) \left(\frac{1 - j1}{1 - j1 + j1}\right) = 14.14 \underline{/-25^\circ}$$

The impedance looking into the dead network is

$$Z_{eq} = \frac{j(1 - j)}{j + (1 - j)} = 1 + j$$

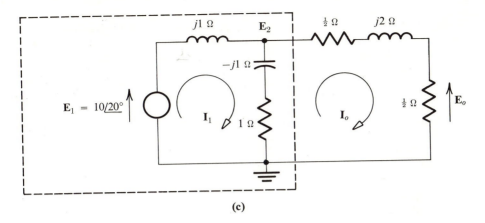

(c)

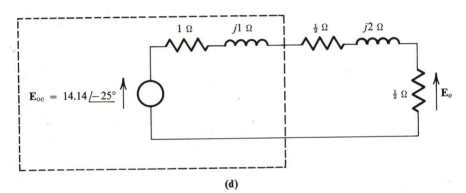

(d)

Figure 5.3-3

The complete Thévenin equivalent circuit is shown in Fig. 5.3-3d. By the voltage divider rule,

$$\mathbf{E}_o = (14.14\underline{/-25°})\left(\frac{0.5}{2+j3}\right) = 1.96\underline{/-81.3°}$$

Example 5.3-2 illustrates the three steps in determining the a-c steady-state response of a fixed, linear circuit. Instead of working directly with sinusoidal functions of time, the first step is to characterize voltages and currents by their phasors, and passive elements by their impedances, as in Fig. 5.3-3b. Figure 5.3-3a is sometimes called the *time-domain circuit*, since voltages and currents are functions of time. Phasors and impedances are not functions of time, but their values do depend upon the angular frequency of the source. (Table 5.3-1 clearly shows that the impedances, and hence the relationship between the phasors, depend in general upon ω.) For this reason, Fig. 5.3-3b is often called the *frequency-domain circuit*.

The second step, which is the only one requiring extensive calculations, is to determine the output phasor by any of the methods of Chapter 2. Finally, the transition is made from the frequency domain back to the time domain, and the steady-state output is written as a sinusoidal function of time.

The steady-state analysis of circuits containing controlled sources is illustrated by Examples 5.3-3 and 5.3-6. Additional examples will be found in Chapter 6, and in the problems at the end of this chapter.

Example 5.3-3.   Figure 5.3-4a is the model used under certain circumstances for a transistor amplifier. The numerical values given for the elements are not typical, but are chosen to simplify the calculations. Find the output current $i_o(t)$ in the steady state at very low frequencies, at very high frequencies, and for $\omega = 2$ rad per sec.

Solution.   In the frequency-domain circuit of Fig. 5.3-4b, the impedance of the parallel R_3C_3 combination is

$$Z_3 = \frac{R_3(1//j\omega C_3)}{R_3 + 1//j\omega C_3} = \frac{R_3}{1 + j\omega R_3 C_3}$$

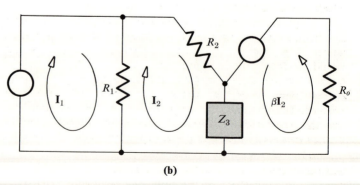

$$R_1 = 4\ \Omega,\ R_2 = R_3 = R_o = 1\ \Omega,\ C_3 = 1\ \text{f},\ \beta = 20$$

(a)

(b)

Figure 5.3-4

Using the loop currents shown in the figure, and writing a voltage-law equation around the loop that does not contain a current source, we obtain

$$R_1(\mathbf{I}_2 - \mathbf{I}_1) + R_2\mathbf{I}_2 + Z_3(\mathbf{I}_2 + \beta\mathbf{I}_2) = 0$$

or

$$\mathbf{I}_2 = \frac{R_1\mathbf{I}_1}{R_1 + R_2 + (\beta + 1)Z_3}$$

When the previous expression for Z_3 and the given numerical values are inserted,

$$\mathbf{I}_o = -\beta\mathbf{I}_2 = \left[\frac{-\beta R_1(1 + j\omega R_3 C_3)}{(R_1 + R_2)(1 + j\omega R_3 C_3) + (\beta + 1)R_3} \right]\mathbf{I}_1 = \left[\frac{-80(1 + j\omega)}{26 + j5\omega} \right]\mathbf{I}_1$$

where $\mathbf{I}_1 = 1\underline{/0°}$. In the notation of Eq. 5.2-6, $\mathbf{I}_o = \mathbf{Y}$ (the output phasor), $\mathbf{I}_1 = \mathbf{X}$ (the input phasor), and the quantity in the brackets is the a-c steady-state network function $H(j\omega)$.

At very low frequencies, where $\omega \ll 1$, $\mathbf{I}_o = -\frac{80}{26} = -3.08$ and

$$i_o(t) = -3.08 \cos \omega t$$

For $\omega \gg \frac{26}{5}$, $\mathbf{I}_o = -j80\omega/j5\omega = -16$ and

$$i_o(t) = -16 \cos \omega t$$

For $\omega = 2$ rad per sec,

$$\mathbf{I}_o = \frac{-80(1 + j2)}{26 + j10} = \frac{-80(2.24\underline{/63.5°})}{27.9\underline{/21.0°}} = -6.43\underline{/42.5°}$$

$$i_o(t) = -6.43 \cos (2t + 42.5°)$$

Notice that the *current gain* $|\mathbf{I}_o|/|\mathbf{I}_1|$ depends upon the angular frequency of the input. For sufficiently high frequencies, the impedance $1/j\omega C_3$ of the capacitance approaches zero, and the element looks like a short circuit. Then the current $i_2(t)$, found by the current divider rule, is $i_2(t) = [R_1/(R_1 + R_2)]i_1(t)$, and

$$i_o(t) = -\beta i_2(t) = \frac{-\beta R_1}{(R_1 + R_2)}i_1(t) = -16 \cos \omega t$$

which agrees with the previous answer for $\omega \gg \frac{26}{5}$.

As ω approaches zero, the capacitance's impedance approaches infinity, so at very low frequencies this element can be approximated by an open circuit. The voltage across R_3 is then $R_3(1 + \beta)i_2(t)$, which is the voltage that the current $i_2(t)$ *alone* would produce if the resistance were $R_3(1 + \beta)$. At low frequencies, therefore, the current $i_2(t)$ sees a much larger total resistance, so a greater proportion of the source current flows through R_1, and the size of $i_2(t)$ and $i_o(t) = -\beta i_2(t)$ is considerably reduced. This conclusion is consistent with the previous solution for $\omega \ll 1$. The fact that the steady-state response of most circuits depends upon the angular frequency of the input is considered at greater length in Section 5.4 and in Chapter 6.

Input Impedance. In Chapter 2, it was demonstrated that any combination of resistances and controlled sources that has only two external terminals is equivalent to a single resistance, as far as the calculation of external voltages and currents is concerned. If the combination also contains one or more independent sources, it may be replaced by a Thévenin or Norton equivalent circuit. In the a-c steady state, any two-terminal frequency-domain circuit that contains no independent sources may be reduced to a single impedance, which is called the *input impedance*. The input impedance, which is the ratio of the external voltage and current phasors, is independent of the magnitude and angle of the external voltage or current, but does depend upon the angular frequency. Since the input impedance Z and the *input admittance* $Y = 1/Z$ are complex quantities, they may be expressed in rectangular form as

$$Z = R + jX, \qquad Y = G + jB \tag{5.3-10}$$

The quantities R, X, G, and B are called the input *resistance*, *reactance*, *conductance*, and *susceptance*, respectively. Although $Z = 1/Y$, the student should be warned that in general $R \neq 1/G$ and $X \neq 1/B$. To find the relationship between R, X, G, and B, note that

$$Y = \left(\frac{1}{R + jX}\right)\left(\frac{R - jX}{R - jX}\right) = \frac{R}{R^2 + X^2} + j\frac{-X}{R^2 + X^2}$$

so

Similarly,

$$G = \frac{R}{R^2 + X^2}, \qquad B = \frac{-X}{R^2 + X^2}$$

$$R = \frac{G}{G^2 + B^2}, \qquad X = \frac{-B}{G^2 + B^2}$$

It will be shown in Chapter 8 that for circuits with no controlled sources, R and G must be nonnegative. The reactance and susceptance may have either sign, although the sign of B is always opposite to that of X.

When the rectangular form for Z or Y is used, a circuit with two external terminals may be replaced by two series or two parallel elements *at any one frequency*. However, R, X, G, and B are functions of frequency, and even the sign of X and B may be different at different frequencies. When Eqs. 5.3-10 are converted to polar form,

$$Z = R + jX = \sqrt{R^2 + X^2} \;\Big/\!\tan^{-1}\frac{X}{R} = |Z| \,\underline{/\theta}$$

and

$$Y = G + jB = \sqrt{G^2 + B^2} \;\Big/\!\tan^{-1}\frac{B}{G} = |Y| \,\underline{/\phi}$$

which suggest the impedance and admittance triangles of Fig. 5.3-5.

Example 5.3-4. Find expressions for the input resistance and reactance in Fig. 5.3-6.

Solution. By the rules for series-parallel combinations,

$$Z = 2 + \frac{(1 + j2\omega)(-j2/\omega)}{1 + j2\omega - j2/\omega} = 2 + \frac{4\omega - j2}{\omega + j2(\omega^2 - 1)}$$

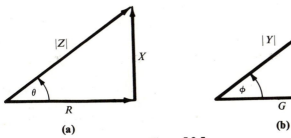

Figure 5.3-5

Rationalizing the second of the two terms, we obtain

$$Z = 2 + \frac{1 + j(-8\omega^3 + 6\omega)}{\omega^2 + 4(\omega^2 - 1)^2} = \left[\frac{8\omega^4 - 14\omega^2 + 9}{\omega^2 + 4(\omega^2 - 1)^2}\right] + j\left[\frac{\omega(-8\omega^2 + 6)}{\omega^2 + 4(\omega^2 - 1)^2}\right]$$

The quantities inside the two brackets are the input resistance and reactance, respectively. Notice that R is positive at all frequencies. For $\omega < \sqrt{3}/2$, X is positive, while for $\omega > \sqrt{3}/2$, X is negative. This result should be expected from an inspection of the circuit, since at low frequencies the impedance of the capacitance is very large and the circuit looks inductive, while at high frequencies the circuit looks capacitive. At the angular frequency $\omega = \sqrt{3}/2$, $R = 3$ and $X = 0$, indicating that the circuit can then be replaced by a 3 Ω resistance.

Example 5.3-5. Find the input impedance of Fig. 5.3-7a for an angular frequency $\omega = 2$ rad per sec.

Solution. The frequency-domain circuit is shown in Fig. 5.3-7b. Since the desired answer cannot be obtained by combining elements in series and in parallel, we use the definition of the input impedance as the ratio of the external voltage and current phasors. If the terminal voltage phasor is **E**, the three loop equations are

$$(2 + j5)\mathbf{I}_1 - (j1)\mathbf{I}_2 - (j4)\mathbf{I}_3 = \mathbf{E}$$

$$(-j1)\mathbf{I}_1 + \mathbf{I}_2 - \mathbf{I}_3 = 0$$

$$(-j4)\mathbf{I}_1 - \mathbf{I}_2 + (1 + j2)\mathbf{I}_3 = 0$$

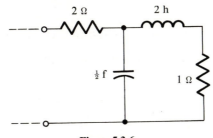

Figure 5.3-6

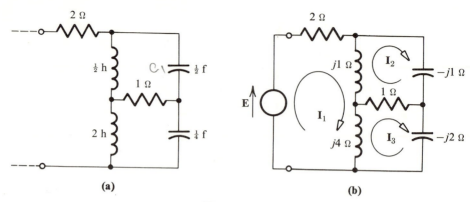

(a)

(b)

Figure 5.3-7

Solving simultaneously for I_1, we obtain

$$I_1 = \frac{E}{3 - j7.5}$$

so

$$Z = \frac{E}{I_1} = 3 - j7.5 = 8.08\underline{/-68.2°}$$

The input resistance and reactance are $R = 3\ \Omega$ and $X = -7.5\ \Omega$.

   Any two-terminal frequency-domain circuit that contains independent sources can be replaced by a Thévenin or Norton equivalent circuit, as in Fig. 2.3-1. The dead network can be reduced to a single equivalent impedance Z_{eq}.

   Example 5.3-6.   For Fig. 5.3-8a, find the Thévenin equivalent circuit that can be used in a-c steady-state analysis.

   Solution.   The frequency-domain circuit for the calculation of the open-circuit voltage is given in Fig. 5.3-8b. For the left-hand loop,

$$E_1 = I_1 - (2I_1)$$

so $I_1 = -E$, $I_2 = -2E_1 = -2\underline{/-90°} = j2$. Then,

$$E_{oc} = E_1 - (-j)I_2 = -2 - j = -\sqrt{5}\underline{/26.5°}$$

The impedance of the dead network is found from Fig. 5.3-8c.

$$I_1 = -jE, \qquad I_2 = -j2E, \qquad I = I_2 - I_1 = -jE$$

so

$$Z_{eq} = \frac{E}{I} = j\,1\ \Omega$$

Notice that the dead network looks inductive even though the circuit contains no

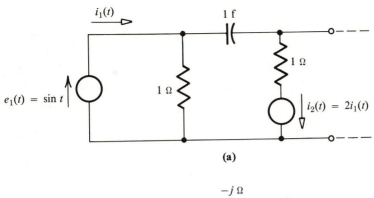

(a)

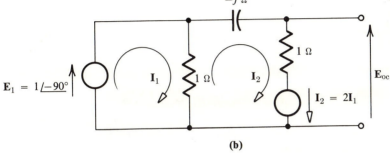

(b)

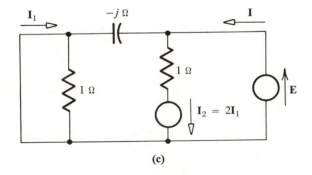

(c)

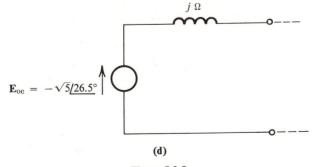

(d)

Figure 5.3-8

inductance. This effect results from the presence of the controlled source. The complete frequency-domain Thévenin equivalent circuit is given in Fig. 5.3-8d.

Circuits with Sources at Different Frequencies.   A single frequency-domain circuit cannot usually be drawn for a time-domain circuit that contains independent sources at different frequencies, because the impedances of the inductances and capacitances are different at different frequencies. Furthermore, there is no way of combining phasors corresponding to sinusoidal functions with different frequencies. Suppose that the functions are expressed as the real-axis projection of their CCW rotating phasors, as in Eq. 5.2-5. $F_1 \angle \omega t$

$$f_1(t) = \text{Re} \ (\mathbf{F}_1 \epsilon^{j\omega_1 t}) = \text{Re} \ (F_{m1}\epsilon^{j\phi_1}\epsilon^{j\omega_1 t})$$

$$f_2(t) = \text{Re} \ (\mathbf{F}_2 \epsilon^{j\omega_2 t}) = \text{Re} \ (F_{m2}\epsilon^{j\phi_2}\epsilon^{j\omega_2 t})$$

If $\omega_2 = \omega_1$, then

$$f_1(t) + f_2(t) = \text{Re} \ [(\mathbf{F}_1 + \mathbf{F}_2)\epsilon^{j\omega_1 t}]$$

and the resulting phasor $\mathbf{F}_1 + \mathbf{F}_2$ is simply the sum of the individual phasors. If $\omega_2 \neq \omega_1$,

$$f_1(t) + f_2(t) = \text{Re} \ (\mathbf{F}_1 \epsilon^{j\omega_1 t} + \mathbf{F}_2 \epsilon^{j\omega_2 t})$$

and there is no way in which the two phasors $\mathbf{F}_1$ and $\mathbf{F}_2$ can be combined. From a graphical point of view, the discussion associated with Fig. 5.2-3 pointed out that the angle between two different CCW phasors remains constant only if they rotate at the same angular frequency.

When a circuit contains independent sources at different frequencies, the steady-state response to each independent source should be found as a separate function of time and the entire response written down by superposition. One important special case occurs in Chapter 9, where certain nonsinusoidal inputs are represented as the sum of sinusoidal components at different frequencies. The steady-state response is then the sum of the responses to the individual components.

Example 5.3-7.   Find the steady-state output voltage $e_o(t)$ in Fig. 5.3-9a.

Solution.   The steady-state response to each of the three components in $e_1(t)$ may be found separately. The response to the constant (or d-c) component may be found by replacing the capacitance and inductance by an open and short circuit, respectively, as in Table 4.5-1. Then, by the voltage divider rule,

$$e_o(t) = \frac{\frac{1}{2}}{4 + \frac{1}{2}} \ (3) = \frac{1}{3}$$

If $e_1(t)$ consisted of a single sinusoidal component at an angular frequency ω, the frequency-domain circuit of Fig. 5.3-9b would apply. A node equation gives

$$\frac{\mathbf{E}_o - \mathbf{E}_1}{4 + j\omega} + 2\mathbf{E}_o + j\omega\mathbf{E}_o = 0$$

or

$$\mathbf{E}_o = \left[\frac{1}{1 + (4 + j\omega)(2 + j\omega)} \right] \mathbf{E}_1 = \left(\frac{1}{9 + j6\omega - \omega^2} \right) \mathbf{E}_1$$

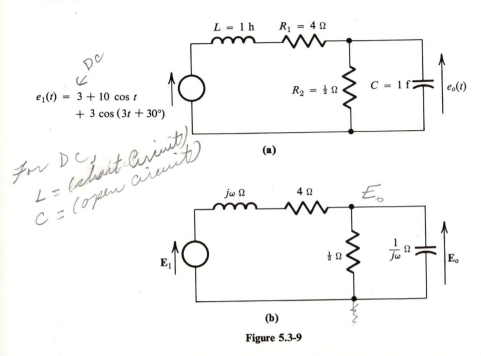

(a)

(b)

Figure 5.3-9

The quantity in the brackets in the above equation is the a-c steady-state network function. If $e_1(t) = 10 \cos t$, then $\omega = 1$, $\mathbf{E}_1 = 10\underline{/0°}$, and

$$\mathbf{E}_o = \frac{10\underline{/0°}}{8 + j6} = 1\underline{/-36.9°}$$

$$e_o(t) = \cos(t - 36.9°)$$

If $e_1(t) = 3 \cos(3t + 30°)$, then $\omega = 3$, $\mathbf{E}_1 = 3\underline{/30°}$, and

$$\mathbf{E}_o = \frac{3\underline{/30°}}{j18} = \frac{1}{6}\underline{/-60°}$$

$$e_o(t) = \frac{1}{6}\cos(3t - 60°)$$

For the input $e_1(t) = 3 + 10 \cos t + 3 \cos(3t + 30°)$, the entire steady-state response is

$$e_o(t) = \frac{1}{3} + \cos(t - 36.9°) + \frac{1}{6}\cos(3t - 60°)$$

Since the sinusoidal input $x(t) = X_m \cos \omega t$ reduces to a constant when $\omega = 0$, the d-c steady state may be regarded as a special case of the a-c steady state. This is consistent with Table 5.3-1, because when ω approaches zero the impedance of a capacitance or inductance approaches infinity (an open circuit) or zero (a short circuit), respectively. The network function $H(j\omega) = |H(j\omega)|\underline{/\theta}$ reduces to the real quantity $H(0)$. By Eqs. 5.2-8 through 5.2-10, the steady-state response to $x(t)$ $= X_m \cos \omega t$ is $y_P(t) = |H(j\omega)| X_m \cos (\omega t + \theta)$. By letting $\omega = 0$, we find that the response to $x(t) = X_m$ is $y_P(t) = H(0)X_m$. In Example 5.3-7,

$$H(0) = \left[\frac{1}{9 - j6\omega - \omega^2}\right]_{\omega=0} = \frac{1}{9}$$

so the steady-state response to $e_1(t) = 3$ volts is $e_o(t) = \frac{1}{3}$ volt.

It is frequently necessary to determine how the magnitude and phase angle of the steady-state response vary when the angular frequency of the input (or the value of one of the circuit elements) is varied. The techniques developed in Sections 5.4, 6.4, and 6.5 can be very helpful in such problems.

5.4

Phasor Diagrams

As discussed in Sections 5.2 and 5.3, the addition of two sinusoidal functions with the same angular frequency can be accomplished by adding their phasors. Phasors are complex numbers that can be represented by directed lines, so the voltage phasors for elements in series, and the current phasors for elements in parallel can be added vectorially.

A typical phasor diagram for the series connection in Fig. 5.4-1a is shown in Fig. 5.4-1b. Since the purpose of such a diagram is to portray the relative magnitudes and angles of the input and output phasors, the magnitude and angle of any one phasor may be selected arbitrarily. The resulting phasor diagram may always be adjusted later to give the response to an input with a specified magnitude and angle, as was done in Method II of Example 5.3-2. In constructing Fig. 5.4-1b, it is convenient to assume that $\mathbf{I} = |\mathbf{I}|/\underline{0°}$. The voltage phasors are $\mathbf{E}_R = R|\mathbf{I}|$, $\mathbf{E}_L = \omega L|\mathbf{I}|/\underline{90°}$, $\mathbf{E}_C = (|\mathbf{I}|/\omega C)/\underline{-90°}$, and $\mathbf{E} = \mathbf{E}_C + \mathbf{E}_L + \mathbf{E}_R$.

For a resistance, the angles of $\mathbf{E}$ and $\mathbf{I}$ are identical; i.e., the voltage and current are "in phase." For an inductance, the angle of $\mathbf{E}$ is 90° more positive than that of $\mathbf{I}$; i.e., the voltage leads the current by 90°, and the current lags the voltage. The current through a capacitance leads the voltage by 90°. More generally, for a RL combination with two external terminals and no controlled sources, the external current *lags* the external voltage by an angle between 0° and 90°. The current into a two-terminal RC circuit *leads* the voltage by an angle between 0° and 90°, while for a RLC circuit or a circuit with controlled sources, either the voltage or the current may lead the other.[†]

In the phasor diagram in Fig. 5.4-1b it is assumed that $|\mathbf{E}_L| > |\mathbf{E}_C|$, i.e., that

---

[†] Although it should not be necessary, some students like to use a mnemonic aid such as "ELI the ICE man." In an inductive (L) circuit, the voltage phasor $\mathbf{E}$ leads the current phasor $\mathbf{I}$, while in a capacitive (C) circuit, the current leads the voltage.

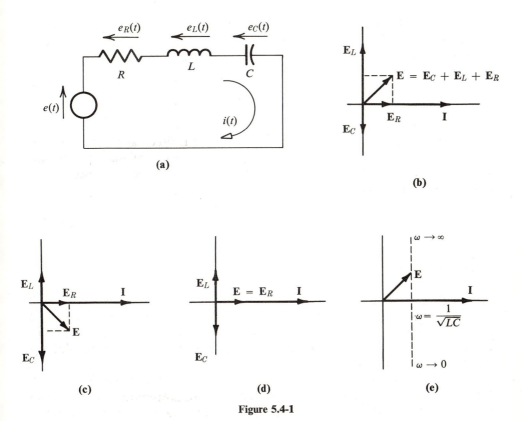

Figure 5.4-1

$\omega L > 1/\omega C$ or $\omega^2 > 1/LC$. If $\omega^2 < 1/LC$, then $|\mathbf{E}_L| < |\mathbf{E}_C|$ as in Fig. 5.4-1c, while for $\omega^2 = 1/LC$, $|\mathbf{E}_L| = |\mathbf{E}_C|$ and $\mathbf{E} = \mathbf{E}_R$, as in part (d) of the figure. For fixed values of R, L, and C and for a fixed current phasor $\mathbf{I}$, the locus traced out by the voltage phasor $\mathbf{E}$, as the angular frequency is varied, is the dashed line in Fig. 5.4-1e. The ratio $|\mathbf{E}|/|\mathbf{I}|$ will be a minimum, and the ratio $|\mathbf{I}|/|\mathbf{E}|$ a maximum when $\omega = 1/\sqrt{LC}$. A sinusoidal voltage source at this angular frequency will produce the largest steady-state current. This conclusion also follows from the equation

$$\mathbf{I} = \frac{\mathbf{E}}{Z} = \frac{\mathbf{E}}{R + j(\omega L - 1/\omega C)}$$

The dual of Fig. 5.4-1a is the parallel combination in Fig. 5.4-2a. For a given phasor $\mathbf{E}$ at an angle of $0°$, the current phasor $\mathbf{I}$ traces out the vertical locus indicated in Fig. 5.4-2b. At low frequencies where Z_L is small, most of the source current flows through the inductance, so the circuit appears inductive. At high frequencies, most of the current flows through the capacitance, while for $\omega^2 = 1/LC$, $\mathbf{I}_C$ and $\mathbf{I}_L$ cancel and $\mathbf{I} = \mathbf{I}_R$. The minimum value of $|\mathbf{I}|/|\mathbf{E}|$ and, hence, the maximum value of $|\mathbf{E}|/|\mathbf{I}|$ occur when $\omega = 1/\sqrt{LC}$.

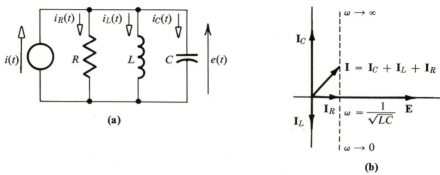

Figure 5.4-2

A phasor is uniquely described by its length and its angle with respect to the positive real axis and can be drawn anywhere in the complex plane. It may not be convenient to show all the phasors starting from the origin. In constructing a typical phasor diagram for Fig. 5.4-3a, the voltage phasors are drawn "head-to-tail" in Fig. 5.4-3b. Similarly, the current phasors may be drawn as in part (c) of the figure.

A complete phasor diagram for Method II of Example 5.3-2, using $\mathbf{E}_o = 0.5 \underline{/0°}$ as an arbitrary reference, is shown in Fig. 5.4-4. A number of other examples are found in Chapter 8.

Example 5.4-1.   Phasor diagrams are often useful in analyzing phase-shift circuits, where the angle but not the magnitude of the output depends upon the value of some circuit parameter. For the circuit in Fig. 5.4-5a, show by a phasor diagram that $|\mathbf{E}_o| = \frac{1}{2}|\mathbf{E}_1|$, but that $\mathbf{E}_o$ may be made to lag $\mathbf{E}_1$ by any angle between 0° and 180°, by means of the proper choice of R_1.

Solution.   Arbitrarily, let the angle of the input voltage phasor $\mathbf{E}_1$ be zero. Half of the input voltage appears across each of the resistances of R_2 ohms, locating the points A, B, and F in Fig. 5.4-5b. The current phasor $\mathbf{I}_C$ through the capacitive

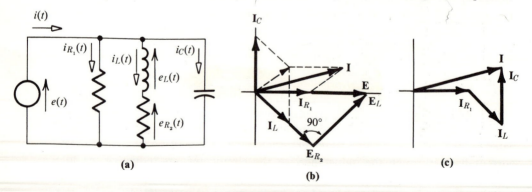

Figure 5.4-3

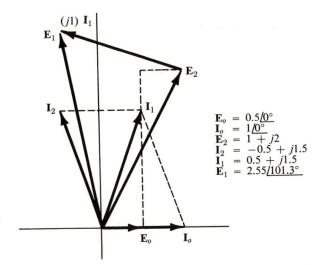

Figure 5.4-4

branch leads the input voltage by an angle between $0°$ and $90°$. Then $\mathbf{E}_C$ lags $\mathbf{I}_C$ by $90°$, while $\mathbf{E}_{R_1}$ is in phase with $\mathbf{I}_C$, producing the right triangle ADF. Finally, the output voltage phasor $\mathbf{E}_o$ is drawn from point B to point D.

By using the geometrical fact that a semicircle (shown dotted) is uniquely defined by the three corners of a right triangle, we see that point D is on a semicircle whose center is at B. Thus $|\mathbf{E}_o| = \frac{1}{2}|\mathbf{E}_1|$, and $\mathbf{E}_o$ lags $\mathbf{E}_1$ by the angle α. As R_1 is varied from zero to infinity, the point D moves from F to A, and α varies from $0°$ to $180°$. Notice, specifically, that

$$\tan\frac{\alpha}{2} = |\mathbf{E}_{R_1}|/|\mathbf{E}_C| = \frac{R_1}{1/\omega C} = \omega C R_1$$

so $\alpha = 2\tan^{-1}\omega C R_1$.

The above solution, which relies heavily upon elementary geometry, may be

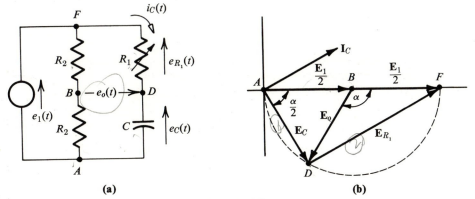

(a)   (b)

Figure 5.4-5

checked analytically as follows. For many students the analytical solution is the easier of the two.

$$\mathbf{I}_C = \frac{\mathbf{E}_1}{R_1 + 1/j\omega C}$$

$$\mathbf{E}_o = -\frac{\mathbf{E}_1}{2} + \frac{\mathbf{I}_C}{j\omega C} = -\frac{\mathbf{E}_1}{2} + \frac{\mathbf{E}_1}{1 + j\omega C R_1}$$

$$\mathbf{E}_o = \frac{\mathbf{E}_1}{2}\left(\frac{1 - j\omega C R_1}{1 + j\omega C R_1}\right) = \frac{\mathbf{E}_1}{2}(1\underline{/-2\tan^{-1}\omega C R_1})$$

which agrees with the previous result.

The circuit may be modified so that $|\mathbf{E}_o| = \frac{1}{2}|\mathbf{E}_1|$, but so $\mathbf{E}_o$ *leads* $\mathbf{E}_1$ by any angle between $0°$ and $180°$. This is done either by interchanging R_1 and C, or else by replacing the capacitance with an inductance. The first choice would probably be preferred, since it is easier to physically approximate a pure capacitance than a pure inductance.

> **Example 5.4-2.**   For fixed values of R_1, R_2, and C in Fig. 5.4-5a, how does the magnitude and angle of $\mathbf{E}_o$ vary as a function of the angular frequency of the source?
>
> **Solution.**   From either the graphical or analytical solution of the previous example, $|\mathbf{E}_o| = \frac{1}{2}|\mathbf{E}_1|$. As ω varies from zero to infinity, the angle $\alpha = 2\tan^{-1}\omega C R_1$ by which $\mathbf{E}_o$ lags $\mathbf{E}_1$ again varies from $0°$ to $180°$.

The phasor diagram of Fig. 5.4-5b is sometimes called a *locus diagram*, because the locus traced out by point D as R_1 increases from zero to infinity shows how the magnitude and angle of the response vary. A useful theorem states that when the value of any one passive circuit element is varied, the locus of the response is always circular.[†] When this theorem applies, the locus can be plotted by locating two or three points on the circumference of the circle.

> **Example 5.4-3.**   Sketch the locus of the current phasor $\mathbf{I}$ as the resistance in Fig. 5.4-6a is varied from zero to infinity.
>
> **Solution.**   The voltage phasor $\mathbf{E} = E_m\underline{/0°}$ is taken as the reference phasor. For $R = 0$ and ∞, $\mathbf{I} = \mathbf{E}/j\omega L$ and 0, respectively, locating the two end points of the locus in Fig. 5.4-6b. The locus must be confined to the fourth quadrant, since the current of an inductive branch lags the voltage by an angle between $0°$ and $90°$. The locus is a semicircle of radius $E_m/2\omega L$, as shown.

[†] Straight lines are considered special, limiting cases of circles. The statement is *not necessarily* true when the angular frequency of the source is varied. The locus for Fig. 5.4-5a must be circular even when the angular frequence is varied, because for that particular circuit varying ω is equivalent to varying the value of one of the passive circuit elements, namely, C. For the same reason, the locus when ω is varied is always circular for any circuit containing only one reactive element, i.e., one capacitance or one inductance. The response of a circuit as a function of the angular frequency of the source is a general topic of considerable importance and is considered in the next chapter.

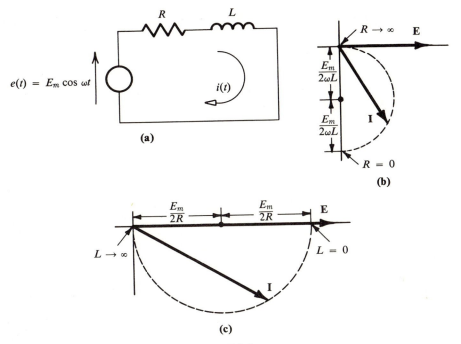

Figure 5.4-6

The correctness of the locus may be verified, if desired, by noting that

$$\mathbf{I} = \left(\frac{E_m \underline{/0^\circ}}{R + j\omega L} \right) \left(\frac{R - j\omega L}{R - j\omega L} \right) = \frac{RE_m - j\omega L E_m}{R^2 + (\omega L)^2} = x + jy$$

where

$$x = \frac{RE_m}{R^2 + (\omega L)^2} \quad \text{and} \quad y = \frac{-\omega L E_m}{R^2 + (\omega L)^2}$$

It is straightforward to show that these expressions for x and y satisfy the equation

$$x^2 + \left(y + \frac{E_m}{2\omega L} \right)^2 = \left(\frac{E_m}{2\omega L} \right)^2$$

which describes the semicircle in Fig. 5.4-6b. Part (c) of the figure shows the locus for **I** when L is varied, and is a semicircle of radius $E_m/2R$. An alternative method for constructing these loci is suggested in the problems at the end of the chapter.

Example 5.4-4.   Sketch the locus of the current phasor **I** as the angular frequency of the source in Fig. 5.4-7a is varied.

Solution.   Varying ω is equivalent to varying C, so the locus will be circular. When $\omega = 0$ and ∞, $\mathbf{I}_1 = 0$ and E/R_1, respectively, where $\mathbf{E} = E_m\underline{/0^\circ}$. Since the current leads the voltage in a capacitive circuit, the locus of $\mathbf{I}_1$ is a semicircle in the first quadrant of the complex plane, as shown in Fig. 5.4-7b. The phasor $\mathbf{I}_2$ has the constant value E_m/R_2, so the locus of $\mathbf{I} = \mathbf{I}_1 + \mathbf{I}_2$ is the semicircle in part (c) of the figure.

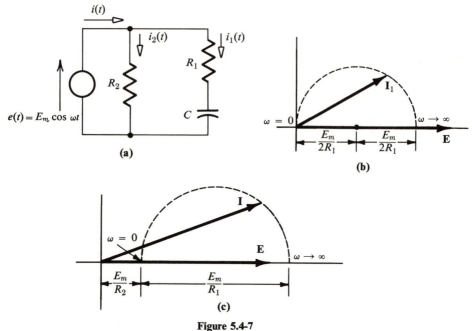

Figure 5.4-7

Combining Trigonometric Terms by Adding Phasors.   The sum of trigonometric terms having the same angular frequency can be expressed as a single trigonometric term with a phase angle. By the use of appropriate identities, suggested after Eq. 5.2-3, it can be shown that

$$A \cos \omega t + B \sin \omega t = \sqrt{A^2 + B^2} \cos \left(\omega t - \tan^{-1} \frac{B}{A} \right)$$

$$= \sqrt{A^2 + B^2} \sin \left(\omega t + \tan^{-1} \frac{A}{B} \right)$$

(5.4-1)

Not only are such relationships difficult to remember, but there is some ambiguity, since $\tan^{-1} B/A$ is not a unique angle.

Each of the trigonometric terms to be combined may be represented by a phasor. If A and B are both positive constants, the phasors representing $A \cos \omega t$ and $B \sin \omega t = B \cos (\omega t - 90°)$ are those shown in Fig. 5.4-8a. The phasor representing the sum of the two terms is the sum of the individual phasors. From the figure,

$$A \cos \omega t + B \sin \omega t = \sqrt{A^2 + B^2} \cos \left(\omega t - \tan^{-1} \frac{B}{A} \right)$$

$$= \sqrt{A^2 + B^2} \sin \left(\omega t + \tan^{-1} \frac{A}{B} \right)$$

agreeing with Eq. 5.4-1.

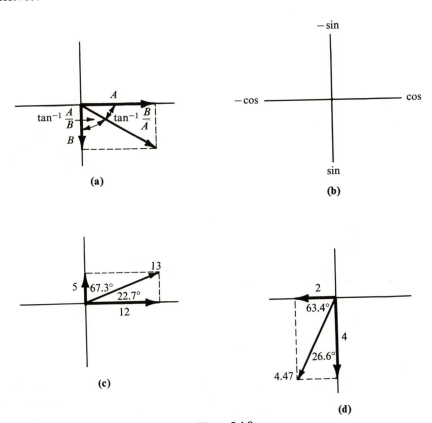

Figure 5.4-8

The axes of the complex plane are sometimes labelled as in Fig. 5.4-8b, indicating where the phasors for positive and negative cosine and sine terms lie. By the construction in part (c) of the figure, for example,

$$-5 \sin \omega t + 12 \cos \omega t = 13 \cos (\omega t + 22.7°)$$

$$= -13 \sin (\omega t - 67.3°)$$

From part (d),

$$4 \sin \omega t - 2 \cos \omega t = -4.47 \cos (\omega t + 63.4°)$$

$$= 4.47 \sin (\omega t - 26.6°)$$

Mechanically, combining trigonometric terms of the same frequency is similar to converting complex numbers from rectangular to polar form. The same rules apply when using the slide rule, but it is advisable to sketch a rough phasor diagram in order to determine the proper quadrant for the resulting phase angle.

5.5

Summary

In this chapter, sinusoidal functions of time are represented by their phasors. The phasor corresponding to $f(t) = F_m \cos(\omega t + \phi)$ is the complex number $\mathbf{F} = F_m\underline{/\phi}$. Thus the input $x(t) = X_m \cos(\omega t + \phi_1)$ and the steady-state output $y_P(t) = Y_m \cos(\omega t + \phi_2)$ are represented by the phasors $\mathbf{X} = X_m\underline{/\phi_1}$ and $\mathbf{Y} = Y_m\underline{/\phi_2}$.

The a-c steady-state network function $H(j\omega) = |H(j\omega)|\underline{/\theta}$ may be found by an inspection of the differential equation relating $x(t)$ and $y(t)$. Then the equation

$$\mathbf{Y} = H(j\omega)\mathbf{X}$$

may be used to calculate the output phasor, from which the output as a function of time immediately follows. Normally, however, $H(j\omega)$ is not determined by first finding the circuit's differential equation. Instead, a frequency-domain circuit is drawn, in which the voltages and currents are characterized by their phasors, and passive elements by their impedances (using Table 5.3-1). The output phasor is then found by any of the methods of Chapter 2. Any combination of elements that has only two external terminals and no independent sources may be represented by an input impedance. A two-terminal circuit containing independent sources may be replaced by a Thevenin or Norton equivalent circuit.

If the circuit contains sources at different frequencies or a source with several different components, the steady-state response may be found by superposition. As discussed after Example 5.3-7, the steady-state response to a constant may be regarded as a special case of the a-c steady-state response, with $\omega = 0$.

The network function $H(j\omega)$ and the magnitude and angle of the output phasor depend upon the frequency of the source. One application of the phasor diagrams in Section 5.4 is to show this dependency graphically. Other methods are developed in Chapter 6.

This chapter is restricted to finding the *forced* response to a *sinusoidal* input. In the next chapter, other types of inputs and the relationship of the network function to the circuit's free response are considered.

PROBLEMS

5.1 For the circuit in Fig. P5.1, $i_1(t) = 2 \cos 100t$, $i_2(t) = 2\sqrt{2} \cos(100t - 45°)$, and $i_3(t) = 5 \cos(100t + 90°)$. By the use of appropriate trigonometric identities, such as Eq. 5.2-3, find an expression for $i(t)$ that has the form $i(t) = A \cos(100t + \theta)$. Express the complex quantity $2\underline{/0°} + 2\sqrt{2}\,\underline{/-45°} + 5\underline{/90°}$ in polar form, and compare the result with your expression for $i(t)$.

5.2 The complex numbers in parts (a) through (h) might represent current phasors for a particular circuit. Change the numbers that are in polar form to rectangular form and those that are in rectangular form to polar form. In parts (i) through (m), carry

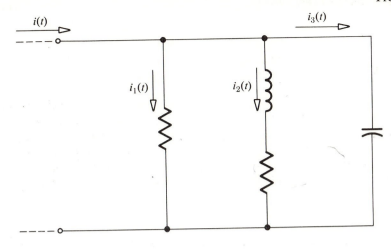

Figure P5.1

out the indicated operations and express the answers in polar form. (The mathematical operations in this problem are typical of the manipulations needed for the solution of subsequent circuit problems.)

(a) $10\underline{/-30°}$

(b) $6\underline{/150°}$

(c) $-8\underline{/60°}$

(d) $20\underline{/-225°}$

(e) $-5 + j12$

(f) $16 - j12$

(g) $-1 - j15$

(h) $-10 + j0.5$

(i) $10\underline{/135°} - 12\underline{/-60°} + 6\underline{/-120°}$

(j) $\dfrac{(1 + j)^2}{1 - j}$

(k) $\dfrac{2\underline{/30°}}{j10} + \dfrac{1}{-8 + j6}$

(l) $\dfrac{(3 - j8)^2(-1 + j2)}{j(-1 - j)^2}$

(m) $\left(\dfrac{3 - j}{1 + j} - \dfrac{j2}{2 + j3}\right)\left(\dfrac{j2}{9 - j12}\right)$

5.3   Find the following quantities.

(a) the cube roots of j

(b) j^9

(c) ϵ^j

(d) j^j

(e) $|(3 - j4)(10\underline{/20°})\epsilon^{j\pi/8}|$

(f) the real and imaginary parts of $z^2/|z|^2$ if $z = x + jy$

5.4   Show the regions in the complex plane that are described by the following inequalities.

(a) $|z - 3| < 2$

(b) $\text{Im } z < 4$

(c) $\text{Re } \dfrac{1}{z} > \dfrac{1}{4}$

5.5   Find the differential equation relating $e_1(t)$ and $e_o(t)$ in Fig. P5.5. Find the steady-state response to $e_1(t) = 2 \cos 2t$ by the following methods.

(a) The classical solution of differential equations, as in Example 5.2-1.

(b) The use of the a-c steady-state network function, as in Example 5.2-2.

5.6   Find $H(j\omega)$ for the circuit in Fig. P5.6 by the following methods.

(a) First find the differential equation relating $e_1(t)$ and $i_o(t)$.

(b) Characterize the passive elements by their impedances.

5.7   For the circuit in Fig. P5.7, find the steady-state output voltage $e_o(t)$ by using phasors and impedances.

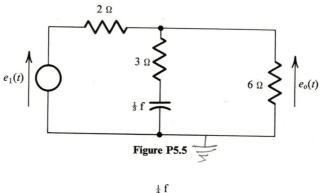

Figure P5.5

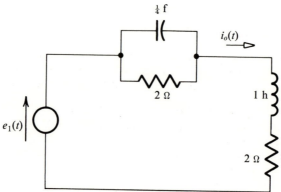

Figure P5.6

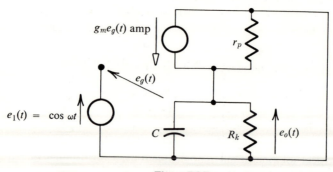

Figure P5.7

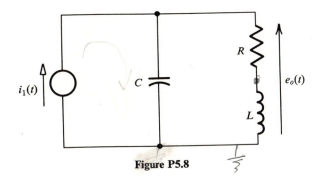

Figure P5.8

5.8 If $L = 1$ mh, $C = 1$ nf, $R = 2$ kΩ, and $i_1(t) = \cos 10^6 t$ amp in Fig. P5.8, find $e_o(t)$ in the steady state. Find the complete response for all $t > 0$, if $i_1(t) = (\cos 10^6 t)U_{-1}(t)$.

5.9 Find $e_o(t)$ in the steady state for the circuit in Fig. P5.9 when $e_1(t) = 10 \cos (2t + 30°)$.

5.10 Find $e_o(t)$ in the steady state for each of the circuits in Fig. P5.10.

5.11 Find the steady-state current $i_2(t)$ for the circuit in Fig. P5.11.

5.12 Find $e_o(t)$ in the steady state for the circuit in Fig. P5.12.

5.13 Three circuits that are sometimes used for the compensation of feedback control systems are shown in Fig. P5.13. For each circuit, determine the values of ω for which the output voltage $e_o(t)$ lags the input voltage $e_1(t)$ in the a-c steady state.

5.14 For the circuit in Fig. P5.14, $e_1(t) = 25 \cos (100t - 30°)$, $i_1(t) = 10 \sin 100t$, and $i_3(t) = 10 \cos 100t$. The impedance of N_1 and N_2 is $Z_1 = Z_2 = 5 + j8.66$ Ω when $\omega = 100$. Find the impedance of N_3, and the steady-state voltage across the current source. From these results, find the impedance looking to the right at the dashed line.

5.15 Find the input impedance for each of the circuits in Fig. P5.15. For what values of ω can each circuit be replaced by a single resistance?

5.16 Find a parallel combination of two elements that is equivalent to Fig. P5.15b when the angular frequency is $\omega = \frac{1}{2}$ rad per sec. Repeat when $\omega = 2$ rad per sec.

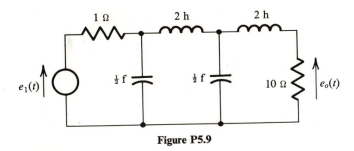

Figure P5.9

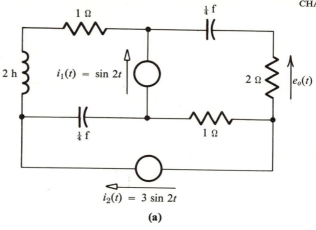

(a)

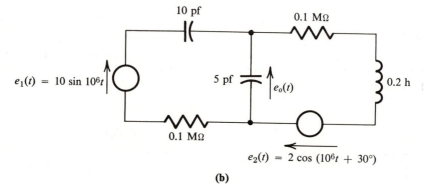

(b)

Figure P5.10

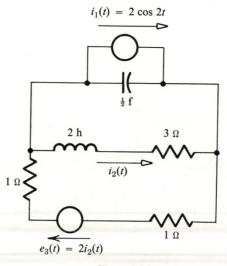

Figure P5.11

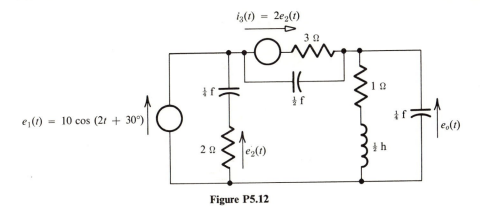

Figure P5.12

5.17 Use Thévenin's or Norton's theorem to find the steady-state current through the inductance in Fig. P5.17.

5.18 What value of C_o in Fig. P5.18 will result in the maximum steady-state current through R_o? (HINT: Apply Thévenin's theorem and a little thought to the portion of the circuit within the dashed lines.)

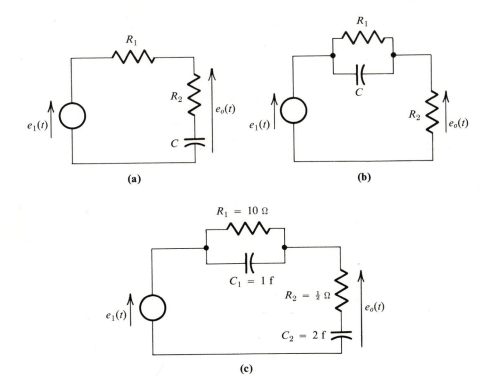

Figure P5.13

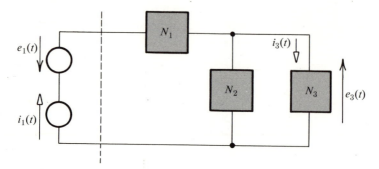

Figure P5.14

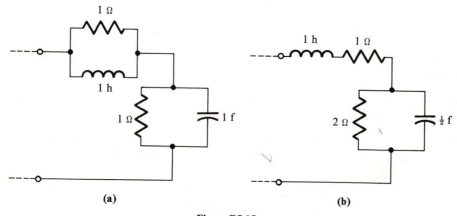

(a)

(b)

Figure P5.15

5.19 Find $e_o(t)$ in the steady state for each of the circuits in Fig. P5.19.

5.20 The circuit in Fig. P5.20 has been operating in the steady state with the switch K closed. The source voltage $e_1(t)$ is sinusoidal and has a maximum value of 10 volts and an angular frequency of 1 rad per sec. At an instant when the waveshape of the

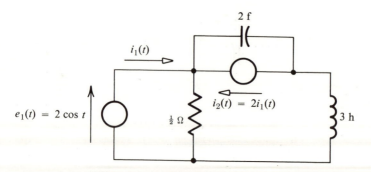

Figure P5.17

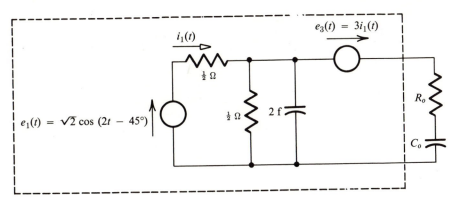

Figure P5.18

current through the switch passes through zero, the switch opens. Find $e_o(t)$ for all time after the switch opens.

5.21 After having been in position ① for a long period of time, switch K in Fig. P5.21 is moved to position ② at $t = 0$. Find $e_o(t)$ for all $t > 0$.

5.22 The circuit in Fig. P5.22 has been operating in the steady state with the switch K closed. If the switch opens at $t = 0$, find the values of $e_o(t)$ and de_o/dt at $t = 0+$.

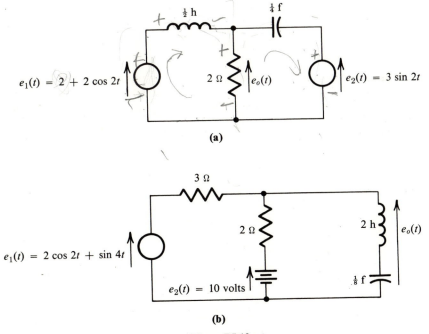

(a)

(b)

Figure P5.19

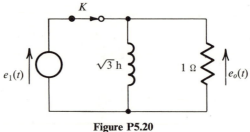

Figure P5.20

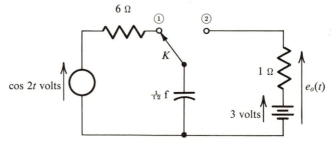

Figure P5.21

5.23 The frequency-domain circuit for an "impedance bridge" is shown in Fig. P5.23a. Prove that $\mathbf{E}_o = 0$ if $Z_1 Z_4 = Z_2 Z_3$. The bridge in Fig. P5.23b may be used to measure the resistance R_4 and the inductance L_4 of an inductor whose impedance is unknown. The resistances R_2 and R_3 are adjusted until $e_o(t) = 0$, i.e., until the bridge is "balanced." Find expressions for R_4 and L_4 in terms of the other circuit elements when the bridge is balanced.

5.24 Two other circuits that may be used to measure the resistance and inductance of an inductor are shown in Fig. P5.24. In each case, find expressions for R_4 and L_4 when the bridge is balanced. Discuss any advantages or disadvantages that these circuits may have.

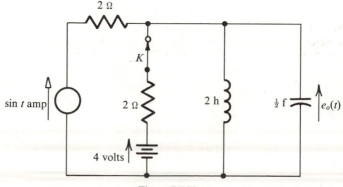

Figure P5.22

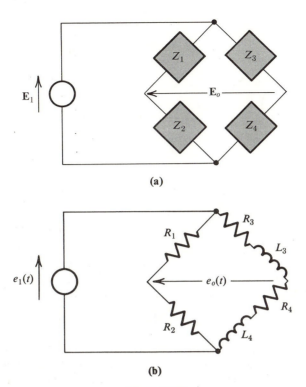

(a)

(b)

Figure P5.23

5.25 Draw a phasor diagram that shows the voltage and current of every element in Fig. P5.23b when the bridge is not balanced. Show graphically that the bridge may be balanced by varying R_2 and R_3, but not by varying only R_1 and R_2.

5.26 Draw complete phasor diagrams for the circuits in Fig. P5.24 when the bridge is unbalanced, and when the bridge is balanced.

5.27 The magnitude but not the angle of an input impedance may be determined from voltmeter and ammeter readings, as discussed in Section 8.2. One method of experimentally determining the complete input impedance of a network is suggested in Fig. P5.27. Denote the input impedances of the circuits in parts (a) and (b) of the figure by Z_a and Z_b, respectively. Then $Z_a = Z_b - Z_C$, where $Z_C = 1/j\omega C$. If $|Z_a| = 100\,\Omega$, $|Z_b| = 50\,\Omega$, $\omega = 100$ rad per sec, and $C = 100\,\mu$f, find Z_a. Represent each of the impedances by a directed line in the complex plane, and solve the problem graphically. Check the answer analytically.

5.28 Draw a phasor diagram for the circuit in Fig. P5.28. If the angular frequency of the source is 10 rad per sec, and if $e_3(t)$ leads $e_2(t)$ by 120°, use the phasor diagram to determine the value of C.

5.29 If $z = |z|\underline{/\theta}$, prove that the equations $|z| = 2a\,|\cos\theta|$ and $|z| = 2a\,|\sin\theta|$ describe the circles of radius a in Figs. P5.29a and P5.29b, respectively.

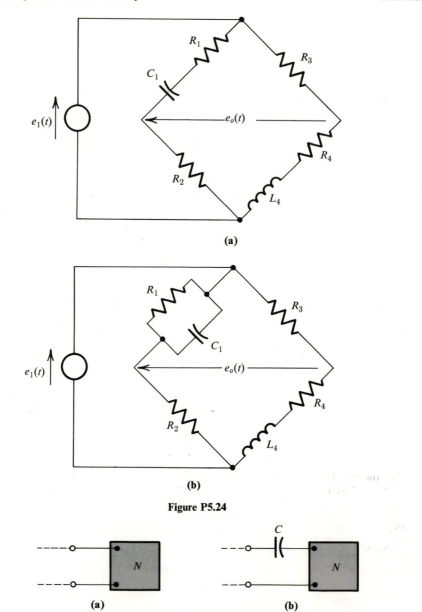

(a)

(b)

Figure P5.24

(a)

(b)

Figure P5.27

5.30 For the circuit in Fig. P5.30, the voltage phasor is $\mathbf{E} = E_m\underline{/0^\circ}$. Show that $|\mathbf{I}| = (E_m/R)\cos\phi = \omega C E_m \sin\phi$, where $\phi = \tan^{-1}(1/\omega CR)$. Use the results of Problem 5.29 to plot the locus of the current phasor under the following conditions:

(a) As R varies from zero to infinity.   (b) As ω varies from zero to infinity.

Compare the second answer with Fig. 5.4-7b.

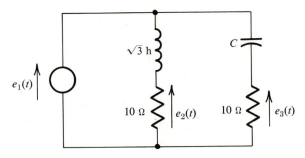

Figure P5.28

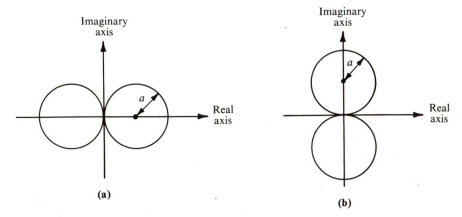

Figure P5.29

5.31 Use the results of Problem 5.29 to plot the locus of the current phasor for the circuit in Fig. P5.31 under the following conditions:

(a) As R varies from zero to infinity.

(b) As L varies from zero to infinity.

Compare the results with Fig. 5.4-6.

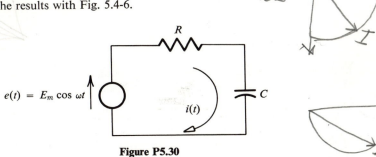

Figure P5.30

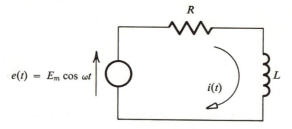

Figure P5.31

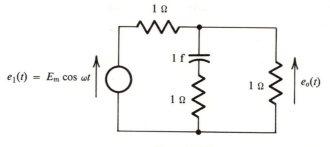

Figure P5.32

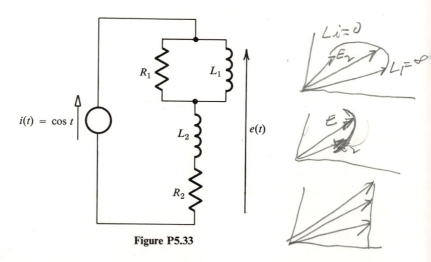

Figure P5.33

5.32 Sketch the locus of the a-c steady-state output phasor $\mathbf{E}_o$ in the circuit of Fig. P5.32, as the angular frequency of the source is varied from zero to infinity.

5.33 Sketch to scale and completely label the locus of the voltage phasor $\mathbf{E}$ for the circuit in Fig. P5.33 when

(a) $L_2 = 2$ h, $R_1 = 1 \ \Omega$, $R_2 = \frac{1}{2} \ \Omega$, L_1 varies from zero to infinity.

(b) $L_1 = 1$ h, $L_2 = 2$ h, $R_2 = \frac{1}{2} \ \Omega$, R_1 varies from zero to infinity.

(c) $L_1 = 1$ h, $R_1 = 1 \ \Omega$, $R_2 = \frac{1}{2} \ \Omega$, L_2 varies from zero to infinity.

5.34 If the voltage phasor is $\mathbf{E} = E_m\underline{/0°}$ for the circuit in Fig. 5.4-1a, sketch the locus of the current phasor as the angular frequency of the source varies from zero to infinity.

5.35 If $\omega = 1$ rad per sec for the circuit in Fig. P5.35, sketch the locus of the current phasor $\mathbf{I}$ as C varies from zero to infinity. For how many values of C are the voltage and current in phase?

5.36 If $C = 1f$ for the circuit in Fig. P5.35, sketch the locus of the current phasor $\mathbf{I}$ as ω varies from zero to infinity. Is the locus circular? Are the voltage and current in phase for any finite, nonzero values of ω?

5.37 Express each of the following functions as a single cosine term with a phase angle.

  (a) $-5 \cos 3t + 3 \sin 3t$.

  (b) $2 \cos (t + 45°) - 2 \sin t$.

  (c) $3 \sin (2t - 120°) - 4 \cos 2t$.

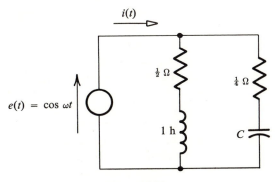

Figure P5.35

$$I_L = \frac{2}{5} - j\,\frac{4}{5}$$

locus of I is semicircle.

6

Complex Frequency

In the last chapter, we developed methods for calculating the a-c steady-state response of a circuit. The steady-state response to a constant input, which was previously discussed in Section 4.5, corresponds to the special case of $\omega = 0$. We now seek techniques for calculating the forced response to other types of inputs, still without explicitly having to find and solve a differential equation. These techniques are developed in Sections 6.1 and 6.2 by generalizing the results of the last chapter, including the concepts of impedance and admittance. The approach used in these sections is quite similar to that in Sections 5.2 and 5.3. In order to develop a coherent and meaningful view of circuit theory, a-c steady-state theory should be retained as an important special case of any new results.

Although in this chapter we do extend the techniques of Chapter 5 to include a few more inputs, that is not our primary purpose. One principal aim is to shed more light upon the important question of how the a-c steady-state response varies when the angular frequency of the input is changed. This question is considered at length in Sections 6.4 and 6.5. A still more important goal is to devise methods of calculating the complete response to any input, without having to explicitly solve a differential equation, or to evaluate a set of arbitrary constants of integration. Some progress toward this goal is made in Section 6.3, but it will not be fully achieved until Chapter 10. The present chapter helps to form a bridge that relates the previous three chapters to the transform techniques that are developed in Chapters 9 and 10.

6.1

Representing Decaying and Growing Oscillations

Recall from Eq. 5.2-4 that a sinusoidal function of constant amplitude may be written

$$f_1(t) = F_m \cos (\omega t + \phi) = \tfrac{1}{2}(F_m \epsilon^{j\phi} \epsilon^{j\omega t} + F_m \epsilon^{-j\phi} \epsilon^{-j\omega t}) \qquad (6.1\text{-}1)$$

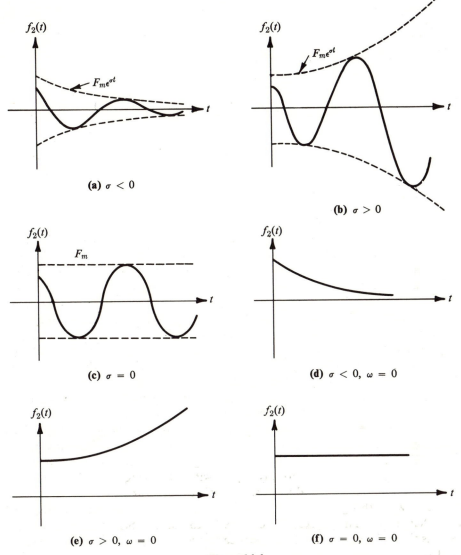

Figure 6.1-1

Consider next the function

$$f_2(t) = F_m\epsilon^{\sigma t}\cos{(\omega t + \phi)} = \tfrac{1}{2}(F_m\epsilon^{j\phi}\epsilon^{(\sigma+j\omega)t} + F_m\epsilon^{-j\phi}\epsilon^{(\sigma-j\omega)t}) \qquad (6.1\text{-}2)$$

which reduces to $f_1(t)$ when the parameter σ (which may take on any real value) is zero. The function $f_2(t)$ is sketched in Fig. 6.1-1 for several different values of σ and ω. It is seen that $f_2(t)$ can represent decaying, growing, or constant-amplitude oscillations, and also exponential and constant functions. Note especially that it retains the a-c and d-c steady-state waveshapes as special cases.

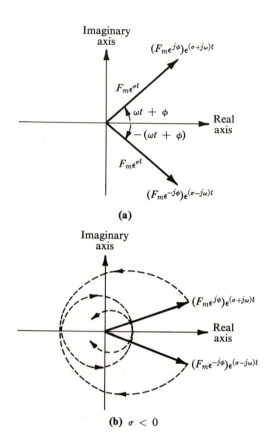

(a)

(b) $\sigma < 0$

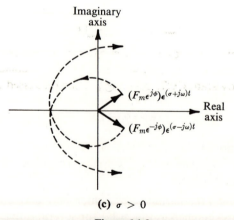

(c) $\sigma > 0$

Figure 6.1-2

The sinusoidal function $f_1(t)$ was seen to be one-half the sum of the two counter-rotating phasors $(F_m\epsilon^{j\phi})\epsilon^{j\omega t}$ and $(F_m\epsilon^{-j\phi})\epsilon^{-j\omega t}$, which move with an angular frequency ω, as illustrated in Fig. 5.2-3b. Equivalently, $f_1(t)$ is the projection of the CCW rotating phasor on the real axis.

The more general function $f_2(t)$ also can be expressed as one-half the sum of the counter-rotating phasors $(F_m\epsilon^{j\phi})\epsilon^{(\sigma+j\omega)t}$ and $(F_m\epsilon^{-j\phi})\epsilon^{(\sigma-j\omega)t}$ in Fig. 6.1-2a, or as the real-axis projection of the CCW one. The rotating phasors again move with an angular velocity ω, but unless $\sigma = 0$ their length also changes with time. If $\sigma < 0$, the phasors rotate and shrink with increasing t, as suggested by the dotted lines that spiral in toward the origin in Fig. 6.1-2b. If $\sigma > 0$, the rotating phasors spiral outward as in Fig. 6.1-2c. (For the special case of $\omega = 0$, they do not rotate, but just shrink or grow in size.)

In the CCW rotating phasor $F_m\epsilon^{j\phi}\epsilon^{(\sigma+j\omega)t}$, ω is the angular velocity (i.e., the rate of change of angle) measured in units of radians per second, while σ is the logarithmic rate of decay or growth in size, measured in units of nepers per second. Both nepers and radians are dimensionless quantities, so the units of both σ and ω are reciprocal seconds. The values of σ and ω determine the nature of the CCW rotating phasor and the nature of the real function $f_2(t)$ in Eq. 6.1-2. Thus we define the *complex frequency* s as

$$s = \sigma + j\omega \qquad (6.1\text{-}3)$$

The name "complex frequency" emphasizes that s is a complex quantity and that the present discussion is a generalization of the a-c steady-state response previously considered.

The complex quantity

$$\mathbf{F} = F_m\epsilon^{j\phi} = F_m\underline{/\phi} \qquad (6.1\text{-}4)$$

is the value of the CCW rotating phasor at $t = 0$ and is again called the phasor, with no modifying adjective. Thus,

$$F_m\epsilon^{j\phi}\epsilon^{(\sigma+j\omega)t} = \mathbf{F}\epsilon^{st} \qquad (6.1\text{-}5)$$

If s is specified separately, there is a one-to-one correspondence between the phasor $\mathbf{F}$ and the function in Eq. 6.1-2. When $\sigma = 0$, i.e., when s is replaced by $j\omega$, Eq. 6.1-5 reduces to the a-c steady-state CCW rotating phasor discussed in Section 5.2.

By Eq. 4.6-1, the forced response to an input $x(t) = \mathbf{X}\epsilon^{st}$ is $y_P(t) = \mathbf{Y}\epsilon^{st}$, where

$$\mathbf{Y} = H(s)\mathbf{X} \qquad (6.1\text{-}6)$$

From Eq. 4.6-3, the network function $H(s)$ can be written down by an inspection of the circuit's differential equation, so the forced response to the CCW rotating phasor $\mathbf{X}\epsilon^{st}$ may be easily found.

If the complex quantities in Eq. 6.1-6 are expressed in polar form,

$$\mathbf{X} = X_m\epsilon^{j\phi_1}, \qquad H(s) = |H(s)|\epsilon^{j\theta}, \qquad \mathbf{Y} = Y_m\epsilon^{j\phi_2} \qquad (6.1\text{-}7)$$

Then the magnitude and angle of the output phasor $\mathbf{Y}$ are given by

$$Y_m = |H(s)|X_m, \qquad \phi_2 = \theta + \phi_1 \qquad (6.1\text{-}8)$$

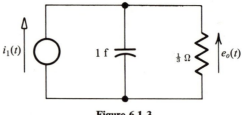

Figure 6.1-3

When using these equations, we must evaluate $H(s)$ for the value of $s = \sigma + j\omega$ that corresponds to the input signal $x(t)$. Notice the similarity of Eqs. 6.1-6 and 6.1-8 to Eqs. 5.2-6 and 5.2-8. Since functions having the form of Eq. 6.1-2 may be represented by the projection of their CCW rotating phasors on the real axis, the forced response to

$$x(t) = X_m \epsilon^{\sigma t} \cos(\omega t + \phi_1) = \text{Re }(\mathbf{X}\epsilon^{st}) \tag{6.1-9}$$

is

$$y_P(t) = Y_m \epsilon^{\sigma t} \cos(\omega t + \phi_2) = \text{Re }(\mathbf{Y}\epsilon^{st}) \tag{6.1-10}$$

where Y_m and ϕ_2 may be found from Eq. 6.1-8. Notice that when $\sigma = 0$, i.e., when $s = \sigma + j\omega$ is replaced by $j\omega$, Eqs. 6.1-9 and 6.1-10 describe the a-c steady-state response. When $\sigma = \omega = 0$, i.e., when s is replaced by zero, $x(t)$ and $y_P(t)$ become constants and describe the d-c steady-state behavior.

Example 6.1-1.   Find the forced response of the circuit in Fig. 6.1-3 to the input $i_1(t) = \epsilon^{-t} \cos t$.

Solution.   The circuit is described by the node equation

$$\frac{de_o}{dt} + 3e_o = i_1$$

The network function, which can be written down by an inspection of the differential equation, is $H(s) = 1/(s + 3)$. The input current has the form of Eq. 6.1-9, with $\sigma = -1$ neper per sec and $\omega = 1$ rad per sec. When s is replaced by $-1 + j$,

$$H(-1 + j) = \frac{1}{2 + j} = \frac{1}{\sqrt{5}} \underline{/-26.5°}$$

The input phasor is $\mathbf{I}_1 = 1\underline{/0°}$, so by Eq. 6.1-6 or 6.1-8 the output phasor is

$$\mathbf{E}_o = \frac{1}{\sqrt{5}} \underline{/-26.5°}$$

Thus the forced response is

$$(e_o)_P = \frac{1}{\sqrt{5}} \epsilon^{-t} \cos(t - 26.5°)$$

which is equivalent to the particular solution in Example 4.1-3.

Examples of determining the network function and the forced response for more complicated circuits are found in Section 6.2.

6.2

Impedance and Admittance

The network function $H(s)$, which appears in the equation

$$\mathbf{Y} = H(s)\mathbf{X} \tag{6.2-1}$$

is not normally determined by first finding the circuit's differential equation. By analogy with Section 5.3, we let the input be the CCW rotating phasor

$$x(t) = \mathbf{X}\epsilon^{st}$$

In the calculation of the forced response, the voltage and current of a typical two-terminal element must have a similar form, namely,

$$e(t) = \mathbf{E}(s)\epsilon^{st}, \qquad i(t) = \mathbf{I}(s)\epsilon^{st}$$

Although the phasors $\mathbf{E}$ and $\mathbf{I}$ are not functions of time, their values do depend upon the value of s, which in turn is determined by the source. In order to emphasize this dependency, we have added an s in parentheses and have written $\mathbf{E}(s)$ and $\mathbf{I}(s)$.

If the typical element is a resistance, the equation $e(t) = Ri(t)$ becomes

$$\mathbf{E}(s)\epsilon^{st} = R\mathbf{I}(s)\epsilon^{st}$$

so that

$$\frac{\mathbf{E}(s)}{\mathbf{I}(s)} = R \tag{6.2-2}$$

For a capacitance, $i = C(de/dt)$ becomes

$$\mathbf{I}(s)\epsilon^{st} = sC\mathbf{E}(s)\epsilon^{st}$$

so

$$\frac{\mathbf{E}(s)}{\mathbf{I}(s)} = \frac{1}{sC} \tag{6.2-3}$$

Finally, for an inductance, $e = L(di/dt)$ and

$$\mathbf{E}(s)\epsilon^{st} = sL\mathbf{I}(s)\epsilon^{st}$$

so

$$\frac{\mathbf{E}(s)}{\mathbf{I}(s)} = sL \tag{6.2-4}$$

The *impedance*

$$Z(s) = \frac{\mathbf{E}(s)}{\mathbf{I}(s)} \tag{6.2-5}$$

is again defined as the ratio of the voltage and current phasors, while the *admittance* is

$$Y(s) = \frac{1}{Z(s)} = \frac{\mathbf{I}(s)}{\mathbf{E}(s)} \tag{6.2-6}$$

TABLE 6.2-1

Element	General expressions		The a-c steady state	The d-c steady state
Resistance	$Z(s) = R$	$Y(s) = \dfrac{1}{R}$	$Z(j\omega) = R$	$Z(0) = R$
Capacitance	$Z(s) = \dfrac{1}{sC}$	$Y(s) = sC$	$Z(j\omega) = \dfrac{1}{j\omega C}$	$Z(0) \to \infty$ (open circuit)
Inductance	$Z(s) = sL$	$Y(s) = \dfrac{1}{sL}$	$Z(j\omega) = j\omega L$	$Z(0) = 0$ (short circuit)

Although the impedance is not a function of time, it does depend in general upon the value of the complex frequency s, as indicated by Eqs. 6.2-2 through 6.2-4.

The impedance and admittance for the three types of passive elements are summarized in Table 6.2-1. When s is replaced by $j\omega$, the expressions for the impedances reduce to those for the a-c steady state in Table 5.3-1, as expected. When s is replaced by zero, the expressions agree with Table 4.5-1 for the d-c steady state.

Equation 6.2-5 is similar to Ohm's law, except that the quantities are in general complex. It can again be proved, by the approach in Section 5.3, that the voltage and current phasors $E(s)$ and $I(s)$ obey Kirchhoff's laws. Thus, if voltages and currents are characterized by their phasors, and passive elements by their impedances, all the procedures for resistive circuits may be used in solving for the unknown phasors. Once the relationship between the input and output phasors is found, the network function $H(s)$ follows from Eq. 6.2-1.

Example 6.2-1. Find the network function $H(s)$ for the circuit in Fig. 5.3-3a, which is repeated in Fig. 6.2-1a.

Solution. The *complex-frequency-domain circuit*, in which voltages and currents are characterized by their phasors and passive elements by their impedances, is given in Fig. 6.2-1b. While any of the five methods of Example 5.3-2 may be used to find the output phasor $E_o(s)$, we choose to write a single node equation at node 2.

$$\frac{E_2(s) - E_1(s)}{s/10} + \frac{E_2(s)}{1 + 10/s} + \frac{E_2(s)}{1 + s/5} = 0$$

Solving algebraically for $E_2(s)$, we obtain

$$E_2(s) = \left[\frac{10(s + 5)(s + 10)}{s^3 + 20s^2 + 200s + 500} \right] E_1(s)$$

Then,

$$E_o(s) = \left(\frac{1/2}{1 + s/5} \right) E_2(s) = \left[\frac{25(s + 10)}{s^3 + 20s^2 + 200s + 500} \right] E_1(s)$$

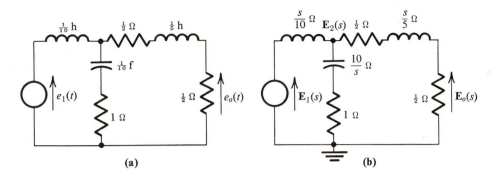

Figure 6.2-1

By Eq. 6.2-1, the network function is

$$H(s) = \frac{\mathbf{E}_o(s)}{\mathbf{E}_1(s)} = \frac{25(s + 10)}{s^3 + 20s^2 + 200s + 500}$$

Example 6.2-2.   Find the forced response of Fig. 6.2-1a to each of the following inputs:

$$e_1(t) = 5\epsilon^{-3t} \cos (t - 10°)$$
$$e_1(t) = 10 \cos (10t + 20°)$$
$$e_1(t) = 10\epsilon^{-t}$$
$$e_1(t) = 10$$

Solution.   All four of the given inputs are special cases of the expression $e_1(t) = E_m \epsilon^{\sigma t} \cos (\omega t + \phi_1)$, as in Eq. 6.1-9. Corresponding to

$$e_1(t) = 5\epsilon^{-3t} \cos (t - 10°)$$

$\sigma = -3$ nepers per sec and $\omega = 1$ rad per sec, so $s = -3 + j1$. For this value of s, the network function $H(s)$ reduces to

$$H(-3 + j1) = \left[\frac{25(s + 10)}{s^3 + 20s^2 + 200s + 500} \right]_{s=-3+j1} = 1.55\underline{/-60.3°}$$

In this case, considerable manipulation with complex numbers is necessary before obtaining the given result, so it is probably not worthwhile for the student to verify it. Since the input phasor $\mathbf{E}_1(s)$ is $5\underline{/-10°}$,

$$\mathbf{E}_o = (1.55\underline{/-60.3°})(5\underline{/-10°}) = 7.75\underline{/-70.3°}$$

The forced response, having the form of Eq. 6.1-10, now can be written down:

$$(e_o)_P = 7.75\epsilon^{-3t} \cos (t - 70.3°)$$

The above steps may be interpreted in the light of Theorem 4.6-1 as follows. The

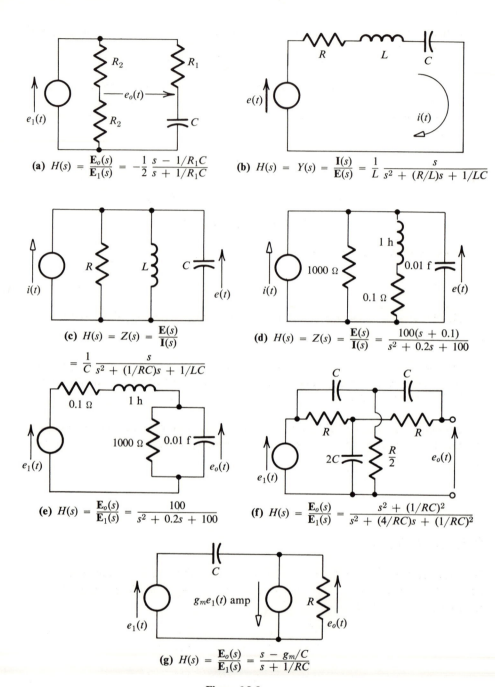

(a) $H(s) = \dfrac{\mathbf{E}_o(s)}{\mathbf{E}_1(s)} = -\dfrac{1}{2}\dfrac{s - 1/R_1C}{s + 1/R_1C}$

(b) $H(s) = Y(s) = \dfrac{\mathbf{I}(s)}{\mathbf{E}(s)} = \dfrac{1}{L}\dfrac{s}{s^2 + (R/L)s + 1/LC}$

(c) $H(s) = Z(s) = \dfrac{\mathbf{E}(s)}{\mathbf{I}(s)}$

$= \dfrac{1}{C}\dfrac{s}{s^2 + (1/RC)s + 1/LC}$

(d) $H(s) = Z(s) = \dfrac{\mathbf{E}(s)}{\mathbf{I}(s)} = \dfrac{100(s + 0.1)}{s^2 + 0.2s + 100}$

(e) $H(s) = \dfrac{\mathbf{E}_o(s)}{\mathbf{E}_1(s)} = \dfrac{100}{s^2 + 0.2s + 100}$

(f) $H(s) = \dfrac{\mathbf{E}_o(s)}{\mathbf{E}_1(s)} = \dfrac{s^2 + (1/RC)^2}{s^2 + (4/RC)s + (1/RC)^2}$

(g) $H(s) = \dfrac{\mathbf{E}_o(s)}{\mathbf{E}_1(s)} = \dfrac{s - g_m/C}{s + 1/RC}$

Figure 6.2-2

forced response to $e_1(t) = \mathbf{E}_1(s)\epsilon^{st}$ is $e_o(t) = \mathbf{E}_o(s)\epsilon^{st}$. Letting $\mathbf{E}_1(s) = 5\underline{/-10°}$ and $s = -3 + j1$, we find

$$(e_o)_P = (7.75\underline{/-70.3°})\epsilon^{(-3+j1)t}$$

Taking the real part of both the input and forced response, we obtain the above answer.

For the input $e_1(t) = 10 \cos(10t + 20°)$, $\sigma = 0$, $\omega = 10$ rad per sec, $s = j10$, and $\mathbf{E}_1(s) = 10\underline{/20°}$. Then,

$$H(j10) = \left[\frac{25(s+10)}{s^3 + 20s^2 + 200s + 500}\right]_{s=j10} = 0.196\underline{/-101.3°}$$

$$\mathbf{E}_o = (0.196\underline{/-101.3°})(10\underline{/20°}) = 1.96\underline{/-81.3°}$$

$$(e_o)_P = 1.96 \cos(10t - 81.3°)$$

which agrees with the answer to Example 5.3-2.

For $e_1(t) = 10\epsilon^{-t}$, $s = -1$ and $\mathbf{E}_1(s) = 10$. Then $H(-1) = 0.705$, $\mathbf{E}_o = 7.05$, and

$$(e_o)_P = 7.05\epsilon^{-t}$$

Finally, for $e_1(t) = 10$, $s = 0$ and $\mathbf{E}_1(s) = 10$. Since $H(0) = 0.5$, $\mathbf{E}_o = 5$ and

$$(e_o)_P = 5$$

The last answer also may be found by an inspection of the circuit. In the d-c steady state, the inductances and capacitance behave like short and open circuits, respectively. The source then sees two $\frac{1}{2}\ \Omega$ resistances in series.

The network functions for several other circuits, some of which are used later in the chapter, are given in Fig. 6.2-2. The student is urged to check these results, preferably using a variety of methods.

6.3

Poles and Zeros

The network function $H(s)$ is the quotient of two polynomials in s, as indicated in Eq. 4.6-3. Suppose that each of the two polynomials $A(s)$ and $B(s)$ are expressed in factored form as follows:

$$H(s) = \frac{B(s)}{A(s)} = \frac{b_m s^m + \cdots + b_1 s + b_0}{a_n s^n + \cdots + a_1 s + a_0}$$

$$= K\frac{(s - z_1)(s - z_2) \cdots (s - z_m)}{(s - p_1)(s - p_2) \cdots (s - p_n)} \tag{6.3-1}$$

where $K = b_m/a_n$. The quantities $z_1, z_2, \ldots, z_m$ are the roots of the equation

$B(s) = 0$, and they are called the *zeros* of the network function. The roots of the equation $A(s) = 0$ are denoted by $p_1, p_2, \ldots, p_n$, and are called the *poles* of the network function.

When the forced response is calculated by noting that the output phasor is $\mathbf{Y}(s) = H(s)\mathbf{X}(s)$, the value of s is determined by the form of the input. Whenever the value of s is equal to one of the zeros, then $H(s) = 0$, and the forced response is zero. Whenever s is equal to one of the poles, the denominator of $H(s)$ vanishes, causing the function to "blow up." In summary, zeros or poles are values of s for which the network function is zero or approaches infinity, respectively.[†]

As seen from Eq. 6.3-1, the network function is completely described by its poles, its zeros, and the multiplying constant K. The poles and zeros together are often called the *critical frequencies* and are most important to the rest of this chapter. Since the poles and zeros are in general complex numbers, they can be represented by points in a complex plane, called the *s-plane* or the *complex-frequency plane*. Since $s = \sigma + j\omega$, the real and imaginary axes can be labelled σ and $j\omega$, respectively. Poles are indicated by crosses, and zeros by circles.

The network function for the circuit in Fig. 6.3-1a is

$$H(s) = \frac{1/sC}{R + 1/sC} = \frac{1}{RCs + 1} = K\frac{1}{s - p_1} \qquad (6.3\text{-}2)$$

where $K = 1/RC$ and $p_1 = -1/RC$. There is a single pole on the negative real axis of the s-plane, as shown in Fig. 6.3-1b. Although $H(s)$ is not zero for any finite value of s, it does approach zero when s approaches infinity. The network function is therefore said to have a zero at infinity.

If the polynomials in the numerator and denominator of the network function are quadratic, they can be easily factored by the quadratic formula. Occasionally, one of them may be a cubic or higher-order polynomial. Although there are standard techniques for finding the factors of such a polynomial, they are not included in this book. On the few occasions when high-order polynomials are involved, we

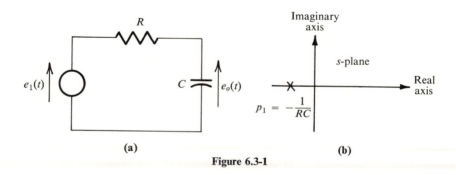

(a)   (b)

Figure 6.3-1

[†] A pole is defined more precisely in Section 10.4 in terms of complex-variable theory.

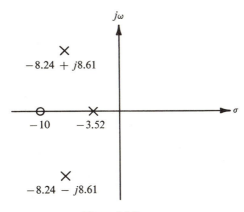

Figure 6.3-2

shall always give their factors. The network function in Examples 6.2-1 and 6.2-2 is

$$H(s) = \frac{25(s + 10)}{s^3 + 20s^2 + 200s + 500}$$

$$= \frac{25(s + 10)}{(s + 3.52)(s + 8.24 - j8.61)(s + 8.24 + j8.61)} \qquad (6.3\text{-}3)$$

It has a zero at $s = -10$, and poles at $s = -3.52$, $-8.24 + j8.61$, and $-8.24 - j8.61$, as indicated in Fig. 6.3-2.

If the numerator or denominator of $H(s)$ has repeated factors, the network function is said to have higher-order zeros or poles, respectively. For example,

$$H(s) = \frac{7s(s + 3)^2}{(s + 1 - j)^2(s + 1 + j)^2} = \frac{7(s - 0)(s + 3)^2}{(s + 1 - j)^2(s + 1 + j)^2} \qquad (6.3\text{-}4)$$

has a first-order (or "simple") zero at $s = 0$, a second-order (or "double") zero at

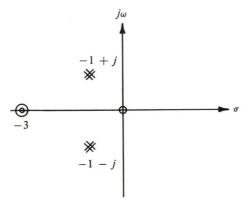

Figure 6.3-3

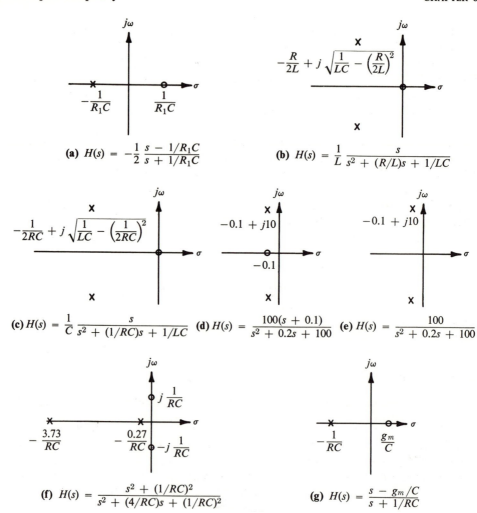

Figure 6.3-4

$s = -3$, and second-order poles at $s = -1 + j$ and $s = -1 - j$. Figure 6.3-3 shows one method of indicating the higher-order poles and zeros in the s-plane.

In Eq. 6.3-3, $H(s) \doteq 25/s^2$ as $|s|$ becomes large and is said to have a second-order zero at infinity. In Eq. 6.3-4, $H(s) \doteq 7/s$ as $|s|$ becomes large and has a first-order zero at infinity. The network function

$$H(s) = \frac{2s^2 + 3s + 1}{s + 3}$$

approaches $2s$ as $|s|$ becomes large and has a first-order pole at infinity. If the poles and zeros at infinity are included and if each pole and zero is counted according to its order, the number of poles always equals the number of zeros. Of course, when the pole-zero pattern is plotted in the s-plane, only finite-plane poles and zeros can be shown explicitly.

The pole-zero patterns for the network functions of Fig. 6.2-2 are shown in Fig. 6.3-4.[†] Note that any complex poles or zeros always occur in complex-conjugate pairs. A standard mathematical theorem states that, if $A(s)$ is a polynomial with real coefficients, any complex roots of $A(s) = 0$ must occur in complex-conjugate pairs.

A circuit's differential equation may be reconstructed once its network function is known, as in Eqs. 4.6-2 and 4.6-3, so $H(s)$ must contain all the information of the differential equation. Thus, we should be able to find the complete response when the network function, the input, and the initial conditions are given. Example 6.2-2 demonstrated how the forced response (particular solution) can be obtained from $H(s)$ for a variety of inputs. To see how the free response (complementary solution) can be obtained, recall that the network function of Eq. 6.3-1 corresponds to the differential equation

$$a_n \frac{d^n y}{dt^n} + \cdots + a_1 \frac{dy}{dt} + a_0 y = b_m \frac{d^m x}{dt^m} + \cdots + b_1 \frac{dx}{dt} + b_0 x \qquad (6.3\text{-}5)$$

Classically, the complementary solution is found by examining the characteristic equation

$$a_n r^n + \cdots + a_1 r + a_0 = 0 \qquad (6.3\text{-}6)$$

If the roots, denoted by $r_1, r_2, \ldots, r_n$, are distinct,

$$y_H(t) = K_1 \epsilon^{r_1 t} + K_2 \epsilon^{r_2 t} + \cdots + K_n \epsilon^{r_n t} \qquad (6.3\text{-}7)$$

The characteristic polynomial in Eq. 6.3-6 is identical with the denominator of $H(s)$ in Eq. 6.3-1, except that the symbol r is replaced by s. Thus, the poles of the network function are the roots of the characteristic equation, and the form of the free response can be written down once the poles are found. For the network functions of Eqs. 6.3-2 and 6.3-3, respectively, the form of the free response is

$$K_1 \epsilon^{-t/RC}$$

and

$$K_2 \epsilon^{-3.52t} + K_3 \epsilon^{(-8.24+j8.61)t} + K_4 \epsilon^{(-8.24-j8.61)t}$$

$$= K_2 \epsilon^{-3.52t} + K_5 \epsilon^{-8.24t} \cos(8.61t + \phi_5)$$

Higher-order poles correspond to repeated roots of the characteristic equation, which are discussed in Chapter 4. For the network function in Eq. 6.3-4, the free response has the form

$$K_1 \epsilon^{-t} \cos(t + \phi_1) + K_3 t \epsilon^{-t} \cos(t + \phi_3)$$

[†] The poles in Figs. 6.3-4b and 6.3-4c may be either real or complex, depending upon the values of the circuit elements. For the poles to be complex, $1/LC > (R/2L)^2$ and $1/LC > (1/2RC)^2$, respectively.

Since the poles of $H(s)$ are the roots of the characteristic equation, the remarks of Section 4.4 apply to the poles. The student should review Section 4.4, especially the discussion associated with Figs. 4.4-1, 4.4-2, and 4.2-5, replacing the phrase "roots of the characteristic equation" by "poles of the network function." A circuit is unstable if the network function has poles in the open right half of the s-plane or higher-order poles on the imaginary axis. Since circuits containing only resistances, capacitances, and inductances are always stable, the pole pattern for such circuits is confined to the left half plane, with only first-order poles allowed on the imaginary axis.

In summary, we have seen how the forced response and the form of the free response can be determined from the network function. The arbitrary constants in the free response depend upon the circuit, the source, and any initial stored energy. Although an easier method is developed in Chapter 10, at present these constants must be evaluated as in Chapter 4. Examples in which the complete response is required may be found in the problems at the end of the chapter.

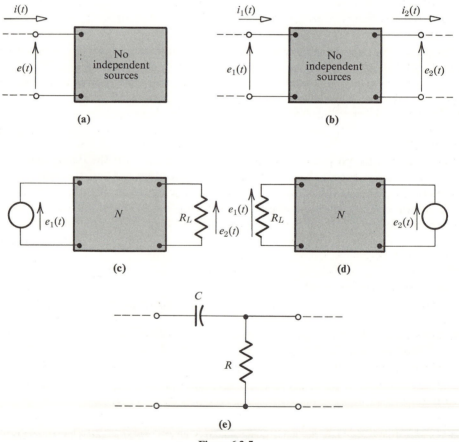

Figure 6.3-5

Driving-Point and Transfer Functions. A circuit having two external terminals and containing no independent sources is shown in Fig. 6.3-5a. If the input is a current source $i(t)$, the response must be the voltage $e(t)$, and the network function is $H(s) = \mathbf{E}(s)/\mathbf{I}(s) = Z(s)$, the input impedance. For a voltage source $e(t)$ and a response $i(t)$, $H(s) = \mathbf{I}(s)/\mathbf{E}(s) = Y(s)$, the input admittance. In either case, the network function relates the voltage and current phasors at the same pair of terminals and is called a *driving-point* or input function. The units of the driving-point function may be either ohms or mhos.

The forced response and the form of the free response in Fig. 6.3-5a can be found from the appropriate driving-point function. For a current source, the nature of the free response (which is the entire response when the source is replaced by an open circuit) is determined by the poles of $Z(s)$. For a voltage source, the free response (which is the entire response when the source is replaced by a short circuit) depends upon the poles of $Y(s)$. Since $Y(s) = 1/Z(s)$, these statements can be summarized as follows. The form of the free response under open- and short-circuit conditions is determined by the poles and zeros of $Z(s)$, respectively.

A circuit with no controlled sources is stable under all circumstances. For such a circuit, therefore, both the poles and zeros of the input impedance must be confined to the left half of the s-plane, with only first-order poles and zeros permitted on the imaginary axis. Example 4.4-1 showed how the pole locations of $Z(s)$ for a series RLC circuit varied, as the value of one of the elements was changed. For positive values of R, L, and C, the poles remained in the left half plane.

Figure 6.3-5b shows a circuit having two pairs of external terminals and containing no independent sources. In many cases, a voltage or current input is applied to the left-hand pair of terminals, and a combination of passive elements, called the *load*, is connected to the right-hand pair. The response is taken to be either the voltage or the current at the output terminals. Depending upon the choice of input and response, the network function could be any one of the following four quantities: $\mathbf{E}_2(s)/\mathbf{I}_1(s)$, $\mathbf{E}_2(s)/\mathbf{E}_1(s)$, $\mathbf{I}_2(s)/\mathbf{I}_1(s)$, and $\mathbf{I}_2(s)/\mathbf{E}_1(s)$. The first quantity has units of ohms, the next two are dimensionless, and the last has units of mhos. In each case, the network function relates the voltage and current phasors at different pairs of terminals, and is called a *transfer* function.

In contrast to Fig. 6.3-5a, the transfer functions of Fig. 6.3-5b do depend upon the external connections. Changing the load connected to the output terminals will change all four of the transfer functions discussed in the previous paragraph. Thus, $H(s)$ does not completely describe the external behavior of Fig. 6.3-5b for arbitrary terminations. (The discussion of the complete characterization of such a circuit is deferred until Chapters 7 and 12.)

The poles of the transfer function again determine the form of the free response. For a stable circuit, including all circuits composed only of passive elements, $H(s)$ cannot have right half plane poles or repeated poles on the imaginary axis. A similar statement is not valid for the zeros of $H(s)$, because $1/H(s)$ is *not*, in general, another transfer function for the same circuit. The transfer functions for Figs. 6.3-5c and 6.3-5d are $H_c(s) = \mathbf{E}_2(s)/\mathbf{E}_1(s)$ and $H_d(s) = \mathbf{E}_1(s)/\mathbf{E}_2(s)$, but $H_d(s) \neq 1/H_c(s)$

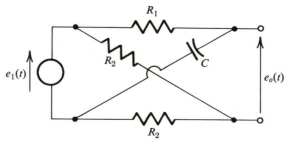

Figure 6.3-6

because the terminations are different. The fact that $H_d(s)$ is not the reciprocal of $H_e(s)$ can be easily seen if the network N is, for example, the simple circuit in Fig. 6.3-5e. The fact that the transfer function of a passive circuit can have right half plane zeros is demonstrated by Fig. 6.2-2a, which is redrawn in Fig. 6.3-6. The transfer function

$$H(s) = \frac{\mathbf{E}_o(s)}{\mathbf{E}_1(s)} = -\frac{1}{2}\left(\frac{s - 1/R_1 C}{s + 1/R_1 C}\right)$$

has a zero on the positive real axis of the s-plane.

It is sometimes, but not always, possible to see how the zeros of the network function are caused by a certain group of elements in the circuit. This knowledge is particularly helpful when we try to devise a circuit that has a specified network function. In Fig. 6.2-2d, for example, the capacitance acts like a short circuit at very high frequencies, so the output phasor must vanish when s becomes infinite. The impedance of the RL branch is $s + 0.1$, which is zero when $s = -0.1$. Thus $H(s)$ must have zeros at $s = -0.1$ and infinity, which agrees with the expression given in the figure.

In Fig. 6.2-2e, the capacitance behaves like a short circuit, and the inductance like an open circuit when s approaches infinity. Either of these two facts would be sufficient to make $\mathbf{E}_o(s)$ zero, so $H(s)$ has a double zero at infinity. Notice that the network functions in Figs. 6.2-2d and 6.2-2e have the same poles, which should be expected, since the poles determine the form of the free response. When the independent sources are made dead (when the current and voltage sources are replaced by open and short circuits, respectively), the two circuits are identical.

6.4

Vectors in the s-Plane

A network function may be evaluated for a particular value of complex frequency s (corresponding to a particular input) either analytically or graphically. In the analytical approach, which was used in Example 6.2-2, every s is replaced by its given value, and the result is simplified using the rules of complex algebra. The graphical method is based upon the pole-zero pattern of $H(s)$.

Consider a typical network function in factored form.

$$H(s) = \frac{K(s - z_1)(s - z_2) \cdots}{(s - p_1)(s - p_2) \cdots} \tag{6.4-1}$$

K is a real number, but the poles, zeros, and complex frequency generally are complex quantities. Like any complex quantity, they may be represented by points in a complex plane as in Fig. 6.4-1a, or by directed lines as in Fig. 6.4-1b. The value of s corresponding to the input is indicated by a solid dot. The directed lines could be called either phasors or vectors, so to distinguish them from the phasors representing voltages and currents we shall use the term *vector*. From the vector subtraction in Fig. 6.4-1c, a vector drawn from z_1 to the point s represents the factor $s - z_1$ in the numerator of $H(s)$. The magnitude and angle of this factor are equal to the length and angle (measured with respect to the positive real axis) of the corresponding vector in the *s*-plane. In a similar way, the remaining factors in $H(s)$ can be represented by vectors drawn from the zeros and poles to the point s, as in Fig. 6.4-1e.

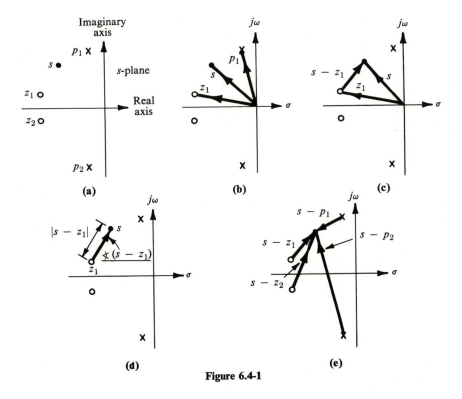

Figure 6.4-1

The network function of Eq. 6.4-1 involves the product and quotient of complex quantities. From the rules of complex algebra,

$$|H(s)| = \frac{|K|\,|s - z_1|\,|s - z_2| \cdots}{|s - p_1|\,|s - p_2| \cdots}$$

$$\angle H(s) = \angle K + [\angle(s - z_1) + \angle(s - z_2) + \cdots] \tag{6.4-2}$$
$$- [\angle(s - p_1) + \angle(s - p_2) + \cdots]$$

K is a real constant, so $\angle K$ is zero if K is positive, and is $\pm 180°$ if K is negative. The other factors in Eqs. 6.4-2 can be evaluated by examining the vectors drawn from the poles and zeros.

Example 6.4-1.   The transfer function $H(s)$ for Example 6.2-2 is

$$H(s) = \frac{25(s + 10)}{(s + 3.52)(s + 8.24 - j8.61)(s + 8.24 + j8.61)}$$

as in Eq. 6.3-3. Evaluate $H(s)$ graphically for $s = -3 + j1$ and for $s = j10$.

Solution.   The pole-zero pattern for this transfer function is shown in Fig. 6.4-2a. In Fig. 6.4-2b, vectors are drawn from the poles and zeros to the point $s = -3 + j1$. The magnitude and angle of each vector, which can be found with a ruler and protractor, are indicated in the sketch. Thus,

$$|H(-3 + j1)| = \frac{(25)(7.07)}{(1.13)(9.24)(10.92)} = 1.55$$

$$\angle H(-3 + j1) = 8.1° - (62.5° - 55.5° + 61.4°) = -60.3°$$

$$H(-3 + j1) = 1.55\underline{/-60.3°}$$

which agrees with Example 6.2-2.

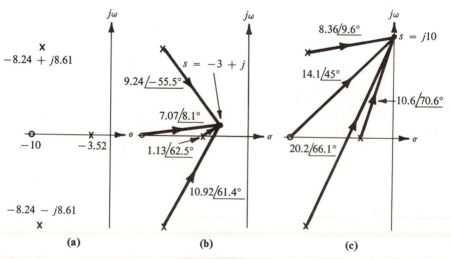

Figure 6.4-2

The vectors in the s-plane for $s = j10$ are given in Fig. 6.4-2c. Thus,

$$|H(j10)| = \frac{(25)(14.1)}{(10.6)(8.36)(20.2)} = 0.196$$

$$\measuredangle\, H(j10) = 45° - (70.6° + 9.6° + 66.1°) = -101.3°$$

$$H(j10) = 0.196\underline{/-101.3°}$$

which again agrees with Example 6.2-2.

There are certain inherent advantages and limitations to any graphical approach. Graphical methods are most useful when an approximate answer, obtained with a minimum of labor, is desired. Furthermore, they often give some insight (which cannot be easily obtained from an analytical method, devoid of sketches) into the role and relative importance of the various factors. The above method clearly indicates, for example, that if the point s is near a zero, the magnitude of $H(s)$ is relatively small; if s is close to a pole, $|H(s)|$ is relatively large. Thus, in Example 6.4-1, the fact that $|H(-3 + j1)|$ is considerably larger than $|H(j10)|$ can be anticipated by a quick look at the pole-zero pattern. Graphical methods are particularly helpful in discovering how a circuit's response varies as some parameter, such as the value of s, is varied. Their importance cannot be fully appreciated from Example 6.4-1, but it will become evident from many of the examples in the rest of this chapter.

Frequency Response Curves.   The steady-state response to sinusoidal inputs can be found by replacing s by $j\omega$ in the network function. For the input and output signals

$$x(t) = X_m \cos(\omega t + \phi_1) \quad \text{and} \quad y_P(t) = Y_m \cos(\omega t + \phi_2) \qquad (6.4\text{-}3)$$

we know from Eq. 6.1-8 that

$$Y_m = |H(j\omega)|X_m \quad \text{and} \quad \phi_2 = \measuredangle\, H(j\omega) + \phi_1 \qquad (6.4\text{-}4)$$

In practice, signals often consist of a number of sinusoidal components having different frequencies. In some cases (as in telephones and phonographs), the output and input waveshapes should have the same shape. In those cases, the relative amplitudes and phase angles of the different sinusoidal components must be the same in the two signals, which (as discussed in Chapter 9) requires that $|H(j\omega)|$ be constant and $\measuredangle\, H(j\omega)$ be proportional to ω over the frequency range of interest. In other cases (such as selecting one radio signal from those available in the atmosphere), the circuit should discriminate between signals at different frequencies. For these reasons, it is important to examine the variation of the a-c steady-state network function as a function of the angular frequency of the source. This information is normally summarized by two frequency response curves: $|H(j\omega)|$ versus ω, and $\measuredangle\, H(j\omega)$ versus ω.

The frequency response curves, of course, can be constructed by forming $H(j\omega)$, calculating the magnitude and angle for a sufficient number of different values of ω, and drawing curves through the plotted points. This procedure is not only quite laborious, but it does not give an intuitive feeling for the factors responsible for the shapes of the curves. Recall that the magnitude and angle of an a-c steady-state transfer function were found graphically in Example 6.4-1 for one particular angular frequency ($\omega = 10$ rad per sec). By an extension of this approach, it is often possible to sketch the frequency response curves to scale, using little or no numerical calculation.

Example 6.4-2. Sketch the frequency response curves for the circuit in Fig. 6.3-1a, which is repeated in Fig. 6.4-3a.

Solution. From Eq. 6.3-2, the transfer function is

$$H(s) = \frac{1}{RC}\left(\frac{1}{s - p_1}\right)$$

where $p_1 = -1/RC$. Since s is replaced by $j\omega$ in forming the a-c steady-state transfer function, the point s lies on the imaginary axis of the complex plane, a distance ω above the origin. As the angular frequency ω increases from zero toward infinity, the point s moves up the imaginary axis, starting from the origin. As in Fig. 6.4-1e, the factor $s - p_1$ is represented by the vector shown in Fig. 6.4-3b, which is equal to $(1/RC)\underline{/0°}$ at $\omega = 0$, and $(\sqrt{2}/RC)\underline{/45°}$ at $\omega = 1/RC$. As ω approaches infinity, the length of the vector increases without limit and its angle approaches 90°. Thus, the frequency response curves in Figs. 6.4-3c and 6.4-3d follow directly from an inspection of the pole-zero pattern.

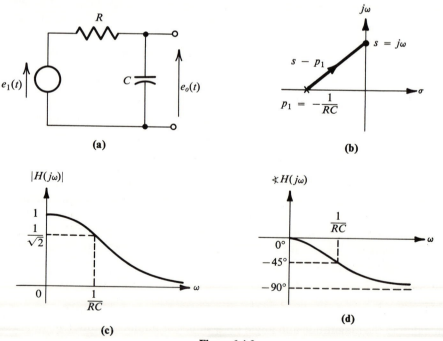

(a)

(b)

(c)

(d)

Figure 6.4-3

The circuit in Fig. 6.4-3a is called a *low-pass* filter, since the magnitude of the output diminishes as the frequency of the source increases. The *bandwidth* is defined as the frequency range over which $|H(j\omega)|$ is at least $1/\sqrt{2}$ times its maximum value. The filter in Fig. 6.4-3 has a bandwidth extending from d-c up to $1/RC$ radians per second. For reasons discussed in Chapter 8, the limit of the bandwidth (in this case, $\omega = 1/RC$) is called the *half-power point*.

Example 6.4-3.   Sketch the frequency response curves for the circuit in Fig. 6.2-2a, which is repeated in Fig. 6.4-4a.

Solution.   The transfer function is

$$H(s) = -\frac{1}{2}\left(\frac{s - z_1}{s - p_1}\right)$$

where $z_1 = 1/R_1C$ and $p_1 = -1/R_1C$, and its pole-zero pattern is shown in Fig. 6.4-4b. The vectors representing the factors $s - z_1$ and $s - p_1$ have the same length for all values of ω, so $|H(j\omega)| = \frac{1}{2}$ at all frequencies. Because the magnitude of the a-c steady-state transfer function does not depend upon ω, the circuit is called an *all-pass* network.

To determine the angle curve, first notice that the factor $-\frac{1}{2}$ has a fixed angle of $-180°$. For $s = j\omega = 0$, $\measuredangle(s - z_1) = 180°$ and $\measuredangle(s - p_1) = 0°$, so $\measuredangle H(j\omega) = 0°$. As ω increases, $\measuredangle(s - z_1)$ becomes less positive and $\measuredangle(s - p_1)$ becomes more

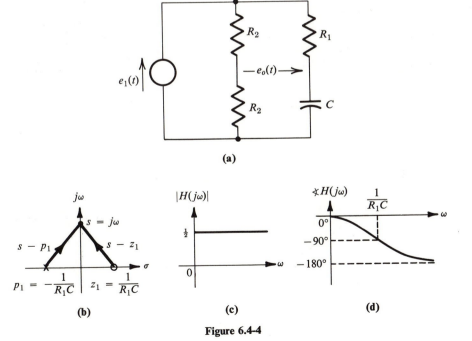

Figure 6.4-4

positive, producing the curve in Fig. 6.4-4d. When ω becomes very large, $\not\prec (s - z_1)$ $\doteq \not\prec (s - p_1) \doteq 90°$, and $\not\prec H(j\omega) = -180°$. The information contained in the frequency response curves is equivalent to that given by the locus diagram in Fig. 5.4-5b, but the frequency response curves usually are preferred.

Example 6.4-4. The transfer function for the circuit in Fig. 6.4-5a is readily found to be

$$H(s) = \frac{1}{LC} \left(\frac{1}{s^2 + (R/L)s + 1/LC} \right) = \frac{\omega_0^2}{s^2 + 2\alpha s + \omega_0^2}$$

where $\alpha = R/2L$ and $\omega_0^2 = 1/LC$. The denominator is similar to the characteristic polynomial in Eq. 4.4-1. The location of the poles (and hence the nature of the frequency response curves) depends upon the element values, as was illustrated in Fig. 4.4-2. Sketch the frequency response curves when $\alpha = \omega_0$ and when $\alpha \ll \omega_0$.

Solution. When $\alpha = \omega_0$,

$$H(s) = \frac{\alpha^2}{(s + \alpha)^2}$$

The transfer function has a double pole at $s = -\alpha$, and its denominator is represented by the *two* identical vectors shown in Fig. 6.4-5b. The magnitude curve $|H(j\omega)|$, which is shown in Fig. 6.4-5c, falls off more rapidly with ω than for a single pole at $s = -\alpha$. The phase angle $\not\prec H(j\omega)$, which is sketched in Fig. 6.4-5d, is twice what it would be for a single pole (which was given in Fig. 6.4-3d).

When $\alpha < \omega_0$, the poles are located at $s = -\alpha \pm j\omega_d$, where

$$\omega_d = \sqrt{\omega_0^2 - \alpha^2}$$

as discussed in Example 4.4-1. When $\alpha \ll \omega_0$, $\omega_d \doteq \omega_0$ and the poles are relatively close to the imaginary axis, as shown in part (e) of the figure. For $s = 0$, the vectors $s - p_1$ and $s - p_2$ have the same length but angles of opposite sign. As the point $s = j\omega$ moves up the imaginary axis, $\not\prec (s - p_1)$ and $\not\prec (s - p_2)$ both increase in the positive direction, causing $\not\prec H(j\omega)$ to become negative. The value of $|H(j\omega)|$ at first changes very slowly, since $|s - p_1|$ decreases and $|s - p_2|$ increases. As ω increases further, however, the percentage change in $|s - p_1|$ becomes considerably greater than the percentage change in $|s - p_2|$, and the curve of $|H(j\omega)|$ rises more rapidly. The magnitude of the transfer function is a maximum when the point s is opposite the pole p_1, because the vector $s - p_1$ then has its shortest length. For $s = j\omega_0$, $s - p_1 = \alpha\underline{/0°}$ and $s - p_2 \doteq 2\omega_0\underline{/90°}$, so $H(j\omega_0) \doteq (\omega_0/2\alpha)\underline{/-90°}$.

As long as the point $s = j\omega$ is in the vicinity of p_1, the percent variation in $s - p_1$ is very large compared to that in $s - p_2$. When studying the frequency response curves in this range, therefore, the vector $s - p_2$ can be regarded as essentially constant. An inspection of Fig. 6.4-5e reveals that at $s = j(\omega_0 \pm \alpha)$, $s - p_1 = \sqrt{2}\alpha\underline{/\pm 45°}$, and $|H(j\omega)|$ is $1/\sqrt{2}$ times its maximum value. This locates the half-power points at $\omega = \omega_0 \pm \alpha$ for the curves in parts (f) and (g). Finally, at very large frequencies, the magnitude and angle of both $s - p_1$ and $s - p_2$ approach infinity and $90°$, respectively, so the magnitude and angle of the transfer function approach zero and $-180°$.

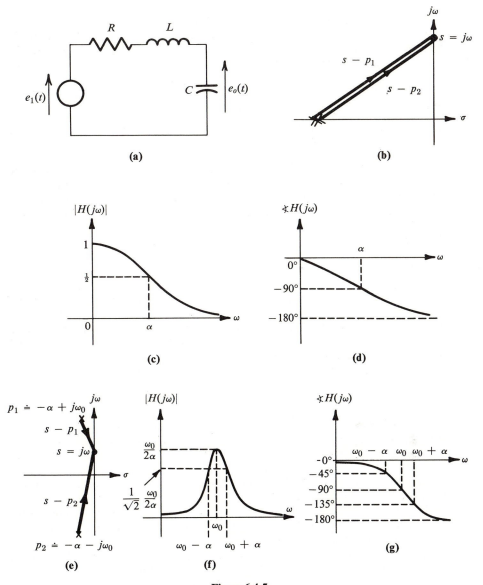

Figure 6.4-5

The student may have some initial difficulty following all the arguments in Example 6.4-4, particularly since the solution was carried out in literal rather than numerical terms. With a little practice, however, the frequency response can be quickly sketched to scale, without any written arguments or computational work. Normally, even the vectors in the *s*-plane are not explicitly shown, but merely visualized.

A circuit whose behavior is described by the curve in Fig. 6.4-5f is called a *band-pass filter*. The frequency at which the response is the greatest is known as the *resonant frequency*. In the last example, the resonant angular frequency is approximately ω_0 radians per second. Also in the last example, the bandwidth is seen to be 2α radians per second, extending from $\omega_0 - \alpha$ to $\omega_0 + \alpha$. In many applications, a sharp peak is desired in the $|H(j\omega)|$ curve, so that only a very narrow band of frequencies will be passed by the circuit. A measure of the sharpness of the resonance phenomenon is the *selectivity* Q, a dimensionless quantity defined as

$$Q = \frac{\text{resonant frequency}}{\text{bandwidth}} \tag{6.4-5}$$

In this equation, the units for the resonant frequency and bandwidth are both cycles per second, or both radians per second. In the last example, $Q \doteq \omega_0/2\alpha = \omega_0 L/R$.[†]

Students are often troubled by the question of what approximations are reasonable in a given situation. Hence, it is instructive to examine more closely the degree of approximation in the construction of Figs. 6.4-5f and 6.4-5g. The actual resonant frequency is slightly less than ω_0 for two reasons, each of about equal importance. First, the pole $p_1 = -\alpha + j\omega_d$ is slightly lower than the one shown in Fig. 6.4-5e, since $\omega_d < \omega_0$. Second, consider the approximation that the length of $s - p_2$ is essentially constant in the vicinity of the resonant frequency. Since $|s - p_2|$ does increase slightly as the point $s = j\omega$ moves up past the pole p_1, the maximum response occurs slightly *below* p_1. The effect is magnified by the fact that at the resonant frequency, the tip of the vector $s - p_1$ is moving perpendicularly to the vector itself.

Example 6.4-5.   Find an expression for the *exact* resonant frequency when $\alpha \ll \omega_0$ in Example 6.4-4.

Solution.   The magnitude squared of the a-c steady-state transfer function

$$H(j\omega) = \frac{\omega_0^2}{(-\omega^2 + \omega_0^2) + j2\alpha\omega}$$

is

$$|H(j\omega)|^2 = \frac{\omega_0^4}{(-\omega^2 + \omega_0^2)^2 + (2\alpha\omega)^2} = \frac{\omega_0^4}{\omega^4 - (2\omega_0^2 - 4\alpha^2)\omega^2 + \omega_0^4}$$

$|H(j\omega)|$ is a maximum when the denominator of the last expression is minimized. Setting the derivative of the denominator with respect to ω equal to zero, we obtain

$$\omega[4\omega^2 - (4\omega_0^2 - 8\alpha^2)] = 0$$

[†] The "Q of a coil" is sometimes defined as $\omega_0 L/R$, where R is the resistance of the wire, since in the special case of a circuit containing only the coil and a capacitance, this is also the Q of the circuit.

so the exact resonant angular frequency is

$$\omega = \sqrt{\omega_0^2 - 2\alpha^2}$$

which, as expected, is slightly less than ω_0.

If, for example, $\omega_0 = 100$ rad per sec and $\alpha = 1$ neper per sec, the approximate and exact resonant frequencies become 100 and 99.99 rad per sec, respectively. It is noteworthy that our graphical approach, basically designed to give a rough idea of the frequency response characteristics, yields numerical results that are so close to the exact answers.

Example 6.4-6. The curves in parts (c) and (f) of Fig. 6.4-5 represent two special choices for the circuit elements: $\alpha = \omega_0$, and $\alpha \ll \omega_0$. If the value of α lies between these two extreme cases, the $|H(j\omega)|$ curve might then have the shape in Fig. 6.4-6a. It is reasonable to expect that there is some value of α for which this curve just fails to have a resonant peak. Find this value of α.

Solution. From the previous example, the angular frequency at which the peak occurs is $\omega = \sqrt{\omega_0^2 - 2\alpha^2}$, which decreases as α increases. For $2\alpha^2 = \omega_0^2$, i.e., for $\alpha = \omega_d = \omega_0/\sqrt{2}$, the largest response occurs at $\omega = 0$. The pole-zero plot and the magnitude curve for this special case are shown in Figs. 6.4-6b and 6.4-6c. Notice that

$$|H(j\omega)|^2 = \frac{\omega_0^4}{\omega^4 + \omega_0^4}$$

so that the half-power point is at $\omega = \omega_0$. The frequency response is seen to be quite flat over a fairly broad frequency range. Such a circuit is called a *Butterworth* (or "maximally flat") *filter.*

Example 6.4-7. Sketch the frequency response curves for the circuits in Figs 6.4-7a and 6.4-7b.

Solution. The network function for Fig. 6.4-7a is the input admittance

$$Y_a(s) = \frac{I(s)}{E(s)} = \frac{1}{sL + R + 1/sC} = \frac{1}{L}\left(\frac{s}{s^2 + (R/L)s + 1/LC}\right)$$

$$= K\frac{s}{s^2 + 2\alpha s + \omega_0^2} = K\frac{s - 0}{(s - p_1)(s - p_2)}$$

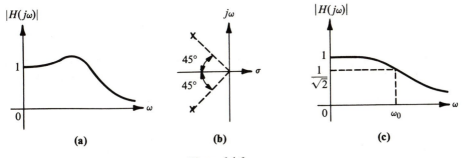

Figure 6.4-6

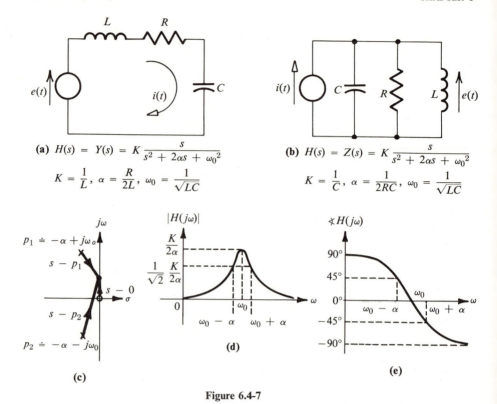

(a) $H(s) = Y(s) = K \dfrac{s}{s^2 + 2\alpha s + \omega_0{}^2}$

$K = \dfrac{1}{L}, \quad \alpha = \dfrac{R}{2L}, \quad \omega_0 = \dfrac{1}{\sqrt{LC}}$

(b) $H(s) = Z(s) = K \dfrac{s}{s^2 + 2\alpha s + \omega_0{}^2}$

$K = \dfrac{1}{C}, \quad \alpha = \dfrac{1}{2RC}, \quad \omega_0 = \dfrac{1}{\sqrt{LC}}$

Figure 6.4-7

where $K = 1/L$, $\alpha = R/2L$, and $\omega_0{}^2 = 1/LC$. The network function for Fig. 6.4-7b is the impedance

$$Z_b(s) = \frac{E(s)}{I(s)} = \frac{1}{sC + 1/R + 1/sL} = K \frac{s}{s^2 + 2\alpha s + \omega_0{}^2}$$

where $K = 1/C$, $\alpha = 1/2RC$, and $\omega_0{}^2 = 1/LC$. Since the form of the network function $H(s)$ is the same for the two circuits,[†] they may be conveniently considered together.

We shall consider only the most interesting case, where $\alpha \ll \omega_0$. As in Example 6.4-4, the poles are located at $-\alpha \pm j\omega_d$, where $\omega_d = \sqrt{\omega_0{}^2 - \alpha^2}$ is approximately equal to ω_0. The pole-zero pattern is shown in Fig. 6.4-7c. The variation of the vectors from the poles to the point $s = j\omega$ is similar to that in Fig. 6.4-5e. The additional vector from the zero at the origin causes $H(j\omega)$ to vanish when ω approaches zero. At the resonant frequency ω_0, the network function, found from the length and angle of the vectors in the s-plane, is

$$\frac{K(\omega_0 / \underline{90°})}{\alpha(2\omega_0 / \underline{90°})} = \frac{K}{2\alpha}$$

Near the resonant frequency, the vectors $s - 0$ and $s - p_2$ can be regarded as essentially constant. At $s = j(\omega_0 \pm \alpha)$, the length of the vector $s - p_1$ is $\sqrt{2}$ times its

---

[†] This result should be expected, since the circuits are duals.

value at $s = j\omega_0$. The bandwidth therefore extends from $\omega_0 - \alpha$ to $\omega_0 + \alpha$, and is 2α radians per second. The selectivity is $Q = \omega_0/2\alpha$. Since the angle of the vector $s - 0$ is 90° for all values of ω, the angle of $H(j\omega)$ varies from 90° to $-90°$ (instead of from 0° to $-180°$ as in Example 6.4-4). The complete frequency response curves are sketched in Figs. 6.4-7d and 6.4-7e.

In Example 6.4-7, two approximations, similar to those in the previous example, were used in determining the resonant frequency. First, since $\omega_d < \omega_0$, the pole $p_1 = -\alpha + j\omega_d$ is slightly *lower* than indicated in the figure. Second, as the point $s = j\omega$ moves up past p_1, the lengths of $s - 0$ and $s - p_2$ are not constant but increase. Since the percentage increase in $|s - 0|$ is twice that of $|s - p_2|$, the maximum response occurs slightly *above* p_1. The two approximations are in opposing directions, and happen to exactly cancel. Therefore, in this example ω_0 is the exact resonant frequency.

For the series *RLC* circuit, $|H(j\omega)| = |Y(j\omega)|$ has a maximum value of $1/R$ at $\omega_0 = 1/\sqrt{LC}$. If R is very small, the admittance and hence the current are extremely large at this frequency. The bandwidth is $2\alpha = R/L$ radians per second, and the selectivity is $\omega_0 L/R$. The frequency response curves are consistent with the phasor diagrams in Fig. 5.4-1. At the angular frequency $1/\sqrt{LC}$ radians per second, $\mathbf{E}_L + \mathbf{E}_C = 0$ and $\mathbf{E} = \mathbf{E}_R = R\mathbf{I}$. For frequencies below and above the resonant frequency, the circuit looks capacitive (a positive angle for the admittance and a negative angle for the impedance) and inductive, respectively.

It is interesting to compare the frequency response curves for the circuits in Figs. 6.4-7a and 6.4-5a, since the only difference is the quantity chosen as the response. The maximum a-c steady-state current occurs when ω is exactly $1/\sqrt{LC}$ radians per second. Since $|Z_C(j\omega)| = 1/\omega C$ decreases with increasing ω, the maximum voltage across the capacitance occurs at a slightly lower angular frequency, given in Example 6.4-5. In the d-c steady-state, when $\omega = 0$, the capacitance is an open circuit, and no current flows. The voltage across the capacitance then equals the source voltage.

For the parallel *RLC* circuit, $|Z(j\omega)|_{\max} = R$ at $\omega_0 = 1/\sqrt{LC}$, and the bandwidth and selectivity are $2\alpha = 1/RC$ radians per second and $Q = \omega_0 RC$. The frequency response curves agree with the conclusions that may be drawn from the phasor diagram in Fig. 5.4-2.

Example 6.4-8.   The circuit in Fig. 6.4-8a is a very common example of parallel resonance. Discuss the frequency response curves when the value of R is small.

Solution.   The network function is the impedance

$$Z(s) = \frac{(sL + R)(1/sC)}{sL + R + 1/sC} = \frac{sL + R}{s^2 LC + sRC + 1} = \frac{1}{C}\left(\frac{s + R/L}{s^2 + 2\alpha s + \omega_0^2}\right)$$

where $\alpha = R/2L$ and $\omega_0^2 = 1/LC$. The pole-zero pattern, which is given in Fig. 6.4-8b, differs from Fig. 6.4-7c only in the location of the zero. If the resistance R is

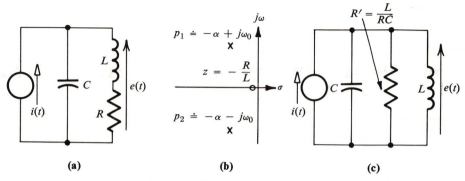

Figure 6.4-8

sufficiently small, the vector drawn from the zero to the point $s = j\omega$ will be approximately the same in the two cases, except at very low frequencies. From another point of view,

$$Y(j\omega) = \frac{-\omega^2 LC + j\omega RC + 1}{j\omega L + R}$$

is approximately

$$\frac{-\omega^2 LC + j\omega RC + 1}{j\omega L} = j\omega C + \frac{RC}{L} + \frac{1}{j\omega L}$$

in the frequency range for which $\omega L \gg R$. The last expression for the admittance suggests the circuit in Fig. 6.4-8c, which is similar to Fig. 6.4-7b. Thus, the magnitude and phase angle curves of $Z(j\omega)$ are similar to the curves in Figs. 6.4-7d and 6.4-7e for $\omega L \gg R$. The behavior of the two circuits in Fig. 6.4-8 is not, of course, the same at low frequencies. In the d-c steady-state, when $\omega = 0$, the impedances for parts (a) and (c) become R and zero, respectively.

Example 6.4-9. Discuss the forced response of the circuits in Figs. 6.4-7a and 6.4-7b when the resistances are omitted.

Solution. Although the frequency response of a parallel LC combination can be found from Figs. 6.4-7d and 6.4-7e by letting R approach infinity and α approach zero, consider the impedance for the circuit in Fig. 6.4-9a:

$$Z(s) = \frac{1}{C}\left(\frac{s}{s^2 + 1/LC}\right)$$

The pole-zero pattern is confined to the imaginary axis, as shown in Fig. 6.4-9b, and leads to the curves in parts (c) and (d). When the point $s = j\omega$ coincides with the pole p_1, the impedance becomes infinite, and the circuit acts like an open circuit. Since $\angle Z(j\omega)$ is always $\pm 90°$, the a-c steady-state impedance is purely imaginary. Since $Z(j\omega) = R + jX = jX$, we can replace the two curves in parts (c) and (d) by the single curve in part (e) of the figure. Notice that the circuit looks inductive for $\omega < 1/\sqrt{LC}$ and capacitive for $\omega > 1/\sqrt{LC}$.

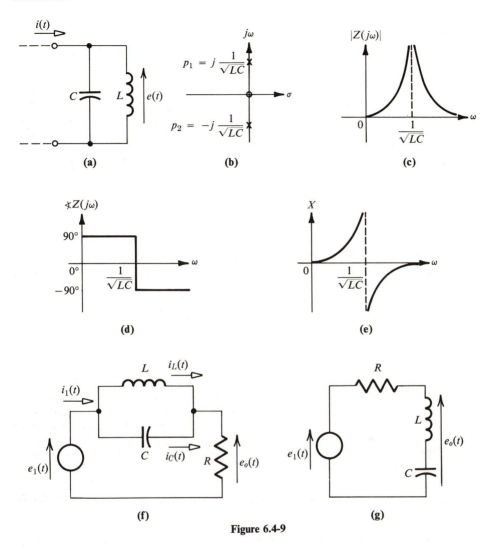

Figure 6.4-9

The parallel *LC* combination in Fig. 6.4-9f may be used to eliminate from the output a particular frequency in the input. At $\omega = 1/\sqrt{LC}$, the *LC* combination acts like an open circuit, and $\mathbf{E}_0 = 0$. The total current $\mathbf{I}_1 = \mathbf{I}_L + \mathbf{I}_C$ is zero, even though the individual values of $\mathbf{I}_L$ and $\mathbf{I}_C$ are nonzero. Because there is some resistance associated with an inductor, however, the complete rejection of a specified frequency by this simple circuit is not possible in practice.

The pole-zero pattern and frequency response curves for the admittance of a series *LC* combination are similar to those in Figs. 6.4-9b through 6.4-9d. At the angular frequency $\omega = 1/\sqrt{LC}$, such a combination acts like a short circuit. Thus, Fig. 6.4-9g also may be used to prevent a specified frequency from appearing in the output.

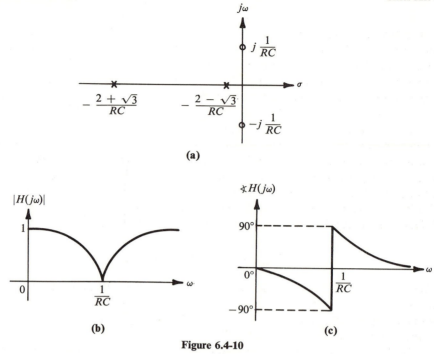

(a)

(b)   **(c)**

Figure 6.4-10

Example 6.4-10. The transfer function for the twin-T circuit in Fig. 6.2-2f is

$$H(s) = \frac{s^2 + (1/RC)^2}{s^2 + (4/RC)s + (1/RC)^2}$$

Sketch the frequency response curves.

Solution. The pole-zero pattern for $H(s)$, which was given in Fig. 6.3-4f, is repeated in Fig. 6.4-10a. The frequency response curves, sketched directly from the pole-zero plot, are shown in parts (b) and (c) of the figure. Notice that the steady-state response is zero at $\omega = 1/RC$, because of the zero on the imaginary axis. The twin-T circuit is often used as a sharp rejection filter, especially in low-frequency applications. A major asset is that no inductances are needed, in contrast to Figs. 6.4-9f and 6.4-9g. It is more difficult to physically approximate a pure inductance than it is to approximate a pure resistance or capacitance.

6.5

Bode Diagrams

The magnitude of an a-c steady-state transfer function is often expressed in *decibels*, abbreviated db, as follows.

$$|H(j\omega)|_{\text{db}} = 20 \log_{10} |H(j\omega)| \qquad (6.5\text{-}1)$$

Notice that the decibel gain $|H(j\omega)|_{db}$ is positive when $|H(j\omega)| > 1$, and negative when $|H(j\omega)| < 1$. When the magnitude of the transfer function is 1, 10, 100, and 1000, respectively, the corresponding decibel gain is 0, 20, 40, and 60. The student should realize that the definition of Eq. 6.5-1 is meaningful only when $H(j\omega)$ is a dimensionless quantity, i.e., when the output and input signals are both voltages, or both currents.

There is some confusion surrounding the above definition, because historically the decibel was first defined as a unit of power gain. The number of bels was $\log_{10}$ (output power/input power), and the number of decibels was 10 $\log_{10}$ (output power/input power). Power relationships in the a-c steady state are discussed in Chapter 8, where the power is shown to be proportional to the square of the magnitude of the voltage or current phasor, if the resistance seen by the source equals the load resistance. Under these conditions, the number of decibels is

$$10 \log_{10} \left|\frac{Y(j\omega)}{X(j\omega)}\right|^2 = 20 \log_{10} |H(j\omega)|$$

where $Y(j\omega)$ and $X(j\omega)$ are the output and input phasors. Frequently, however, Eq. 6.5-1 is used even when the power gain does *not* equal the square of the voltage or current gain. In the present section, Eq. 6.5-1 is taken as an arbitrary definition.

Bode diagrams are an alternative method of portraying the variation of the transfer function with the angular frequency of the input. They are equivalent to the frequency response curves of Section 6.4 and consist of two curves: $|H(j\omega)|_{db}$ versus ω, and $\angle H(j\omega)$ versus ω. In both cases, the angular frequency is plotted on a logarithmic scale.

Plotting ω on a logarithmic scale permits us to include a wide frequency range. Bode diagrams are most often used when the frequency range of interest is large, and when the poles and zeros of $H(s)$ are on or near the real axis of the s-plane. Then the decibel curve can be quickly approximated by a series of straight lines, using the technique developed in the following examples.

Another use of Bode diagrams is illustrated by the cascade connection of two circuits in Fig. 6.5-1. Suppose that the presence of the second circuit does not change the transfer function of the first, or that $H_1(s) = E_2(s)/E_1(s)$ is calculated with any loading effect of the second circuit included. (Such an assumption was discussed briefly in Section 3.5 and will be examined again in Chapter 12.) Then,

$$H(s) = \frac{E_o(s)}{E_1(s)} = \frac{E_2(s)}{E_1(s)} \frac{E_o(s)}{E_2(s)} = H_1(s)H_2(s)$$

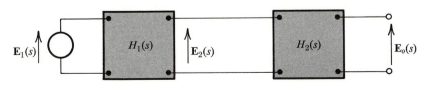

Figure 6.5-1

so

$$|H(j\omega)|_{db} = 20 \log_{10} |H_1(j\omega)H_2(j\omega)| = |H_1(j\omega)|_{db} + |H_2(j\omega)|_{db}$$
$$\angle H(j\omega) = \angle H_1(j\omega) + \angle H_2(j\omega)$$
(6.5-2)

Therefore the individual decibel and angle curves may be added to obtain the curves for the cascaded combination. This is one of the reasons why Bode diagrams are frequently used in the analysis of control systems and electronic circuits.

Example 6.5-1. Construct the decibel curve for the following transfer functions, if K is a real, positive constant. These results will be needed in all of the subsequent examples in this chapter.

(a) $H(j\omega) = K$

(b) $H(j\omega) = j\omega K$

(c) $H(j\omega) = \dfrac{K}{j\omega}$

(d) $H(j\omega) = \dfrac{K}{(j\omega)^n}$

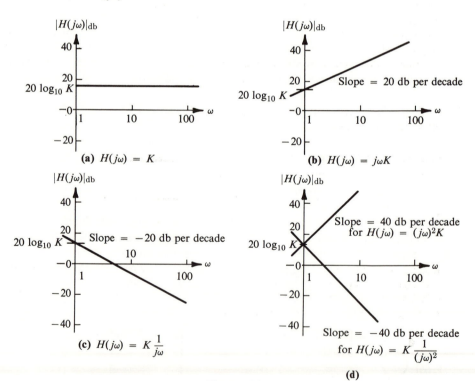

Figure 6.5-2

Solution.   For $H(j\omega) = K$, $|H(j\omega)|_{db} = 20 \log_{10} K$, which gives the horizontal line in Fig. 6.5-2a. The angle of $H(j\omega)$ is identically zero.

For the second transfer function, $|H(j\omega)|_{db} = 20 \log_{10} (\omega K) = 20 \log_{10} K + 20 \log_{10} \omega$, which describes the straight line in Fig. 6.5-2b. The slope of the line is usually expressed in units of decibels per decade of frequency or decibels per octave. A *decade* corresponds to increasing the angular frequency by a factor of 10, and an *octave* to doubling the angular frequency. Thus, the frequency range $\omega = 1$ rad per sec to $\omega = 10$ rad per sec or $\omega = 2.86$ rad per sec to $\omega = 28.6$ rad per sec is one decade, while $\omega = 2.86$ rad per sec to $\omega = 5.72$ rad per sec is one octave. For the example under consideration, when the angular frequency is increased from ω_a to $10\omega_a$ the resulting decibel change is

$$20 \log_{10} (10\omega_a K) - 20 \log_{10} (\omega_a K) = 20 \log_{10} 10 = 20$$

If we double the angular frequency, we obtain a decibel change of

$$20 \log_{10} (2\omega_a K) - 20 \log_{10} (\omega_a K) = 20 \log_{10} 2 = 6$$

The slope of the line in Fig. 6.5-2b is, therefore, 20 db per decade or 6 db per octave. The curve of $\angle H(j\omega)$ is $90°$ for all values of ω.

For $H(j\omega) = K/j\omega$, $|H(j\omega)|_{db} = 20 \log_{10} (K/\omega) = 20 \log_{10} K - 20 \log_{10} \omega$, which yields a straight line with a slope of -20 db per decade, as in Fig. 6.5-2c. The angle is a constant $-90°$.

Finally, for $H(j\omega) = K/(j\omega)^n$, $|H(j\omega)|_{db} = 20 \log_{10} (K/\omega^n) = 20 \log_{10} K - 20n \log_{10} \omega$, which describes a line with a slope of $-20n$ db per decade. The angle is a constant $-90n$ degrees. Similarly, the decibel curve for $H(j\omega) = K(j\omega)^n$ has a slope of $+20n$ db per decade, while the angle is $+90n$ degrees.

Example 6.5-2.   Draw the Bode diagrams for the circuit of Fig. 6.4-3a, which is repeated in Fig. 6.5-3a.

Solution.   The transfer function is

$$H(s) = \frac{1}{RCs + 1} = \frac{1}{RC} \left(\frac{1}{s + 1/RC} \right)$$

so

$$H(j\omega) = \frac{1}{j\omega RC + 1}$$

Instead of calculating $|H(j\omega)|_{db}$ for a number of different values of ω, we usually draw a straight-line approximation first. For $\omega \ll 1/RC$, the $j\omega RC$ term is negligible compared to the 1 in the denominator of $H(j\omega)$, while for $\omega \gg 1/RC$, the 1 can be neglected. As very rough approximations,

$$H(j\omega) \doteq 1 \quad \text{and} \quad |H(j\omega)|_{db} \doteq 20 \log_{10} 1 = 0 \qquad \text{for } \omega < \frac{1}{RC}$$

$$H(j\omega) \doteq \frac{1}{j\omega RC} \quad \text{and} \quad |H(j\omega)|_{db} \doteq -20 \log_{10} \omega RC \qquad \text{for } \omega > \frac{1}{RC}$$

The two approximations describe, respectively, a horizontal line and a line with a

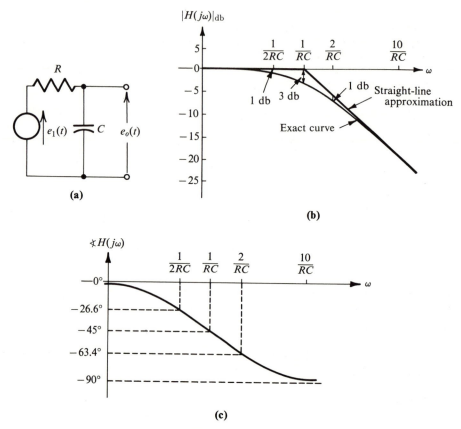

Figure 6.5-3

slope of -20 db per decade. Both lines are shown in Fig. 6.5-3b. Note that both approximations give the same result at $\omega = 1/RC$, which is called the *break frequency*.

The two straight lines really constitute the low- and high-frequency asymptotes for the exact curve. The straight-line approximation is least accurate in the vicinity of the break point.

$$\left| H\left(j\frac{1}{RC} \right) \right|_{\text{db}} = 20 \log_{10} \frac{1}{\sqrt{2}} = -3$$

so the exact curve deviates from the approximation by 3 db at the break point. In a similar way, the student may show that one octave above and below the break frequency, the deviation is about 1 db. Then the $|H(j\omega)|_{\text{db}}$ curve can be drawn with considerable accuracy, as is done in the figure.

Since reducing $|H(j\omega)|_{\text{db}}$ by 3 db is equivalent to multiplying $|H(j\omega)|$ by a factor of $1/\sqrt{2}$, the break point in the straight-line approximation marks the limit of the band-width. In this example, the bandwidth extends from d-c to $1/RC$ radians per second, a conclusion that was also reached in Example 6.4-2 by examining the vectors in the *s*-plane.

In the construction of the angle curve, note first that for $\omega \ll 1/RC$ and $\omega \gg 1/RC$, respectively, $\angle H(j\omega) = 0°$ (the low-frequency asymptote) and $-90°$ (the high-frequency asymptote), respectively. At the break point, $\angle H(j\omega) = -45°$, while one octave away from the break frequency the deviation of the angle from the asymptote is $26.6°$. The complete curve is sketched in Fig. 6.5-3c.

If RC is replaced by T (which has the units of seconds and which is intended to suggest a time constant), the curves in Figs. 6.5-3b and 6.5-3c may be used to describe any factor of the form $1/(j\omega T + 1)$. If $H(j\omega) = j\omega T + 1$, these same curves would apply except for a sign change in the ordinates. The form of a typical transfer function is

$$H(s) = \frac{K(s - z_1)(s - z_2) \cdots}{(s - p_1)(s - p_2) \cdots} = \frac{K'(1 + sT_1)(1 + sT_3) \cdots}{(1 + sT_2)(1 + sT_4) \cdots} \qquad (6.5\text{-}3)$$

in which case

$$H(j\omega) = \frac{K'(1 + j\omega T_1)(1 + j\omega T_3) \cdots}{(1 + j\omega T_2)(1 + j\omega T_4) \cdots}$$

If all the T's are real numbers, each individual factor could be handled as in the last example. In the next few examples, we consider the Bode diagrams for transfer functions that contain several factors.

Example 6.5-3. Sketch the Bode diagrams for the circuit in Fig. 6.4-5a, which is repeated in Fig. 6.5-4a, when $R = 21 \ \Omega$, $L = 1$ h, and $C = \frac{1}{20}$ f.

Solution. The transfer function is

$$H(s) = \frac{1}{LC}\left[\frac{1}{s^2 + (R/L)s + 1/LC}\right] = \frac{20}{s^2 + 21s + 20}$$

$$= \frac{20}{(s + 1)(s + 20)} = \frac{1}{(s + 1)(s/20 + 1)}$$

$$H(j\omega) = \frac{1}{(j\omega + 1)(j\omega/20 + 1)}$$

One possible approach is based upon the relationships

$$\left|H(j\omega)\right|_{\text{db}} = 20 \log_{10}\left|\frac{1}{(j\omega + 1)(j\omega/20 + 1)}\right|$$

$$= 20 \log_{10}\left|\frac{1}{j\omega + 1}\right| + 20 \log_{10}\left|\frac{1}{j\omega/20 + 1}\right|$$

$$\angle H(j\omega) = \angle\left(\frac{1}{j\omega + 1}\right) + \angle\left(\frac{1}{j\omega/20 + 1}\right)$$

Individual curves for $1/(j\omega + 1)$ and $1/(j\omega/20 + 1)$ can be drawn as in the previous example. These curves then may be added to yield the desired answer.

Most students, however, prefer a slightly different approach. For each factor of the

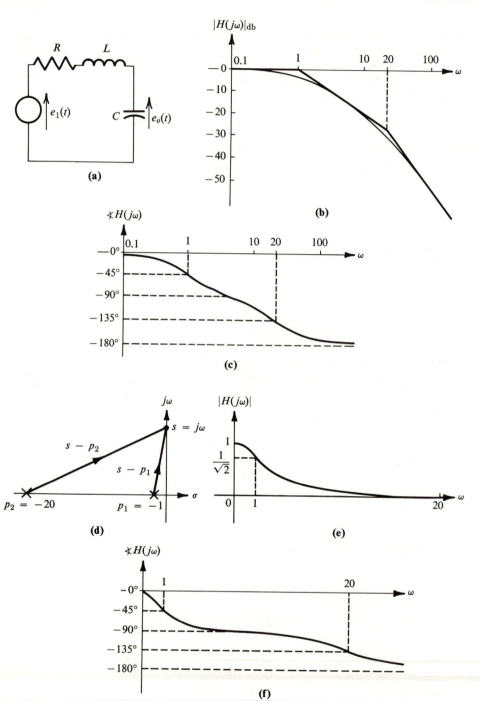

Figure 6.5-4

form $j\omega T + 1$, either the real or imaginary term, whichever is the smaller, is again neglected when the decibel curve is approximated. The angular frequency $\omega = 1/T$ locates the break point, where one approximation is replaced by another. It is usually advisable to first list all the break points, which in this example are at $\omega = 1$ and 20 rad per sec. Next, determine the low-frequency asymptote

$$|H(j\omega)|_{\mathrm{db}} \doteq 20 \log_{10} \frac{1}{(1)(1)} = 0$$

which is used for frequencies below the lowest break frequency. For $1 < \omega < 20$, the approximation used is

$$|H(j\omega)|_{\mathrm{db}} \doteq 20 \log_{10} \left| \frac{1}{(j\omega)(1)} \right|$$

which by Example 6.5-1 describes a lines with a slope of -20 db per decade. Finally, for $\omega > 20$, the approximation is

$$|H(j\omega)|_{\mathrm{db}} \doteq 20 \log_{10} \left| \frac{1}{(j\omega)(j\omega/20)} \right|$$

yielding a line with a slope of -40 db per decade. In general, at each break point in a *denominator* factor, the db approximation breaks *downward*, with the slope suddenly becoming 20 db per decade more negative.

If the break points are separated sufficiently, i.e., if they are a decade or more apart, only one of the factors in the transfer function differs appreciably from its approximation in the vicinity of any given break point. The exact curve then deviates from the straight-line approximation by 3 db at the break points and by 1 db an octave away. Both the approximate and exact curves are shown in Fig. 6.5-4b. The bandwidth extends from d-c to 1 rad per sec.

For the construction of the angle curve, note that the angle of each $j\omega T + 1$ factor varies from $0°$ to $90°$, is $45°$ at the break point, and is $26.6°$ or $63.4°$ one octave away. In general, each *denominator* factor corresponds to a *decrease* of $90°$ in the angle, with one-half of the decrease occurring before and one-half after the break point. The complete angle curve is sketched in part (c) of the figure.

It is interesting to relate the Bode diagrams in Figs. 6.5-4b and 6.5-4c to the frequency response curves obtained from the vectors in the s-plane. The pole-zero pattern for the transfer function

$$H(s) = \frac{20}{(s + 1)(s + 20)} = \frac{20}{(s - p_1)(s - p_2)}$$

is shown in Fig. 6.5-4d. For small values of ω, $s - p_2 \doteq 20\underline{/0}$, but the vector $s - p_1$ changes relatively rapidly. At $\omega = 1$ rad per sec, $s - p_1 = \sqrt{2}\underline{/45°}$, so $H(j1) = (1/\sqrt{2})\underline{/-45°}$. At $\omega = 20$ rad per sec, $s - p_1 \doteq 20\underline{/90°}$ and $s - p_2 = 20\sqrt{2}\underline{/45°}$, so $H(j20) = (1/20\sqrt{2})\underline{/-135°}$. For large values of ω, the length of both vectors increases without limit, causing $|H(j\omega)|$ to be inversely proportional to ω^2. The frequency response curves are shown in parts (e) and (f). The values $1/\sqrt{2}$ and $1/20\sqrt{2}$ correspond to -3 db and -29 db, respectively.

Example 6.5-4. Repeat Example 6.5-3 with $R = 2\,\Omega$, $L = 1$ h, and $C = 1$ f.

Solution. The transfer function becomes

$$H(s) = \frac{1}{s^2 + 2s + 1} \quad \text{and} \quad H(j\omega) = \frac{1}{(j\omega + 1)^2}$$

Because of the double break point at $\omega = 1$ rad per sec, the slope of the straight-line approximation changes from 0 to -40 db per decade, as shown in Fig. 6.5-5a. The exact curve now deviates from the approximation by 6 db at the break point and by 2 db one octave away. Therefore, the double break point does *not* mark the limit of the bandwidth, which now extends from d-c to 0.64 rad per sec. The angle curve is given in Fig. 6.5-5b.

Example 6.5-5. This example demonstrates the effect on the Bode diagrams of a pole at the origin of the s-plane, of a zero on the negative or positive real axis, and of changing the multiplying constant in $H(s)$. Sketch the Bode diagrams for

$$H_1(s) = \frac{10(s + 2)}{s(s + 40)}$$

$$H_2(s) = \frac{10(s - 2)}{s(s + 40)}$$

$$H_3(s) = \frac{100(s + 2)}{s(s + 40)}$$

$$H_4(s) = \frac{100s}{(10s + 1)^2}$$

Solution. In the first case, the a-c steady-state transfer function is

$$H_1(j\omega) = \frac{10(j\omega + 2)}{j\omega(j\omega + 40)} = \frac{1}{2}\left[\frac{1 + j\omega/2}{j\omega(1 + j\omega/40)}\right]$$

In constructing the straight-line approximation to the decibel curve, we neglect the

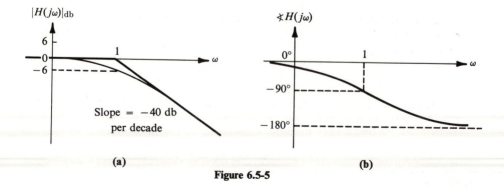

(a)

(b)

Figure 6.5-5

smaller of the real and imaginary terms in each linear factor. The break points, where the slope of the approximation changes, are at $\omega = 2$ and 40 rad per sec.

$$\left| H_1(j\omega) \right|_{db} \doteq 20 \log_{10} \frac{1}{2\omega} \qquad \text{for } \omega < 2 \text{ rad per sec}$$

$$\doteq 20 \log_{10} \frac{1}{4} = -12 \qquad \text{for } 2 < \omega < 40 \text{ rad per sec}$$

$$\doteq 20 \log_{10} \frac{10}{\omega} \qquad \text{for } \omega > 40 \text{ rad per sec}$$

which yields the straight-line curve in Fig. 6.5-6a. The low-frequency asymptote can be drawn by realizing that its slope is -20 db per decade and by locating one point on this straight line. At $\omega = 0.1$ rad per sec, for example, its value is $20 \log_{10} 5 = 14$ db. Another convenient method is to note that the asymptote (extended if necessary) crosses the 0 db line when $1/2\omega = 1$; i.e., when $\omega = \frac{1}{2}$ rad per sec.

In practice, only one of the three approximations given above is written down. If the low-frequency asymptote is drawn first, the slope is *increased* by 20 db per decade at each numerator break point and *decreased* by 20 db per decade at each denominator break point. (When the straight-line approximation contains a horizontal segment, as in this example, the student may prefer to draw that portion of the curve first.) The exact decibel curve is included in Fig. 6.5-6a.

The phase angle curve appears in Fig. 6.5-6b. Before sketching the curve, it is advisable to estimate quickly the angle at the break points and at very low and very high frequencies. Notice also that when the straight-line decibel curve breaks up or down, respectively, the slope of the angle curve is positive or negative, respectively.[†] This is always true, as long as $H(s)$ does not contain any poles or zeros in the right half of the s-plane.

Since the magnitudes of the factors $j\omega - 2$ and $j\omega + 2$ are equal, $|H_2(j\omega)|_{db} = |H_1(j\omega)|_{db}$ and is again described by the curve in Fig. 6.5-6a. While the angle of $j\omega + 2$ varies from $0°$ to $90°$, however, the angle of $j\omega - 2$ varies from $180°$ to $90°$ as ω is increased. The complete sketch of $\measuredangle H_2(j\omega)$ is given in Fig. 6.5-6c.

Since $H_3(s) = 10H_1(s)$, these two transfer functions will have the same phase angle curve. Furthermore,

$$|H_3(j\omega)|_{db} = 20 \log_{10} |10H_1(j\omega)| = 20 + 20 \log_{10} |H_1(j\omega)|$$

giving the decibel curve in Fig. 6.5-6d. Multiplying a transfer function by the positive constant K simply raises the decibel curve by $20 \log_{10} K$. Notice in the figure that the low-frequency asymptote

$$\left| H_3(j\omega) \right|_{db} \doteq 20 \log_{10} \frac{5}{\omega}$$

still has a slope of -20 db per decade, but that its extension now crosses 0 db at $\omega = 5$ rad per sec.

[†] A more accurate plot of the angle curve can be obtained directly from the straight-line decibel curve with a phase ruler or template, which is often called a *Bode plotter* and which is commercially available.

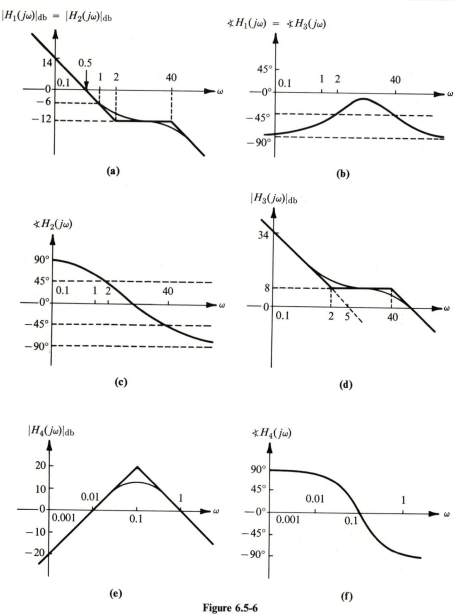

Figure 6.5-6

For $\omega < \frac{1}{10}$ rad per sec in the fourth and last transfer function,

$$|H_4(j\omega)|_{db} \doteq 20 \log_{10}(100\omega)$$

which describes a straight line passing through 0 db at $\omega = 0.01$ rad per sec and having a slope of 20 db per decade. As ω approaches zero, $H_4(j\omega)$ vanishes, and the decibel gain approaches minus infinity. Since ω is plotted on a logarithmic scale, the point $\omega = 0$ cannot be shown. There is a double break point at $\omega = \frac{1}{10}$ rad per sec, giving the decibel and angle curves in parts (e) and (f) of Fig. 6.5-6.

Example 6.5-6.   The transfer function of a linear circuit may be reconstructed from the straight-line decibel curve by means of the rules developed in the previous examples. Find the transfer function corresponding to the experimentally determined curve in Fig. 6.5-7a.

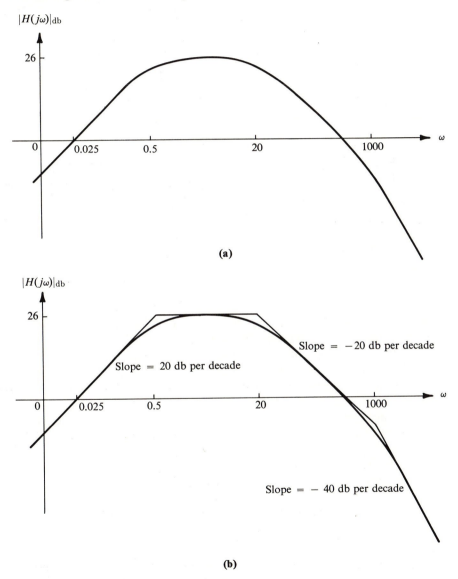

(a)

(b)

Figure 6.5-7

Solution. The first step is to fit a straight-line approximation to the given curve, which usually requires some trial and error. The slope of the approximation always must be an integral multiple of 20 db per decade, and the deviation from the exact curve should be about 3 db at each break point (unless there is another break point nearby). It is usually best to draw the low- and high-frequency asymptotes first, before attempting to determine the rest of the approximation.

The straight-line approximation for this example is shown in Fig. 6.5-7b. It has an initial slope of 20 db per decade, and denominator break points at $\omega = 0.5, 20$, and 1000 rad per sec. Therefore,

$$H(j\omega) = \frac{K(j\omega)}{(1 + j2\omega)(1 + j\omega/20)(1 + j\omega/1000)}$$

There are several ways to determine the gain constant K. For $\omega \ll 0.5$ rad per sec,

$$|H(j\omega)|_{db} \doteq 20 \log_{10} K\omega$$

which must be zero when $\omega = 0.025$ rad per sec. Thus, $K = 1/0.025 = 40$. Finally,

$$H(s) = \frac{40s}{(1 + 2s)(1 + s/20)(1 + s/1000)}$$

If $H(s)$ can be reconstructed from the magnitude of $H(j\omega)$, then $\measuredangle H(j\omega)$ can be determined. Example 6.5-6 therefore implies that the variation of the phase angle is fixed, once $|H(j\omega)|$ is specified. This is indeed the case, if all the poles and zeros are known to lie in the left half of the s-plane. In Example 6.5-5, $|H_1(j\omega)| = |H_2(j\omega)|$ and $\measuredangle H_1(j\omega) \neq \measuredangle H_2(j\omega)$, but $H_2(s)$ had a zero on the positive real axis. In Example 6.5-6, the transfer function

$$H(s) = \frac{40s}{(1 - 2s)(1 + s/20)(1 + s/1000)}$$

would also have the decibel curve given in Fig. 6.5-7, but it contains a pole on the positive real axis. Such a transfer function represents an unstable circuit, so if the circuit is known to be stable, the transfer function must be the one given in the example.[†] If the magnitude *and* angle of a transfer function were independently specified at *all* frequencies in the a-c steady state, a circuit having the desired properties could not, in general, be found.

The a-c steady-state output of some circuits, including many electronic amplifiers, is directly proportional to the input signal over a fairly wide frequency range, but falls off at very low and very high frequencies. Usually, the reduced output at low frequencies is caused by one group of circuit elements, and the reduction at high frequencies is caused by a different group. Then the Bode diagram can be constructed by examination of the low- and high-frequency behavior separately. In this technique, any circuit elements that are not important in the frequency range under consideration are ignored.

[†] A relatively trivial variation could be obtained by replacing the multiplying constant 40 by −40. This gives an added phase shift of 180° and corresponds to changing the direction of the reference arrow representing the output voltage or current.

Example 6.5-7. Figure 6.5-8a is a simplified model of one stage of a vacuum-tube amplifier, including some of the elements which limit its low- and high-frequency performance. An amplifier may consist of several such stages in cascade. The part of the circuit within the dashed lines is the equivalent circuit of the vacuum tube itself. As mentioned in Section 1.5, C_1, C_2, and C_3 represent the interelectrode and wiring capacitances. Their values are quite small (perhaps less than 100 pf), so their impedance, $1/j\omega C$, is very large except at high frequencies. At low and intermediate frequences, these elements may be replaced by open circuits.

The value of the external resistance R_2 is chosen to establish suitable quiescent conditions, a matter to be discussed in Chapter 12. The symbols R_5 and C_5 represent the input resistance and capacitance of the following stage, or of some other load. Normally, C_5 is small enough to be treated as an open circuit except at high frequencies. Therefore, at low and intermediate frequencies the original circuit may be replaced by Fig. 6.5-8b.

The only purpose of C_4 is to prevent any d-c voltage across R_2 from being transmitted to the following stage, as explained in Chapter 12. The value of C_4 is made relatively large, and in practice is limited only by its allowable physical size. Except at low frequencies, the impedance $1/j\omega C_4$ is so small that the element may be treated as a short circuit. A circuit model that is valid at intermediate and high frequencies is shown in Fig. 6.5-8c. When the frequency of the input is large enough so that C_4 acts like a short circuit and yet small enough to allow C_1, C_2, C_3, and C_5 to be treated as open circuits, the circuit reduces to the one in part (d) of the figure.

Draw the complete Bode diagrams.

Solution.   At intermediate frequencies, where Fig. 6.5-8d is valid,

$$\mathbf{E}_o = -(g_m\mathbf{E}_1)R \quad \text{and} \quad H(j\omega) = \frac{\mathbf{E}_o}{\mathbf{E}_1} = -g_mR$$

where R is the parallel combination of r_p, R_2, and R_5. The decibel curve in this frequency range is a straight horizontal line, and the phase angle is $-180°$. To study the low-frequency behavior, we apply the current divider rule to Fig. 6.5-8b.

$$\mathbf{I}_o = -(g_m\mathbf{E}_1)\frac{R_{p2}}{R_{p2} + R_5 + 1/j\omega C_4} = -g_m\mathbf{E}_1\frac{j\omega C_4 R_{p2}}{j\omega C_4(R_{p2} + R_5) + 1}$$

where R_{p2} is the resistance of r_p and R_2 in parallel. Since $\mathbf{E}_o = \mathbf{I}_oR_5$ and $R = R_{p2}R_5/(R_{p2} + R_5)$,

$$H(j\omega) = \frac{\mathbf{E}_o}{\mathbf{E}_1} = -g_m\frac{j\omega C_4 R_{p2}R_5}{j\omega C_4(R_{p2} + R_5) + 1} = -g_mR\frac{j\omega/\omega_1}{1 + j\omega/\omega_1}$$

where $\omega_1 = 1/C_4(R_{p2} + R_5)$. Notice that when $\omega C_4(R_{p2} + R_5) \gg 1$, i.e., when $\omega \gg \omega_1$, this expression becomes $-g_mR$, which agrees with the previous result at intermediate frequencies. The decibel curve has a slope of 20 db per decade at very low frequencies, and a break point at $\omega = \omega_1$.

Figure 6.5-8c is used to determine the high-frequency behavior. The a-c steadystate node equation is

$$(\mathbf{E}_o - \mathbf{E}_1)(j\omega C_3) + g_m\mathbf{E}_1 + \frac{\mathbf{E}_o}{R} + j\omega(C_2 + C_5)\mathbf{E}_o = 0$$

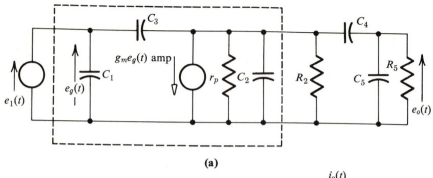

(a)

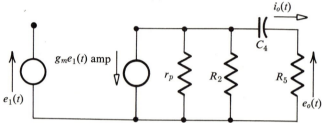

(b) Model valid at low and intermediate frequencies

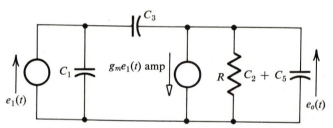

(c) Model valid at high and intermediate frequencies
$$1/R = 1/r_p + 1/R_2 + 1/R_5$$

from which

$$H(j\omega) = \frac{\mathbf{E}_o}{\mathbf{E}_1} = \frac{-g_m + j\omega C_3}{j\omega(C_2 + C_3 + C_5) + 1/R} = -g_m R \frac{1 - j\omega(C_3/g_m)}{1 + j\omega(C_2 + C_3 + C_5)R}$$

When the frequency is sufficiently low, this expression again reduces to $-g_m R$. Notice the break point in the denominator at $\omega = 1/R(C_2 + C_3 + C_5)$, and in the numerator at $\omega = g_m/C_3$. In practice, the latter break point occurs at a much higher frequency than the first. The complete decibel and angle curves are sketched in parts (e) and (f) of Fig. 6.5-8.

For television and other wide-band amplifiers, the horizontal section of the decibel curve, where all sinusoidal signals receive the same amplification, should extend over a wide frequency range. The lower limit to the bandwidth at $f_1 = \omega_1/2\pi = 1/2\pi C_4(R_{p2} + R_5)$ might be less than 100 cycles per second. At low frequencies, C_4 tends to block any current flow to the resistance R_5. The upper limit to the bandwidth at $f_2 = 1/2\pi R(C_2 + C_3 + C_5)$ may have to be several megacycles. At high

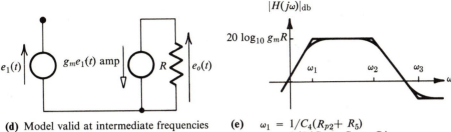

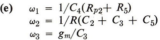

(d) Model valid at intermediate frequencies

(e) $\omega_1 = 1/C_4(R_{p2} + R_5)$
$\omega_2 = 1/R(C_2 + C_3 + C_5)$
$\omega_3 = g_m/C_3$

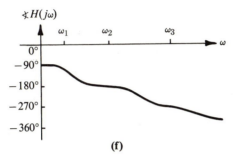

(f)

Figure 6.5-8

frequencies, the low impedance of the shunt capacitances C_2, C_3, and C_5 reduces the current flow through R_5. The capacitance C_1 has no effect upon the transfer function, since it is directly in parallel with the input voltage. It does, however, affect the total impedance seen by the source. In the model for the stage preceding the one under consideration, it will contribute to the value of C_5.

For a wide bandwidth, the product $R(C_2 + C_3 + C_5)$ should be small. Although efforts can be made to limit the size of these capacitances, a certain minimum capacitance is unavoidable. A small value of R increases the bandwidth, but at the same time reduces the size of the mid-frequency gain $-g_m R$.

The model in Fig. 6.5-8a is an oversimplification of a typical amplifier, where still other components must be considered. The above procedure for constructing the Bode diagram is, however, frequently used for both vacuum-tube and transistor amplifiers. Further discussion of such circuits will be found in Chapter 12.

The preceding examples have developed a general method for rapidly sketching the Bode diagrams when the poles and zeros are real numbers. Consider the transfer function

$$H(s) = \frac{\omega_0^2}{s^2 + 2\alpha s + \omega_0^2} = \frac{\omega_0^2}{s^2 + 2\zeta\omega_0 s + \omega_0^2} \tag{6.5-4}$$

where $\alpha = \zeta\omega_0$. Some discussion concerning the general significance of the symbols ω_0, α, and ζ may be found in Examples 4.4-1 and 6.4-4. (The transfer function for Fig. 6.5-4a has the form of Eq. 6.5-4, with $\zeta = \frac{1}{2}R\sqrt{C/L}$). When the damping ratio ζ is less than unity, the poles of $H(s)$ are complex.

In contrast to the method of Section 6.4, no attempt is made to factor quadratic factors which have complex factors. Instead, we consider

$$H(j\omega) = \frac{\omega_0^2}{-\omega^2 + j(2\zeta\omega_0)\omega + \omega_0^2} \tag{6.5-5}$$

The low- and high-frequency asymptotes are, respectively,

$$|H(j\omega)|_{db} \doteq 20 \log_{10} \frac{\omega_0^2}{\omega_0^2} = 0$$

$$|H(j\omega)|_{db} \doteq 20 \log_{10} \frac{\omega_0^2}{\omega^2}$$

which describe the two straight lines in Fig. 6.5-9. The approximate decibel curve consists of only these two asymptotes and has a double break point at $\omega = \omega_0$. In effect, the $j(2\zeta\omega_0)\omega$ term is ignored and only the larger of the ω^2 and ω_0^2 terms is retained. This may be an extremely bad approximation in the vicinity of the break point, since $|H(j\omega_0)|_{db}$ may have any value between -6 db for $\zeta = 1$ and ∞ for $\zeta = 0$.

The exact decibel curves for several different values of ζ are given in Fig. 6.5-10a. For $\zeta = 1$, $H(s) = \omega_0^2/(s + \omega_0)^2$, and the deviation of the exact curve at the double break point is the expected 6 db. For $\zeta = 1/\sqrt{2}$, the exact curve just fails to rise above the approximation. As ζ falls below $1/\sqrt{2}$, the exact curve possesses a progressively higher peak, until for $\zeta = 0.05$ it is 20 db above the approximation. These results are consistent with Example 6.4-4. As ζ decreases, the complex-conjugate poles move closer to the imaginary axis of the s-plane, as was first discussed in connection with Fig. 4.4-2b. The pole pattern of Fig. 6.4-5b corresponds to $\zeta = 1$, and the magnitude curve of Fig. 6.4-5c continually decreases with increasing ω. When $\zeta \ll 1$, the pole pattern and magnitude curve of Figs. 6.4-5e and 6.4-5f apply. As the values of α and hence ζ decrease, the peak in the resonance curve becomes higher and sharper.

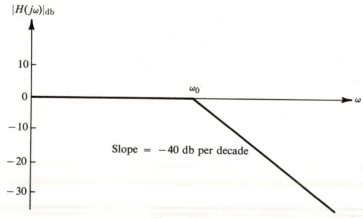

Figure 6.5-9

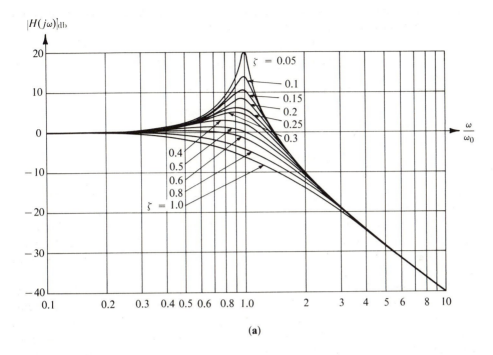

(a)

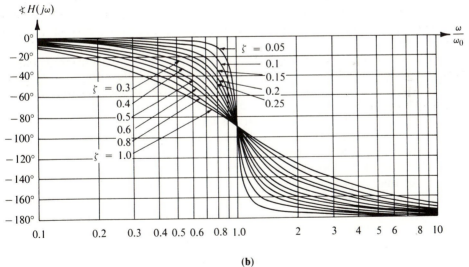

(b)

Figure 6.5-10

From H. Chestnut and R. W. Mayer, *Servomechanisms and Regulating System Design*, vol. 1, 2nd ed. (New York: Wiley, 1959).

Although the straight-line approximation is usually satisfactory when ζ is larger than say $\frac{1}{2}$, it is certainly very poor for small values of ζ. The method of Section 6.4 is, however, particularly well suited to the latter case. Bode diagrams are rarely drawn for transfer functions having complex poles or zeros near the imaginary axis and are most frequently used when the poles and zeros are real.

An inspection of Eq. 6.5-5 reveals that $\measuredangle H(j\omega) = 0°$ at very low frequencies, $-90°$ at $\omega = \omega_0$, and $-180°$ at very high frequencies, regardless of the value of ζ. The complete angle curves for different values of ζ are shown in Fig. 6.5-10b. The curves for $0.5 \leq \zeta \leq 1$ are fairly close together, but they differ appreciably from the one for $\zeta = 0.05$. Notice that these phase angle curves are consistent with Fig. 6.4-5. From part (g) of that figure, the angle change becomes increasingly rapid in the vicinity of ω_0 when α and hence ζ are reduced.

If all the values on the ordinates of Fig. 6.5-10 are multiplied by -1, the curves then represent the reciprocal of Eq. 6.5-4, namely,

$$\frac{1}{\omega_0{}^2}(s^2 + 2\zeta\omega_0 s + \omega_0{}^2) = \left(\frac{1}{\omega_0{}^2}\right)s^2 + \left(\frac{2\zeta}{\omega_0}\right)s + 1$$

Therefore, the figure may be used when either the numerator or denominator of the transfer function contains a quadratic with complex factors.

6.6

Summary

The forced response to the input $x(t) = X_m\epsilon^{\sigma t} \cos(\omega t + \phi_1)$ is $y_P(t) = Y_m\epsilon^{\sigma t} \cos(\omega t + \phi_2)$. Depending upon the value of the complex frequency $s = \sigma + j\omega$, these functions may describe decaying, growing, or constant-amplitude signals. The corresponding phasors $\mathbf{X} = X_m\underline{/\phi_1}$ and $\mathbf{Y} = Y_m\underline{/\phi_2}$ are related by the equation $\mathbf{Y} = H(s)\mathbf{X}$. If the network function $H(s)$ is known, the output phasor (and hence the forced response) may be calculated when the values of X_m, ϕ_1, and s are given.

The network function is usually found from a complex-frequency-domain circuit, in which the voltages and currents are represented by their phasors and passive elements by their impedances (using Table 6.2-1). The values of s for which $H(s)$ is zero or approaches infinity are called zeros or poles, respectively. The poles of $H(s)$ are identical with the roots of the characteristic equation that was discussed in Chapter 4, and they determine the form of the free response. If there are any poles in the right half of the s-plane or any higher-order poles on the imaginary axis, the circuit is unstable.

The forced response to a sinusoidal input of constant amplitude may be studied by replacing s by $j\omega$ in the network function. Curves of $|H(j\omega)|$ and $\measuredangle H(j\omega)$ versus ω describe a circuit's a-c steady-state behavior as the angular frequency of the input is varied. In many cases, these "frequency response" curves may be sketched by an inspection of the pole-zero plot of $H(s)$. The method, which is described in Section

6.4, is especially useful when $H(s)$ has poles or zeros that are close to the imaginary axis of the s-plane. When there is a pair of complex poles very close to the imaginary axis, the curve of $|H(j\omega)|$ versus ω has a high, narrow peak, and the circuit discriminates against all but a narrow band of frequencies.

The Bode diagrams of Section 6.5 are another way of presenting the variation of the a-c steady-state network function with the angular frequency of the input. The decibel curve, defined by Eq. 6.5-1, usually can be quickly approximated by a series of straight lines. When $H(s)$ has a pair of complex poles or zeros that are close to the imaginary axis, however, the curves in Fig. 6.5-10 must be used in place of the straight-line approximation, and the method loses much of its simplicity. Bode diagrams are most frequently used when the poles and zeros are real, and when the frequency range of interest is large.

The emphasis in the first six chapters of this book has been upon the development of methods for finding the output voltage or current of a linear circuit. More complicated circuits and inputs and more sophisticated analysis techniques have been gradually introduced. This task is not yet completed and is pursued further in Chapters 9 and 10. Chapters 7 and 8 constitute a major digression to investigate some important related topics.

PROBLEMS

6.1 The approximate transfer function for a certain circuit is known to be $H(s) = -4s$. If there is no stored energy for $t < 0$ and if the input has the waveshape in Fig. P6.1, sketch the output waveshape.

6.2 If a circuit's transfer function is $H(s) = 3s/(s + 2)$, find the forced response to

(a) $x(t) = 1$ (b) $x(t) = \epsilon^{-t}$ (c) $x(t) = \epsilon^{-2t} \cos 2t$

(d) $x(t) = \epsilon^{-2t}$ (e) $x(t) = t$

6.3 Verify the expressions for $H(s)$ in Fig. 6.2-2.

6.4 Find the transfer function $H(s)$ for each of the circuits in Fig. P6.4.

6.5 Find $E_2(s)/I_1(s)$, $E_2(s)/E_1(s)$, $I_2(s)/I_1(s)$, and $I_2(s)/E_1(s)$ for the circuit in Fig. P6.5.

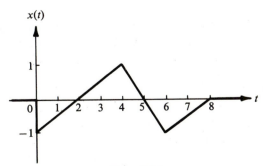

Figure P6.1

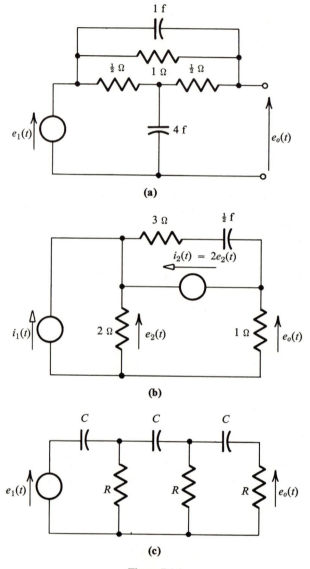

Figure P6.4

6.6   Find the transfer function $H(s) = E_o(s)/E_1(s)$ for the symmetrical lattice shown in Fig. P6.6. Plot the pole-zero pattern in the s-plane. Under what conditions can the transfer function have a zero in the right half plane?

6.7   Consider a symmetrical lattice terminated by a resistance, as shown in Fig. P6.7. If $Z_A(s)Z_B(s) = R^2$, prove that

$$\frac{E_o(s)}{E_1(s)} = \frac{R - Z_A(s)}{R + Z_A(s)} \quad \text{and} \quad \frac{E_1(s)}{I_1(s)} = R$$

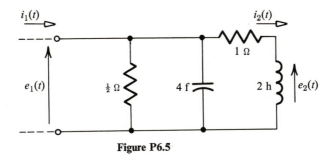

Figure P6.5

Because the input impedance is a real constant, the circuit is called a *constant resistance network*. Prove that, if the four branches of the lattice contain only inductances and capacitances, the circuit must be an all-pass network (a term defined in Example 6.4-3).

6.8 Use the results of the previous problem to find a circuit whose transfer function is

(a) $\dfrac{E_o(s)}{E_1(s)} = \dfrac{s-3}{s+3} = \dfrac{1-3/s}{1+3/s}$

(b) $\dfrac{E_o(s)}{E_1(s)} = \dfrac{s^2-s+1}{s^2+s+1} = \dfrac{1-s/(s^2+1)}{1+s/(s^2+1)}$

(c) $\dfrac{E_o(s)}{E_1(s)} = \dfrac{(s-3)(s^2-s+1)}{(s+3)(s^2+s+1)}$

In part (c), make use of the fact that the circuit's input impedance is a real constant.

6.9 Find the network N in Fig. P6.9, if $E_o(s)/E_1(s) = K(s^2 - s + 1)/(s^2 + s + 1)$. The 1 Ω and 2 Ω resistances might represent a load resistance and part of a Thévenin equivalent circuit, respectively. What is the value of the real constant K for your circuit?

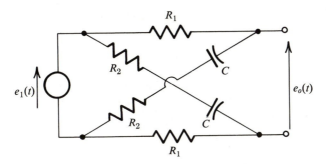

Figure P6.6

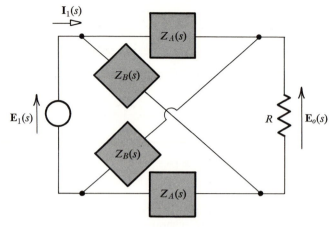

Figure P6.7

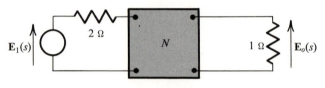

Figure P6.9

6.10 If $Z_A(s) = R + R^2 Y_B(s)$ for the circuit in Fig. P6.10, show that

$$\frac{E_o(s)}{E_1(s)} = \frac{R}{R + Z_B(s)} \quad \text{and} \quad \frac{E_1(s)}{I_1(s)} = R$$

Then find a circuit for which

(a) $\dfrac{E_o(s)}{E_1(s)} = \dfrac{s}{s+3} = \dfrac{1}{1+3/s}$

(b) $\dfrac{E_o(s)}{E_1(s)} = \dfrac{s+1}{2s+4} = \dfrac{1}{1+[(s+3)/(s+1)]} = \dfrac{1}{1+[1+2/(s+1)]}$

(c) $\dfrac{E_o(s)}{E_1(s)} = \dfrac{s(s^2-s+1)}{(s+3)(s^2+s+1)}$

The results of Problem 6.8 are useful in part (c).

6.11 Find the transfer function $H(s) = E_o(s)/E_1(s)$ for the circuit in Fig. P6.11, and plot the pole-zero pattern. Is the circuit stable? If the controlled source were changed so that $i_2(t) = g_m e_o(t)$, what would be the largest value of g_m for which the circuit is stable?

6.12 The circuit in Fig. P6.12 is a simple model for a phase shift oscillator. Notice that

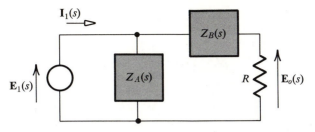

Figure P6.10

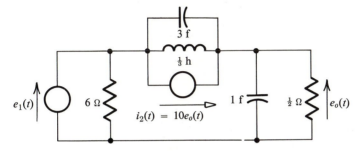

Figure P6.11

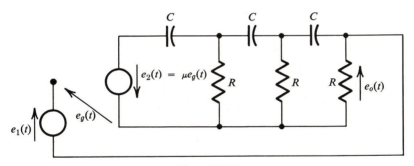

Figure P6.12

part of the circuit is similar to Fig. P6.4c. Find $H(s) = E_o(s)/E_1(s)$. For what value of μ will the free response contain a constant-amplitude sinusoid? What will be the frequency of this oscillation? (HINT: Assume that one factor in the denominator of the transfer function has the form $s^2 + \omega_o^2$. By long division, determine the conditions under which such an assumption is valid. An alternative procedure is suggested in Example 10.6-3.)

6.13 For Fig. P6.12, consider the quantity $E_o(s)/E_g(s)$ when s is replaced by $j\omega$. For what value of ω does the angle of this quantity equal zero? Compare the answer with the angular frequency of oscillation in Problem 6.12. Explain why you should expect the two results to be identical.

6.14 Which of the following functions of s could be the transfer function of a stable

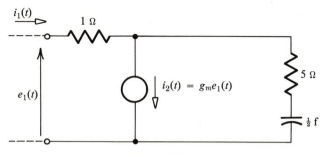

Figure P6.15

circuit? Could any of the functions be the input impedance of a two-terminal network that contains no sources?

$$H_1(s) = \frac{s - 2}{(s + 1)(s + 20)}$$

$$H_2(s) = \frac{s + 3}{s^2(s^2 + 2s + 2)}$$

$$H_3(s) = \frac{s(s + 1)}{(s^2 + 3)^2}$$

6.15 Find the input impedance $Z(s)$ for the circuit in Fig. P6.15. For what positive or negative values of g_m will the circuit be stable when it is connected to an external current source? For what values of g_m (if any) can the circuit be replaced by a single resistance?

6.16 A circuit is described by the differential equation $dy/dt + 3y = x$, where $x(t) = \epsilon^{-2t} \cos 2t$ for $t > 0$. If $y(0+) = -1$, find $y(t)$ for all $t > 0$, by using the network function $H(s)$.

6.17 If the network function is $H(s) = 10(s + 3)/(s^2 + 2s + 2)$, find the output as a function of time when

(a) The input is $\sin t$ for $t > 0$.

(b) The input is t^2 for $t > 0$.

Assume that the output and its first derivative are zero at $t = 0+$.

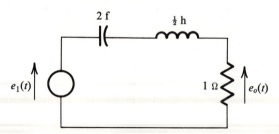

Figure P6.18

6.18 In Fig. P6.18, find $e_o(t)$ for all $t > 0$ when

(a) $e_1(t) = 10U_{-1}(t)$.

(b) $e_1(t) = \epsilon^{-2t}$ for $t > 0$.

(c) $e_1(t) = 8\epsilon^{3t} \cos (4t + 60°)$ for $t > 0$.

Assume that the circuit contains no stored energy for $t < 0$.

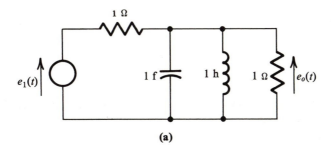

(a)

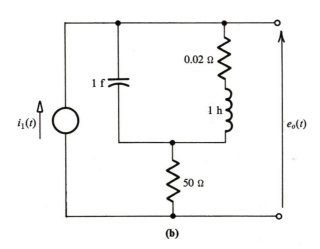

(b)

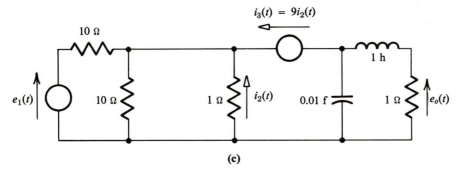

(c)

Figure P6.20

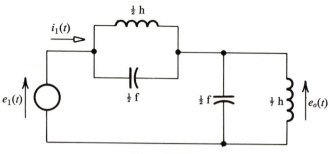

Figure P6.21

6.19 For each of the network functions listed below, plot the pole-zero pattern. Sketch and fully label the frequency response curves $|H(j\omega)|$ and $\angle H(j\omega)$ versus ω.

(a) $H_1(s) = \dfrac{100(s + 1)}{s^2 + 2s + 400}$

(b) $H_2(s) = \dfrac{100s(s + 1)}{s^2 + 2s + 400}$

(c) $H_3(s) = \dfrac{100(s^2 - 2s + 400)}{s^2 + 2s + 400}$

(d) $H_4(s) = \dfrac{s^2 + 2s + 400}{s(s + 1)}$

(e) $H_5(s) = \dfrac{100s^2}{(s^2 + 2s + 100)(s^2 + 2s + 400)}$

(f) $H_6(s) = \dfrac{100s^2}{(s^2 + 2s + 100)^2}$

(g) $H_7(s) = \dfrac{s^2 + 4}{s(s^2 + 9)}$

(h) $H_8(s) = \dfrac{1}{(s + 10)^3}$

For parts (a), (b), and (f), determine numerical values for the bandwidth and for the selectivity Q. Also determine the bandwidth for part (h). If the transfer function of a circuit is given by $H_2(s)$, what is the forced response to the input $x(t) = 10 + 5 \cos (10t + 20°) + 2 \cos (20t + 40°)$?

6.20 Find the transfer function $H(s)$ for each of the circuits in Fig. P6.20. Plot the pole-zero pattern, and sketch to scale the frequency response curves.

6.21 For the circuit in Fig. P6.21, find the input impedance $Z(s) = \mathbf{E}_1(s)/\mathbf{I}_1(s)$ and the transfer function $H(s) = \mathbf{E}_o(s)/\mathbf{E}_1(s)$. In each case, plot the pole-zero pattern, and sketch the frequency response curves. Explain how the location of the poles and zeros can be checked by an inspection of the circuit, without specifically examining the expressions for $Z(s)$ and $H(s)$.

6.22 For the circuit in Fig. 6.4-7a, show that $|\mathbf{E}_L| = |\mathbf{E}_C| = Q|\mathbf{E}|$ when $\omega = 1/\sqrt{LC}$. For the circuit in Fig. 6.4-7b, show that $|\mathbf{I}_L| = |\mathbf{I}_C| = Q|\mathbf{I}|$ at the resonant frequency.

6.23 The transfer function $H(s)$ for the circuit in Fig. P6.23a is readily found to be

$$H(s) = \frac{\mathbf{E}_o(s)}{\mathbf{I}_1(s)} = \frac{Ks^2}{(s^2 + 2\alpha_1 s + \omega_1{}^2)(s^2 + 2\alpha_2 s + \omega_2{}^2)}$$

where $\alpha_1 = 1/2R_1C_1$, $\alpha_2 = 1/2R_2C_2$, $\omega_1{}^2 = 1/L_1C_1$, $\omega_2{}^2 = 1/L_2C_2$, and $K = g_m/C_1C_2$.

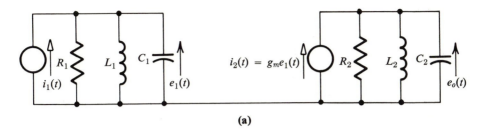

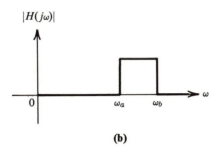

(b)

Figure P6.23

Assume that $\alpha_1 \ll \omega_1$ and $\alpha_2 \ll \omega_2$, so that

$$H(s) \doteq \frac{Ks^2}{(s + \alpha_1 + j\omega_1)(s + \alpha_1 - j\omega_1)(s + \alpha_2 + j\omega_2)(s + \alpha_2 - j\omega_2)}$$

Assume further that $R_1 = R_2$ and $C_1 = C_2$, so that $\alpha_1 = \alpha_2 = \alpha$. Define $\omega_0 = \frac{1}{2}(\omega_1 + \omega_2)$.

(a) When $\omega_1 = \omega_2$, the two *RLC* combinations in Fig. 6.23a are said to be synchronously tuned. For this case, plot the pole-zero pattern, sketch the frequency response curves, and find the bandwidth. Remember that $\alpha \ll \omega_0$.

(b) When $\omega_1 \neq \omega_2$, the two *RLC* combinations are said to be staggered tuned. Sketch the frequency response curves for $\omega_2 - \omega_1 = \alpha$, for $\omega_2 - \omega_1 = 2\alpha$, and for $\omega_2 - \omega_1 = 5\alpha$.

(c) What is the largest value of $\omega_2 - \omega_1$ for which the curve of $|H(j\omega)|$ versus ω has a single peak? Compare the bandwidth for this case with the bandwidth when $\omega_1 = \omega_2$. The results of Example 6.4-6 are helpful.

(d) Discuss how the value of $\omega_2 - \omega_1$ might be chosen in order to best approximate an idealized band-pass filter having the characteristics shown in Fig. P6.23b.

6.24 Find a transfer function which corresponds to the angle curve in Fig. P6.24.

6.25 Sketch to scale curves of $|H(j\omega)|$ and $\angle H(j\omega)$ versus ω for the following transfer functions. Also draw (on semilog paper) the Bode diagrams.

(a) $H_1(s) = \dfrac{s}{s + 2}$

(b) $H_2(s) = \dfrac{10(1 + 10s)}{s(s + 50)}$

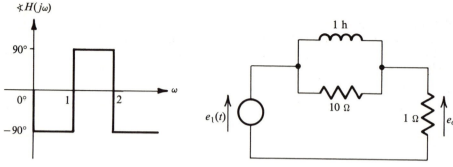

Figure P6.24   Figure P6.27

6.26 Draw the Bode diagrams (decibel and angle curves drawn on semilog paper) for the following transfer functions.

(a) $H_1(s) = \dfrac{100(10s + 1)}{(s + 1)(s + 100)}$

(b) $H_2(s) = \dfrac{10s}{(s + 1)(s + 100)}$

(c) $H_3(s) = \dfrac{1250s}{(s + 0.4)(s + 50)^2}$

(d) $H_4(s) = \dfrac{200s(s + 100{,}000)}{(s + 10)(s + 1000)^2}$

(e) $H_5(s) = \dfrac{10(s - 1)}{(s + 100)^2}$

(f) $H_6(s) = \dfrac{s^2(20 - s)}{(1 + 50s)(s + 1)^2}$

(g) $H_7(s) = \dfrac{100(10s + 1)}{s^2 + 2s + 100}$

(h) $H_8(s) = \dfrac{s}{s^2 + s + 1}$

(i) $H_9(s) = \dfrac{s^2 + 4s + 100}{(10s + 1)(s + 100)}$

6.27 Sketch the Bode diagrams for the phase-lag circuit in Fig. P6.27.

6.28 For each of the circuits in Fig. P5.13, plot the pole-zero patterns for the transfer function, and draw the Bode diagrams. In parts (a) and (b), assume that the break points are at least a decade apart. Verify the answers to Problem 5.13.

6.29 Sketch the Bode diagrams for the transfer function $H(s) = 5/s(1 + 10s)(s + 4)$. To what value should the multiplying constant of 5 be changed, if $\angle H(j\omega)$ should be approximately $-135°$ when $|H(j\omega)|_{db} = 0$?

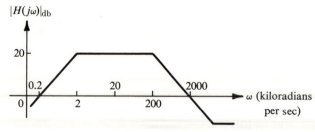

Figure P6.30

6.30 For a certain amplifier, the straight-line approximation to the decibel curve is given in Fig. P6.30. What is the bandwidth? Suppose that two such amplifiers are connected in cascade and that the second one does not load down the first. Sketch the decibel curve, and determine the bandwidth for the combination.

6.31 Find the transfer function $H(s)$ corresponding to each of the straight-line approximations shown in Fig. P6.31. Assume that the transfer function has no zeros in the open right half plane.

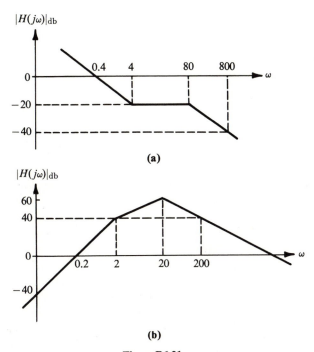

Figure P6.31

7

Transformers and

Equivalent Circuits

Inductance (or self-inductance) is one of the passive, two-terminal elements defined in Section 1.3 and is a simple model for a current-carrying coil of wire. The model accounts for the fact that a changing current produces a changing magnetic field, which in turn induces a voltage in the coil. A model that includes the effect of the wire's resistance and the capacitance between adjacent turns of the coil was shown in Fig. 1.5-1c.

In this chapter we consider models representing two or more coils of wire, which are so situated that the magnetic flux produced by the current in one coil induces a voltage in the other coils. Such a device is usually called a *transformer*. Although the student already may be acquainted with the effects of magnetic flux, this subject is briefly reviewed in Section 7.1, in order to justify the models that are used.

The basic model of a transformer is characterized by a mutual inductance and two self-inductances, and the analysis of circuits containing such a model is considered in Section 7.2. If all the flux links all the turns of each coil and if the self- and mutual inductances are large, then a simpler model, called the *ideal transformer*, often can be used. The development of this simpler model and a discussion of when it may be used is included in Section 7.3.

In Section 7.4 it is shown how any transformer may be adequately represented by a model composed only of those circuit elements that were defined in Chapter 1. Therefore, it might seem that a separate treatment of the transformer is unnecessary. The device is very widely used, however, and is sufficiently important to warrant consideration as a basic circuit component. For the same reason, Chapter 12 is devoted to vacuum tubes and transistors, even though their models are also composed of the elements in Chapter 1.

The equivalence of different circuits, each having two external terminals, was first considered in Sections 2.1 and 2.3. The models in Section 7.4 lead into a general discussion of the equivalence of circuits that have more than two external terminals.

7.1

Properties of the Two-Winding Transformer

The self-inductance shown in Fig. 7.1-1a is characterized by the relationship

$$e = L\frac{di}{dt} \tag{7.1-1}$$

regardless of what might be connected to the two external terminals. In the physical sketch in Fig. 7.1-1b, the current $i(t)$ produces a magnetic flux $\phi(t)$, whose direction is given by the familiar right-hand rule.† Assume that the coil has N turns, all of which are linked by the common flux $\phi(t)$. If the current and hence the flux change with time, then a voltage

$$e = N\frac{d\phi}{dt} = \left(N\frac{d\phi}{di}\right)\frac{di}{dt} \tag{7.1-2}$$

is induced in the coil. The voltage has such a polarity as to tend to oppose the change in current that was originally responsible for it. A comparison of the last two equations indicates that

$$L = N\frac{d\phi}{di} \tag{7.1-3}$$

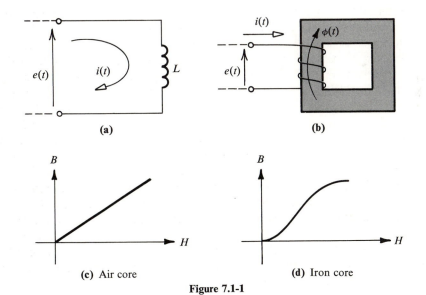

(a)

(b)

(c) Air core

(d) Iron core

Figure 7.1-1

---

† If the fingers of the closed right hand point in the direction of the current flow, the thumb indicates the direction of the magnetic flux.

In some physics books, a minus sign is inserted on the right side of Eqs. 7.1-1 and 7.1-2 to indicate that the induced voltage opposes the change in current. We can use, however, either a plus or a minus sign, depending upon how we choose the direction of the voltage and current reference arrows. The fact that our choice of the plus sign corresponds to the reference arrows in Fig. 7.1-1a can be physically justified by letting $i(t) = U_{-2}(t)$, the unit ramp function. Then by Eq. 7.1-1, $e(t) = LU_{-1}(t)$ and the power absorbed by the inductance is $p(t) = e(t)i(t) = Lt$ for all $t > 0$. A minus sign would indicate that the inductance supplies increasingly large amounts of power to the rest of the circuit, which is not physically possible.

The value of L depends upon the magnetic characteristics of the core on which the coil is wound. Assume for simplicity that all of the flux in Fig. 7.1-1b is confined to the core and links all of the N turns. The *magnetic flux density B* and the *magnetic intensity H* are given by

$$B = \frac{\phi}{A} \text{ webers per meter}^2$$

$$H = \frac{Ni}{l} \text{ ampere-turns per meter}$$

where l and A are the length and cross-sectional area of the core. When these expressions are substituted into Eq. 7.1-3,

$$L = \frac{N^2 A}{l} \frac{dB}{dH} \tag{7.1-4}$$

where dB/dH is the slope of a curve of B versus H.

The relationship between B and H depends only upon the material of the core, and two typical curves are shown in Figs. 7.1-1c and 7.1-1d. For the air core, dB/dH equals a constant denoted by μ, which is called the *permeability*, so

$$L = \frac{\mu N^2 A}{l}$$

Since the inductance has a constant value, independent of the size of the current that flows, it is a linear circuit element. For the iron core, the value of dB/dH depends upon H and therefore upon the current, so the inductance is a nonlinear element. Even for the iron core, however, an equivalent linear inductance often can be used with good results. In all the problems in this book, L is assumed to be a constant.

It has been convenient to assume that the flux is confined to the core on which the coil is wound, so that the same flux links all the turns. Equation 7.1-1 (and Eqs. 7.1-5 through 7.1-11) is valid whether or not this is the case. If the flux linking different turns of the same coil is not the same, Eqs. 7.1-2 and 7.1-3 should be replaced by $e = d\lambda/dt$ and $L = d\lambda/di$, where $\lambda(t)$ denotes the flux linkages. When the same flux links all the turns in a coil, $\lambda(t) = N\phi(t)$. When the portion of the flux linking some of the turns is reduced, these turns do not contribute as much to the flux linkages and to the value of L. For a linear inductance, $\lambda(t)$ is directly proportional to $i(t)$, and L is a constant.

A sketch of a two-winding transformer is given in Fig. 7.1-2a, where it is assumed that at least part of the flux produced by one winding links the turns of the other winding.† The symbol used in circuit diagrams is shown in Fig. 7.1-2b. In order to distinguish the two windings, the one on the left side of the diagram is often called the *primary winding*, and the other one the *secondary winding*. Let $\phi_1(t)$ denote the total flux linking the primary winding and $\phi_2(t)$ the flux linking the secondary winding. Since each flux is completely determined by the two currents, which in turn are functions of t,

$$d\phi_1 = d\phi_1(i_1, i_2) = \frac{\partial\phi_1}{\partial i_1}\, di_1 + \frac{\partial\phi_1}{\partial i_2}\, di_2$$

$$d\phi_2 = d\phi_2(i_1, i_2) = \frac{\partial\phi_2}{\partial i_1}\, di_1 + \frac{\partial\phi_2}{\partial i_2}\, di_2$$

Then,

$$e_1 = N_1\frac{d\phi_1}{dt} = \left(N_1\frac{\partial\phi_1}{\partial i_1}\right)\frac{di_1}{dt} + \left(N_1\frac{\partial\phi_1}{\partial i_2}\right)\frac{di_2}{dt}$$

$$e_2 = N_2\frac{d\phi_2}{dt} = \left(N_2\frac{\partial\phi_2}{\partial i_1}\right)\frac{di_1}{dt} + \left(N_2\frac{\partial\phi_2}{\partial i_2}\right)\frac{di_2}{dt}$$

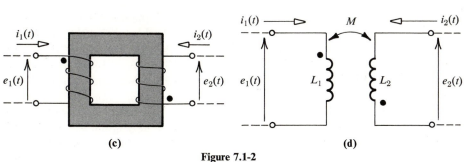

Figure 7.1-2

---

† Some of the problems at the end of the chapter consider transformers with three magnetically coupled windings.

These equations have the form

$$e_1 = L_1 \frac{di_1}{dt} + M \frac{di_2}{dt}$$

$$e_2 = M \frac{di_1}{dt} + L_2 \frac{di_2}{dt}$$

(7.1-5)

where L_1 and L_2 are self-inductances, and where M is called the *mutual inductance*. In the above derivation, $M = N_1(\partial\phi_1/\partial i_2) = N_2(\partial\phi_2/\partial i_1)$. A proof that these two expressions for M give the same result is suggested in the problems at the end of the chapter. The mutual inductance M is a constant for an air-core transformer, but depends upon the size of the currents for an iron core. Some further comments about iron-core transformers appear in a later section, but we shall always assume that M may be treated as a constant.

If $i_2(t)$ is identically zero, i.e., if the secondary winding in Fig. 7.1-2a is open circuited, Eqs. 7.1-5 become $e_1 = L_1(di_1/dt)$ and $e_2 = M(di_1/dt)$. Clearly, the mutual-inductance term represents the voltage induced in the second winding by a changing current in the first. If $i_1(t)$ is identically zero, $e_1 = M(di_2/dt)$.

If the current arrows are drawn into the terminals toward which the voltage arrows are directed, as is done in Fig. 7.1-2, the sign of the self-inductance terms in Eqs. 7.1-5 is positive. Since the sign of $\partial\phi_1/\partial i_2$ and $\partial\phi_2/\partial i_1$ is not immediately clear, consider the sketches of Figs. 7.1-2a and 7.1-2c, where the flux is assumed to be substantially confined to the core. In part (a), the fluxes produced by the currents $i_1(t)$ and $i_2(t)$ are *both* clockwise, so that the mutual coupling increases the size of the induced voltages. Then the signs of the mutual- and self-inductance terms must be the same. In part (c), $i_1(t)$ produces a clockwise flux, and $i_2(t)$ produces a counterclockwise flux. Since the two components of flux are in opposite directions, the mutual coupling decreases the size of the induced voltages, so that the self- and mutual-inductance terms have opposite signs.

A simple method of determining the sign of the mutual-inductance term *without* reference to a detailed sketch of the windings is provided by polarity markings (the "dot convention"). Dots are placed at one end of each coil to indicate that, when positive currents flow into the dotted end of each coil, the fluxes produced by the currents are in the same direction. The student should verify that the dots in Figs. 7.1-2a and 7.1-2c agree with this convention. The dot convention leads to the following rule for writing equations. *If both current references are directed into (or both out of) the dotted end of the coil, the mutual- and self-inductance terms have the same sign. If one reference arrow is into and the other out of a dot, the mutual- and self-inductance terms have opposite signs.*[†]

It is convenient to regard M, along with L_1 and L_2, as a nonnegative quantity.

[†] There are other, equivalent ways of stating the dot convention. For example, a positively increasing current flowing into the dotted end of one winding induces a voltage in the other winding that is positive at the dotted end. Since most students have no difficulty with the sign of the self-inductance term, they often prefer the previous statement of the convention.

Equations 7.1-5 are then valid only for the polarity markings shown in Figs. 7.1-2a and 7.1-2b. For Figs. 7.1-2c and 7.1-2d,

$$e_1 = L_1 \frac{di_1}{dt} - M \frac{di_2}{dt}$$

$$e_2 = -M \frac{di_1}{dt} + L_2 \frac{di_2}{dt}$$

$$(7.1\text{-}6)$$

Notice that part (d) of the figure is essentially identical with part (b), except that the reference arrows for $e_2(t)$ and $i_2(t)$ have been reversed with respect to the polarity dot. Thus, Eqs. 7.1-6 can be obtained directly from Eqs. 7.1-5 by replacing $e_2(t)$ by $-e_2(t)$ and $i_2(t)$ by $-i_2(t)$. Note also that, in the above and in later equations, changing one polarity dot is equivalent to replacing M by $-M$.

Stored Energy. As indicated in Section 1.3, the stored energy contained in the magnetic field of a self-inductance is

$$w(t) = \tfrac{1}{2}Li^2(t) \qquad\qquad (7.1\text{-}7)$$

We now wish to derive an expression for the energy stored in the two-winding transformer of Fig. 7.1-2b, which is described by Eqs. 7.1-5. By Eqs. 1.2-2 and 1.2-4, the net energy supplied *to* the transformer from time t_1 to time t_2 is

$$\int_{t_1}^{t_2} e_1 i_1 \, dt + \int_{t_1}^{t_2} e_2 i_2 \, dt$$

The net energy obtained *from* the transformer between these two instants is the negative of this expression.

Suppose that at the instant t_0 the values of the two currents are denoted by i_{1_0} and i_{2_0}. To determine the energy stored in the magnetic field, we collapse the field to zero in two steps. First, reduce $i_1(t)$ to zero while keeping $i_2(t) = i_{2_0}$. Since $di_2/dt = 0$, $e_1 = L_1(di_1/dt)$ and $e_2 = M(di_1/dt)$. Then $e_1 i_1 \, dt = L_1 i_1 \, di_1$ and $e_2 i_2 \, dt = M i_{2_0} \, di_1$. The energy obtained from the transformer during this step is

$$-\left(\int_{i_{1_0}}^{0} L_1 i_1 \, di_1 + \int_{i_{1_0}}^{0} M i_{2_0} \, di_1 \right) = \tfrac{1}{2}L_1 i_{1_0}^2 + M i_{1_0} i_{2_0}$$

Next, leave the first winding open circuited, so that $i_1(t)$ remains equal to zero, and reduce $i_2(t)$ to zero. Then $e_1 i_1 \, dt = 0$ and $e_2 i_2 \, dt = L_2(di_2/dt)i_2 \, dt = L_2 i_2 \, di_2$. The energy removed from the transformer is

$$-\int_{i_{2_0}}^{0} L_2 i_2 \, di_2 = \tfrac{1}{2}L_2 i_{2_0}^2$$

Since the energy in the magnetic field has been reduced to zero, the initial energy at the time t_0 must be the sum of the above two expressions. Since t_0 may be any instant of time,

$$w(t) = \tfrac{1}{2}L_1 i_1^2(t) + M i_1(t) i_2(t) + \tfrac{1}{2}L_2 i_2^2(t) \qquad (7.1\text{-}8)$$

For the polarity markings in Fig. 7.1-2d,

$$w(t) = \tfrac{1}{2}L_1 i_1^2(t) - M i_1(t) i_2(t) + \tfrac{1}{2}L_2 i_2^2(t) \qquad (7.1\text{-}9)$$

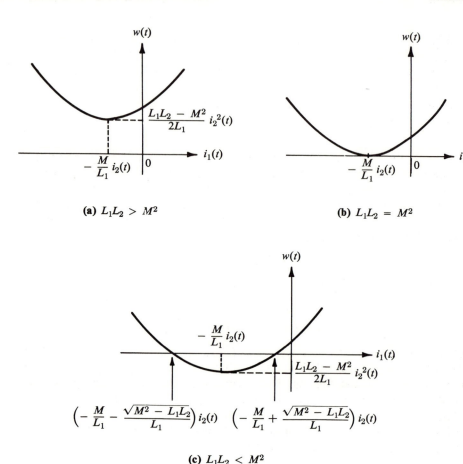

(a) $L_1L_2 > M^2$

(b) $L_1L_2 = M^2$

(c) $L_1L_2 < M^2$

Figure 7.1-3

The currents $i_1(t)$ and $i_2(t)$ in Fig. 7.1-2 depend upon the external connections and can take on any real values. Since the energy stored within any passive component is always a nonnegative quantity, the above expressions for $w(t)$ must be nonnegative for all possible choices of $i_1(t)$ and $i_2(t)$. This fact restricts the size of the mutual inductance, since the $Mi_1(t)i_2(t)$ term in Eq. 7.1-8 is negative if $i_1(t)$ and $i_2(t)$ have opposite signs. In order to examine this restriction more carefully, rewrite Eq. 7.1-8 as follows, by completing the square in the variable $i_1(t)$.

$$w(t) = \frac{1}{2}\left\{\left[\sqrt{L_1}\,i_1(t) + \frac{M}{\sqrt{L_1}}\,i_2(t)\right]^2 + \left(L_2 - \frac{M^2}{L_1}\right)i_2^2(t)\right\}$$

or

$$\left[\sqrt{L_1}\,i_1(t) + \frac{M}{\sqrt{L_1}}\,i_2(t)\right]^2 = 2\left[w(t) - \left(\frac{L_1L_2 - M^2}{2L_1}\right)i_2^2(t)\right]$$

Remember that L_1, L_2, and M are nonnegative constants. Suppose that we now assign an arbitrary, fixed value to $i_2(t)$, and plot $w(t)$ versus $i_1(t)$ for all real values of $i_1(t)$. The last equation is the standard form for a parabola opening upward, with a vertex at $i_1(t) = -(M/L_1)i_2(t)$ and $w(t) = (L_1L_2 - M^2)(1/2L_1)i_2^2(t)$. Typical sketches are shown in Fig. 7.1-3. It is seen that the stored energy is always nonnegative if and only if

$$L_1L_2 \geq M^2 \tag{7.1-10}$$

If the student is not familiar with the standard equation for a parabola, the following reasoning may be used to justify the above inequality. For any arbitrary, fixed value of $i_2(t)$, $w(t)$ in Eq. 7.1-8 is certainly positive for extremely large positive or negative values of $i_1(t)$, since $\frac{1}{2}L_1i_1^2(t)$ is a positive term that increases as the *square* of $i_1(t)$. As the size of $i_1(t)$ is decreased, $w(t)$ can become negative only by first passing through zero. Setting

$$(\tfrac{1}{2}L_1)i_1^2(t) + [Mi_2(t)]i_1(t) + [\tfrac{1}{2}L_2i_2^2(t)] = 0$$

and solving for $i_1(t)$ by the quadratic formula, we obtain

$$i_1(t) = \left(-\frac{M}{L_1} \pm \frac{1}{L_1} \sqrt{M^2 - L_1L_2} \right)i_2(t)$$

which has two distinct, real values if $M^2 > L_1L_2$, yielding the curve in Fig. 7.1-3c. If $M^2 = L_1L_2$, the quadratic in $i_1(t)$ has a *double* zero at $-(M/L_1)i_2(t)$ as shown in part (b) of the figure. For $M^2 < L_1L_2$, the zeros are complex, so that $w(t) \neq 0$ for any real value of $i_1(t)$ [except for the trivial case $i_1(t) = i_2(t) = 0$].

While Eq. 7.1-10 was derived using Eq. 7.1-8, the same result can be obtained from Eq. 7.1-9. Another method of proving the inequality is discussed in Example 7.2-2. The *coefficient of coupling* k is defined by the relationship

$$M = k\sqrt{L_1L_2} \tag{7.1-11}$$

where $0 \leq k \leq 1$ by Eq. 7.1-10. A *unity-coupled* transformer is one for which k has its maximum value of unity. Physically, this corresponds to the case where all the flux links all the turns of both windings. If $k = 0$, none of the flux produced by a current in one winding links any of the turns in the other winding.

7.2
Circuits Containing Transformers

The two-winding transformer is a four-terminal component characterized by three parameters (L_1, L_2, and M) and is formally defined by Eqs. 7.1-5 and 7.1-6. The complete solution of circuits containing transformers can be carried out using only these equations, and occasionally the expressions for stored energy in Eqs. 7.1-8 and 7.1-9. Since the mathematical definition of the transformer expresses the voltages in terms of the currents, it is usually easier to sum voltages around a loop than to sum currents leaving a node. Unless there are compelling reasons for doing otherwise, loop equations rather than node equations are used in analyzing transformer circuits.

Example 7.2-1. When a transformer's windings are connected in series, as in Fig. 7.2-1a or 7.2-1b, it is equivalent to a single inductance. Find an expression for the equivalent inductance in each part of the figure.

Solution. The same current $i(t)$ flows through both windings of the transformer. In Fig. 7.2-1a, the current reference arrow enters the dotted end of both windings, so

$$e = e_1 + e_2 = \left(L_1 \frac{di}{dt} + M \frac{di}{dt} \right) + \left(M \frac{di}{dt} + L_2 \frac{di}{dt} \right) = (L_1 + L_2 + 2M) \frac{di}{dt}$$

which describes an inductance $L_a = L_1 + L_2 + 2M$. In part (b) of the figure, the current leaves the dotted end of the second winding, so the mutual-inductance terms are negative. The equivalent inductance is $L_b = L_1 + L_2 - 2M$.

Notice that $(L_a - L_b)/4 = M$. This is an interesting result, because if L_a and L_b can be accurately measured on an impedance bridge, then M can be calculated.

In Chapters 5 and 6 we considered the forced response of circuits to inputs of the form ϵ^{st}, $\epsilon^{j\omega t}$, and $\cos(\omega t + \phi)$. In order to be able to carry over the concepts developed in these chapters, assume that the transformer in Fig. 7.2-2a is part of a circuit whose input is ϵ^{st}. In the calculation of the forced response, the voltages and currents for the transformer have the form

$$e_1(t) = \mathbf{E}_1(s)\epsilon^{st}, \qquad e_2(t) = \mathbf{E}_2(s)\epsilon^{st}$$
$$i_1(t) = \mathbf{I}_1(s)\epsilon^{st}, \qquad i_2(t) = \mathbf{I}_2(s)\epsilon^{st}$$

When these expressions are substituted into Eqs. 7.1-5,

$$\mathbf{E}_1(s)\epsilon^{st} = sL_1\mathbf{I}_1(s)\epsilon^{st} + sM\mathbf{I}_2(s)\epsilon^{st}$$
$$\mathbf{E}_2(s)\epsilon^{st} = sM\mathbf{I}_1(s)\epsilon^{st} + sL_2\mathbf{I}_2(s)\epsilon^{st}$$

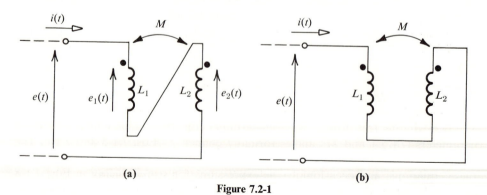

(a)   **(b)**

Figure 7.2-1

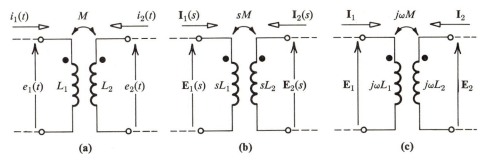

Figure 7.2-2

or

$$\mathbf{E}_1(s) = sL_1\mathbf{I}_1(s) + sM\mathbf{I}_2(s)$$
$$\mathbf{E}_2(s) = sM\mathbf{I}_1(s) + sL_2\mathbf{I}_2(s) \tag{7.2-1}$$

$\mathbf{E}_1(s)$, $\mathbf{E}_2(s)$, $\mathbf{I}_1(s)$, and $\mathbf{I}_2(s)$ are voltage and current phasors that depend upon the value of the complex frequency s. The quantities sL_1 and sL_2 are the self-impedances, and sM is the mutual impedance of the transformer. In a complex-frequency-domain circuit, the transformer is labelled with phasors and impedances, as in Fig. 7.2-2b.

When the a-c steady-state response is calculated, s is replaced by $j\omega$, which gives

$$\mathbf{E}_1 = j\omega L_1\mathbf{I}_1 + j\omega M\mathbf{I}_2$$
$$\mathbf{E}_2 = j\omega M\mathbf{I}_1 + j\omega L_2\mathbf{I}_2 \tag{7.2-2}$$

The self- and mutual impedances are shown in the frequency-domain circuit of Fig. 7.2-2c. Reversing one of the polarity dots in Fig. 7.2-2 would, of course, change the sign of the mutual-impedance terms in Eqs. 7.2-1 and 7.2-2. For the d-c steady-state response, s is replaced by zero, and both windings of the transformer become short circuits.

Example 7.2-2. Find the transfer function $H(s) = \mathbf{E}_o(s)/\mathbf{E}_1(s)$ for Fig. 7.2-3a, and comment upon the stability of the circuit.

Solution. The two loop equations for the complex-frequency-domain circuit in Fig. 7.2-3b are

$$(sL_1 + R_1)\mathbf{I}_1(s) - sM\mathbf{I}_2(s) = \mathbf{E}_1(s)$$

$$-sM\mathbf{I}_1(s) + (sL_2 + R_2)\mathbf{I}_2(s) = 0$$

Solving simultaneously for $\mathbf{I}_2(s)$, and noting that $\mathbf{E}_o(s) = R_2\mathbf{I}_2(s)$, we obtain

$$H(s) = \frac{\mathbf{E}_o(s)}{\mathbf{E}_1(s)} = \frac{sM\,R_2}{(L_1L_2 - M^2)s^2 + (L_1R_2 + L_2R_1)s + R_1R_2}$$

If $L_1L_2 - M^2 \geq 0$, the poles of the transfer function are in the left half of the s-plane, and the circuit is stable. If $L_1L_2 - M^2$ is negative, one of the poles is in the right half

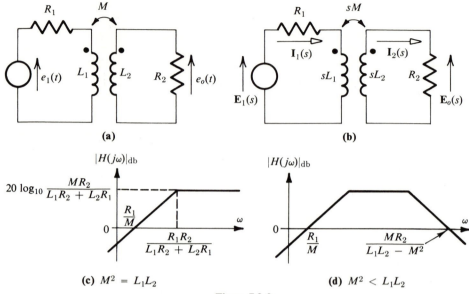

Figure 7.2-3

plane, and the free response increases without limit. Since the transformer and the two resistances are passive elements, however, the free response cannot possibly increase without limit. We therefore conclude that $M^2 \leq L_1L_2$, which agrees with Eq. 7.1-10.

Example 7.2-3.   Sketch and discuss the decibel voltage gain as a function of ω in the a-c steady-state for the circuit in Fig. 7.2-3a.

Solution.   For the special case of $M^2 = L_1L_2$,

$$H(j\omega) = \frac{j\omega MR_2}{(L_1R_2 + L_2R_1)j\omega + R_1R_2}$$

Figure 7.2-3c is a Bode diagram showing the straight-line approximation to the decibel curve (plotted on semilog paper). Among other applications, a transformer can be used to provide voltage amplification. Notice from the figure that if $MR_2 > (L_1R_2 + L_2R_1)$, the output voltage is larger than the input voltage for all $\omega \dot> R_1/M$. To achieve a large voltage gain over a wide frequency range, R_1 should be small, as should have been anticipated. Then the high-frequency asymptote is $H(j\omega) \doteq M/L_1 = \sqrt{L_1L_2}/L_1 = \sqrt{L_2/L_1}$, so the ratio L_2/L_1 should be large. Equation 7.1-4 indicates that the secondary winding should have more turns than the primary winding.

A typical Bode diagram for $M^2 < L_1L_2$ is shown in part (d) of the figure. The low- and high-frequency asymptotes

$$H(j\omega) \doteq \frac{j\omega M}{R_1}$$

$$H(j\omega) \doteq \frac{MR_2}{j\omega(L_1L_2 - M^2)}$$

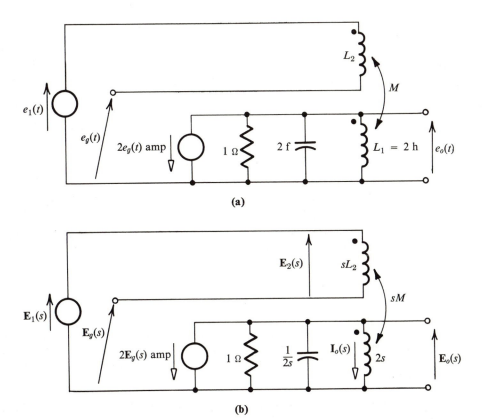

Figure 7.2-4

cross the zero-decibel axis at $\omega = R_1/M$ and $MR_2/(L_1L_2 - M^2)$, respectively. To achieve a large voltage gain over a wide frequency range, R_1 should be small, R_2 large, $L_2 \gg L_1$, and M close to L_1L_2. Some numerical values are given in the problems at the end of the chapter.

Example 7.2-4. A suitably scaled model of a vacuum-tube circuit is given in Fig. 7.2-4a. Find the transfer function $H(s) = E_o(s)/E_1(s)$. For what value of M will the circuit act like an oscillator, i.e., produce a constant-amplitude sinusoidal output when $e_1(t) = 0$? What will be the angular frequency of this oscillation?

Solution. No current can flow through the top winding of the transformer, so, for the frequency-domain circuit in Fig. 7.2-4b,

$$E_2(s) = sM\,I_o(s) \quad \text{and} \quad E_o(s) = sL_1I_o(s)$$

The current-law equation at node o is

$$\left(1 + 2s + \frac{1}{2s}\right) E_o(s) = -2E_g(s)$$

Since $\mathbf{E}_g(s) = \mathbf{E}_1(s) - \mathbf{E}_2(s) = \mathbf{E}_1(s) - sM[\mathbf{E}_o(s)/2s]$,

$$\left(1 - M + 2s + \frac{1}{2s}\right) \mathbf{E}_o(s) = -2\mathbf{E}_1(s)$$

$$H(s) = \frac{\mathbf{E}_o(s)}{\mathbf{E}_1(s)} = \frac{-s}{s^2 + \frac{1}{2}(1 - M)s + \frac{1}{4}}$$

The free response is a constant-amplitude sinusoid if and only if the poles of $H(s)$ are on the imaginary axis of the s-plane. We therefore choose $M = 1$. The poles are then at $s = \pm j\frac{1}{2}$, so the angular frequency of oscillation is $\frac{1}{2}$ rad per sec.

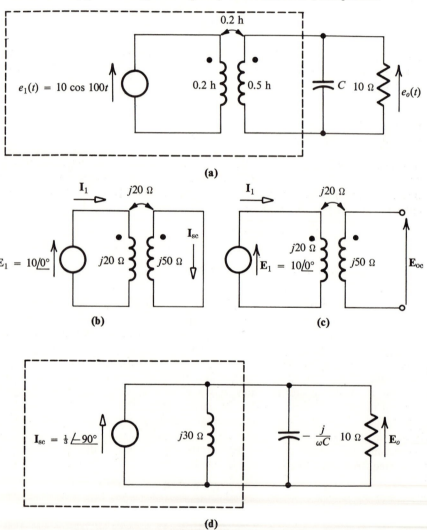

Figure 7.2-5

Example 7.2-5. In the a-c steady state, find the Norton equivalent circuit for the part of the circuit in Fig. 7.2-5a enclosed by the dashed line. What value of C will produce the largest steady-state output voltage?

Solution. The short-circuit current phasor $\mathbf{I}_{sc}$ can be found by writing two loop equations for Fig. 7.2-5b.

$$10 = j20\mathbf{I}_1 - j20\mathbf{I}_{sc}$$
$$0 = -j20\mathbf{I}_1 + j50\mathbf{I}_{sc}$$

from which $\mathbf{I}_{sc} = 1/j3$ amp. To find the equivalent impedance of the dead network, we first calculate the open-circuit voltage phasor from part (c) of the figure.

$$\mathbf{I}_1 = \frac{10}{j20}, \qquad \mathbf{E}_{oc} = j20\mathbf{I}_1 = 10 \text{ volts}$$

Then $Z_{eq} = \mathbf{E}_{oc}/\mathbf{I}_{sc} = j30 \ \Omega$, and the complete Norton equivalent circuit is shown in part (d). From the discussion of the parallel RLC circuit in Figs. 5.4-2 and 6.4-7, $|\mathbf{E}_o|$ will be a maximum when $\mathbf{I}_L + \mathbf{I}_C = 0$, i.e., when $30 = 1/\omega C$ and $C = 1/3000$ f. For this choice of C, $\mathbf{E}_o = \frac{10}{3}\underline{/-90°}$ and $e_o(t) = \frac{10}{3} \sin 100t$ volts.

Example 7.2-6. In the circuit in Fig. 7.2-6a, the switch K is open for $t < 0$, and all currents are zero. If K closes at $t = 0$, find $i_2(t)$ for all $t > 0$.

Solution. The transfer function with the switch closed can be found from Fig. 7.2-6b. The set of loop equations is usually easier to write if only one loop current is drawn through each transformer winding, as is done in the figure. When like terms are collected in the two loop equations,

$$\mathbf{E}_1(s) = (s + 2)\mathbf{I}_1(s) + (2s + 1)\mathbf{I}_2(s)$$
$$\mathbf{E}_1(s) = (2s + 1)\mathbf{I}_1(s) + (6s + 3)\mathbf{I}_2(s)$$

Solving simultaneously for $\mathbf{I}_2(s)$, we obtain

$$H(s) = \frac{\mathbf{I}_2(s)}{\mathbf{E}_1(s)} = \frac{-s + 1}{2s^2 + 11s + 5} = \frac{-s + 1}{(2s + 1)(s + 5)}$$

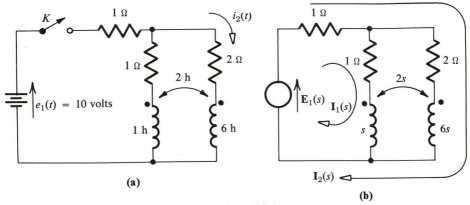

(a)

(b)

Figure 7.2-6

The steady-state response to a 10 volt d-c source is $10(\frac{1}{5}) = 2$, as can be seen from the transfer function or from the circuit. Because $H(s)$ has poles at $s = -\frac{1}{2}$ and -5, the complete response for $t > 0$ is

$$i_2(t) = 2 + K_1\epsilon^{-t/2} + K_2\epsilon^{-5t}$$

In order to evaluate the arbitrary constants, the values of $i_2(t)$ and di_2/dt at $t = 0+$ are needed. Since no impulse functions are present, the stored energy of the transformer cannot instantaneously change from its initial value of zero, and so $i_1(0+) = i_2(0+) = 0$. Therefore, there can be no voltage across the resistances at $t = 0+$, so 10 volts must appear across each inductance.

$$10 = \frac{di_1}{dt}(0+) + 2\frac{di_2}{dt}(0+)$$

$$10 = 2\frac{di_1}{dt}(0+) + 6\frac{di_2}{dt}(0+)$$

from which

$$\frac{di_2}{dt}(0+) = -5$$

A routine evaluation of the arbitrary constants then gives $K_1 = -\frac{10}{3}$, $K_2 = \frac{4}{3}$, and

$$i_2(t) = 2 - \tfrac{10}{3}\epsilon^{-t/2} + \tfrac{4}{3}\epsilon^{-5t}$$

Example 7.2-7.   The circuit in Fig. 7.2-7 contains no stored energy for $t < 0$. If the switch K suddenly closes at $t = 0$, find $i_2(t)$ for all $t > 0$.

Solution.   The transfer function with K closed is

$$H(s) = \frac{I_2(s)}{E_1(s)} = \frac{-sM}{(L_1L_2 - M^2)s^2 + (L_1R_2 + L_2R_1)s + R_1R_2} = \frac{-2s}{5s + 1}$$

so, for $t > 0$,

$$i_2(t) = 0 + K\epsilon^{-0.2t}$$

The student might first assume that the currents would remain equal to zero at $t = 0+$, as in the previous example. But this leads to the physically unreasonable conclusion that $i_2(t) \equiv 0$. It also leads to a mathematical contradiction in the two loop equations

$$1 = i_1 + \frac{di_1}{dt} + 2\frac{di_2}{dt}, \qquad 0 = i_2 + 2\frac{di_1}{dt} + 4\frac{di_2}{dt}$$

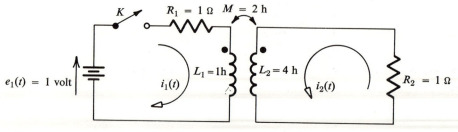

Figure 7.2-7

Subtraction of the second equation from twice the first equation gives $2 = 2i_1(t) - i_2(t)$, which is not satisfied by $i_1(0+) = i_2(0+) = 0$.

This complication arises only because $M^2 = L_1L_2$, i.e., because the coefficient of coupling has its maximum value of unity. If $M^2 < L_1L_2$, the expression for $i_2(t)$ would have two arbitrary constants, and $i_2(t)$ could remain zero at $t = 0+$ without being zero for all values of t.

Since the two windings of a transformer are linked by magnetic flux, the entire transformer should be considered as a single circuit element, whose stored energy $w(t)$ is given by Eq. 7.1-8. Figure 7.1-3 indicates that if $M^2 < L_1L_2$, $w(t)$ is zero only if $i_1(t) = i_2(t) = 0$, but that for $L_1L_2 = M^2$, $w(t)$ remains zero if $i_1(t) = -(M/L_1)i_2(t)$. Thus, only in a unity-coupled transformer can the currents change instantaneously even when the stored energy remains zero, and when no impulses are present.

In this example, the equations

$$2 = 2i_1(0+) - i_2(0+)$$

$$i_1(0+) = -\frac{M}{L_1} i_2(0+) = -2i_2(0+)$$

give $i_2(0+) = -0.4$ amp, $K = -0.4$, and

$$i_2(t) = -0.4\epsilon^{-0.2t} \qquad \text{for all } t > 0$$

7.3

Reflected Impedance; Unity-Coupled and Ideal Transformers

Consider the a-c steady-state analysis of the circuit in Fig. 7.3-1, where a voltage source is connected to the primary winding of a transformer, and a load impedance Z_2 is attached to the secondary winding. The two loop equations are

$$\mathbf{E}_1 = (j\omega L_1)\mathbf{I}_1 - (j\omega M)\mathbf{I}_2$$
$$0 = -(j\omega M)\mathbf{I}_1 + (j\omega L_2 + Z_2)\mathbf{I}_2$$

Solving these equations for $\mathbf{I}_1$ and $\mathbf{I}_2$, and noting that $\mathbf{E}_2 = Z_2\mathbf{I}_2$, we find that

$$\begin{aligned}
Z_1 &= \frac{\mathbf{E}_1}{\mathbf{I}_1} = j\omega L_1 + \frac{\omega^2 M^2}{j\omega L_2 + Z_2} \\[2mm]
\frac{\mathbf{E}_2}{\mathbf{E}_1} &= \frac{(j\omega M)Z_2}{-\omega^2(L_1L_2 - M^2) + (j\omega L_1)Z_2} \\[2mm]
\frac{\mathbf{I}_2}{\mathbf{I}_1} &= \frac{j\omega M}{j\omega L_2 + Z_2}
\end{aligned} \qquad (7.3\text{-}1)$$

The input impedance Z_1 seen by the source is the sum of the self-impedance $j\omega L_1$ and the *reflected impedance*

$$Z_f = \frac{\omega^2 M^2}{j\omega L_2 + Z_2} \qquad (7.3\text{-}2)$$

The reflected impedance involves the mutual coupling between the coils and the

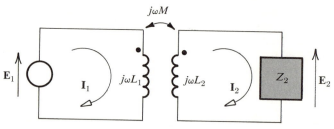

Figure 7.3-1

total self-impedance $j\omega L_2 + Z_2$ around the right-hand loop. Since the load impedance is, in general, a complex quantity, let

$$Z_2 = R_2 + jX_2$$

so that

$$Z_f = \left[\frac{\omega^2 M^2}{R_2 + j(X_2 + \omega L_2)}\right]\left[\frac{R_2 - j(X_2 + \omega L_2)}{R_2 - j(X_2 + \omega L_2)}\right]$$

$$= \left[\frac{\omega^2 M^2 R_2}{R_2{}^2 + (X_2 + \omega L_2)^2}\right] + j\left[\frac{-\omega^2 M^2(X_2 + \omega L_2)}{R_2{}^2 + (X_2 + \omega L_2)^2}\right] = R_f + jX_f$$

For a passive load impedance, $R_2 \geq 0$ but X_2 may have either sign. The last equation indicates that $R_f \geq 0$ as expected, but that the sign of X_f is *opposite* to that of $(X_2 + \omega L_2)$. Thus, an inductive load will "reflect" a capacitive reactance back into the left-hand loop.[†]

It is interesting to note that if $X_2 + \omega L_2 = 0$ at some frequency, then the impedance $Z_f = \omega^2 M^2 / R_2$, and the reflected resistance is inversely proportional to the load resistance. If, on the other hand, $X_2 + \omega L_2 \gg R_2$, the reflected resistance is directly proportional to the load resistance.

We next examine Eqs. 7.3-1 for certain important special cases. The *unity-coupled* transformer is one in which M has its maximum value of $\sqrt{L_1 L_2}$, i.e., in which all of the flux links all of the turns of both windings. In this case,

$$\frac{\mathbf{E}_2}{\mathbf{E}_1} = \sqrt{\frac{L_2}{L_1}} = N$$

By Eq. 7.1-4, the self-inductance is proportional to the square of the number of turns in the winding, so the constant N is called the *turns ratio*. Furthermore,

$$\frac{\mathbf{I}_2}{\mathbf{I}_1} = \frac{\sqrt{L_1/L_2}}{1 + Z_2/j\omega L_2} = \frac{1/N}{1 + Z_2/j\omega L_2}$$

$$Z_1 = j\omega L_1 + \frac{\omega^2 L_1 L_2}{j\omega L_2 + Z_2} = \frac{(L_1/L_2)Z_2}{1 + Z_2/j\omega L_2} = \frac{Z_2/N^2}{1 + Z_2/j\omega L_2}$$

[†] This statement is valid at any one particular frequency, but since the expression for X_f does not have the form $-(1/\omega C)$, the variation of X_f with the frequency of the source will not be the same as for a capacitance.

In the last expression for the input impedance, the separate identity of the reflected impedance Z_f has been lost.

The *ideal transformer* is a further idealization of the unity-coupled transformer, in which the self- and mutual inductances approach infinity. These characteristics can be physically approximated by winding both coils on a common, high-permeability iron core. The above equations then reduce to

$$\frac{E_2}{E_1} = N, \qquad \frac{I_2}{I_1} = \frac{1}{N}, \qquad Z_1 = \frac{Z_2}{N^2} \qquad (7.3\text{-}3)$$

If $N > 1$, the output voltage is larger than the input voltage, and the component is called a *step-up transformer*. A *step-down ideal transformer* is one for which $N < 1$. Notice that the quotient of the first two of Eqs. 7.3-3 gives

$$\frac{E_2 I_1}{I_2 E_1} = N^2 \quad \text{or} \quad \frac{Z_2}{Z_1} = N^2$$

which yields the third equation. The quantity Z_2/N^2 is sometimes called the reflected impedance, but such a definition is not consistent with Eqs. 7.3-1 and 7.3-2, where Z_f is only one part of the input impedance.

Equations 7.3-3 indicate that the two voltage phasors have the same phase angle, as do the two current phasors. In the steady state, therefore,

$$e_2(t) = N e_1(t) \quad \text{and} \quad i_2(t) = \frac{1}{N} i_1(t) \qquad (7.3\text{-}4)$$

where the voltages and currents are expressed as functions of time. Notice that

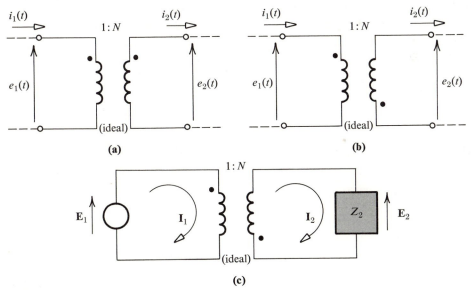

(a)   (b)

(c)

Figure 7.3-2

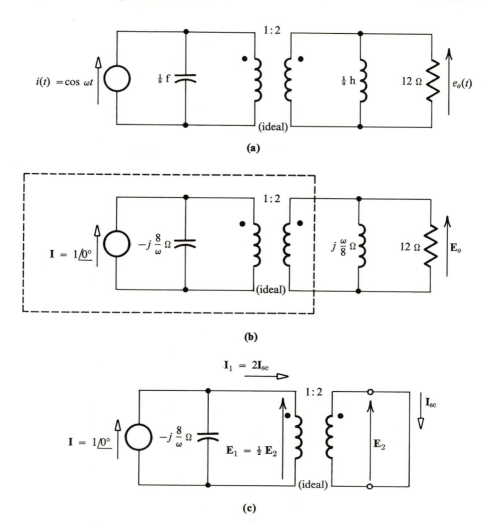

(a)

(b)

(c)

$e_2(t)i_2(t) = e_1(t)i_1(t)$, so that all the power delivered by the source is transmitted to the load, and no energy is dissipated or stored in the transformer itself.

Equations 7.3-4 are taken as the formal definition of the *ideal transformer*, whose symbol is shown in Fig. 7.3-2a. The definition applies even if the external connections are not similar to those in Fig. 7.3-1, and even if the analysis is not restricted to the a-c steady state. In the a-c steady-state analysis of Fig. 7.3-1, it is fairly easy to decide whether the ideal transformer is a suitable model for a physical component. Essentially all of the flux should link all the turns of both windings, and ωL_2 should be large compared to the magnitude of Z_2. The model certainly would not be satisfactory in the d-c steady state, because ω is then zero. In other applications, it may be difficult to decide whether or not the ideal transformer is a suitable model.

The ordinary transformer is characterized by three parameters: L_1, $L_2 = N^2 L_1$,

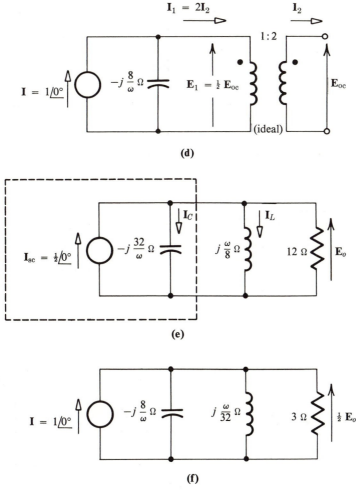

Figure 7.3-3

and $M = k\sqrt{L_1 L_2}$. For the ideal transformer, k equals unity and the inductances approach infinity, so the turns ratio N is the only additional information needed. Because the inductances become infinite, many of the remarks of Sections 7.1 and 7.2, including the previous statement of the dot convention, are difficult to apply to the ideal transformer. In circuit analysis, it is best to regard the ordinary and the ideal transformer as two separate and distinct types of components.[†]

[†] Some students may be confused by an apparent inconsistency in the dot conventions for the ordinary and ideal transformers. In the circuit of Fig. 7.3-1, the current reference arrows enter the terminal of the primary winding that has a dot, and leave the terminal of the secondary winding that has a dot. In the loop equations, the signs of the mutual-inductance terms are minus. The solution given in Eqs. 7.3-1, however, indicates that, if $|Z_2| \ll \omega L$, $I_2 = (M/L_2)I_1$ and the secondary current does flow in the clockwise direction.

Reversing one of the polarity dots on the ideal transformer is equivalent to replacing $e_2(t)$ by $-e_2(t)$, and $i_2(t)$ by $-i_2(t)$, so for Fig. 7.3-2b

$$e_2(t) = -Ne_1(t) \quad \text{and} \quad i_2(t) = -\frac{1}{N} i_1(t) \qquad (7.3\text{-}5)$$

In the frequency-domain circuit of Fig. 7.3-2c,

$$\mathbf{E}_2 = -N\mathbf{E}_1, \qquad \mathbf{I}_2 = -\frac{1}{N}\mathbf{I}_1, \qquad Z_1 = \frac{Z_2}{N^2} \qquad (7.3\text{-}6)$$

Notice that reversing one polarity dot replaces N by $-N$ in all the equations but does not affect the sign of the input impedance seen by the source.

In many applications, the ideal transformer divides the circuit into two parts. The following example develops the method that is usually used for circuits falling into this category.

Example 7.3-1.   For what value of ω is the steady-state output voltage in Fig. 7.3-3a a maximum? Find the steady-state output voltage for this value of ω.

Solution.   The frequency-domain circuit is shown in Fig. 7.3-3b. One approach, similar to the solution of Example 7.2-5, is to replace the part of the circuit within the dashed lines by its Norton equivalent. In part (c) of the figure, $\mathbf{E}_2 = 0$, so $\mathbf{E}_1 = 0$ and all the source current flows into the transformer. Then $\mathbf{I}_1 = 1\underline{/0^\circ}$ and $\mathbf{I}_{sc} = \frac{1}{2}\underline{/0^\circ}$. In part (d), $\mathbf{I}_2 = 0$, so $\mathbf{I}_1 = 0$ and all the source current flows into the capacitance. Then $\mathbf{E}_{oc}/2 = (1\underline{/0^\circ})(-j8/\omega)$, and the equivalent impedance of the dead network is $Z_{eq} = \mathbf{E}_{oc}/\mathbf{I}_{sc} = -j32/\omega \ \Omega$. As a check on this result, the last of Eqs. 7.3-3 gives

$$Z_{eq} = \frac{-j8/\omega}{(\frac{1}{2})^2} = \frac{-j32}{\omega}$$

For the Norton equivalent circuit in part (e), $|\mathbf{E}_o|$ is a maximum when $\mathbf{I}_C + \mathbf{I}_L = 0$, i.e., when $32/\omega = \omega/8$ or $\omega = 16$ rad per sec. For this value of ω, $\mathbf{E}_o = 12(\frac{1}{2}\underline{/0^\circ})$ and $e_o(t) = 6 \cos 16t$.

An alternative approach is to notice that the impedance looking into the primary of the transformer in Fig. 7.3-3b is $Z_1 = \frac{1}{4}Z_2$, where Z_2 is the impedance of the parallel RL combination. Therefore, the total impedance seen by the source must be given by the circuit in Fig. 7.3-3f. The voltage across the RL combination is now really the voltage across the primary of the transformer in the original circuit, and is labelled $\frac{1}{2}\mathbf{E}_o$. The magnitude of the voltage phasor $\mathbf{E}_o$ is a maximum when $8/\omega = \omega/32$ or $\omega = 16$ rad per sec. Then $\mathbf{E}_o/2 = 3\underline{/0^\circ}$ and $e_o(t) = 6 \cos 16t$.

A generalization of the procedure used in Example 7.3-1 leads to the following rule. The ideal transformer may be removed if all the circuit elements on one side are transferred to the other side. When transferring elements, multiply all voltages by n, all currents by $1/n$, and all impedances by n^2, where n is the ratio of the number of turns on the new side to the turns on the side where the elements were originally found. Since $Z_R = R$, $Z_L = j\omega L$, and $Z_C = 1/j\omega C$, multiplying the impedances by

n^2 is equivalent to multiplying a resistance or inductance by n^2, and dividing a capacitance by n^2.

The procedure of transferring all the elements to one side of an ideal transformer is normally used when the transformer divides the circuit into two parts. The method is not valid, however, when there are external connections between the two windings of the transformer, as in the following example. One convenient method that is always valid is to replace the ideal transformer by two controlled sources. The defining equations for the ideal transformer in Fig. 7.3-4a are

$$e_2(t) = Ne_1(t) \quad \text{and} \quad i_1(t) = Ni_2(t)$$

and suggest the two circuits in parts (b) and (c) of the figure. The latter two circuits are said to be *equivalent* to the ideal transformer, because they all have the same relationship between the external voltages and currents. If one of the polarity dots in Fig. 7.3-4a were reversed, N would be replaced by $-N$ in parts (b) and (c) of the figure.

Example 7.3-2.   Find the steady-state output voltage in Fig. 7.3-5a.

Solution.   In the frequency-domain circuit of Fig. 7.3-5b, the ideal transformer has been replaced by the model in Fig. 7.3-4b. When the two node equations

$$\frac{E_1 - E}{2} + \frac{E_1 - 2E_1}{2} + 2I_2 = 0$$

$$\frac{2E_1 - E_1}{2} + \frac{2E_1}{1 - j} - I_2 = 0$$

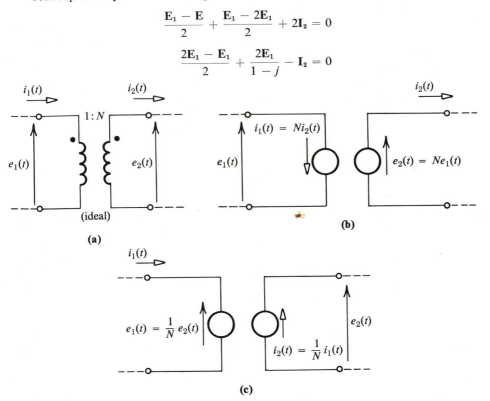

(a)

(b)

(c)

Figure 7.3-4

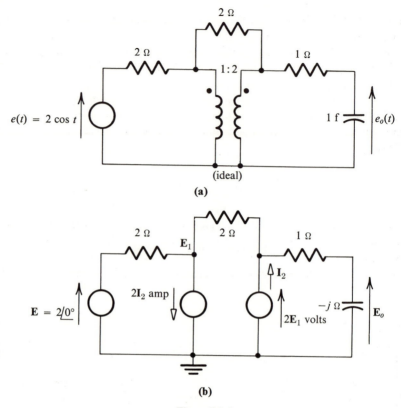

Figure 7.3-5

are solved simultaneously, there results $E_1 = E/(6 + j4)$. By the voltage divider rule,

$$E_o = \frac{(-j)(2E_1)}{1 - j} = \frac{(-j)(2\underline{/0°})}{(1 - j)(3 + j2)} = 0.39\underline{/-79°}$$

so $e_o(t) = 0.39 \cos (t - 79°)$.

7.4

Equivalent Circuits

Two circuits are said to be equivalent if the relationships between their external voltages and currents are always identical, i.e., if the circuits cannot be distinguished from one another by external electrical measurements. Any linear network with two external terminals can be replaced by its Thévenin or Norton equivalent circuit, when the external voltages and currents are calculated. Two-terminal networks are equivalent, therefore, if and only if they have the same Thévenin equivalent circuit. If a two-terminal network does not contain independent sources, its external behavior is completely described by its input impedance $Z(s)$, as discussed in connection with Fig. 6.3-5a.† Two such networks are equivalent if and only if they have the same impedance.

An equivalent circuit was developed in Fig. 7.3-4 for the ideal transformer, and it is logical to seek equivalent circuits for other components with more than two external terminals. Consider the ordinary transformer in Fig. 7.4-1a, which is described by the equations

$$e_1 = L_1 \frac{di_1}{dt} + M \frac{di_2}{dt}, \qquad e_2 = M \frac{di_1}{dt} + L_2 \frac{di_2}{dt}$$

These equations immediately suggest the combination of self-inductances and controlled sources shown in Fig. 7.4-1b. The values of the controlled sources are proportional to the derivative of a current instead of the current itself, but they are otherwise similar to the controlled sources found throughout the previous chapters.

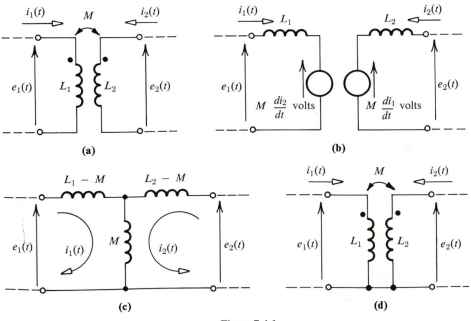

Figure 7.4-1

---

† Any initial stored energy within the network is considered as an independent source, but the network may contain controlled sources.

The equations for the transformer may be rewritten as

$$e_1 = (L_1 - M)\frac{di_1}{dt} + M\frac{d}{dt}(i_1 + i_2)$$

$$e_2 = M\frac{d}{dt}(i_2 + i_1) + (L_2 - M)\frac{di_2}{dt}$$

which suggest the *T* equivalent circuit in Fig. 7.4-1c. It was pointed out in Section 7.1 that reversing one polarity dot on a transformer is equivalent to replacing M by $-M$ in the circuit equations. This change would also have to be made in Figs. 7.4-1b and 7.4-1c. When numerical values are inserted into the T equivalent circuit, either $L_1 - M$ or $L_2 - M$ may turn out to be negative and hence not individually realizable, but this does not invalidate the equivalent circuit. The parts of an equivalent circuit need not necessarily have any physical significance when they are considered individually.

There is, however, another reason for questioning the validity of the proposed T equivalent circuit. The voltage between the two bottom terminals in parts (a) and (b) of Fig. 7.4-1 is not necessarily zero, but the corresponding voltage in part (c) is zero. Since the latter circuit can be distinguished from the other two by measuring

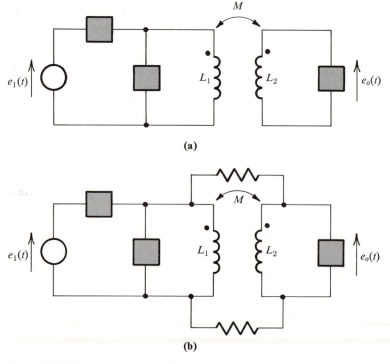

(a)

(b)

Figure 7.4-2

the resistance between the two bottom terminals, it is not completely equivalent to them. If the bottom terminals of the original transformer were joined, as in part (d) of the figure, then the circuit of part (c) would be a valid equivalent circuit, and the configuration of part (b) would be correct only if its bottom terminals were also connected.

The transformer in Fig. 7.4-2a divides the circuit into two parts. Joining the bottom ends of the transformer windings would not change the relationship between $e_1(t)$ and $e_o(t)$, because no current would flow through the added connecting wire. The T equivalent circuit could then be used in the calculation of $e_o(t)$. An example (which rarely occurs) where the T equivalent circuit may not be used is given in Fig. 7.4-2b.

Example 7.4-1.   Find the input impedance $Z(s)$ in Fig. 7.4-3a by using the T equivalent circuit for the transformer.

Solution.   From the equivalent circuit in Fig. 7.4-3b,

$$Z(s) = \frac{(2s + 2/s)(s + 4/s)}{(2s + 2/s) + (s + 4/s)} = \frac{2(s^2 + 1)(s^2 + 4)}{3s(s^2 + 2)}$$

The same result can be obtained by writing and solving two loop equations for the original circuit, as suggested in one of the problems at the end of the chapter.

We next consider any component or circuit which has three external terminals and which contains no independent sources. The relationships between the external voltages and currents may be given by a set of differential equations or, equivalently, by algebraic equations in the complex-frequency parameter s. In Fig. 7.4-4a, only the relationships between $e_1(t)$, $e_2(t)$, $i_1(t)$, and $i_2(t)$ need be considered, since these quantities uniquely determine the remaining voltage and current. All the following remarks will be valid if the component in part (a) of the figure is redrawn as in part (b). The two bottom terminals are at the same potential, so there are still only three independent terminals. As illustrated by the discussion associated with

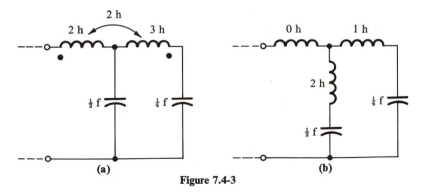

(a)   (b)

Figure 7.4-3

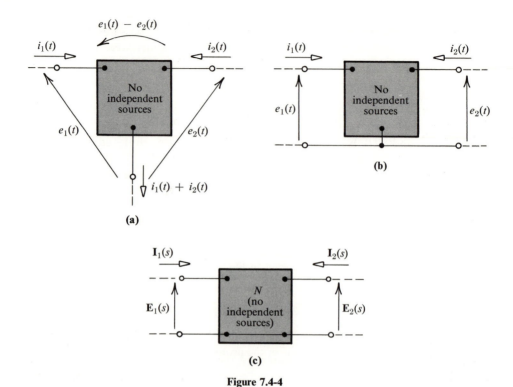

Figure 7.4-4

Fig. 7.4-1, however, some care must be exercised when trying to extend the results to a component with four independent terminals.

Suppose that all the voltages and currents are represented by their phasors and the passive elements by their impedances (as functions of s). If the network N in Fig. 7.4-4c is described by a set of node equations with the bottom terminal chosen as the reference node, $E_1(s)$ and $E_2(s)$ are two of the node voltages. The voltages at all the other nodes inside N also must be considered, and may be denoted by $E_3(s)$, $E_4(s)$, ... The set of node equations will have the form of Eqs. 2.5-2, except that phasors will be used and G_{ij} will be replaced by the admittance $Y_{ij}(s)$. The terms $I_1(s)$ and $I_2(s)$ represent the currents entering from the outside and may be treated as sources. $I_3(s) = I_4(s) = \cdots = 0$, because there are no independent sources inside the network. The value of any controlled sources may be expressed in terms of the node voltages and included on the left-hand side of the equals sign. The node equations will have the form

$$Y_{11}(s)E_1(s) - Y_{12}(s)E_2(s) - Y_{13}(s)E_3(s) - \cdots = I_1(s)$$

$$- Y_{21}(s)E_1(s) + Y_{22}(s)E_2(s) - Y_{23}(s)E_3(s) - \cdots = I_2(s)$$

$$- Y_{31}(s)E_1(s) - Y_{32}(s)E_2(s) + Y_{33}(s)E_3(s) - \cdots = 0$$

$$\cdots\cdots\cdots\cdots\cdots\cdots\cdots\cdots\cdots\cdots\cdots\cdots\cdots\cdots\cdots\cdots\cdots\cdots\cdots$$

and may be solved for $\mathbf{E}_1(s)$ and $\mathbf{E}_2(s)$ in terms of $\mathbf{I}_1(s)$ and $\mathbf{I}_2(s)$, giving

$$\mathbf{E}_1(s) = z_{11}(s)\mathbf{I}_1(s) + z_{12}(s)\mathbf{I}_2(s)$$
$$\mathbf{E}_2(s) = z_{21}(s)\mathbf{I}_1(s) + z_{22}(s)\mathbf{I}_2(s)$$

(7.4-1)

The quantities $z_{11}(s)$, $z_{12}(s)$, $z_{21}(s)$, and $z_{22}(s)$ can be expressed in terms of the coefficients $Y_{ij}(s)$ in the original node equations. If, for example, the node equations are solved by determinants and Crammer's rule,

$$z_{11}(s) = \frac{\Delta_{11}}{\Delta}, \qquad z_{12}(s) = \frac{\Delta_{21}}{\Delta}, \qquad z_{21}(s) = \frac{\Delta_{12}}{\Delta}, \qquad z_{22}(s) = \frac{\Delta_{22}}{\Delta}$$

where

$$\Delta = \begin{vmatrix} Y_{11}(s) & -Y_{12}(s) & -Y_{13}(s) & \cdots \\ -Y_{21}(s) & Y_{22}(s) & -Y_{23}(s) & \cdots \\ -Y_{31}(s) & -Y_{32}(s) & Y_{33}(s) & \cdots \\ \cdots & \cdots & \cdots & \cdots \end{vmatrix}$$

and where Δ_{ij} denotes the ijth cofactor, i.e., $(-1)^{i+j}$ times the determinant formed by deleting the ith row and jth column of Δ. The four z parameters therefore depend only upon the values of the circuit elements within the network N and upon the complex frequency s. In order to simplify the form of the equations, the s in parentheses usually will be omitted in the remainder of this chapter. The student should keep in mind, however, that all impedances and phasors are understood to be functions of s, unless otherwise stated.

Equations 7.4-1 indicate that any three-terminal network may be replaced by the equivalent circuit in Fig. 7.4-5a, once its z parameters are known. These parameters may be determined by inspection in a few cases. The transformer in Fig. 7.4-1d is described by Eqs. 7.2-1, which are

$$\mathbf{E}_1 = sL_1\mathbf{I}_1 + sM\mathbf{I}_2$$
$$\mathbf{E}_2 = sM\mathbf{I}_1 + sL_2\mathbf{I}_2$$

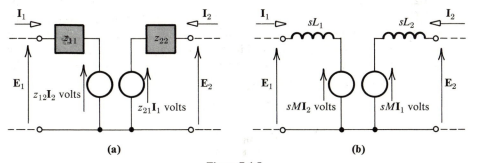

(a)   (b)

Figure 7.4-5

Thus,

$$z_{11} = sL_1, \qquad z_{12} = z_{21} = sM, \qquad z_{22} = sL_2$$

The equivalent circuit in Fig. 7.4-5b is, of course, similar to Fig. 7.4-1b.

Even if the values of the z parameters are not immediately evident, it is still not necessary to write and solve a complete set of node equations. Recall that the parameters do not depend upon the external connections to the network N, and that Eqs. 7.4-1 are always valid. If the right-hand side of the network is left open circuited, so that $\mathbf{I}_2 = 0$,

$$z_{11} = \left[\frac{\mathbf{E}_1}{\mathbf{I}_1}\right]_{\mathbf{I}_2=0} \quad \text{and} \quad z_{21} = \left[\frac{\mathbf{E}_2}{\mathbf{I}_1}\right]_{\mathbf{I}_2=0} \tag{7.4-2}$$

Similarly, if the left-hand side is open circuited, the equations give

$$z_{12} = \left[\frac{\mathbf{E}_1}{\mathbf{I}_2}\right]_{\mathbf{I}_1=0} \quad \text{and} \quad z_{22} = \left[\frac{\mathbf{E}_2}{\mathbf{I}_2}\right]_{\mathbf{I}_1=0} \tag{7.4-3}$$

Equations 7.4-2 and 7.4-3 suggest a procedure for determining the z parameters analytically for a given circuit model or for obtaining them experimentally by voltage and current measurements. They also explain why the parameters are often called the *open-circuit impedance parameters*.

Example 7.4-2.   Find the z parameters for the π (or delta) circuit in Fig. 7.4-6a.

Solution.   When a source is connected to the left-hand side of the circuit and when $\mathbf{I}_2 = 0$, as in Fig. 7.4-6b,

$$\mathbf{I}_1 = \frac{\mathbf{E}_1}{Z_1} + \frac{\mathbf{E}_1}{Z_2 + Z_3}$$

or

$$\mathbf{E}_1 = \frac{Z_1(Z_2 + Z_3)}{Z_1 + Z_2 + Z_3} \mathbf{I}_1$$

Also,

$$\mathbf{E}_2 = \frac{Z_2}{Z_2 + Z_3} \mathbf{E}_1 = \frac{Z_1 Z_2}{Z_1 + Z_2 + Z_3} \mathbf{I}_1$$

Thus,

$$z_{11} = \frac{Z_1(Z_2 + Z_3)}{Z_1 + Z_2 + Z_3} \quad \text{and} \quad z_{21} = \frac{Z_1 Z_2}{Z_1 + Z_2 + Z_3}$$

The other z parameters can be found in a similar way from Fig. 7.4-6c, and are

$$z_{12} = \frac{Z_1 Z_2}{Z_1 + Z_2 + Z_3} \quad \text{and} \quad z_{22} = \frac{Z_2(Z_1 + Z_3)}{Z_1 + Z_2 + Z_3}$$

Example 7.4-3.   Determine the z parameters for the transistor model in Fig. 7.4-7.

Solution.   When $\mathbf{I}_2 = 0$, $\mathbf{E}_1 = r_n \mathbf{I}_1$ and $\mathbf{E}_2 = -r_o(\beta \mathbf{I}_1)$, so $z_{11} = r_n$ and $z_{21} = -\beta r_o$. When $\mathbf{I}_1 = 0$, $\mathbf{E}_1 = 0$ and $\mathbf{E}_2 = r_o \mathbf{I}_2$, so $z_{12} = 0$ and $z_{22} = r_o$.

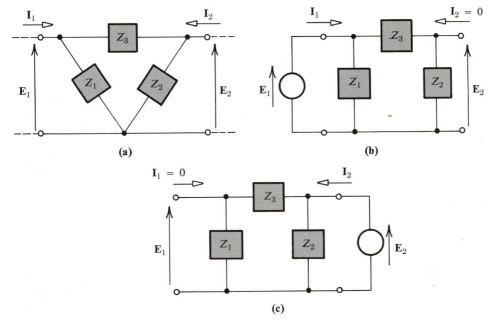

(a)

(b)

(c)

Figure 7.4-6

Notice that $z_{12} = z_{21}$ for the circuit in Example 7.4-2, which contained no controlled sources, but that $z_{12} \neq z_{21}$ in Example 7.4-3. In order to be able to draw
some general conclusions about z_{12} and z_{21}, reconsider the set of node equations
from which Eqs. 7.4-1 were derived. From the discussion associated with Eqs. 2.5-2,
we see that $Y_{ij}(s) = Y_{ji}(s)$ for circuits without controlled sources. Then the
determinant Δ will be symmetrical, so $\Delta_{21} = \Delta_{12}$ and $z_{12} = z_{21}$. We conclude that
z_{12} and z_{21} are always identical for circuits that are composed only of resistances,
capacitances, self-inductances, and mutual inductance.[†]

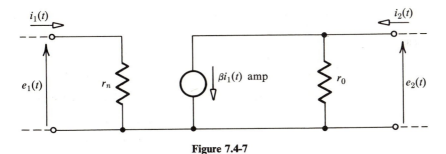

Figure 7.4-7

[†] We shall not bother to formally prove that $Y_{ij}(s) = Y_{ji}(s)$ even when mutual inductance is
present. A plausibility argument can be based upon the fact that Fig. 7.4-1d is equivalent to the
three self-inductances in Fig. 7.4-1c.

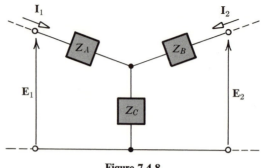

Figure 7.4-8

Whenever $z_{12} = z_{21}$, Eqs. 7.4-1 may be rewritten as

$$\mathbf{E}_1 = (z_{11} - z_{21})\mathbf{I}_1 + z_{21}(\mathbf{I}_1 + \mathbf{I}_2)$$

$$\mathbf{E}_2 = z_{21}(\mathbf{I}_2 + \mathbf{I}_1) + (z_{22} - z_{21})\mathbf{I}_2$$

which describe the T (or wye) equivalent circuit in Fig. 7.4-8, where

$$Z_A = z_{11} - z_{21}, \qquad Z_C = z_{21}, \qquad Z_B = z_{22} - z_{21}$$

Figure 7.4-1c is one example of the T equivalent circuit. Using the results of Example 7.4-2, we may replace the π circuit of Fig. 7.4-6a by an equivalent T, in which

$$Z_A = \frac{Z_1 Z_3}{Z_1 + Z_2 + Z_3}$$

$$Z_B = \frac{Z_2 Z_3}{Z_1 + Z_2 + Z_3} \qquad (7.4\text{-}4)$$

$$Z_C = \frac{Z_1 Z_2}{Z_1 + Z_2 + Z_3}$$

Equations 7.4-4 are known as the π-T (*or delta-wye*) *transformation*, and may be used to replace a π network by an equivalent T. Notice that the numerator of Z_A is the product of the two impedances in the π that are connected to node A.

Example 7.4-4. Use the delta-wye transformation to find the input impedance for the frequency-domain circuit in Fig. 7.4-9a.

Solution. The transformation enables the circuit to be replaced by the one in Fig. 7.4-9b. The input impedance is

$$Z = 3 + \frac{(j5)(-j3)}{j5 - j3} = 3 - j7.5 \ \Omega$$

which agrees with the answer to Example 5.3-5.

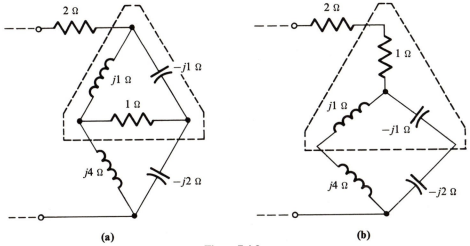

(a)   (b)

Figure 7.4-9

The z parameters constitute only one of several common ways of describing the external behavior of three-terminal networks. When Eqs. 7.4-1 are solved simultaneously for the two currents,

$$\mathbf{I}_1(s) = y_{11}(s)\mathbf{E}_1(s) + y_{12}(s)\mathbf{E}_2(s)$$
$$\mathbf{I}_2(s) = y_{21}(s)\mathbf{E}_1(s) + y_{22}(s)\mathbf{E}_2(s)$$

(7.4-5)

where

$$y_{11} = \frac{z_{22}}{|z|}, \quad y_{12} = \frac{-z_{12}}{|z|}, \quad y_{21} = \frac{-z_{21}}{|z|}, \quad y_{22} = \frac{z_{11}}{|z|}, \quad |z| = z_{11}z_{22} - z_{12}z_{21}$$

(7.4-6)

The four y parameters depend only upon the elements within the network and upon the complex frequency s. Equations 7.4-5 could also have been obtained by writing and solving a set of loop equations for the network N in Fig. 7.4-4c. The equations describe the equivalent circuit in Fig. 7.4-10.

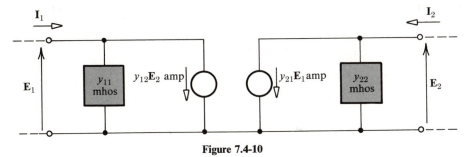

Figure 7.4-10

The usual method of evaluating the y parameters is to first short circuit the right-hand side of the network, so that $\mathbf{E}_2 = 0$. Equations 7.4-5 then reduce to

$$y_{11} = \left[\frac{\mathbf{I}_1}{\mathbf{E}_1}\right]_{\mathbf{E}_2=0} \quad \text{and} \quad y_{21} = \left[\frac{\mathbf{I}_2}{\mathbf{E}_1}\right]_{\mathbf{E}_2=0} \tag{7.4-7}$$

When the left-hand terminals are shorted, $\mathbf{E}_1 = 0$.

$$y_{12} = \left[\frac{\mathbf{I}_1}{\mathbf{E}_2}\right]_{\mathbf{E}_1=0} \quad \text{and} \quad y_{22} = \left[\frac{\mathbf{I}_2}{\mathbf{E}_2}\right]_{\mathbf{E}_1=0} \tag{7.4-8}$$

Because of Eqs. 7.4-7 and 7.4-8, the y parameters are called the *short-circuit admittance parameters*. Notice that, in general, $y_{12} \neq 1/z_{12}$ and $y_{21} \neq 1/z_{21}$.

Example 7.4-5. Find the y parameters for the T circuit in Fig. 7.4-8.

Solution. When the two right-hand terminals are shorted,

$$\mathbf{E}_1 = \left(Z_A + \frac{Z_B Z_C}{Z_B + Z_C}\right) \mathbf{I}_1$$

$$\mathbf{I}_1 = \frac{Z_B + Z_C}{Z_A Z_B + Z_A Z_C + Z_B Z_C} \mathbf{E}_1$$

$$\mathbf{I}_2 = \frac{-Z_C}{Z_B + Z_C} \mathbf{I}_1 = \frac{-Z_C}{Z_A Z_B + Z_A Z_C + Z_B Z_C} \mathbf{E}_1$$

so

$$y_{11} = \frac{Z_B + Z_C}{Z_A Z_B + Z_A Z_C + Z_B Z_C}, \quad y_{21} = \frac{-Z_C}{Z_A Z_B + Z_A Z_C + Z_B Z_C}$$

By examining the T network when the left-hand terminals are shorted, it is found that $y_{12} = y_{21}$ and that

$$y_{22} = \frac{Z_A + Z_C}{Z_A Z_B + Z_A Z_C + Z_B Z_C}$$

Example 7.4-6. Determine the y parameters for the transistor model in Fig. 7.4-7.

Solution. When $\mathbf{E}_2 = 0$, $\mathbf{I}_1 = (1/r_n)\mathbf{E}_1$ and $\mathbf{I}_2 = (\beta/r_n)\mathbf{E}_1$, so $y_{11} = 1/r_n$ and $y_{21} = \beta/r_n$. When $\mathbf{E}_1 = 0$, $\mathbf{I}_1 = 0$ and $\mathbf{I}_2 = (1/r_o)\mathbf{E}_2$, so $y_{12} = 0$ and $y_{22} = 1/r_o$. The same results can be obtained from Example 7.4-3 and Eqs. 7.4-6.

For networks composed only of resistances, capacitances, and self- and mutual inductances, $z_{12} = z_{21}$ so $y_{21} = y_{12}$. In such cases, Eqs. 7.4-5 may be rewritten as

$$\mathbf{I}_1 = (y_{11} + y_{21})\,\mathbf{E}_1 - y_{21}\,(\mathbf{E}_1 - \mathbf{E}_2)$$

$$\mathbf{I}_2 = -y_{21}(\mathbf{E}_2 - \mathbf{E}_1) + (y_{22} + y_{21})\mathbf{E}_2$$

which describe the π equivalent circuit in Fig. 7.4-11b, where

$$Y_1 = y_{11} + y_{21}, \qquad Y_2 = y_{22} + y_{21}, \qquad Y_3 = -y_{21}$$

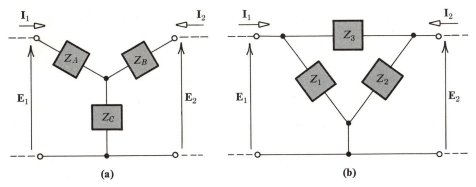

(a)   (b)

Figure 7.4-11

If the T circuit of Fig. 7.4-11a is to be replaced by a π equivalent circuit, then, by the results of Example 7.4-5,

$$Z_1 = \frac{Z_A Z_B + Z_A Z_C + Z_B Z_C}{Z_B}$$

$$Z_2 = \frac{Z_A Z_B + Z_A Z_C + Z_B Z_C}{Z_A} \qquad (7.4\text{-}9)$$

$$Z_3 = \frac{Z_A Z_B + Z_A Z_C + Z_B Z_C}{Z_C}$$

These equations also could have been obtained by solving simultaneously Eqs. 7.4-4. Notice that the denominator of Z_1 is Z_B, which corresponds to the terminal to which Z_1 is not connected. One application of this transformation is indicated by the following examples.

Example 7.4-7. Replace the transformer's T equivalent circuit of Fig. 7.4-1c by a π equivalent circuit.

Solution. The $T - \pi$ transformation yields the circuit of Fig. 7.4-12, if $M^2 \neq L_1 L_2$. Such a circuit may be useful when writing a set of node equations. Unlike the T equivalent circuit, it does not introduce an additional internal node. If one of the

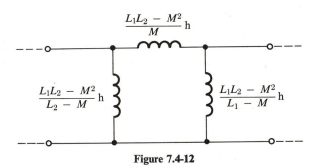

Figure 7.4-12

polarity dots on the original transformer were changed, M would be replaced by $-M$ in this equivalent circuit.

Example 7.4-8.   For the circuit in Fig. 7.4-13a, what value of C will produce the largest steady-state output voltage?

Solution.   The transformer is replaced by a π equivalent circuit in Fig. 7.4-13b. Although the frequency-domain circuit in Fig. 7.2-5d now follows by inspection, for variety we instead write a node equation. In the a-c steady state, with $\omega = 100$ rad per sec,

$$\left(\frac{1}{j30} + j100C + \frac{1}{10} \right) \mathbf{E}_o = \left(\frac{1}{j30} \right) \mathbf{E}_1$$

The magnitude of the output phasor will be a maximum when $-j/30 = j100C$, i.e., when $C = 1/3000$ f. For this value of C, $\mathbf{E}_o = (10/j30)\mathbf{E}_1 = (1/j3)(10\underline{/0°}) = \frac{10}{3}\underline{/-90°}$ and $e_o(t) = \frac{10}{3} \sin 100t$. These answers agree with those for Example 7.2-5.

Once the z or y parameters have been found, either analytically or experimentally, a three-terminal network may be replaced by one of the equivalent circuits we have developed. When the external connections are made, the network function $H(s)$ can be found by using the equivalent circuit. In some cases, however, an analysis using Eqs. 7.4-1 or 7.4-5 may be easier. Suppose, for example, that the voltage gain $\mathbf{E}_2/\mathbf{E}_1$ in Fig. 7.4-14 is desired when the right-hand terminals are left open circuited. When $\mathbf{I}_2 = 0$ in Eqs. 7.4-1, $\mathbf{E}_1 = z_{11}\mathbf{I}_1$ and $\mathbf{E}_2 = z_{21}\mathbf{I}_1$, so $\mathbf{E}_2/\mathbf{E}_1 = z_{21}/z_{11}$. Alternatively, by letting $\mathbf{I}_2 = 0$ in Eqs. 7.4-5, $\mathbf{E}_2/\mathbf{E}_1 = -y_{21}/y_{22}$.

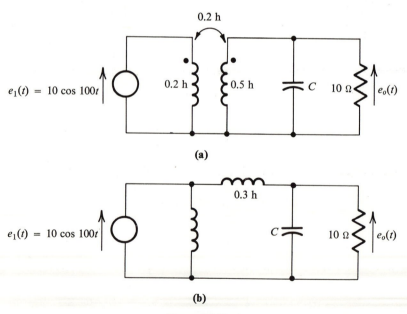

(a)

(b)

Figure 7.4-13

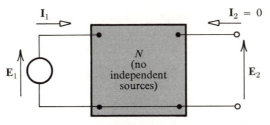

Figure 7.4-14

The equivalent circuits in this section have been based upon the z and y parameters. In Chapter 12 we discuss two other methods of characterizing three-terminal networks that are particularly convenient for transistors and vacuum tubes.

It is sometimes useful to develop equivalent circuits in which the individual parts have some physical significance. The behavior of large power transformers, for example, roughly approximates that of an ideal transformer. It would be useful to have a model for the power transformer that includes an ideal transformer plus elements that represent the causes of the deviation from the desired behavior. One

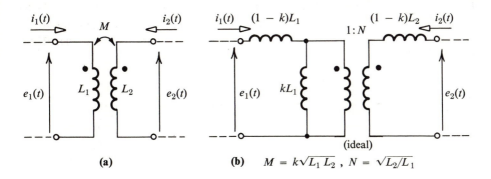

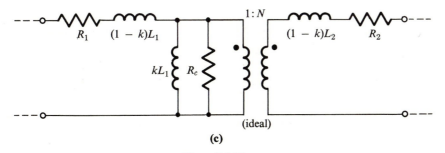

Figure 7.4-15

such equivalent circuit that is often used is suggested by the development of Eqs. 7.3-3 from Eqs. 7.3-1, and is shown in Fig. 7.4-15b. The student should show that in the a-c steady state the figure is described by the equations

$$\mathbf{E}_1 = j\omega L_1\mathbf{I}_1 + j\omega M\mathbf{I}_2$$

$$\mathbf{E}_2 = j\omega M\mathbf{I}_1 + j\omega L_2\mathbf{I}_2 \tag{7.4-10}$$

so that it is an equivalent circuit for the ordinary transformer in Fig. 7.4-15a. If all the flux links all the turns of both windings, $k = 1$, and $1 - k = 0$. The inductances $(1 - k)L_1$ and $(1 - k)L_2$ are therefore related to the flux leakage, and are sometimes called *leakage inductances*. They become short circuits for a unity-coupled transformer. Notice that even if the secondary terminals are left open circuited, current may still flow through the inductance kL_1 and will establish a magnetic field in the core. Thus, kL_1 is called the *magnetizing inductance*. If $k = 1$ and if the inductances become infinite (corresponding to an ideal core with infinite permeability), then the entire equivalent circuit reduces to an ideal transformer, as it should.

There is some resistance in the windings of any physical transformer and also hysteresis and eddy-current losses within any iron core. In Fig. 7.4-15c, the winding resistances are represented by R_1 and R_2 and the core loss by R_c.

7.5

Summary

The ordinary two-winding transformer is a four-terminal component that is characterized by three parameters: the self-inductances L_1 and L_2 and the mutual inductance M. The defining equations are given in Eqs. 7.1-5, 7.2-2 (for frequency-domain circuits), and 7.2-1 (for complex-frequency-domain circuits). The polarity markings that are placed at one end of each winding depend upon the way in which the individual coils are wound and are used to determine the proper sign for the mutual-inductance terms. Changing one polarity dot, without changing the voltage and current reference arrows, is equivalent to replacing M by $-M$ in all the equations.

The ideal transformer can be regarded as a very special case of the ordinary transformer and is characterized by a single parameter: the turns ratio N. Reversing a polarity dot is equivalent to replacing N by $-N$ in the defining equations. Whenever an ideal transformer divides the circuit into two parts, all the elements on one side may be transferred to the other side. The voltages of the elements that are transferred are multiplied by n, the currents by $1/n$, and the impedances by n^2, where n is the ratio of the number of turns on the new side to the number of turns on the original side.

Any network containing no independent sources and having three external terminals (or four external terminals if two of them can be connected together without affecting the behavior) can be completely characterized by four parameters. These parameters may be the z parameters defined in Eqs. 7.4-1 or the y parameters in Eqs. 7.4-5. One convenient way of determining the parameters is by analyzing the network under open-circuit or short-circuit conditions. If the network contains no controlled sources, $z_{12} = z_{21}$ and $y_{12} = y_{21}$, and there are only three independent parameters.

One important byproduct of the discussion of equivalent circuits is the delta-wye transformation. By the use of Eqs. 7.4-4 and 7.4-9, either of the circuits in Fig. 7.4-11 may be transformed into the other one.

The z and the y parameters are two of several methods of characterizing three-terminal networks. Other methods, including some that are particularly useful in vacuum-tube and transistor circuits, are discussed in Chapter 12.

PROBLEMS

7.1 In the development of Eqs. 7.1-5, it was assumed that $N_1(\partial \phi_1/\partial i_2) = N_2(\partial \phi_2/\partial i_1)$. If this assumption were not valid, Eqs. 7.1-5 would have the form

$$e_1 = L_1 \frac{di_1}{dt} + M_{12} \frac{di_2}{dt}, \qquad e_2 = M_{21} \frac{di_1}{dt} + L_2 \frac{di_2}{dt}$$

Suppose that at time t_0 the currents are i_{1_0} and i_{2_0}. If $i_1(t)$ is first reduced to zero while $i_2(t) = i_{2_0}$, and if $i_2(t)$ is then reduced to zero, find an expression for the energy obtained from the transformer. Find a similar expression when the currents are reduced to zero in the opposite order, and prove that $M_{12} = M_{21}$.

7.2 A sketch of a three-winding transformer is given in Fig. P7.2a. Assume that at least part of the flux produced by one winding links the turns of the other windings. Such a transformer is characterized by three self-inductances and three mutual inductances, as in Fig. P7.2b.

(a) By a generalization of Eqs. 7.1-5, express the three voltages in terms of the three currents.

(b) Place polarity dots on one end of each of the windings shown in Figs. P7.2c and P7.2d. The dots should be placed so that, when positive currents flow into the dotted end of each coil, the fluxes produced by the currents are in the same direction. Then a generalization of the dot convention for two-winding transformers may be used.

7.3 Is it possible to place three polarity dots on the sketch in Fig. P7.3 so that they will be consistent with the dot convention? If it is possible, do so; if it is not possible, suggest another system of polarity markings. Assume that substantially all the flux is confined to the core.

7.4 Find an expression for the equivalent inductance for each of the circuits in Fig. P7.4.

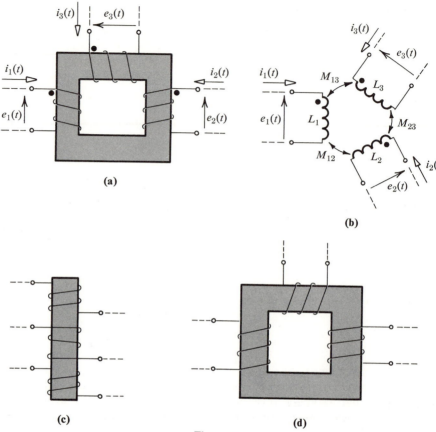

(a)

(b)

(c)

(d)

Figure P7.2

7.5 Let $R_1 = 1 \, \Omega$, $L_1 = 1$ h, and $R_2 = 100 \, \Omega$ for the circuit in Fig. 7.2-3a. Sketch the decibel voltage gain as a function of ω, and determine the approximate bandwidth for each of the following cases.

(a) $L_2 = 10$ h, $k = \frac{1}{2}$

(b) $L_2 = $ 10 h, $k = 1$

(c) $L_2 = 100$ h, $k = \frac{1}{2}$

(d) $L_2 = 100$ h, $k = 1$

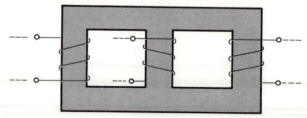

Figure P7.3

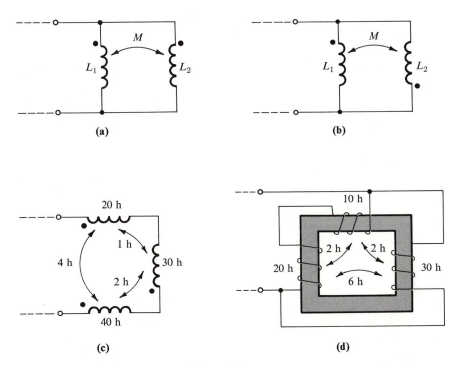

Figure P7.4

7.6 Find the input impedance $Z(s) = \mathbf{E}_1(s)/\mathbf{I}_1(s)$ for the circuit in Fig. P7.6 by writing and solving two loop equations. Plot the pole-zero pattern, and sketch curves of $|Z(j\omega)|$ and $\measuredangle\ Z(j\omega)$ versus ω.

7.7 Find the transfer function $H(s) = \mathbf{E}_o(s)/\mathbf{E}_1(s)$ for the circuit in Fig. P7.7.

7.8 The largest a-c steady-state response for the circuit in Fig. P7.8 occurs at approximately 1000 rad per sec.

(a) If $i_1(t) = \cos 1000t$, find an expression for the output phasor $\mathbf{E}_o$ in terms of the mutual inductance M.

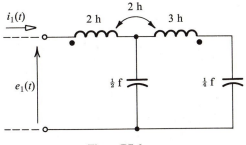

Figure P7.6

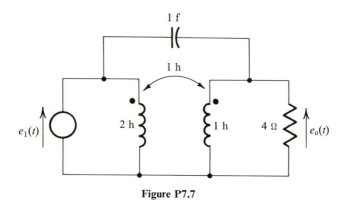

Figure P7.7

Figure P7.8

(b) What value of M results in the largest value of $|\mathbf{E}_o|$?

(c) Calculate the bandwidth when M has the value in part (b), and also when M is very small.

7.9 Verify the calculation of Z_{eq} in Example 7.2-5 by a direct analysis of the dead network.

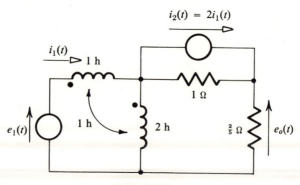

Figure P7.10

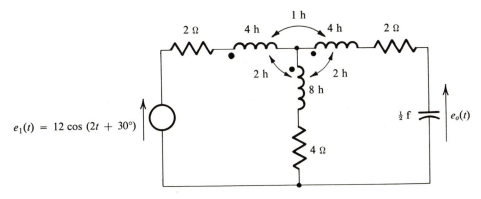

Figure P 7.11

7.10 Use Thévenin's theorem to find the steady-state output voltage $e_o(t)$ for the circuit in Fig. P7.10 when $e_1(t) = \cos t$.

7.11 The circuit in Fig. P7.11 contains a three-winding transformer. Find the voltage $e_o(t)$ in the steady state.

7.12 For the circuit in Fig. P7.12, show that the denominator of the transfer function $H(s) = E_o(s)/E_1(s)$ is $(1 + 2M_{12}M_{13}M_{23} - M_{12}^2 - M_{13}^2 - M_{23}^2)s^3 + (3 - M_{12}^2 - M_{13}^2 - M_{23}^2)Rs^2 + 3R^2s + R^3$. Notice that, even when the values of the mutual inductances do not exceed one (e.g., $M_{12} = M_{13} = 1$ and $M_{23} = 0$), the transfer function may have a pole in the right half of the s-plane. What does this result infer about the permissible values of the mutual inductances in a three-winding transformer?

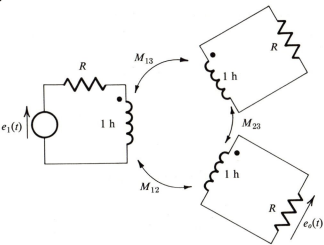

Figure P7.12

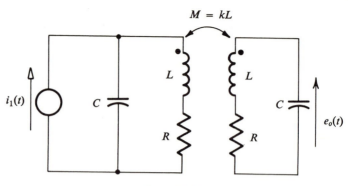

Figure P7.13

7.13 Show that the transfer function for the "doubly tuned" circuit in Fig. P7.13 is

$$H(s) = \frac{E_o(s)}{I_1(s)} = \frac{M}{C^2 L^2} \frac{s}{(s^2 + 2\alpha s + \omega_0^2)^2 - k^2 s^4}$$

where $\alpha = R/2L$ and $\omega_0^2 = 1/LC$, and where k is the coefficient of coupling.

(a) If $\alpha \ll \omega_0$ and if $k \ll 1$, show that

$$H(s) \doteq \frac{M}{C^2 L^2} \frac{s}{(s + \alpha + j\omega_1)(s + \alpha - j\omega_1)(s + \alpha + j\omega_2)(s + \alpha - j\omega_2)}$$

where $\omega_1 = \omega_0(1 - k/2)$ and $\omega_2 = \omega_0(1 + k/2)$. Notice the similarity of this transfer function with the one in Problem 6.23.

(b) Use the expression for the reflected impedance Z_f in Section 7.3 to explain how the coupling of two identical RLC circuits can produce an effect similar to that for the cascading of two different RLC circuits.

7.14 In each of the circuits in Fig. P7.14, steady-state conditions exist at $t = 0-$, with the switch K open. If the switch closes at $t = 0$, find the output voltage or current for all $t > 0$.

7.15 Replace all of the circuit in Fig. P7.15, except for the resistance R_o, by a Thévenin equivalent circuit. Use the results of Problem 2.16 to find the value of R_o that will result in the maximum power delivered to R_o. Then calculate the power delivered to R_o as a function of time, and also find the average value of this power.

7.16 For the circuit in Fig. P7.16, transfer all the elements to the right side of the right-hand transformer. Then calculate $e_o(t)$ in the steady state.

7.17 The two circuits in Fig. P7.17 are sometimes called *autotransformers*. The inductances L_1 and L_2 may represent two separate coils or a single coil with an additional contact between the two ends.

(a) In the a-c steady state, find expressions for the input impedance $\mathbf{E}_1/\mathbf{I}_1$, and for the transfer functions $\mathbf{E}_2/\mathbf{E}_1$ and $\mathbf{I}_2/\mathbf{I}_1$.

(b) If the coefficient of coupling k is unity and if the inductances are then allowed to become indefinitely large, to what do the previous results reduce? Draw the equivalent circuit that is suggested by your answers.

7.18 The circuit in Fig. P7.18 is operating in the a-c steady state. Find an expression for the output phasor $\mathbf{E}_o$ in terms of the turns ratio of the ideal transformer.

(a) For what value of N will the output voltage be in phase with the input voltage?

(b) What value of N will result in the largest value of $|\mathbf{E}_o|$?

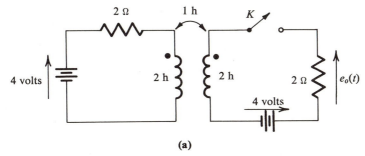

(a)

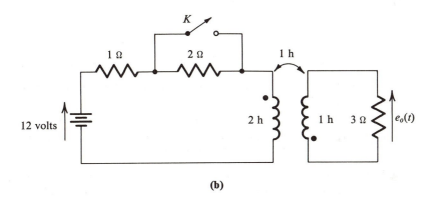

(b)

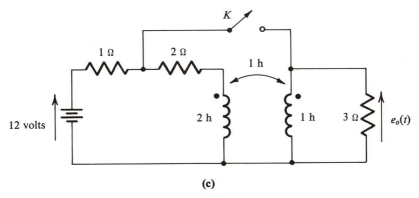

(c)

Figure P7.14

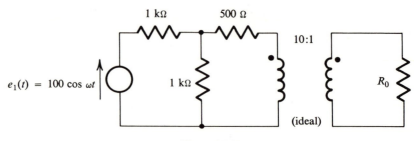

Figure P7.15

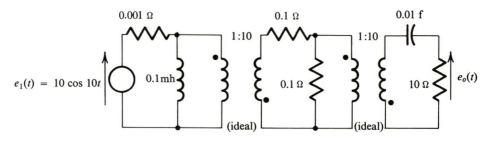

Figure P7.16

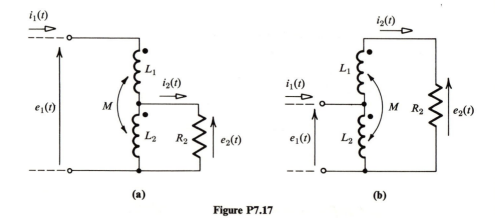

(a)   (b)

Figure P7.17

7.19 At the angular frequency of 200 rad per sec, the transformer in Fig. P7.19 can be regarded as an ideal transformer. Its primary to secondary turns ratio is $1:2$. Find $i_o(t)$ in the steady state.

7.20 Find the transfer function $H(s) = E_o(s)/E_1(s)$ for each of the circuits in Fig. P7.20. Discuss the relative advantages of using a controlled source (as in Fig. 3.5-7a) and of using a transformer to provide isolation, i.e., to prevent one part of the circuit from "loading down" a previous part.

7.21 Solve Problems 7.10 and 7.14 by using a T or π equivalent circuit for the transformer.

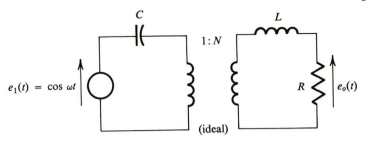

Figure P7.18

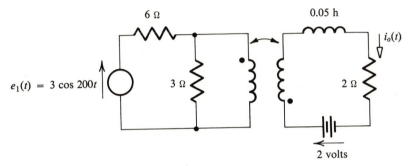

Figure P7.19

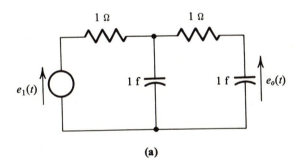

(a)

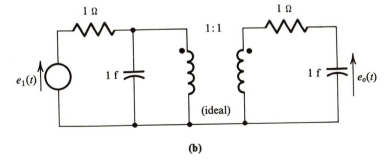

(b)

Figure P7.20

7.22 Find the impedance $Z(s)$ for the portion of the circuit in Fig. 7.2-7 that is to the right of R_1. The result, which may be written as $Z(s) = 1/[(L_2/L_1R_2) + (1/sL_1)]$, suggests the parallel combination of a resistance and inductance. Use this fact to justify the values $i_1(0+) = \frac{4}{5}$ amp and $i_2(0+) = -\frac{2}{5}$ amp in Example 7.2-7. If the transformer is replaced by a T equivalent circuit, show that voltage impulses appear across the branches of the T when the switch is closed, but do not appear across the terminals of the transformer.

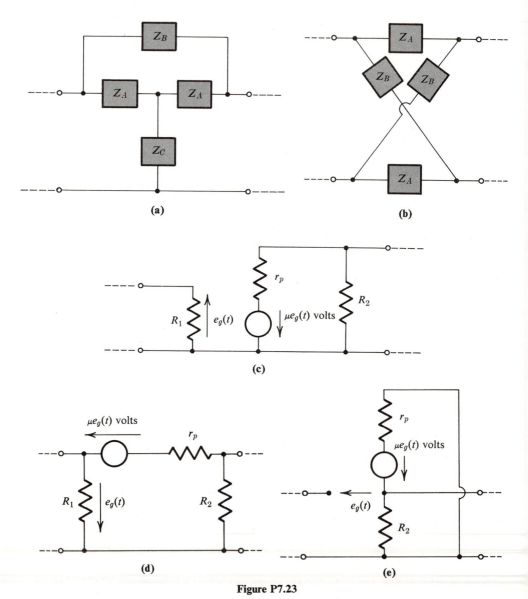

Figure P7.23

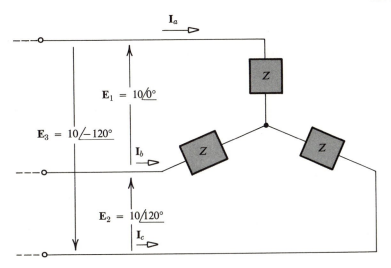

$\mathbf{I}_a$

$\mathbf{E}_1 = 10\underline{/0°}$

$\mathbf{E}_3 = 10\underline{/-120°}$

$\mathbf{I}_b$

$\mathbf{E}_2 = 10\underline{/120°}$

$\mathbf{I}_c$

Z   Z   Z

Figure P7.26

7.23 Find the z and y parameters for each of the circuits in Fig. P7.23.

7.24 Find, if they exist, the z and y parameters for the ideal transformer. If they do not exist, explain what assumptions made in the development of Eqs. 7.4-1 are violated.

7.25 Find the input impedance for the circuit in Fig. 7.4-9 by applying the wye-delta transformation to the 1 Ω resistance and the two capacitances.

7.26 The impedance of each branch in the wye connection in Fig. P7.26 is $Z = 1 + j\sqrt{3}\ \Omega$. Convert the wye connection to an equivalent delta connection, and find the current phasors $\mathbf{I}_a$, $\mathbf{I}_b$, and $\mathbf{I}_c$. Then find the voltage across each of the impedances in the original circuit.

7.27 For the circuit in Fig. P7.27a, find an expression for the transfer function $H(s) = \mathbf{E}_2(s)/\mathbf{I}_1(s)$ in terms of the z parameters for the network N.

(a) Carry out the solution by algebraic manipulation of Eqs. 7.4-1.

(b) Find the transfer function by replacing N by the equivalent circuit in Fig. 7.4-5a.

For the circuit in Fig. P7.27b, find an expression for $H(s) = \mathbf{E}_2(s)/\mathbf{E}_1(s)$ in terms of the y parameters for the network N.

7.28 Find the transfer function $H(s) = \mathbf{E}_o(s)/\mathbf{E}_1(s)$ for the circuit in Fig. P7.28.

7.29 In Section 7.4, the transfer function for the circuit in Fig. 7.4-14 was found to be

$$\frac{\mathbf{E}_2(s)}{\mathbf{E}_1(s)} = \frac{z_{21}}{z_{11}} = \frac{-y_{21}}{y_{22}}$$

For each of the following transfer functions, choose suitable expressions for z_{21} and z_{11}. Then recall that $Z_C = z_{21}$ and $Z_A = z_{11} - z_{21}$ for the circuit in Fig. 7.4-8.

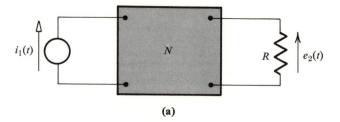

(a)

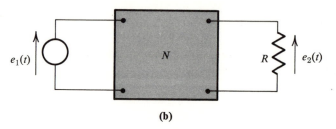

(b)

Figure P7.27

Draw the complete circuit.

(a) $\dfrac{E_2(s)}{E_1(s)} = \dfrac{s+1}{10s+1} = \dfrac{1+1/s}{10+1/s}$

(b) $\dfrac{E_2(s)}{E_1(s)} = \dfrac{s+1}{s+10} = \dfrac{1+1/s}{1+10/s}$

(c) $\dfrac{E_2(s)}{E_1(s)} = \dfrac{(s+1)(10s+1)}{s^2+31s+1} = \dfrac{(s+1)/s}{(s^2+31s+1)/s(10s+1)}$

7.30 The transfer function for the circuit in Fig. P7.27b is $H(s) = E_2(s)/E_1(s) = -y_{21}/(G+y_{22})$ in terms of the y parameters of network N. If

$$\frac{E_2(s)}{E_1(s)} = \frac{2s^2}{2s^2+s+2} = \frac{2s}{1+[(2s^2+2)/s]}$$

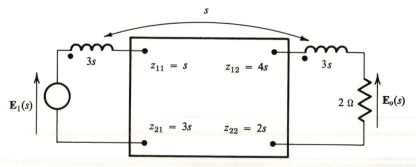

Figure P7.28

choose suitable expressions for $-y_{21}$ and y_{22}. Use the fact that $Y_3 = -y_{21}$ and $Y_2 = y_{22} + y_{21}$ for the configuration in Fig. 7.4-11b in order to find a suitable circuit.

7.31 Prove that the circuit in Fig. 7.4-15b is described by Eqs. 7.4-10.

7.32 The network N in Fig. 7.4-4 is called a *gyrator*, if it is always described by the equations $E_1(s) = I_2(s)$ and $E_2(s) = -I_1(s)$.

   (a) Draw two equivalent circuits for the gyrator, each of which contains two controlled sources.

   (b) Find the input impedance $Z_1(s) = E_1(s)/I_1(s)$, when a load impedance $Z_2(s)$ is connected between the two right-hand terminals.

   (c) How much power is absorbed or stored in the gyrator in the a-c steady state?

   (d) If the gyrator divides a network into two parts, develop a general rule for transferring the elements on one side to the other side, as was done for the ideal transformer after Example 7.3-1.

8

Power and Energy

The primary emphasis in this textbook is on determining the output voltage or current for a given input signal. Very often, however, the output power must be large enough to properly operate some device, such as a loudspeaker, antenna, or motor. Power calculations are encountered most frequently in the analysis of the steady-state behavior of circuits that are excited by a periodic input, especially by a sinusoidal source. Except for Example 8.1-1, the input signals in this chapter are periodic. The discussion in Sections 8.2 through 8.6 is restricted to the a-c steady state.

The effect on a circuit of a loudspeaker, antenna, or amplifier that is connected to the output terminals may be represented by a group of passive elements that is called the *load*. The problem of delivering sufficient power to the load in communication circuits is related to the subject of impedance matching, which is discussed in Section 8.5.

In electric power systems and in the operation of large rotating machines, the voltages are normally sinusoidal waveshapes with a fixed angular frequency, and the only concern is to provide the necessary power at minimum cost. Section 8.3 gives some typical calculations of output power and power losses, and one example of how the cost of supplying power might be reduced. Large amounts of electric power are usually transmitted by "three-phase" circuits, which are examined in Section 8.6.

A study of power also can provide additional insight into some of the topics previously considered. The calculation of the bandwidth in a frequency-selective circuit, for example, is discussed in this chapter as well as in Section 6.4.

8.1

Average Power and RMS Values

From Eqs. 1.2-2 and 1.2-4, the power delivered to the network N in Fig. 8.1-1 is

$$p(t) = e(t)i(t) \quad \text{watts} \tag{8.1-1}$$

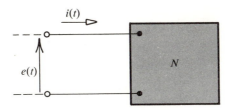

Figure 8.1-1

while the net energy supplied to N from time t_1 to t_2 is

$$w(t_2) - w(t_1) = \int_{t_1}^{t_2} p(t)\, dt \quad \text{joules} \tag{8.1-2}$$

If in a particular problem one of the above expressions is negative, then the network N is actually supplying power or energy, respectively, to the rest of the circuit.

Example 8.1-1.   Consider the circuit in Fig. 3.3-1a, which is repeated in Fig. 8.1-2. If there is no initial stored energy in the capacitance and if $i(t) = U_{-1}(t)$, find the power delivered to the resistance and the capacitance. Also find an expression for the net energy supplied to the capacitance up to the time t.

Solution.   For $t > 0$,

page 369 in (Scott Psch)

See page 129

$$e(t) = R(1 - \epsilon^{-t/RC})$$
$$i_R(t) = 1 - \epsilon^{-t/RC}$$
$$i_C(t) = 1 - i_R(t) = \epsilon^{-t/RC}$$

The power delivered to the resistance and capacitance is

$$p_R(t) = R(1 - \epsilon^{-t/RC})^2$$
$$p_C(t) = R(\epsilon^{-t/RC} - \epsilon^{-2t/RC})$$

The total energy supplied by the source to the two passive components can be found from Eq. 8.1-2. The energy delivered to the resistance is dissipated as heat, while the energy to the capacitance is stored in its electric field. Since no power is supplied for $t < 0$, the net energy received by the capacitance up to time t is

$$w_C(t) = \int_0^t R(\epsilon^{-t/RC} - \epsilon^{-2t/RC})\, dt$$
$$= \tfrac{1}{2}R^2C(1 - 2\epsilon^{-t/RC} + \epsilon^{-2t/RC})$$
$$= \tfrac{1}{2}R^2C(1 - \epsilon^{-t/RC})^2$$

for $t > 0$. Notice that $w_C(t) = \tfrac{1}{2}Ce^2(t)$ as expected.

The power $p(t) = e(t)i(t)$ supplied to the network N in Fig. 8.1-1 is, in general, a function of time. To emphasize that its value depends upon the instant at which it is measured, $p(t)$ is often called the *instantaneous power*. The *average power*

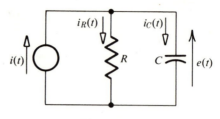

Figure 8.1-2

supplied between times t_1 and t_2 is a constant denoted by p_{avg}. The quantity $(t_2 - t_1)p_{\text{avg}}$ must equal the total energy $\int_{t_1}^{t_2} p(t)\,dt$ delivered between the two instants, so

$$p_{\text{avg}} = \frac{1}{t_2 - t_1} \int_{t_1}^{t_2} p(t)\,dt \tag{8.1-3}$$

which agrees with the usual mathematical definition of an average value.

Circuits with Periodic Inputs.   The remainder of this chapter is restricted to circuits whose voltages and currents are periodic functions of time.† A function $f(t)$ is *periodic* with a period T if it is continually repeated every T seconds, i.e., if $f(t + T) = f(t)$ as in Fig. 8.1-3. If the voltage and current waveshapes both have a period T, then the instantaneous power $p(t) = e(t)i(t)$ is also periodic. The energy supplied to the network N during any one cycle is therefore independent of the cycle selected. For periodic signals, the average power is denoted by P more often than by p_{avg}, and is

$$P = p_{\text{avg}} = \frac{1}{T} \int_0^T p(t)\,dt \tag{8.1-4}$$

In evaluating Eq. 8.1-4, the integration may be carried out over any one full period (e.g., $-T/2$ to $T/2$) or over any integral number of cycles. In some cases the average power may be found by an inspection of the curve of $p(t)$ versus t.

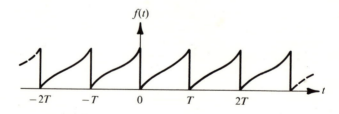

Figure 8.1-3

---

† If a stable circuit is subjected to a periodic input, then in the steady state all voltages and currents are periodic. The general analysis of circuits with periodic inputs is considered in the next chapter.

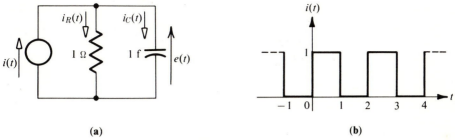

(a)   (b)

Figure 8.1-4

Example 8.1-2.   The circuit in Fig. 8.1-4a is subjected to the square wave of current shown in Fig. 8.1-4b. In Example 9.2-2, the steady-state response is shown to be

$$e(t) = 1 - \left(\frac{\epsilon}{1 + \epsilon}\right) \epsilon^{-t} \qquad \text{for } 0 < t < 1$$

$$= \left(\frac{\epsilon^2}{1 + \epsilon}\right) \epsilon^{-t} \qquad \text{for } 1 < t < 2$$

$$i_C(t) = \left(\frac{\epsilon}{1 + \epsilon}\right) \epsilon^{-t} \qquad \text{for } 0 < t < 1$$

$$= -\left(\frac{\epsilon^2}{1 + \epsilon}\right) \epsilon^{-t} \qquad \text{for } 1 < t < 2$$

Find the average power supplied by the source and delivered to each of the passive circuit elements.

Solution.   The average power delivered by the source is

$$P = \frac{1}{T}\int_0^T ei \, dt = \frac{1}{2}\int_0^1 \left[1 - \left(\frac{\epsilon}{1 + \epsilon}\right)\epsilon^{-t}\right] dt = \frac{1}{1 + \epsilon}.$$

The average power supplied to the resistance is

$$P_R = \frac{1}{T}\int_0^T ei_R \, dt = \frac{1}{T}\int_0^T \frac{e^2}{R} \, dt$$

$$= \frac{1}{2}\left\{\int_0^1 \left[1 - \frac{2\epsilon}{1 + \epsilon}\epsilon^{-t} + \frac{\epsilon^2}{(1 + \epsilon)^2}\epsilon^{-2t}\right] dt + \int_1^2 \frac{\epsilon^4}{(1 + \epsilon)^2}\epsilon^{-2t} \, dt\right\} = \frac{1}{1 + \epsilon}$$

while the average power to the capacitance is

$$P_C = \frac{1}{T}\int_0^T ei_C \, dt = \frac{1}{2}\left\{\int_0^1 \left[\frac{\epsilon}{1 + \epsilon}\epsilon^{-t} - \frac{\epsilon^2}{(1 + \epsilon)^2}\epsilon^{-2t}\right] dt\right.$$

$$\left. - \int_1^2 \frac{\epsilon^4}{(1 + \epsilon)^2}\epsilon^{-2t} \, dt\right\} = 0$$

As expected from the principle of conservation of energy, $P = P_R + P_C$ in Example 8.1-2. Furthermore, in the steady state, the average power supplied to any inductance or capacitance is zero. All the energy supplied to these two elements is stored in their magnetic or electric field and is given by $w_L(t) = \frac{1}{2}Li^2(t)$ or $w_C(t) = \frac{1}{2}Ce^2(t)$. If the steady-state voltages and currents are periodic functions, then their values—and hence the values of $w_L(t)$ and $w_C(t)$—must be the same at the start of each successive cycle. If the stored energy is the same at the start of each cycle, and if no energy can be dissipated as heat, then the average power must be zero. During part of each cycle, energy is supplied to the inductance or capacitance, while in the remainder of the cycle, an equal amount of energy is returned to the rest of the circuit.

 RMS Values. The instantaneous power supplied to the pure resistance in Fig. 8.1-5a is

$$p_R(t) = e(t)i(t) = Ri^2(t) = \frac{e^2(t)}{R} \tag{8.1-5}$$

In these equations, of course, it is assumed that $e(t)$ and $i(t)$ are the voltage across and the current through the resistance. The power supplied to R in Fig. 8.1-5b equals $Ri^2(t)$ but is not $e^2(t)/R$. In part (c) of the figure, $p_R(t) = e^2(t)/R \neq Ri^2(t)$.

For the special case of d-c voltages and currents, Eq. 8.1-5 is a constant, and the instantaneous and average power are identical. The *rms* (or *effective*) value of *any* periodic current is denoted by I_{rms} and is a constant equal to the d-c current which would produce the same average power in a pure resistance. Thus,

$$P_R = p_{\text{avg}} = I_{\text{rms}}{}^2 R$$

The rms voltage is defined in a similar manner, so that for a resistance *only*,

$$P_R = E_{\text{rms}}I_{\text{rms}} = I_{\text{rms}}{}^2 R = \frac{E_{\text{rms}}{}^2}{R} \tag{8.1-6}$$

The instantaneous and average power produced by a current with period T flowing through a resistance are given by

$$p_R(t) = Ri^2(t)$$

$$P_R = \frac{1}{T}\int_0^T Ri^2 \, dt = R\left[\frac{1}{T}\int_0^T i^2 \, dt\right]$$

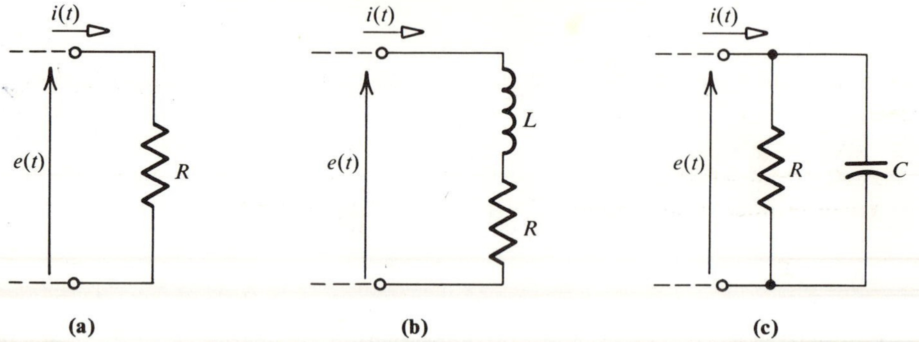

(a)   (b)   (c)

Figure 8.1-5

By Eq. 8.1-6, the quantity in the brackets must be I_{rms}^2, so

$$I_{\text{rms}} = \sqrt{\frac{1}{T}\int_0^T i^2\, dt} = \sqrt{(i^2)_{\text{avg}}} \tag{8.1-7}$$

The letters rms stand for root mean square. As shown in Eq. 8.1-7, the rms current is the square *root* of the average (or *mean*) value of the *square* of the current. Similarly,

$$E_{\text{rms}} = \sqrt{\frac{1}{T}\int_0^T e^2\, dt} = \sqrt{(e^2)_{\text{avg}}} \tag{8.1-8}$$

Equations 8.1-7 and 8.1-8 may be considered as the basic definition of the rms value of a periodic waveshape. It should be emphasized, however, that Eq. 8.1-6 is valid only if $e(t)$ and $i(t)$ are the voltage across and current through a pure resistance.

Example 8.1-3.   Find I_{rms} and E_{rms} for Example 8.1-2.

Solution. $I_{\text{rms}} = 1/\sqrt{2}$ and $E_{\text{rms}} = 1/\sqrt{1 + \epsilon}$. Note that $P_R = E_{\text{rms}}^2/R \neq I_{\text{rms}}^2 R$.

Example 8.1-4.   Find the rms value of $i(t) = I_m \cos(\omega t + \phi)$.

Solution.   By the use of the appropriate trigonometric identity,

$$i^2(t) = I_m^2 \cos^2(\omega t + \phi) = I_m^2[\tfrac{1}{2} + \tfrac{1}{2}\cos(2\omega t + 2\phi)]$$

Since the average value of any sinusoid is zero,

$$(i^2)_{\text{avg}} = \frac{I_m^2}{2}$$

By Eq. 8.1-7,

$$I_{\text{rms}} = \frac{I_m}{\sqrt{2}}$$

It is important to remember that the rms value of any sinusoid is $1/\sqrt{2}$ times the maximum value. Also useful is the fact that the average value of $\cos^2(\omega t + \phi)$ or $\sin^2(\omega t + \phi)$ is $\tfrac{1}{2}$. As illustrated in the following example, the rms values for different voltages or currents cannot be added or subtracted directly. When two elements are in series, for example, the individual voltage phasors (complex numbers) can be added but the individual rms values (real numbers) cannot.

Example 8.1-5.   Find the power supplied to the resistance R in Fig. 8.1-6 when

(a) $e_1(t) = 10 \cos t$ and $e_2(t) = 10 \cos t$

(b) $e_1(t) = 10$ and $e_2(t) = 10 \cos t$

(c) $e_1(t) = 10 \cos t$ and $e_2(t) = 10 \cos(t + 60°)$

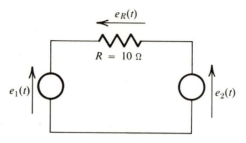

Figure 8.1-6

Solution.   The voltage across the resistance is $e_R(t) = e_1(t) - e_2(t)$, and is zero for part (a) of the example. The power supplied is also zero. In part (b),

$$p_R(t) = \frac{(10 - 10 \cos t)^2}{10} = 10 - 20 \cos t + 10 \cos^2 t$$

Since the average values of $\cos t$ and $\cos^2 t$ are zero and $\frac{1}{2}$, respectively, the average power supplied is 15 watts. In part (c), $e_R(t) = 10 \cos t - 10 \cos (t + 60°) = 10 \cos (t - 60°)$. The last expression follows from a phasor diagram, as discussed at the end of Section 5.4, and also can be obtained by trigonometric identities. The instantaneous power supplied is $p_R(t) = 10 \cos^2 (t - 60°)$, and the average power is 5 watts.

Notice that the rms values of $e_1(t)$, $e_2(t)$, and $e_R(t)$ are $10/\sqrt{2}$, $10/\sqrt{2}$, and zero in part (a); 10, $10/\sqrt{2}$, and $\sqrt{150}$ in part (b); and $10/\sqrt{2}$, $10/\sqrt{2}$, and $10/\sqrt{2}$ in part (c). In parts (b) and (c), $E_{R_{rms}} \neq E_{1_{rms}} - E_{2_{rms}}$ even though $e_R(t) = e_1(t) - e_2(t)$.

Power and Superposition.   In Example 8.1-5a, the average power that would be supplied by either of the voltage sources if the other one were dead is 5 watts. However, when both sources are present, the average power is zero and not 10 watts. In order to determine whether or not a superposition principle ever applies to the calculation of power, consider the more general circuit in Fig. 8.1-7. Let

$i_1(t) =$ the current through R with $e_1(t)$ alive and $e_2(t)$ dead

$i_2(t) =$ the current through R with $e_2(t)$ alive and $e_1(t)$ dead

$p_1(t) =$ the power to R with $e_1(t)$ alive and $e_2(t)$ dead

$p_2(t) =$ the power to R with $e_2(t)$ alive and $e_1(t)$ dead

Since superposition can be used for voltages and currents, $i(t) = i_1(t) + i_2(t)$. The instantaneous power supplied to the resistance R is

$$p(t) = Ri^2(t) = Ri_1^2(t) + 2Ri_1(t)i_2(t) + Ri_2^2(t)$$

$$= p_1(t) + p_2(t) + 2Ri_1(t)i_2(t)$$

Except for the trivial case when $i_1(t)i_2(t)$ is identically zero, $p(t) \neq p_1(t) + p_2(t)$.

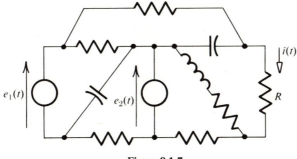

Figure 8.1-7

For periodic waveshapes, the average power is

$$P = \frac{1}{T} \int_0^T R i^2 \, dt$$

$$= \frac{1}{T} \int_0^T R i_1^2 \, dt + \frac{2R}{T} \int_0^T i_1 i_2 \, dt + \frac{1}{T} \int_0^T R i_2^2 \, dt$$

so $P \neq P_1 + P_2$, and superposition does not apply to average power, *unless* $\int_0^T i_1 i_2 \, dt = 0$.

The most important case where superposition is valid for average power is when $i(t)$ consists of sinusoidal components at different frequencies. If

$$i_1(t) = I_{m1} \cos (m\omega_0 t + \phi_1)$$

$$i_2(t) = I_{m2} \cos (n\omega_0 t + \phi_2)$$

where m and n are *different* nonnegative integers (including zero), and where $T = 2\pi/\omega_0$,

$$\int_0^T i_1 i_2 \, dt = I_{m1} I_{m2} \int_0^{2\pi/\omega_0} \cos (m\omega_0 t + \phi_1) \cos (n\omega_0 t + \phi_2) \, dt$$

The evaluation of this integral, which is similar to those encountered in the development of the Fourier series in Chapter 9, does yield zero. Thus, if the voltages and currents consist of several components at different frequencies (which may include a d-c or "zero-frequency" component), the average power at each frequency may be computed individually, and the total average power may be found by superposition.

Example 8.1-6. Find the average power supplied to the network N in Fig. 8.1-1 if $e(t) = 3 + 1.5 \cos 100t$ volts, and $i(t) = 1 + 2 \sin (100t - 30°)$ amperes.

Solution. For the d-c component,

$$P_{\text{d-c}} = 3 \text{ watts}$$

while for the a-c component,

$$p_{\text{a-c}}(t) = 3 \cos 100t \, (\sin 100t \cos 30° - \cos 100t \sin 30°)$$

$$= 1.5 \cos 30° \sin 200t - 3 \sin 30° \cos^2 (100t)$$

Since the average values of $\sin 200t$ and $\cos^2 (100t)$ are zero and $\frac{1}{2}$, respectively,

$$P_{\text{a-c}} = -3(\sin 30°)\tfrac{1}{2} = -\tfrac{3}{4}$$

$$P = P_{\text{d-c}} + P_{\text{a-c}} = 2.25 \text{ watts}$$

Example 8.1-7.   If the current $i(t)$ consists of sinusoidal components at different frequencies (including a possible d-c component), find an expression for I_{rms}.

Solution.   If

$$i(t) = i_1(t) + i_2(t) + \cdots$$

then the average power delivered to a resistance R is

$$P = I_{1_{\text{rms}}}{}^2 R + I_{2_{\text{rms}}}{}^2 R + \cdots$$

Since $P = I_{\text{rms}}{}^2 R$,

$$I_{\text{rms}} = \sqrt{I_{1_{\text{rms}}}{}^2 + I_{2_{\text{rms}}}{}^2 + \cdots}$$

For a voltage $e(t)$ that consists of sinusoidal components at different frequencies (including a possible d-c component),

$$E_{\text{rms}} = \sqrt{E_{1_{\text{rms}}}{}^2 + E_{2_{\text{rms}}}{}^2 + \cdots}$$

If $e(t) = 10 - 10 \cos t$, $E_{\text{rms}} = \sqrt{(10)^2 + (10/\sqrt{2})^2} = \sqrt{150}$, which agrees with the answer to Example 8.1-5b.

8.2
Power in the A-C Steady State

In Example 8.1-4, the rms value of a sinusoidal waveshape was found to be $1/\sqrt{2}$ times the maximum value. Then Eq. 8.1-6 may be used to compute the average power, *provided* that $e(t)$ and $i(t)$ are the voltage across and current through a *resistance*. We now wish to obtain a general expression for the average a-c steady-state power delivered to any two-terminal network whose input impedance is known.

Let the input impedance of the network N in Fig. 8.1-1 be denoted by

$$Z(j\omega) = R + jX = |Z|\underline{/\theta}$$

The relationship between the rectangular and polar forms is summarized by the impedance triangle of Fig. 5.3-5a, which is repeated in Fig. 8.2-1. If

$$e(t) = E_m \cos (\omega t + \phi) = \sqrt{2} \, E_{\text{rms}} \cos (\omega t + \phi)$$

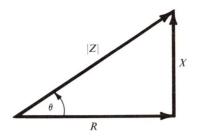

Figure 8.2-1

then the steady-state current is

$$i(t) = I_m \cos (\omega t + \phi - \theta) = \sqrt{2} \, I_{\rm rms} \cos (\omega t + \phi - \theta)$$

where $|Z| = E_m/I_m = E_{\rm rms}/I_{\rm rms}$. By the use of common trigonometric identities, the instantaneous power may be written

$$p(t) = e(t)i(t) = 2E_{\rm rms}I_{\rm rms} \cos (\omega t + \phi)[\cos (\omega t + \phi) \cos \theta + \sin (\omega t + \phi) \sin \theta]$$

$$= [E_{\rm rms}I_{\rm rms} \cos \theta][1 + \cos (2\omega t + 2\phi)] + [E_{\rm rms}I_{\rm rms} \sin \theta] \sin (2\omega t + 2\phi)$$

Since the average value of a sinusoid is zero, the average power is

$$P = E_{\rm rms}I_{\rm rms} \cos \theta \qquad\qquad (8.2\text{-}1)$$

When the angle θ is nearly $\pm 90°$, the instantaneous power may be very large at certain instants, even though the average power is small. As discussed in Section 8.3, the power company may then have to maintain unnecessarily large generating and transmission facilities. For this reason, and because it will be helpful in later power calculations, the coefficient of the second of the two terms in the expression for instantaneous power is given a separate symbol. We define Q (which is not related to the selectivity Q in Section 6.4) as

$$Q = E_{\rm rms}I_{\rm rms} \sin \theta \qquad\qquad (8.2\text{-}2)$$

In the above equations, θ is the angle of Z, i.e., the angle by which the current lags the voltage. Thus, Q is positive for an inductive network and negative for a capacitive network. For passive networks, the average power P is nonnegative.

If the substitutions $E_{\rm rms} = I_{\rm rms}|Z|$, $\cos \theta = R/|Z|$, and $\sin \theta = X/|Z|$ are used, Eqs. 8.2-1 and 8.2-2 may be rewritten as

$$P = I_{\rm rms}^2 R, \qquad Q = I_{\rm rms}^2 X \qquad\qquad (8.2\text{-}3)$$

Since $E_{\rm rms}$ is the rms voltage across the entire network and not across just the resistive part of Z, $P \neq E_{\rm rms}^2/R$ unless $X = 0$. Similarly, $Q \neq E_{\rm rms}^2/X$ unless $R = 0$.

The external behavior of a two-terminal network which contains independent sources cannot be completely described by an input impedance (unless the open-circuit voltage in its Thévenin equivalent circuit is zero). Equations 8.2-3 are therefore not meaningful if there are independent sources within the network N, but

Eqs. 8.2-1 and 8.2-2 remain valid. The angle θ is still interpreted as the angle by which the current lags the voltage, but it is no longer related to an impedance.

Example 8.2-1. Find P if $e(t) = 1.5 \cos 100t$ volts and $i(t) = 2 \sin (100t - 30°)$ amperes in Fig. 8.1-1.

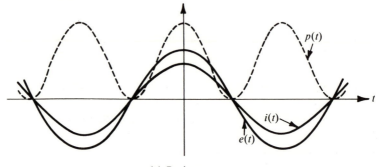

(a) Resistance

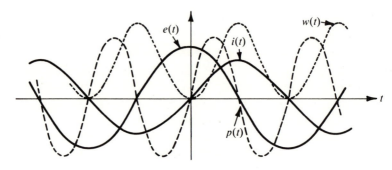

(b) Inductance

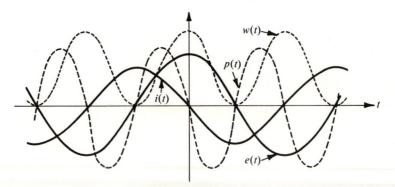

(c) Capacitance

Figure 8.2-2

Solution.   Since $i(t) = 2 \cos (100t - 120°)$ amperes, $\theta = 120°$ and

$$P = \frac{2}{\sqrt{2}} \frac{1.5}{\sqrt{2}} \cos 120° = -\frac{3}{4} \text{ watt}$$

which agrees with Example 8.1-6. The negative sign indicates that the network N is supplying power to the rest of the circuit.

It is instructive to examine the case where the network N consists of a single passive circuit element. Typical curves of $e(t)$, $i(t)$, and $p(t)$ are shown in Fig. 8.2-2. The instantaneous power supplied to the element is positive whenever $e(t)$ and $i(t)$ have the same sign. For the resistance, $p(t)$ is always nonnegative, and oscillates— at twice the angular frequency of $e(t)$ and $i(t)$—about the average value $P = \frac{1}{2}E_m I_m = E_{rms}I_{rms}$. For the inductance or capacitance, $p(t)$ again oscillates at twice the angular frequency of $e(t)$ and $i(t)$, but its average value is zero.

Figure 8.2-2 also shows the energy $w_L(t) = \frac{1}{2}Li^2(t)$ stored in the magnetic field of an inductance and the energy $w_C(t) = \frac{1}{2}Ce^2(t)$ stored in the electric field of a capacitance. During part of each cycle, energy is supplied to the component and temporarily stored in its field. Later in the cycle, the field collapses and the energy is returned to the rest of the circuit. Note that if the voltage waveshape is taken as a common reference, $p_L(t)$ and $p_C(t)$ always have opposite signs, and that $w_L(t)$ or $w_C(t)$ is a maximum when the other stored energy is zero. For an inductance and capacitance in parallel, therefore, energy is continually interchanged between the magnetic field of the inductance and the electric field of the capacitance.

From Eqs. 8.2-3 and the expressions for stored energy,

$$(w_L)_{max} = \frac{1}{2}LI_m{}^2 = LI_{rms}{}^2$$

$$Q_L = I_{rms}{}^2(\omega L)$$

$$(w_C)_{max} = \frac{1}{2}CE_m{}^2 = CE_{rms}{}^2$$

$$Q_C = I_{rms}{}^2 \left(\frac{-1}{\omega C}\right) = -E_{rms}{}^2(\omega C)$$

Thus, Q is directly related to the stored energy in the reactive elements, and because of this is called the *reactive power*. The fact that Q_L is positive and Q_C negative agrees with the above discussion about the sign of $p(t)$ and the interchange of stored energy between the inductance and capacitance.

It is necessary to distinguish carefully between the quantities $p(t)$, P, and Q. The nomenclature most commonly used is summarized below.

$p(t) =$ instantaneous power (watts) $=$ a real function of time

$P =$ power $=$ average power $=$ real power (watts) $=$ a real constant

$Q =$ reactive power (var) $=$ a real constant

Since average power is normally of greater interest than the instantaneous power, "power" is understood to mean average power unless otherwise indicated. The

term "real power" is sometimes used for P to distinguish it from the reactive power Q, although $p(t)$, P, and Q are all real quantities. Although Q is dimensionally equivalent to P, the unit of var (volt-amperes-reactive) is used for Q to emphasize that Q is not associated with the dissipation of energy.

Measuring Instruments.   A-C ammeters or voltmeters are calibrated to read the rms value of a sinusoidal current or voltage, respectively. The usual symbols for these two-terminal instruments are shown in Figs. 8.2-3a and 8.2-3b. In circuit analysis, an ammeter is usually assumed to act like a short circuit, but it does have a coil of low (but nonzero) resistance between its terminals. Although a voltmeter is normally considered to be an open circuit, there is a high (but finite) resistance

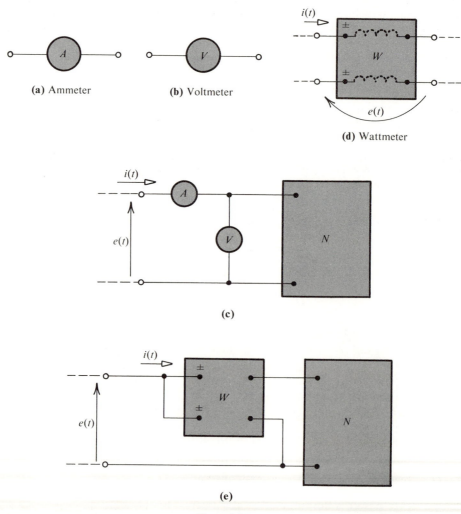

(a) Ammeter   **(b)** Voltmeter

(d) Wattmeter

(c)

(e)

Figure 8.2-3

coil between its terminals. If the meters in Fig. 8.2-3c are ideal, and if the network N is characterized by its input impedance Z, then the ratio of the voltmeter to ammeter readings is $E_{\mathrm{rms}}/I_{\mathrm{rms}} = |Z|$.

A wattmeter is designed to read average power and has the four terminals shown in part (d) of the figure. A low-resistance current-measuring coil is located between the top two terminals, and a high-resistance potential coil between the lower terminals. One end of each coil is marked with a $\pm$ sign (or with some other distinctive mark), while the other end is often labelled with the maximum allowable current or voltage.

A wattmeter reads the average value of the product $e(t)i(t)$, where $i(t)$ is the current flowing into the $\pm$ terminal of the current coil, and $e(t)$ is the voltage of the $\pm$ terminal of the potential coil with respect to the other terminal. To measure the average power supplied to the network N, the wattmeter is connected as in Fig. 8.2-3e. It reads[†]

$$P = \frac{1}{T} \int_0^T ei \, dt = E_{\mathrm{rms}} I_{\mathrm{rms}} \cos \theta$$

where θ is the angle by which the current lags the voltage. A negative reading indicates that the network N is actually supplying power to the rest of the circuit. If the network N is characterized by its input impedance $Z = |Z|\underline{/\theta}$, and if a voltmeter and an ammeter are used to determine E_{rms} and I_{rms}, θ can be calculated from the wattmeter reading. (Another method of experimentally determining the magnitude and angle of Z is suggested in Problem 5.27.) While other meters do exist, including a varmeter to measure Q, they are not discussed here.

Phasors. In Chapter 5, the sinusoidal voltage and current

$$e(t) = E_m \cos (\omega t + \phi) = \sqrt{2}\, E_{\mathrm{rms}} \cos (\omega t + \phi)$$

$$i(t) = I_m \cos (\omega t + \phi - \theta) = \sqrt{2}\, I_{\mathrm{rms}} \cos (\omega t + \phi - \theta)$$

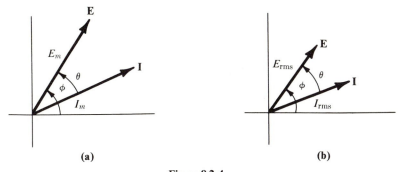

(a)   (b)

Figure 8.2-4

[†] The wattmeter always reads $\frac{1}{T}\int_0^T ei \, dt$, but only for sinusoidal signals is the reading also given by $E_{\mathrm{rms}} I_{\mathrm{rms}} \cos \theta$.

were represented by phasors whose magnitudes were equal to the maximum values of the corresponding functions of time, i.e.,

$$\mathbf{E} = E_m\underline{/\phi}, \qquad \mathbf{I} = I_m\underline{/\phi - \theta}$$

as shown in Fig. 8.2-4a. Voltmeters and ammeters, however, measure the rms value and not the maximum value. Furthermore, Eqs. 8.2-1 through 8.2-3 are expressed in terms of rms values in order to avoid an extra multiplying factor of $\frac{1}{2}$. For these reasons, it is more common to define the magnitude of a phasor as the rms value rather than the maximum value of a sinusoidal function. Then, as shown in Fig. 8.2-4b,

$$\mathbf{E} = E_{\text{rms}}\underline{/\phi} = \frac{E_m}{\sqrt{2}} \underline{/\phi}$$

$$\mathbf{I} = I_{\text{rms}}\underline{/\phi - \theta} = \frac{I_m}{\sqrt{2}} \underline{/\phi - \theta} \qquad (8.2\text{-}4)$$

$$|\mathbf{E}| = E_{\text{rms}}, \qquad |\mathbf{I}| = I_{\text{rms}}$$

Using the rms values changes the length of all phasors by a multiplying factor of $1/\sqrt{2}$, but it does not change their *relative* lengths and their phase angles. The impedance $Z = \mathbf{E}/\mathbf{I}$ is the same whether or not the phasor magnitudes are rms values or maximum values. In the remainder of this chapter, therefore, all phasors will be defined as in Eq. 8.2-4. With this understanding, Eqs. 8.2-1 through 8.2-3 may be rewritten as

$$P = |\mathbf{E}|\ |\mathbf{I}| \cos \theta = |\mathbf{I}|^2\ R$$

$$Q = |\mathbf{E}|\ |\mathbf{I}| \sin \theta = |\mathbf{I}|^2\ X \qquad (8.2\text{-}5)$$

The student should keep in mind that, while the phasor $\mathbf{E}$ is a complex quantity, $|\mathbf{E}| = E_{\text{rms}}$ and $E_m = \sqrt{2}\ E_{\text{rms}}$ are real nonnegative numbers.

Example 8.2-2. The voltage phasors $\mathbf{E}_1$ and $\mathbf{E}_2$ in Fig. 8.2-5a are given. Find the average power consumed by each of the two load impedances and the reading of each wattmeter.

Solution. The current phasors are

$$\mathbf{I}_a = \frac{\mathbf{E}_2}{Z_a} = 12\underline{/60°}, \qquad \mathbf{I}_c = \frac{\mathbf{E}_1}{Z_c} = 12\underline{/-60°}, \qquad \mathbf{I}_b = \mathbf{I}_c - \mathbf{I}_a = 12\sqrt{3}\underline{/-90°}$$

Note for future reference that $\mathbf{E}_3 = \mathbf{E}_1 + \mathbf{E}_2 = 120\ \underline{/60°}$. A complete phasor diagram is shown in Fig. 8.2-5b. The average power consumed by the loads is given by

$$P_a = |\mathbf{E}_2|\ |\mathbf{I}_a| \cos 60° = 720 \text{ watts}$$

$$P_c = |\mathbf{E}_1|\ |\mathbf{I}_c| \cos 60° = 720 \text{ watts}$$

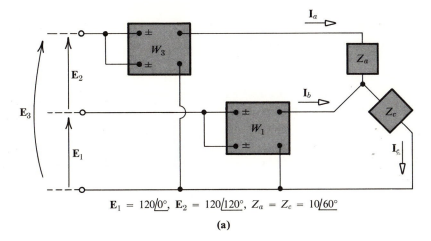

$$\mathbf{E}_1 = 120\underline{/0°}, \ \mathbf{E}_2 = 120\underline{/120°}, \ Z_a = Z_c = 10\underline{/60°}$$

(a)

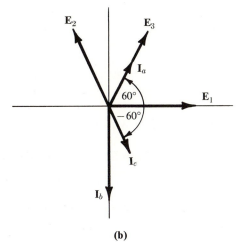

(b)

Figure 8.2-5

or, alternatively,

$$P_a = |\mathbf{I}_a|^2 R_a = (12)^2(10 \cos 60°) = 720 \text{ watts}$$

The wattmeter readings are

$$P_1 = |\mathbf{E}_1| \ |\mathbf{I}_b| \cos 90° = 0$$

$$P_3 = |\mathbf{E}_3| \ |\mathbf{I}_a| \cos 0° = 1440 \text{ watts}$$

It is interesting to notice that although the wattmeters are not connected to read the power delivered to the individual loads, the sum of the wattmeter readings equals the total power delivered to loads A and C. In Section 8.6 we consider whether or not this is a coincidence.

8.3

The Conservation of Real and Reactive Power

Since the instantaneous power is given by $p(t) = e(t)i(t)$, it is logical to inquire whether or not the phasor product **EI** is related to either P or Q. If in Fig. 8.1-1 the current lags the voltage by the angle θ,

$$e(t) = \sqrt{2}\,E_{\text{rms}} \cos{(\omega t + \phi)}$$

$$i(t) = \sqrt{2}\,I_{\text{rms}} \cos{(\omega t + \phi - \theta)}$$

$$\mathbf{E} = E_{\text{rms}}\underline{/\phi} = |\mathbf{E}|\underline{/\phi}$$

$$\mathbf{I} = I_{\text{rms}}\underline{/\phi - \theta} = |\mathbf{I}|\underline{/\phi - \theta}$$

$$\mathbf{EI} = |\mathbf{E}|\,|\mathbf{I}|\underline{/2\phi - \theta}$$

Since neither P nor Q depends upon the phase angle ϕ assumed for the voltage, the product **EI** is *not* a meaningful quantity. The undesirable dependency of **EI** upon the reference angle ϕ can be circumvented, however, by considering

$$\mathbf{EI^*} = (|\mathbf{E}|\underline{/\phi})(|\mathbf{I}|\underline{/-\phi + \theta}) = |\mathbf{E}|\,|\mathbf{I}|\underline{/\theta}$$

$$= |\mathbf{E}|\,|\mathbf{I}| \cos{\theta} + j|\mathbf{E}|\,|\mathbf{I}| \sin{\theta}$$

Thus,

$$\mathbf{EI^*} = P + jQ \tag{8.3-1}$$

This expression for **EI*** (which is sometimes called *complex power*) suggests the power triangle in Fig. 8.3-1a. The sides of the triangle, however, are more commonly labelled with their lengths, as in part (b) of the figure. Note that $|\mathbf{EI^*}| = |\mathbf{E}|\,|\mathbf{I^*}| = |\mathbf{E}|\,|\mathbf{I}|$, which is called the *volt-amperes* (or occasionally the *apparent power*), abbreviated volt-amp. It cannot be emphasized too strongly that the power P, the reactive power Q, and the volt-amperes $|\mathbf{E}|\,|\mathbf{I}|$ are all scalar quantities. For the special case of a pure resistance, $P = |\mathbf{E}|\,|\mathbf{I}|$, but in general P is less than the volt-amperes.

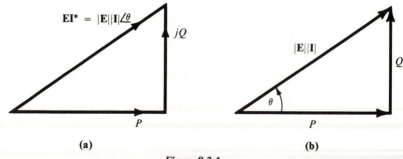

(a)   (b)

Figure 8.3-1

The angle θ, which for a two-terminal network with no independent sources is also the angle of the input impedance, is called the power-factor angle. The *power factor* itself is defined as

$$pf = \cos \theta = \frac{P}{|\mathbf{E}| \, |\mathbf{I}|} \tag{8.3-2}$$

For an inductive load, θ and Q are positive, while for a capacitive load these quantities are negative. Since $\cos(-\theta) = \cos \theta$, specifying the power factor does not uniquely determine the angle θ. To resolve this ambiguity, the word *leading* or the word *lagging* is usually added to the numerical value of the power factor. A lagging power factor means that the current lags the voltage, and that the angle θ is positive.

Large industrial loads are often specified by giving their power and power factor, or equivalently their volt-ampere rating and power factor. If any two of the four quantities P, Q, $|\mathbf{E}| \, |\mathbf{I}|$, and $\cos \theta$ are given, the remaining two can be found from the power triangle of Fig. 8.3-1b. Power companies are concerned about the power factor, and usually increase their rates for an industrial company which operates at a low power factor. For a fixed power P, a low power factor corresponds to large values of Q and $|\mathbf{E}| \, |\mathbf{I}|$. This increases the current that must be delivered to the customer, and requires that the power company maintain the generating and transmission capacity to supply this larger current. A less important factor is that the increased current results in a larger power loss on the transmission lines.

The Conservation of P and Q.   If several components are connected to a source, the power P supplied by the source must be equal to the sum of the power consumed by each component, in accordance with the conservation of energy principle. The student might suspect that a similar conservation law would hold for the reactive power Q, since Q is related to the energy stored in the reactive elements. When Eq. 8.3-1 is applied to the parallel connection in Fig. 8.3-2a, the complex power supplied by the source is

$$P + jQ = \mathbf{E}\mathbf{I}^* = \mathbf{E}(\mathbf{I}_1 + \mathbf{I}_2)^* = \mathbf{E}\mathbf{I}_1^* + \mathbf{E}\mathbf{I}_2^*$$

$$= (P_1 + jQ_1) + (P_2 + jQ_2) = (P_1 + P_2) + j(Q_1 + Q_2)$$

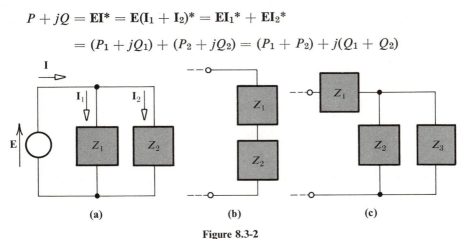

(a)   (b)   (c)

Figure 8.3-2

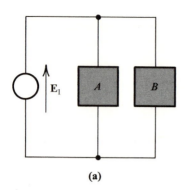

Load	P, kw	Q, kvar	|E||I|, kva	pf = cos θ
A	24	32	40	0.6 lagging
B	8	−6	10	0.8 leading
A and B	32	26	41.23	0.776 lagging

(a)   (b)

Figure 8.3-3

so

$$P = P_1 + P_2, \qquad Q = Q_1 + Q_2 \qquad\qquad (8.3\text{-}3)$$

Thus, the power or reactive power supplied by the source equals the algebraic sum of the power or reactive power received by the individual loads. The student may easily show that this same conclusion applies to the series connection of Fig. 8.3-2b and to the series-parallel connection of Fig. 8.3-2c. In fact, a conservation law applies to both P and Q, no matter how the individual loads might be interconnected. This fact, together with the power triangle of Fig. 8.3-1b, offers a powerful and simple method of solving many problems.

Example 8.3-1. Figure 8.3-3a shows two loads connected across a 220 volt, 60 cycle source. Load A consumes 24 kw and has a power factor of 0.6 lagging. Load B is rated at 8 kw and has a 0.8 leading power factor. Find the power and the current supplied by the source and the power factor seen by the source.

Solution. The information given about each of the two loads is entered in the table of Fig. 8.3-3b. The reactive power and the volt-amperes are then easily obtained from the power triangle of Fig. 8.3-1b. Note that Q is negative for load B because of the leading power factor. The values of P and Q for the combined loads are found by Eqs. 8.3-3, and the total volt-amperes and power factor follow from the power triangle. Notice that the total volt-amperes is *not* the sum of the individual volt-amperes. The source supplies 32 kw at a power factor of 0.776 lagging. The source current is $41,230/220 = 187$ amp. In accordance with the discussion at the end of the previous section, the voltage and current are understood to be rms values.

Example 8.3-2. We wish to increase the overall power factor in Example 8.3-1 to 0.9 without changing the power consumed (presumably to obtain a lower rate from the power company). Find the value of a capacitance that will do this when it is connected in parallel with loads A and B.

Solution. Since the average power supplied to a capacitance is zero, $P = 32$ kw for the entire combination. The total reactive power is $Q = 26 + Q_C$ kvar, where Q_C is the reactive power for the added capacitance. If the overall power factor is to be

0.9 lagging, then from the power triangle, $Q = 15.5$ kvar. This requires $Q_C = -10.5$ kvar and $|E| \, |I_C| = 10,500$ volt-amp. Then,

$$\frac{1}{\omega C} = \frac{|E|}{|I_C|} = \frac{|E|^2}{10,500}$$

$$C = \frac{10,500}{\omega |E|^2} = \frac{10,500}{(377)(220)^2} = 5.8 \times 10^{-4} \, f = 580 \; \mu f$$

When the value of C is impractically large for a physical capacitor, it may be cheaper to simulate the capacitance by a rotating machine called a *synchronous capacitor*.

Example 8.3-3. In Fig. 8.3-4a, the voltage across loads A, B, and C is 2300 volts. The three loads are described as follows:

Load A: 10 kw at 0.707 lagging power factor

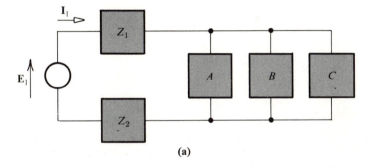

(a)

Load	P, kw	Q, kvar	$\|E\|\|I\|$, kva	$pf = \cos \theta$
A	10	10		0.707 lagging
B	9	-4.4		0.9 leading
C	18	24		0.6 lagging
A, B, and C	37	29.6	47.3	
Z_1 and Z_2	3.4	1.7		
Total	40.4	31.3	51.1	0.79 lagging

(b)

Figure 8.3-4

Load B: 10 kva (kilo-volt-amp) at 0.9 leading power factor
Load C: 18 kw at 0.6 lagging power factor

If $Z_1 = Z_2 = 4 + j2$ Ω (which might represent the impedance of a transmission system), find the voltage, current, power factor, and power supplied by the source.

Solution. The values of P and Q for each load are calculated from the given information and from the power triangle and are entered in Fig. 8.3-4b. The values of P and Q for the combined loads are found by the conservation laws, and the volt-amperes from the power triangle. The magnitude of $\mathbf{I}_1$ is $47,300/2300 = 20.6$ amp. For the impedances Z_1 and Z_2, $P = 2(20.6)^2(4) = 3400$ watts and $Q = 2(20.6)^2(2) = 1700$ watts by Eqs. 8.2-3. The power and reactive power that must be supplied by the source are again determined by the conservation laws and are included in Fig. 8.3-4b. The overall power factor is 0.79 lagging, and $|\mathbf{E}_1| = 51,100/20.6 = 2490$ volts.

8.4

Energy Storage in Resonant Circuits

Frequency-selective circuits that have a large a-c steady-state response over a narrow range of frequencies were discussed in Section 6.4. For a given circuit, the network function $H(s)$ was found, and curves of $|H(j\omega)|$ and $\measuredangle H(j\omega)$ versus ω were sketched. The selectivity Q, which was defined in Eq. 6.4-5, is a measure of the sharpness of the resonant peak in the $|H(j\omega)|$ curve.

We now reconsider the steady-state behavior of circuits that are excited at their resonant frequency. We shall investigate the energy stored and the power dissipated in the circuit, and the power and the reactive power supplied by the external source. One reason for doing this is to obtain additional insight into the resonance phenomenon. Another important result will be the development of another expression for the selectivity—one that is frequently used at microwave frequencies (in the gigacycle per sec range).

Example 8.4-1. The parallel circuit in Fig. 8.4-1 was considered in Example 6.4-7. It was found that the resonant angular frequency is $\omega_0 = 1/\sqrt{LC}$, and that the

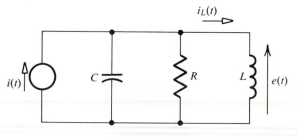

Figure 8.4-1

selectivity is $Q = \omega_0 RC$. Find the energy stored and the average power dissipated within the circuit, when the voltage has the form $e(t) = \sqrt{2}\, E_{rms} \cos \omega t$.

Solution. The energy stored in the capacitance is

$$w_C(t) = \tfrac{1}{2} C e^2(t) = C E_{rms}^2 \cos^2 \omega t$$

The current through the inductance is $i_L(t) = (\sqrt{2}\, E_{rms}/\omega L) \cos (\omega t - 90°) = (\sqrt{2}\, E_{rms}/\omega L) \sin \omega t$, so

$$w_L(t) = \frac{1}{2} L i_L^2(t) = \frac{E_{rms}^2}{\omega^2 L} \sin^2 \omega t$$

The total stored energy is $w_C(t) + w_L(t)$. Notice that $w_C(t)$ has its maximum value of $C E_{rms}^2$ at an instant when $w_L(t)$ is zero. A quarter cycle later the energy in the capacitance is zero, and the energy in the inductance reaches its maximum value of $E_{rms}^2/\omega^2 L$.

The values of $(w_C)_{max}$ and $(w_L)_{max}$ will be identical when $C = 1/\omega^2 L$, i.e., when the angular frequency of the source equals the resonant angular frequency $\omega_0 = 1/\sqrt{LC}$. At this frequency,

$$w_C(t) + w_L(t) = C E_{rms}^2 (\cos^2 \omega t + \sin^2 \omega t) = C E_{rms}^2$$

which is a constant, independent of t. The stored energy is continually transferred between the capacitance and the inductance. No energy need be supplied to the LC combination by the source, which agrees with the fact that the combination acts like an open circuit.

If $\omega < \omega_0$, $(w_L)_{max} > (w_C)_{max}$ and additional energy must be supplied temporarily by the external source. The reactive power delivered by the source is positive, and the circuit looks inductive. If $\omega > \omega_0$, $(w_C)_{max} > (w_L)_{max}$, the source supplies a negative reactive power, and the circuit looks capacitive. Only when $\omega = \omega_0$ is the reactive power of the source zero and the voltage in phase with the current.

The average power dissipated in the circuit is E_{rms}^2/R. Consider the quantity

$$\frac{\omega_0 \ (\text{total energy stored in the circuit})}{\text{average power dissipated in the circuit}}$$

where all factors are to be evaluated at the resonant frequency. This expression gives $\omega_0(C E_{rms}^2)/(E_{rms}^2/R) = \omega_0 RC$, which, according to Example 6.4-7, is the selectivity Q for the circuit.

Consider any network function $H(s)$ that has a single pair of complex-conjugate poles near the imaginary axis of the s-plane.

$$H(s) = \frac{B(s)}{s^2 + 2\alpha s + \omega_0^2} \doteq \frac{B(s)}{(s + \alpha - j\omega_0)(s + \alpha + j\omega_0)} \tag{8.4-1}$$

where the last expression is a good approximation when $\alpha \ll \omega_0$. The numerator $B(s)$ may be a constant or a polynomial in s. In Fig. 8.4-1, for example, $\alpha = 1/2RC$, $\omega_0^2 = 1/LC$, and $B(s) = (1/C)s$. From Section 6.4, the selectivity Q for a circuit described by Eq. 8.4-1 is $Q \doteq \omega_0/2\alpha$.

Suppose that the external source is removed, so that the free response

$$y_H(t) = K\epsilon^{-\alpha t} \cos(\omega_0 t + \phi)$$

is the complete response. Because of the assumption that $\alpha \ll \omega_0$, the value of the exponential factor $\epsilon^{-\alpha t}$ is nearly constant during any one cycle of this damped oscillation. The average energy stored within the circuit during one cycle is $K'(\epsilon^{-\alpha t})^2 = K'\epsilon^{-2\alpha t} = K'\epsilon^{-\omega_0 t/Q}$, where K' is a constant of proportionality.[†] During this cycle, the average *decrease* in the stored energy per unit time, i.e., the average power dissipated, is

$$-\frac{d}{dt}(K'\epsilon^{-\omega_0 t/Q}) = \left(\frac{\omega_0}{Q}\right)(\text{stored energy})$$

Thus,

$$Q = \frac{\omega_0\,(\text{energy stored in the circuit})}{\text{average power dissipated}} \tag{8.4-2}$$

If an external source with an angular frequency ω_0 is used to supply an average power equal to the average power dissipated, then the steady-state response will be a constant-amplitude oscillation.

All the factors in Eq. 8.4-2 should be evaluated at the resonant frequency. When $\alpha \ll \omega_0$, i.e., for high Q circuits, this equation may be used instead of the original definition of selectivity in Eq. 6.4-5. Although, in general, the two equations give the same result only when $Q \gg 1$, they are exactly equivalent in certain special cases, including Examples 8.4-1 and 8.4-2. For high Q circuits, the energy in the capacitance or inductance will be a maximum when the energy in the other reactive element is approximately zero. The stored energy can be calculated, therefore, if an instant is considered when all of it is in the capacitance or when it is all in the inductance.

In microwave circuits, the precise definition of voltage and current becomes difficult, and Eq. 6.4-5 is not used to calculate Q. However, Eq. 8.4-2 is still applicable. (The value of Q for a microwave circuit easily may be several thousand.)

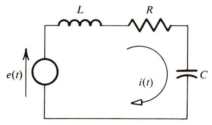

Figure 8.4-2

[†] This derivation is easily seen if, for example, the response is the voltage across a capacitance or the current through an inductance.

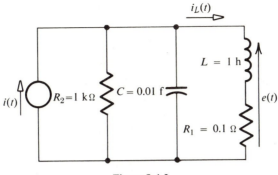

Figure 8.4-3

Example 8.4-2.   Use Eq. 8.4-2 to calculate the selectivity Q for the series circuit in Fig. 8.4-2.

Solution.   At the resonant frequency $\omega_0 = 1/\sqrt{LC}$, the stored energy will be interchanged between the capacitance and inductance. This energy may be calculated by choosing an instant of time at which $w_C(t) = 0$ and $w_L(t)$ has its maximum value of $\tfrac{1}{2}LI_m{}^2 = LI_{\text{rms}}{}^2$. The average power dissipated is $RI_{\text{rms}}{}^2$, so

$$Q = \frac{\omega_0(LI_{\text{rms}}{}^2)}{RI_{\text{rms}}{}^2} = \frac{\omega_0 L}{R}$$

which agrees with Example 6.4-7.

Example 8.4-3.   Find the selectivity Q for the circuit in Fig. 8.4-3.

Solution.   The circuit is a high Q circuit with a resonant angular frequency $\omega_0 = 1/\sqrt{LC} = 10$ rad per sec. This conclusion can be proved by finding the poles of the input impedance $Z(s)$. An easier approach is to formally use Eq. 8.4-2 and to see if the result gives a value of Q that is much larger than 1. The stored energy is $(w_C)_{\text{max}} = \tfrac{1}{2}CE_m{}^2 = CE_{\text{rms}}{}^2 = 0.01E_{\text{rms}}{}^2$. At the angular frequency of 10 rad per sec, $\omega L \gg R_1$ and $(I_L)_{\text{rms}} = E_{\text{rms}}/\omega L = 0.1E_{\text{rms}}$. The power dissipated in R_1 is $(I_L)_{\text{rms}}{}^2R_1 = 0.001E_{\text{rms}}{}^2$, while the power dissipated in R_2 is $E_{\text{rms}}{}^2/R_2 = 0.001E_{\text{rms}}{}^2$. By Eq. 8.4-2,

$$Q = \frac{(10)(0.01E_{\text{rms}}{}^2)}{0.002E_{\text{rms}}{}^2} = 50$$

8.5

Maximum Power Transfer

If the component shown in Fig. 8.5-1 can be described by the impedance $Z = R + jX = |Z|\underline{/\theta}$, the power supplied to it is

$$P = |\mathbf{E}|\,|\mathbf{I}|\cos\theta = |\mathbf{I}|^2 R$$

Frequently, the component represents a load impedance, and we wish to choose

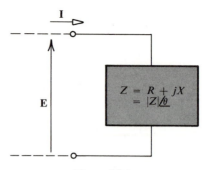

Figure 8.5-1

some parameter γ so that maximum power is delivered from a given source to the load. If the proper choice is not immediately evident, the usual procedure is to find a general expression for P in terms of γ. If the optimum value of the parameter γ is still not evident by inspection, the standard mathematical methods of locating the maximum value of a function can be used. The most common method is to solve the equation $dP/d\gamma = 0$. Several typical problems encountered in electronic circuits are given below as examples.

Example 8.5-1. In Fig. 8.5-2, $Z_2 = R_2 + jX_2$ represents a variable load impedance, while the voltage phasor $\mathbf{E}_1$ and the impedance $Z_1 = R_1 + jX_1$ are fixed. Choose R_2 and X_2 so that maximum power is delivered to the load.

Solution. The load power is

$$P_2 = |\mathbf{I}_2|^2 R_2 = \left| \frac{\mathbf{E}_1}{(R_1 + R_2) + j(X_1 + X_2)} \right|^2 R_2 = \frac{R_2|\mathbf{E}_1|^2}{(R_1 + R_2)^2 + (X_1 + X_2)^2}$$

By inspection, $X_2 = -X_1$ for maximum power transfer. Then

$$P_2 = \frac{R_2|\mathbf{E}_1|^2}{(R_1 + R_2)^2}$$

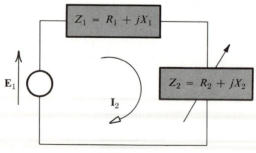

Figure 8.5-2

Since the optimum choice of R_2 is not immediately evident, examine

$$\frac{dP_2}{dR_2} = \frac{(R_1 + R_2)^2 - R_2(2)(R_1 + R_2)}{(R_1 + R_2)^4}\,|E_1|^2 = \frac{R_1{}^2 - R_2{}^2}{(R_1 + R_2)^4}\,|E_1|^2$$

For maximum power, $R_2 = R_1$. In summary, the variable load impedance should be chosen to be the complex conjugate of the source impedance.

By the use of Thévenin's theorem, the result of Example 8.5-1 may be extended to any circuit in which both the real and imaginary parts of the load impedance can be varied independently of each other. In Fig. 8.5-2, E_1 and $R_1 + jX_1$ would represent the Thévenin equivalent circuit for the entire network (with the load impedance removed). For maximum power transfer, the load impedance should be the complex conjugate of Z_{eq}, the impedance of the dead network.

Example 8.5-2.   Only the load resistance R_2 may be varied in Fig. 8.5-3. Choose R_2 so that maximum power is delivered to it.

Solution.   The power delivered to R_2 is

$$P_2 = \frac{|E_1|^2}{(R_1 + R_2)^2 + X_1{}^2}\,R_2$$

$$\frac{dP_2}{dR_2} = \frac{R_1{}^2 - R_2{}^2 + X_1{}^2}{[(R_1 + R_2)^2 + X_1{}^2]^2}\,|E_1|^2$$

so, for maximum power transfer, $R_2 = \sqrt{R_1{}^2 + X_1{}^2} =$ the magnitude of the source impedance.

Example 8.5-3.   In Fig. 8.5-4a, only the turns ratio of the ideal transformer may be varied. Choose N for maximum power transfer to the fixed load impedance $R_2 + jX_2$.

Solution.   The discussion following Example 7.3-1 indicates that Fig. 8.5-4a may be replaced by Fig. 8.5-4b when calculating the load voltage and current. The load power is

$$P_2 = \frac{|NE_1|^2}{(N^2 R_1 + R_2)^2 + (N^2 X_1 + X_2)^2}\,R_2$$

Figure 8.5-4c shows the impedances seen by the voltage source E_1. Notice that the

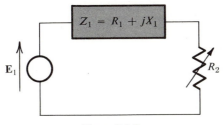

Figure 8.5-3

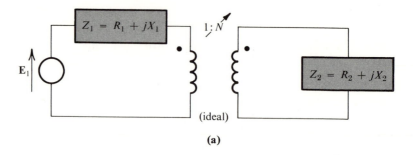

$Z_1 = R_1 + jX_1$ 1: N

E_1

$Z_2 = R_2 + jX_2$

(ideal)

(a)

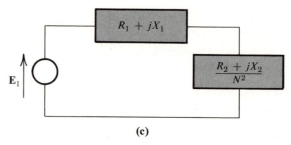

$N^2(R_1 + jX_1)$

$N\mathbf{E}_1$

$R_2 + jX_2$

(b)

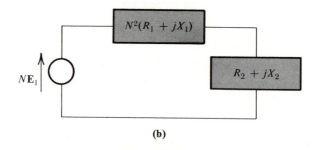

$R_1 + jX_1$

$\mathbf{E}_1$

$\dfrac{R_2 + jX_2}{N^2}$

(c)

Figure 8.5-4

magnitude but not the angle of $(R_2 + jX_2)/N^2$ depends upon N. The power delivered to this impedance is

$$\frac{|\mathbf{E}_1|^2}{(R_1 + R_2/N^2)^2 + (X_1 + X_2/N^2)^2} \frac{R_2}{N^2} = \frac{N^2|\mathbf{E}_1|^2 R_2}{(N^2 R_1 + R_2)^2 + (N^2 X_1 + X_2)^2}$$

and represents the power to the primary winding of the transformer. Since no power is consumed in the ideal transformer, this is also the power supplied to the load, and agrees with the previous expression. Finally,

$$\frac{dP_2}{dN} = \frac{R_2{}^2 + X_2{}^2 - N^4(R_1{}^2 + X_1{}^2)}{[(N^2 R_1 + R_2)^2 + (N^2 X_1 + X_2)^2]^2} R_2|\mathbf{E}_1|^2$$

For the maximum load power, the turns ratio is chosen so that $\sqrt{R_2{}^2 + X_2{}^2} = N^2\sqrt{R_1{}^2 + X_1{}^2}$, i.e., so that the magnitudes of the two impedances in Fig. 8.5-4c are equal.

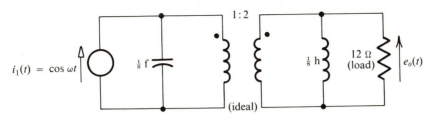

Figure 8.5-5

Suppose that the angle of a load impedance is fixed and that the Thévenin equivalent circuit for the rest of the network has a fixed open circuit voltage phasor $\mathbf{E}_{oc}$ and a fixed Z_{eq}. The analysis of Fig. 8.5-4c shows that for maximum power transfer the magnitude of the load impedance should be made equal to $|Z_{eq}|$. Example 8.5-2 falls into this category, and its answer can be immediately written down. In Example 8.5-3, the ideal transformer is used as an impedance-matching device, so that $\mathbf{E}_1$ sees two impedances of equal magnitude.

Example 8.5-4. For what value of ω is the steady-state output power in Fig. 8.5-5 a maximum? Find the power delivered to the load under these conditions.

Solution. For a fixed load impedance, maximum power occurs when the load voltage is a maximum. From Example 7.3-1, the maximum steady-state voltage occurs at $\omega = 16$ rad per sec and is $e_o(t) = 6 \cos 16t$. The load power is $(6/\sqrt{2})^2/12 = \frac{3}{2}$ watts.

Several other examples of maximizing the load power can be found in the problems at the end of this chapter. Some of the problems can be solved by means of the results of Examples 8.5-1 and 8.5-3. Problems that do not fall into either of these two categories must be solved by basic considerations. Even then, it is usually not necessary to set the derivative of the load power equal to zero in order to obtain the answer.

8.6 SKIP
Three-Phase Circuits

The generation and transmission of large amounts of electrical power is accomplished most efficiently by three-phase circuits. A three-phase generator is a rotating machine that produces three sinusoidal voltages having the same angular frequency and the same amplitude, but displaced in phase by 120°, as shown in Fig. 8.6-1. If the voltage $e_a(t)$ leads $e_b(t)$ by 120°, and $e_b(t)$ leads $e_c(t)$ by 120°, as in part (a) of the figure, the *phase sequence* is said to be a, b, c. The phase sequence for part (b) is a, c, b.

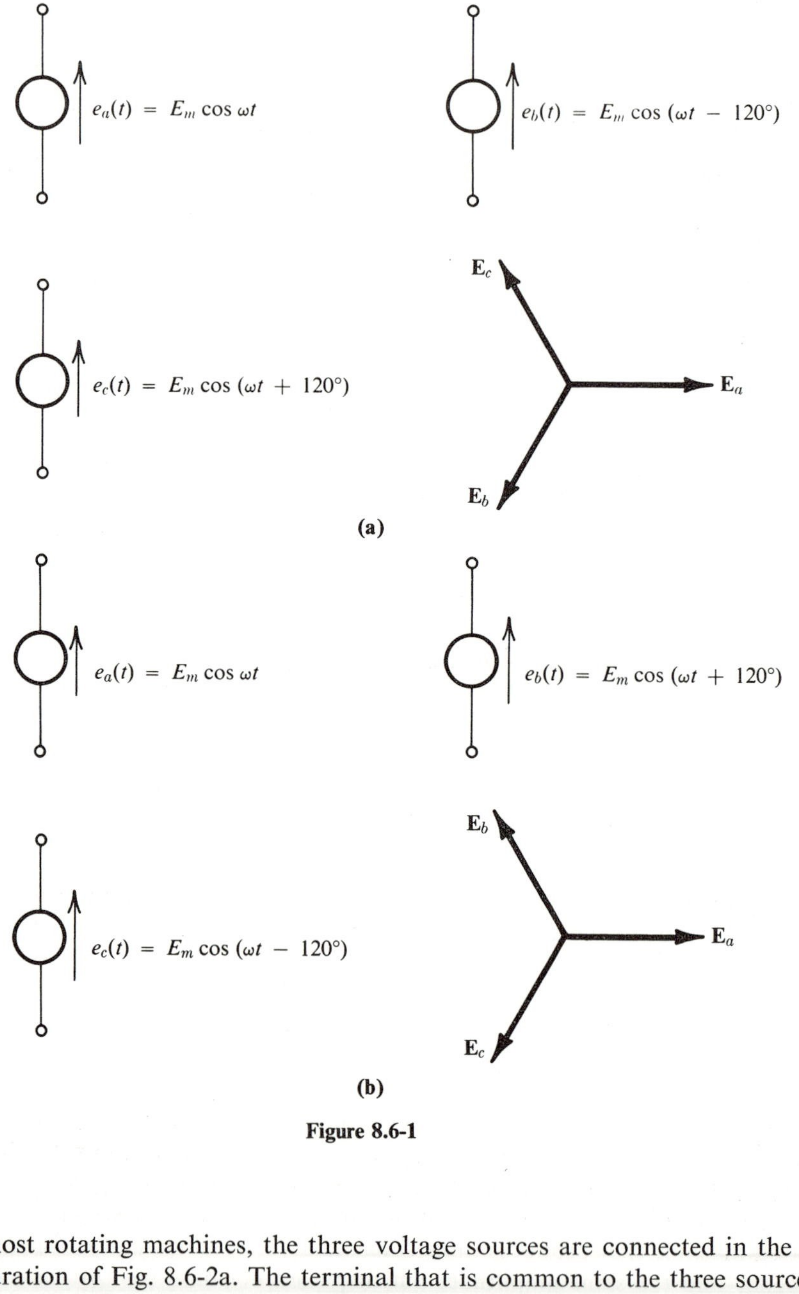
$e_a(t) = E_m \cos \omega t$
$e_b(t) = E_m \cos (\omega t - 120°)$
$e_c(t) = E_m \cos (\omega t + 120°)$
$\mathbf{E}_c$
$\mathbf{E}_a$
$\mathbf{E}_b$
(a)
$e_a(t) = E_m \cos \omega t$
$e_b(t) = E_m \cos (\omega t + 120°)$
$e_c(t) = E_m \cos (\omega t - 120°)$
$\mathbf{E}_b$
$\mathbf{E}_a$
$\mathbf{E}_c$
(b)

Figure 8.6-1

In most rotating machines, the three voltage sources are connected in the wye configuration of Fig. 8.6-2a. The terminal that is common to the three sources is called the *neutral,* is denoted by the letter n, and is normally used as the reference node; it may or may not be externally accessible. The quantities $e_a(t)$, $e_b(t)$, and $e_c(t)$ are called *phase voltages,* while the voltages between terminals a, b, and c are

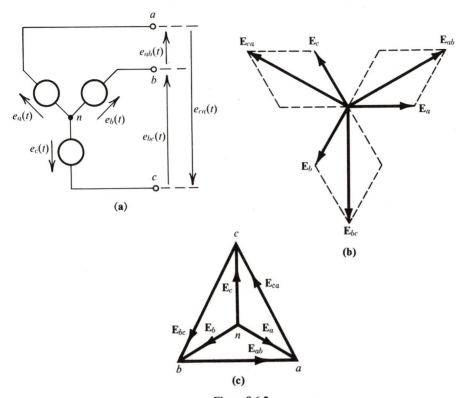

Figure 8.6-2

known as the *line-to-line voltages*. The latter are conveniently expressed by the double subscript notation

$$e_{xy}(t) = e_x(t) - e_y(t) = \text{the voltage of point } x \text{ with respect to point } y$$

Although the voltages

$$e_{ab}(t) = e_a(t) - e_b(t)$$
$$e_{bc}(t) = e_b(t) - e_c(t)$$
$$e_{ca}(t) = e_c(t) - e_a(t)$$

may be found by subtraction of the functions of time, it is easier to subtract the corresponding phasors. The phasor diagram for the phase sequence a, b, c is shown in Fig. 8.6-2b. It is found that the three line-to-line voltages again differ in phase by 120°, and that their magnitudes are $\sqrt{3}$ times the magnitudes of the phase voltages. Specifically,

$$e_{ab}(t) = \sqrt{3}\,E_m \cos{(\omega t + 30°)}$$
$$e_{bc}(t) = \sqrt{3}\,E_m \cos{(\omega t - 90°)}$$
$$e_{ca}(t) = \sqrt{3}\,E_m \cos{(\omega t + 150°)}$$

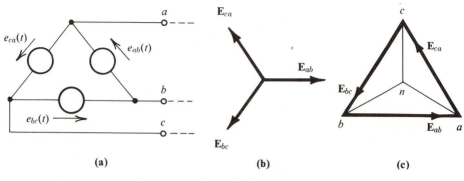

Figure 8.6-3

All the phasors in Fig. 8.6-2b are drawn from a common origin. The phasor diagram may be redrawn in "closed form" as in Fig. 8.6-2c (with $\mathbf{E}_{ab}$ instead of $\mathbf{E}_a$ taken as the reference phasor with an angle of zero degrees, i.e., with the angle of the previous phasors changed by 30°). Notice that the corners of the outside triangle represent the voltages of lines a, b, and c with respect to the neutral point n, which is at the center of the figure.

The use of three-phase transformers sometimes results in a delta-connected source like that in Fig. 8.6-3, which again assumes the phase sequence a, b, c. Although no neutral point is physically present, it is still possible to define one as the center of the closed phasor diagram in part (c) of the figure. Notice that there is no unbalance of voltage around the delta, since $\mathbf{E}_{ab} + \mathbf{E}_{ca} + \mathbf{E}_{bc} = 0$. If, by some mistake, one of the voltage sources were reversed, there would be a net voltage around the delta, in violation of Kirchhoff's voltage law. In actual practice, each voltage source has a small, internal impedance associated with it, so that the net voltage would appear across these small impedances, causing a destructively large circulating current around the delta.

Unbalanced Loads. In three-phase circuits, the load usually consists of three impedances connected in wye or delta. If the three impedances are identical, the load is said to be balanced. We first look at the more general case of unbalanced loads. The voltages and currents may be found by the usual a-c steady-state techniques, and the power may be calculated from the results derived earlier in this chapter.

 Example 8.6-1. For the wye-connected load[†] of Fig. 8.6-4a, the line-to-line voltages are 200 volts rms, and the phase sequence is a, b, c. Find the total real and reactive power delivered to the load and the reading of each wattmeter.

Solution. The line-to-line voltages are shown in the phasor diagram of Fig. 8.6-4b, if $\mathbf{E}_{ab}$ is assigned a phase angle of zero degrees. (Arbitrarily selecting one phase angle

---

[†] In practice, part of each load impedance might represent the impedances of the transmission lines and voltage generator.

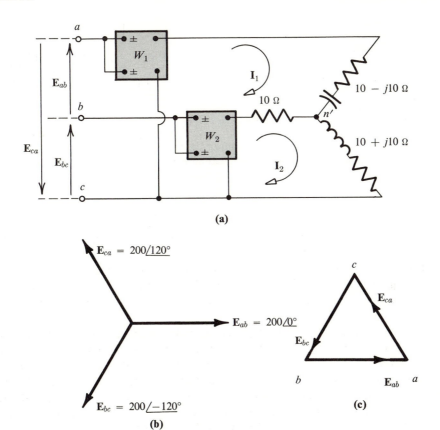

(a)

(b)

(c)

Figure 8.6-4

is equivalent to arbitrarily choosing the time origin, and it does not affect P and Q.)
The currents $\mathbf{I}_1$ and $\mathbf{I}_2$ are found from the two loop equations:

$$200\,\underline{/0°} = (20 - j10)\mathbf{I}_1 - 10\mathbf{I}_2$$

$$200\underline{/-120°} = -10\mathbf{I}_1 + (20 + j10)\mathbf{I}_2$$

which yield

$$\mathbf{I}_1 = 7.5\underline{/5.2°}$$

$$\mathbf{I}_2 = -7.5\underline{/55°}$$

$$\mathbf{I}_1 - \mathbf{I}_2 = 13.6\underline{/30°}$$

By Eqs. 8.2-3, the total real and reactive power is

$$P = (7.5)^2(10) + (13.6)^2(10) + (7.5)^2(10) = 2.97 \text{ kw}$$

$$Q = (7.5)^2(-10) + 0 + (7.5)^2(10) = 0$$

The wattmeter $\boxed{W_1}$ sees the voltage $-\mathbf{E}_{ca} = 200\underline{/-60°}$ and the current $\mathbf{I}_1 = 7.5\underline{/5.2°}$ and reads

$$(200)(7.5) \cos 65.2° = 0.63 \text{ kw}$$

while $\boxed{W_2}$ sees the voltage $\mathbf{E}_{bc} = 200\underline{/-120°}$ and the current $\mathbf{I}_2 - \mathbf{I}_1 = 13.6\underline{/-150°}$ and reads

$$(200)(13.6) \cos 30° = 2.35 \text{ kw}$$

It is interesting to notice that the sum of the two wattmeter readings equals the total power, although neither wattmeter reading has any individual physical significance. The student may also find it worthwhile to show that the neutral point of the load, labelled n', is *not* at the center of the phasor triangle in Fig. 8.6-4c.

Example 8.6-2.   For the delta-connected load of Fig. 8.6-5a, the line-to-line voltages are 100 volts rms, and the phase sequence is a, b, c. Find the total real and reactive power consumed by the load, and the reading of each wattmeter.

Solution.   The voltages are shown in the phasor diagram of Fig. 8.6-5b. Since the voltage across each impedance is known, the currents may be calculated directly.

$$\mathbf{I}_1 = \frac{\mathbf{E}_{ab}}{Z_1} = \frac{100\underline{/0°}}{10\underline{/-70°}} = 10\underline{/70°}$$

$$\mathbf{I}_2 = \frac{\mathbf{E}_{bc}}{Z_2} = \frac{100\underline{/-120°}}{10\underline{/30°}} = 10\underline{/-150°}$$

$$\mathbf{I}_3 = \frac{\mathbf{E}_{ca}}{Z_3} = \frac{100\underline{/120°}}{10\underline{/70°}} = 10\underline{/50°}$$

$$\mathbf{I}_a = \mathbf{I}_1 - \mathbf{I}_3 = 3.51\underline{/149°}$$

$$\mathbf{I}_c = \mathbf{I}_3 - \mathbf{I}_2 = 19.8\underline{/39.6°}$$

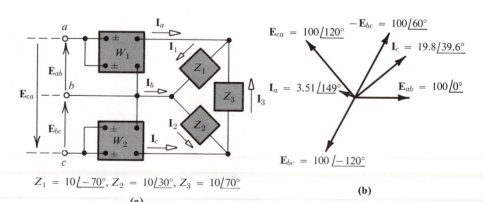

$$Z_1 = 10\underline{/-70°}, \ Z_2 = 10\underline{/30°}, \ Z_3 = 10\underline{/70°}$$

(a)

(b)

Figure 8.6-5

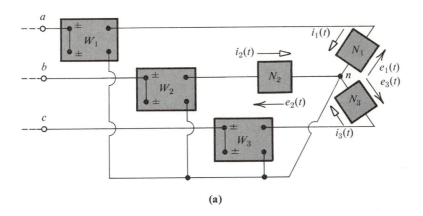

(a)

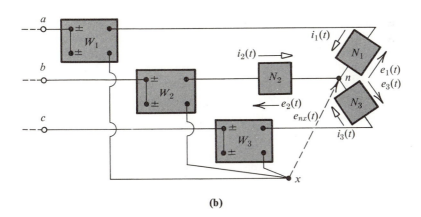

(b)

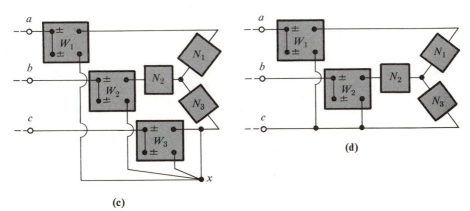

(c)

(d)

Figure 8.6-6

The total real and reactive power is

$$P = (100)(10)\cos 70° + (100)(10)\cos 30° + (100)(10)\cos 70°$$
$$= 342 + 866 + 342 = 1550 \text{ watts}$$
$$Q = -(100)(10)\sin 70° + (100)(10)\sin 30° + (100)(10)\sin 70° = 500 \text{ var}$$

Wattmeter $\boxed{W_1}$ sees the voltage $\mathbf{E}_{ab} = 100\underline{/0°}$ and the current $\mathbf{I}_a = 3.51\underline{/149°}$ and reads

$$(100)(3.51)\cos 149° = -301 \text{ watts}$$

while $\boxed{W_2}$ sees the voltage $-\mathbf{E}_{bc} = 100\underline{/60°}$ and the current $\mathbf{I}_c = 19.8\underline{/39.6°}$ and reads

$$(100)(19.8)\cos 20.4° = 1852 \text{ watts}$$

Once again, the algebraic sum of the wattmeter readings equals the total load power.

If the phase sequence were reversed in the last two examples, the currents and the individual wattmeter readings would be changed, but the total load power would not be affected. In fact, it can be shown that such a conclusion is always valid.

The results of Examples 8.2-2, 8.6-1, and 8.6-2 suggest that the total power delivered to any load with three external terminals can be measured using only two wattmeters. Consider first the general wye-connected load in Fig. 8.6-6a. The wattmeters $\boxed{W_1}$, $\boxed{W_2}$, and $\boxed{W_3}$ measure the power delivered to N_1, N_2, and N_3, respectively. Since P is the average value of the instantaneous power, the individual wattmeter readings are

$$P_1 = \frac{1}{T}\int_0^T e_1 i_1 \, dt$$

$$P_2 = \frac{1}{T}\int_0^T e_2 i_2 \, dt$$

$$P_3 = \frac{1}{T}\int_0^T e_3 i_3 \, dt$$

where T is the period of the voltage and current waveshapes. The sum of the three readings is certainly the total load power, so

$$P_{\text{total}} = \frac{1}{T}\int_0^T (e_1 i_1 + e_2 i_2 + e_3 i_3) \, dt$$

Next, consider the arrangement of wattmeters in Fig. 8.6-6b, where one end of each potential coil is connected to point x instead of to the neutral point of the load. The three wattmeter readings become, respectively,

$$\frac{1}{T}\int_0^T (e_1 + e_{nx})i_1 \, dt$$

$$\frac{1}{T}\int_0^T (e_2 + e_{nx})i_2 \, dt$$

$$\frac{1}{T}\int_0^T (e_3 + e_{nx})i_3 \, dt$$

and their sum is

$$\frac{1}{T}\int_0^T [(e_1i_1 + e_2i_2 + e_3i_3) + e_{nx}(i_1 + i_2 + i_3)]\, dt$$

But $i_1(t) + i_2(t) + i_3(t) = 0$ by Kirchhoff's current law. Thus, even though the individual meter readings have lost their significance, their sum still equals the total load power, no matter where the point x is located.

Suppose that the point x is chosen to be one of the three load terminals, as shown in Fig. 8.6-6c. Since there is no voltage across the potential coil of $|W_3|$, its reading will be zero, and it may just as well be omitted, as in Fig. 8.6-6d. Under these conditions, the algebraic sum of the two remaining readings must still equal the total load power.

Although the validity of the two-wattmeter method has been proved for a wye-connected load, the method is valid for any three-terminal load, no matter what its internal configuration might be. It should be emphasized that the $\pm$ terminal of each current coil should be on the source side of the meter and that the $\pm$ end of the potential coil should be connected to the current coil and the other end to the line that does not pass through the current coil of any wattmeter. The method may be generalized to a load with m external terminals (an "m-wire system"), where the total power may be measured with m-1 wattmeters. The wattmeter configuration for a four-wire system is shown in Fig. 8.6-7.

Balanced Wye-Connected Loads.   All the procedures discussed for unbalanced loads, including the m-1 wattmeter method for measuring total power, remain valid for the special case of a balanced load. The analysis of a balanced load is simplified, however, if we take advantage of its special features. Consider the four-wire system of Fig. 8.6-8a, in which the neutral point of the source is connected to that of the load. The load voltages are identical with the source voltages and are shown in part (b) of the figure for the phase sequence a, b, c.

If the rms magnitude of the phase voltages (across each individual load im-

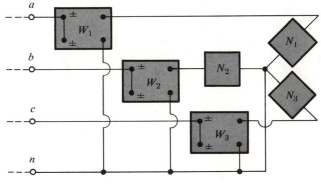

Figure 8.6-7

pedance) is denoted by E_p, and the rms magnitude of the line-to-line voltages by E_L, then, as discussed in connection with Fig. 8.6-2,

$$E_L = \sqrt{3}\,E_p \tag{8.6-1}$$

If I_p and I_L denote the rms magnitudes of the phase currents (through each load impedance) and the line currents, respectively,

$$I_L = I_p \tag{8.6-2}$$

If $Z = |Z|\underline{/\theta}$, the currents $\mathbf{I}_a$, $\mathbf{I}_b$, and $\mathbf{I}_c$ will lag the corresponding phase voltages by the angle θ. The phasor diagram of Fig. 8.6-8b indicates that the vector sum $\mathbf{I}_a + \mathbf{I}_b + \mathbf{I}_c$ is zero, so the current in the neutral wire is $\mathbf{I}_n = \mathbf{I}_a + \mathbf{I}_b + \mathbf{I}_c = 0$. Since no current flows through the neutral wire when the load is balanced, the neutral connection can be omitted (or can contain any arbitrary impedance) without affecting the voltages and currents. It is important to emphasize that even

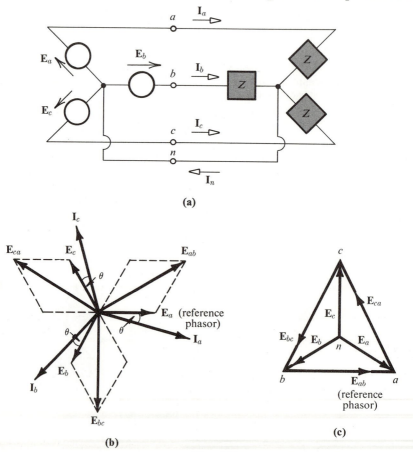

(a)

(b)

(c)

Figure 8.6-8

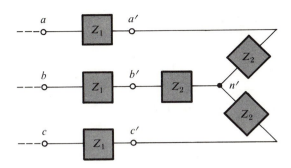

Figure 8.6-9

if the neutral point of the load is not connected to the source, its potential is at the center of the triangle in Fig. 8.6-8c, and that the voltage across each impedance is known.

The real and reactive power delivered to each phase of the load is $P_p = E_p I_p \cos \theta$ and $Q_p = E_p I_p \sin \theta$, so

$$P_{\text{total}} = 3P_p = \sqrt{3}\, E_L I_L \cos \theta$$

$$Q_{\text{total}} = 3Q_p = \sqrt{3}\, E_L I_L \sin \theta$$

$$(\text{volt-amperes})_{\text{total}} = \sqrt{P_{\text{total}}{}^2 + Q_{\text{total}}{}^2} = \sqrt{3}\, E_L I_L$$

$$\text{power factor} = \frac{\sqrt{3}\, E_L I_L \cos \theta}{\sqrt{3}\, E_L I_L} = \cos \theta$$

$$(8.6\text{-}3)$$

Example 8.6-3. In Fig. 8.6-9, a three-phase source with line voltages of 120 volts rms is connected to terminals a, b, c. The impedance of the generator and transmission lines is represented by $Z_1 = 1 + j1$ Ω, and the balanced wye-connected load by $Z_2 = 4 + j3$ Ω. Find the power delivered to the load, and the line voltages between terminals a', b', and c'.

Solution. Even though the neutral point of the load is not connected to the source, the rms voltage between n' and terminal a, b, or c is $120/\sqrt{3}$ volts. The line currents are

$$I_L = \frac{120/\sqrt{3}}{|1 + j1 + 4 + j3|} = \frac{120/\sqrt{3}}{6.4} = 10.8 \text{ amp}$$

The total load power is $3(10.8)^2(4) = 1400$ watts. The rms value of the line voltages between a', b', and c' is $\sqrt{3}(10.8)(5) = 93.5$ volts. By the first of Eqs. 8.6-3, the load power is also given by $\sqrt{3}(93.5)(10.8)(0.8) = 1400$ watts.

Example 8.6-4. Let P_1 and P_2 denote the readings of $\boxed{W_1}$ and $\boxed{W_2}$, respectively, in Fig. 8.6-10. Show that for the phase sequence a, b, c

$$P_1 = E_L I_L \cos (30° - \theta)$$

$$P_2 = E_L I_L \cos (30° + \theta)$$

while for the phase sequence a, c, b the quantities P_1 and P_2 are interchanged in the above expressions.

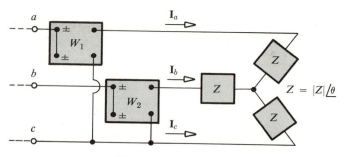

Figure 8.6-10

Solution. The voltages and currents for the phase sequence a, b, c can be represented by the phasor diagram in Fig. 8.6-8b. Wattmeter $\boxed{W_1}$ sees the voltage $-\mathbf{E}_{ca} = E_L \underline{/-30°}$ and the current $\mathbf{I}_a = I_L \underline{/-\theta}$ and reads

$$P_1 = E_L I_L \cos (30° - \theta)$$

The second wattmeter sees $\mathbf{E}_{bc} = E_L \underline{/-90°}$ and $\mathbf{I}_b = I_L \underline{/-120° - \theta}$ and reads

$$P_2 = E_L I_L \cos (30° + \theta)$$

The sum of the two readings is

$$P_2 + P_1 = E_L I_L (\cos 30° \cos \theta - \sin 30° \sin \theta + \cos 30° \cos \theta + \sin 30° \sin \theta)$$

$$= \sqrt{3}\, E_L I_L \cos \theta = P$$

as expected. In a similar way, the difference of the two readings is found to be $P_2 - P_1 = E_L I_L \sin \theta$. The total reactive power is

$$Q = \sqrt{3}\, E_L I_L \sin \theta = \sqrt{3}\, (P_2 - P_1)$$

Division of these expressions for P and Q gives

$$\tan \theta = \sqrt{3}\, \frac{P_2 - P_1}{P_2 + P_1}$$

The student should show that all these results are valid for the phase sequence a, c, b when the roles of P_1 and P_2 are interchanged.

Example 8.6-4 shows that for a balanced load not only the total power, but also the total reactive power and the power factor, may be calculated from the two wattmeter readings. Even if the phase sequence is not known, the magnitude (but not the sign) of the reactive power and of the angle θ can be uniquely determined. Notice that the two wattmeter readings are identical when $\theta = 0$ (unity power factor). When the power factor is less than $\frac{1}{2}$, $|\theta| > 60°$ and one of the two readings is negative.

It is instructive to determine an expression for the instantaneous power delivered to a balanced load. For the wye-connected load in Fig. 8.6-8a (with or without the neutral wire), and for the phase sequence a, b, c, the phase voltages and currents are

$$e_a(t) = E_m \cos \omega t \qquad\qquad i_a(t) = I_m \cos (\omega t - \theta)$$

$$e_b(t) = E_m \cos (\omega t - 120°) \qquad i_b(t) = I_m \cos (\omega t - 120° - \theta)$$

$$e_c(t) = E_m \cos (\omega t + 120°) \qquad i_c(t) = I_m \cos (\omega t + 120° - \theta)$$

By some common trigonometric identities, the power delivered to phase a can be expressed as

$$p_a(t) = e_a(t)i_a(t) = E_m I_m \cos \omega t(\cos \theta \cos \omega t + \sin \theta \sin \omega t)$$

$$= E_m I_m \cos \theta \cos^2 \omega t + E_m I_m \sin \theta \sin \omega t \cos \omega t$$

$$= E_p I_p \cos \theta (1 + \cos 2\omega t) + E_p \dot{I}_p \sin \theta \sin 2\omega t$$

Similarly,

$$p_b(t) = e_b(t)i_b(t)$$

$$= E_p I_p \cos \theta[1 + \cos (2\omega t + 120°)] + E_p I_p \sin \theta \sin (2\omega t + 120°)$$

$$p_c(t) = e_c(t)i_c(t)$$

$$= E_p I_p \cos \theta[1 + \cos (2\omega t - 120°)] + E_p I_p \sin \theta \sin (2\omega t - 120°)$$

so that the total instantaneous power is

$$p(t) = p_a(t) + p_b(t) + p_c(t)$$

$$= 3E_p I_p \cos \theta + E_p I_p \cos \theta[\cos 2\omega t + \cos (2\omega t + 120°) + \cos (2\omega t - 120°)]$$

$$+ E_p I_p \sin \theta[\sin 2\omega t + \sin (2\omega t + 120°) + \sin (2\omega t - 120°)]$$

The quantities in the brackets vanish, as can be easily seen from a phasor diagram, so

$$p(t) = 3E_p I_p \cos \theta = P$$

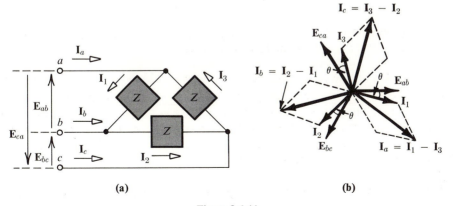

(a)   (b)

Figure 8.6-11

Thus the total instantaneous power is a constant, independent of time.[†] Recall, in contrast, that the power delivered by a single-phase source is pulsating at an angular frequency of 2ω rad per sec. A constant, rather than a pulsating, power flow is one of several advantages of three-phase systems.

Balanced Delta-Connected Loads. Figure 8.6-11a shows a balanced delta-connected load, where $Z = |Z|\underline{/\theta}$. A complete phasor diagram is given in part (b) of the figure for the phase sequence a, b, c. A carefully drawn diagram shows that the three line currents are displaced in phase by 120°, and are $\sqrt{3}$ times as large as the phase currents. Notice that $\mathbf{I}_a + \mathbf{I}_b + \mathbf{I}_c = 0$ as expected. Using the same subscript notation as for the balanced wye load, we have

$$E_L = E_p, \qquad I_L = \sqrt{3}\,I_p \qquad\qquad (8.6\text{-}4)$$

Then,

$$P_{\text{total}} = 3E_pI_p \cos\theta = \sqrt{3}\,E_LI_L \cos\theta$$

$$Q_{\text{total}} = 3E_pI_p \sin\theta = \sqrt{3}\,E_LI_L \sin\theta$$

$$(\text{volt-amperes})_{\text{total}} = \sqrt{P_{\text{total}}^2 + Q_{\text{total}}^2} = \sqrt{3}\,E_LI_L \qquad\qquad (8.6\text{-}5)$$

$$\text{power factor} = \frac{\sqrt{3}\,E_LI_L \cos\theta}{\sqrt{3}\,E_LI_L} = \cos\theta$$

Note that the power formulas in terms of E_L and I_L are identical for balanced delta and wye loads. The equations developed in Example 8.6-4 and the fact that the total instantaneous power is a constant are also valid for both the delta and wye connections.

It is sometimes convenient to transform a delta-connected load into an equivalent wye connection by Eqs. 7.4-4. If each branch of the delta has the impedance Z, each impedance in the equivalent wye is $Z/3$.

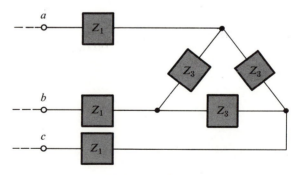

Figure 8.6-12

[†] This conclusion also may be reached graphically if the curves of instantaneous power for the individual phases are sketched and added. The conclusion remains valid if the phase sequence is reversed, but does not hold for unbalanced loads.

Example 8.6-5.   In Fig. 8.6-12, $Z_1 = 1 + j1$ Ω, $Z_3 = 15\underline{/36.9°} = 12 + j9$ Ω, and the line voltages between terminals a, b, and c are 120 volts rms. Find the total power supplied to the three 15 Ω impedances.

Solution.   The delta-wye transformation yields the circuit in Fig. 8.6-9 with $Z_2 = 4 + j3$ Ω. From Example 8.6-3, $P = 1400$ watts.

In other problems, it may be helpful to use Eqs. 7.4-9 to replace a wye-connected load by an equivalent delta. If this were done in Fig. 8.6-4a, the load power could be found without solving two simultaneous loop equations. One of the problems at the end of the chapter asks the student to solve Example 8.6-1 by this method.

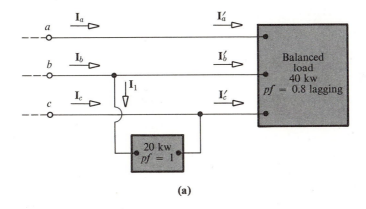

(a)

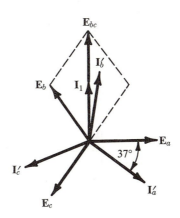

(b)

Figure 8.6-13

Frequently, several loads are connected across one three-phase source. The loads may be balanced or unbalanced and may include single-phase loads.

Example 8.6-6. Determine the rms value of each line current in Fig. 8.6-13a. The phase sequence is a, c, b, and the line voltage is 208 volts.

Solution. The result is the same whether the balanced load is connected in wye or delta. The application of Eqs. 8.6-3 or 8.6-5 to the balanced load gives $40,000 = \sqrt{3}(208)I_L(0.8)$, so $I_L = 139$ amp. From the phasor diagram in Fig. 8.6-13b, where the voltage $\mathbf{E}_a$ has been chosen as a reference,

$$\mathbf{I}'_a = 139\underline{/-37°}, \qquad \mathbf{I}'_b = 139\underline{/83°}, \qquad \mathbf{I}'_c = 139\underline{/-157°}$$

Since $\mathbf{I}_1 = \dfrac{20,000}{208}\underline{/90°}$,

$$\mathbf{I}_b = \mathbf{I}'_b + \mathbf{I}_1 = 234\underline{/86°}$$

$$\mathbf{I}_c = \mathbf{I}'_c - \mathbf{I}_1 = 134\underline{/162°}$$

The rms values of $\mathbf{I}_a$, $\mathbf{I}_b$, and $\mathbf{I}_c$ are 139, 234, and 134 amp, respectively.

8.7

Summary

Nearly all of this chapter is devoted to power calculations in the a-c steady state. The magnitudes of the voltage and current phasors are redefined as the rms instead of the maximum values of the sinusoidal functions of time, in order to simplify the power formulas. The power P is the average value of the instantaneous power $p(t)$. The reactive power Q is related to the energy stored in the reactive elements and is positive for an inductive circuit and negative for a capacitive circuit. The power and reactive power supplied to a two-terminal network characterized by the impedance $Z = R + jX = |Z|\underline{/\theta}$ are given by

$$P = |\mathbf{E}|\,|\mathbf{I}|\cos\theta = |\mathbf{I}|^2 R$$

$$Q = |\mathbf{E}|\,|\mathbf{I}|\sin\theta = |\mathbf{I}|^2 X$$

The power triangle shown in Fig. 8.3-1b relates P, Q, $|\mathbf{E}|\,|\mathbf{I}|$ (the volt-amperes), and $\cos\theta$ (the power factor). A conservation law applies to P and Q, but not to the volt-amperes.

The use of measuring instruments, including wattmeters, is discussed at the end of Section 8.2. In the latter part of this chapter, several important applications of the power formulas are considered: the power-factor correction problem (Section 8.3), the development of an alternative expression for the selectivity of a resonant circuit (Section 8.4), the selection of a circuit element so that maximum power is delivered to the load (Section 8.5), and three-phase circuits (Section 8.6).

The discussion of three-phase circuits provides many illustrations of power calculations, phasor diagrams, and the use of measuring instruments. If a delta-connected load is excited by a three-phase source, the voltage across each load impedance is known, and the currents may be calculated directly. If an unbalanced wye-connected load is supplied by a three-phase, three-wire source, the currents are found by solving two loop equations simultaneously. If the wye-connected load is balanced, however, the neutral point of the load is at the same potential as the neutral point of the source, and the solution of simultaneous equations is not necessary.

The last two chapters have constituted a major digression to consider trans-formers and power calculations. In Chapter 9 we return to the development of additional methods for finding the output voltage or current when the input is given.

PROBLEMS

8.1 Find the rms value of each of the periodic waveshapes shown in Fig. P8.1. The waveshapes in parts (c) and (d) are full-wave and half-wave rectified sinusoids. Most of the answers can be written down with very little computational work.

8.2 Find the average power delivered to the resistance in Fig. P8.2a under the following conditions:

(a) When $e_1(t) = 1$ for all t and when $e_2(t)$ has the waveshape in Fig. P8.2b.

(b) When $e_1(t) = 1$ for all t and when $e_2(t)$ has the waveshape in Fig. P8.2c.

(c) When $e_1(t) = 5 \cos 5t$ and $e_2(t) = 5 \cos (10t + 30°)$.

(d) When $e_1(t) = 10 \cos 25t$ and $e_2(t) = 10 \sin (25t - 30°)$.

8.3 The waveshape of the current flowing through a 1 Ω resistance is given in Fig. P8.3. By the method developed in Section 9.1, the current may be represented by the following infinite series:

$$i(t) = 1 + \frac{2}{\pi} \left(\sin t - \frac{1}{2} \sin 2t + \frac{1}{3} \sin 3t - \cdots \right)$$

Find the average power dissipated in the resistance:

(a) By finding the rms value of the current.

(b) By finding the average power supplied by each of the first four terms in the infinite series.

Compare the answers obtained by the two methods.

8.4 Some of the many types of ammeters are considered in this problem. The thermo-couple, dynamometer, and iron vane meters read the rms value of any periodic current that passes through them. The reading of the D'Arsonval (or d-c) ammeter equals the average value of the current waveshape. The reading of the full-wave rectifier meter is based upon the waveshape of $|i(t)|$ instead of $i(t)$. The meter is

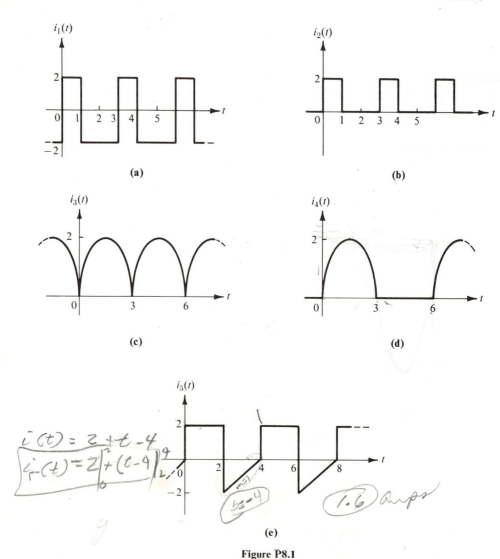

Figure P8.1

calibrated to read 1.11 times the average value of $|i(t)|$. The reading of a half-wave rectifier meter is based upon a waveshape that equals $i(t)$ when $i(t) > 0$ and that is zero when $i(t) < 0$. The reading equals 2.22 times the average value of this modified waveshape. Calculate the reading of each of these ammeters for each of the current waveshapes in Fig. P8.1.

8.5   Assume that $M = \sqrt{L_1 L_2}$ for the transformer in Fig. 7.2-3a. In Example 7.2-3, it was found that the a-c steady-state output voltage can be larger than the input voltage. Derive expressions for the average power dissipated in the resistance R_2 and the average power supplied by the source. Show that the output power is less than the input power if R_1 and R_2 are positive.

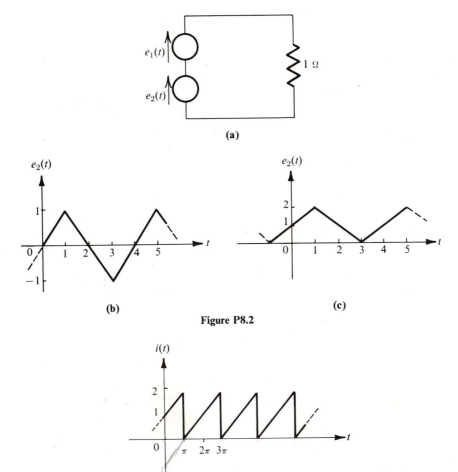

(a)

(b)   **(c)**

Figure P8.2

Figure P8.3

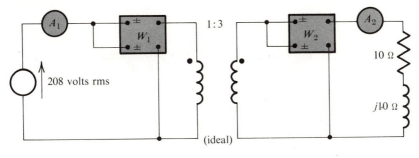

208 volts rms

(ideal)

Figure P8.6

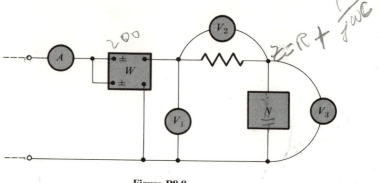

Figure P8.8

8.6 Find the reading of each meter in the frequency-domain circuit shown in Fig. P8.6. What is the power factor seen by the source?

8.7 An inductor that can be described by the model in Fig. 1.5-1b takes 100 watts and 5 amp when it is placed across a 25 volt, 60 cycle per sec source. Find the power consumed in the inductor when it is connected to a 25 volt, 400 cycle per sec source.

8.8 When the circuit in Fig. P8.8 is connected to a d-c voltage source, the steady-state current is zero. When the circuit is connected to a 60 cycle per sec source, the voltmeters V_2 and V_3 read 40 and 100 volts, respectively. The ammeter and wattmeter readings are 2 amp and 200 watts.

(a) If the network N is known to contain only two circuit elements, find the element values.

(b) Calculate the reading of the voltmeter V_1 when the circuit was connected to the 60 cycle per sec source.

8.9 Prove that a conservation law applies to P and Q for the circuits in Figs. 8.3-2b and 8.3-2c.

8.10 The generator in Fig. P8.10 supplies 20 kw of power. All the meter readings are rms values. If Z_1 and Z_2 are known to be capacitive in nature, find X_1, the reactive component of Z_1.

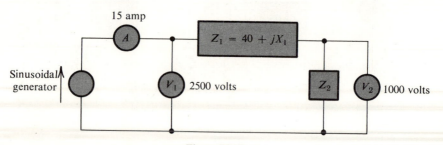

Figure P8.10

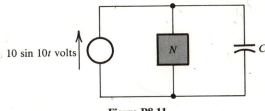

Figure P8.11

8.11 The steady-state power and reactive power supplied to the network N in Fig. P8.11 are 8 watts and 6 var, respectively. Choose the value of the capacitance C so that the source sees a unity power factor load.

8.12 For load A in Fig. P8.12, $P = 10$ kw and the power factor is 0.8 lagging. Load B is rated at 12 kva and has a power factor of 0.7 lagging. What value of resistance, when connected in parallel with the two loads, will result in an overall power factor of 0.9? What is the disadvantage of using a resistance instead of a capacitance to improve the power factor?

8.13 For the circuit in Fig. P8.13, $e_1(t) = 200 \cos \omega t$ and $e_2(t) = 200 \sin \omega t$. Load A consumes 4 kw of power at a power factor of 0.8 lagging, while load B takes 3 kw at a power factor of 0.6 leading. Find the three currents and the reading of each wattmeter.

8.14 Calculate the selectivity for the circuit in Fig. 8.4-3 by plotting the pole-zero pattern of $Z(s) = E(s)/I(s)$. Compare the answer with that in Example 8.4-3.

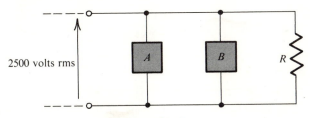

Figure P8.12

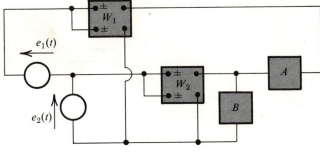

Figure P8.13

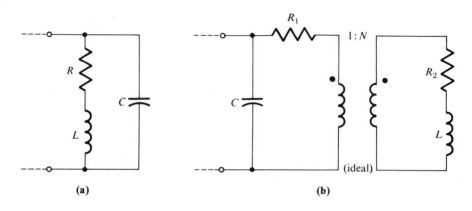

(a)

(b)

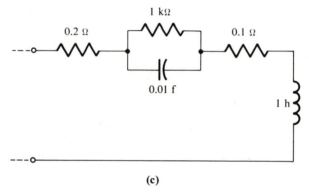

(c)

Figure P8.15

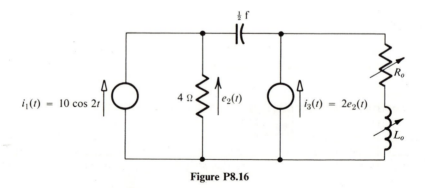

Figure P8.16

8.15 Use Eq. 8.4-2 to find the selectivity Q for each of the circuits in Fig. P8.15. Assume that $Q \gg 1$.

8.16 Choose the values of L_o and R_o in Fig. P8.16 so that maximum power is supplied to the resistance R_o in the steady state.

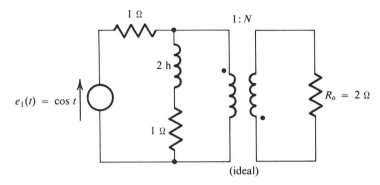

Figure P8.17

8.17 For the circuit in Fig. P8.17, find the value of the turns ratio N that will result in the maximum power delivered to R_o in the steady state. Find the power consumed in R_o for this choice of N.

8.18 In each of the circuits in Fig. P8.18, an arrow is drawn through the elements that may be varied. Choose the values of these elements so that maximum power is delivered to the resistance R_o. (Other problems on maximum power transfer include Problems 2.28, 5.18, and 7.18.)

8.19 The networks N_1 and N_2 in Fig. P8.19 can be characterized by the admittances $Y_1 = G_1 + jB_1 = |Y_1|\underline{/\phi_1}$ and $Y_2 = G_2 + jB_2 = |Y_2|\underline{/\phi_2}$ at the angular frequency of the source. Find the conditions that result in the maximum steady-state power to N_2:

(a) If only G_2 and B_2 may be varied.

(b) If only $|Y_2|$ may be varied.

(c) If only G_1 and G_2 may be varied.

(d) If only $|Y_1|$ may be varied.

8.20 What value of the capacitance C in Fig. P8.20 will allow the maximum steady-state power to be delivered to the resistance R_o?

8.21 The available power P_m for the frequency-domain circuit in Fig. P8.21a is defined as the power that would be supplied to an externally connected, passive load, when the load is chosen to obtain the maximum load power. The power supplied to the resistance R_2 in Fig. P8.21b is denoted by P_{20}. The power supplied to R_2 when a network N is inserted between the two resistances, as in Fig. P8.21c, is denoted by P_2. The insertion power gain in decibels is defined as $10 \log_{10} (P_2/P_{20})$ and the insertion power loss as $10 \log_{10} (P_{20}/P_2)$.

(a) Prove that $P_m = |E_g|^2/4R_g$.

(b) Prove that the insertion power loss in decibels is

$$10 \log_{10} \left(\frac{R_2}{R_g + R_2} \right)^2 \left| \frac{E_g}{E_2} \right|^2$$

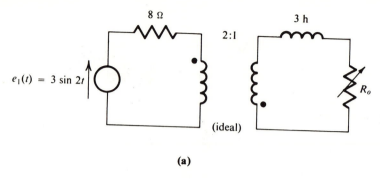

(a)

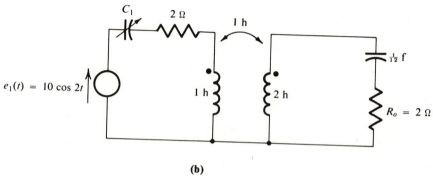

(b)

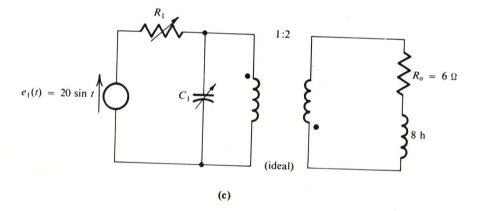

(c)

(c) Prove that

$$\frac{P_2}{P_m} = \frac{4R_g}{R_2} \left| \frac{\mathbf{E}_2}{\mathbf{E}_g} \right|^2$$

8.22 Derive the equations given in the text for the line-to-line voltages in Fig. 8.6-2b by the use of trigonometric identities.

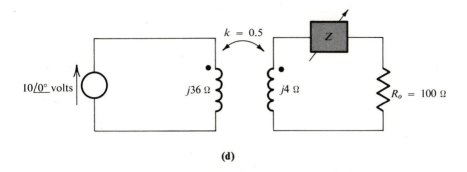

(d)

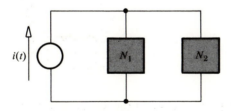

(e)

Figure P8.18

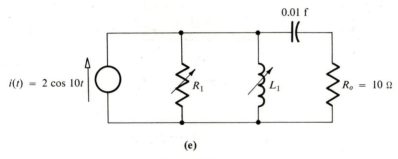

Figure P8.19

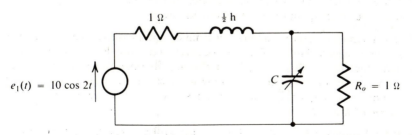

Figure P8.20

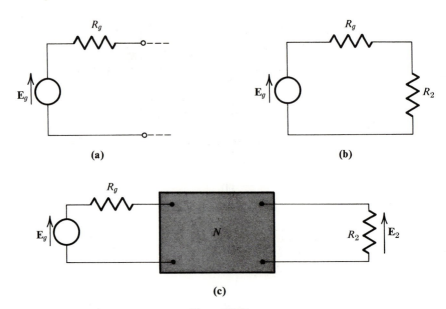

Figure P8.21

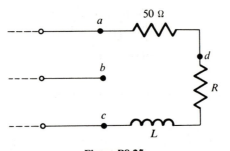

Figure P8.25

8.23 Solve Example 8.6-1 by the use of the wye-delta transformation, which is given by Eqs. 7.4-9. *P 365*

8.24 Prove that reversing the phase sequence in Example 8.6-2 changes the individual wattmeter readings but not the total load power.

8.25 One method of experimentally determining the phase sequence of a three-phase source is suggested in Fig. P8.25. If the rms voltages are $|\mathbf{E}_{ab}| = |\mathbf{E}_{bc}| = |\mathbf{E}_{ca}| = 208$ volts and $|\mathbf{E}_{ad}| = |\mathbf{E}_{bd}| = |\mathbf{E}_{cd}| = 120$ volts, determine the phase sequence. (HINT: Draw phasor diagrams corresponding to each of the two possible sequences.) If, in addition, the frequency of the source is known to be 60 cycles per second, find the values of R and L.

8.26 Solve Example 8.6-1 if the point n' in the load is connected to the neutral point of the three-phase source. Does the sum of the two wattmeter readings still equal the total load power?

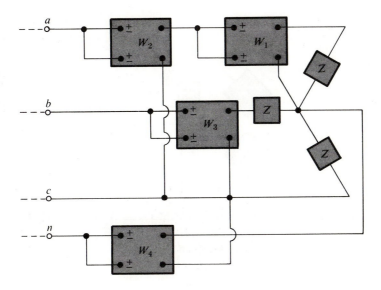

Figure P8.27

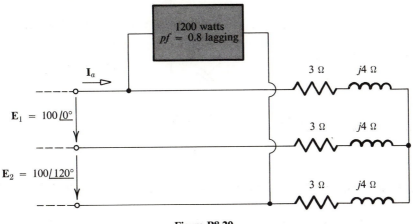

Figure P8.29

8.27 The three impedances in Fig. P8.27 are identical, and are connected to a three-phase source. The readings of $\boxed{W_1}$ and $\boxed{W_2}$ are 10 kw and 12 kw, respectively.

(a) Find the readings of the other two wattmeters.

(b) If the line-to-line voltage is 200 volts rms, find the impedance Z.

8.28 Completely solve Example 8.6-4 for the phase sequence a, c, b.

8.29 Find the current phasor I_a in Fig. P8.29.

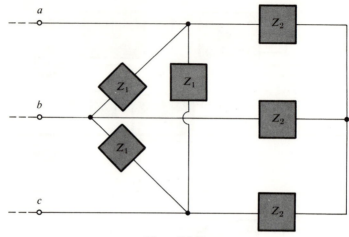

Figure P8.30

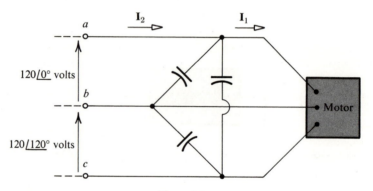

Figure P8.31

8.30 The circuit in Fig. P8.30 is supplied by a three-phase source with line-to-line voltages of 208 volts. If $Z_1 = 6 - j8\ \Omega$ and $Z_2 = 3 + j4\ \Omega$, find the magnitude of the line currents.

8.31 The three-phase motor in Fig. P8.31 is a balanced load that consumes 3 kw of power and has a power factor of 0.6 lagging.

(a) Find the kva rating of the capacitors that are needed to make the overall power factor 0.9 lagging.

(b) Find the current phasors I_1 and I_2.

9

The Fourier Series

and Integral

In Sections 3.4 and 3.5, we showed how an arbitrary input may be expressed as the sum of singularity functions. As soon as a circuit's response to the unit step function or to the unit impulse is known, its response to the arbitrary input can be found by superposition. In Chapter 5, we developed techniques for finding the steady-state response to a sinusoidal input. If an arbitrary input can be decomposed into a set of sinusoidal components, the steady-state response to this input can be found conveniently by superposition.

The student may already know that periodic functions can be expressed as a series of sinusoidal terms known as the Fourier series. Methods for obtaining the series are discussed in Section 9.1. In Section 9.2, the steady-state response to periodic inputs is found by using a Fourier series for the input. Later in the chapter we extend the technique to include nonperiodic inputs.

9.1

The Trigonometric Form of the Fourier Series

A function $f(t)$ is said to be *periodic* with a period T if $f(t + T) = f(t)$ for all values of t, i.e., if the function continually repeats itself every T seconds, as in Fig. 8.1-3. A Fourier series, having the following form, can be found for all periodic functions of practical interest.

$$f(t) = a_0 + a_1 \cos \omega_0 t + a_2 \cos 2\omega_0 t + \cdots + b_1 \sin \omega_0 t + b_2 \sin 2\omega_0 t + \cdots$$

$$= a_0 + \sum_{n=1}^{\infty} (a_n \cos n\omega_0 t + b_n \sin n\omega_0 t) \tag{9.1-1}$$

A set of sufficient conditions (always satisfied in practice) for the existence of the series is that $f(t)$ be single valued, have a finite number of maxima, minima, and discontinuities in any one period, and that the area $\int_0^T |f(t)| \, dt$ be finite. There are,

437

in general, an infinite number of terms in the series, although in certain special cases all but a finite number of these terms may be zero. The quantity ω_0, which is not related to the resonant angular frequency of a circuit, is defined as

$$\omega_0 = \frac{2\pi}{T} \tag{9.1-2}$$

The terms $a_1 \cos \omega_0 t$ and $b_1 \sin \omega_0 t$ together constitute the fundamental component (or the first harmonic), while $a_n \cos n\omega_0 t$ and $b_n \sin n\omega_0 t$ are the nth harmonic. Since the average value of each of the sinusoidal terms in the series is zero, the constant a_0 must be the average value (or d-c component) of $f(t)$.

The terms in the series are said to be *orthogonal* with respect to the period T, i.e., the integral over one period of the product of any two different terms vanishes:

$$\int_0^T \cos n\omega_0 t \; dt = 0$$

$$\int_0^T \sin n\omega_0 t \; dt = 0$$

$$\int_0^T \cos m\omega_0 t \sin n\omega_0 t \; dt = 0$$

$$\int_0^T \cos m\omega_0 t \cos n\omega_0 t \; dt = 0 \qquad (m \neq n)$$

$$\int_0^T \sin m\omega_0 t \sin n\omega_0 t \; dt = 0 \qquad (m \neq n)$$

It can also be shown that

$$\int_0^T \cos^2 n\omega_0 t \; dt = \int_0^T \sin^2 n\omega_0 t \; dt = \frac{\pi}{\omega_0} = \frac{T}{2}$$

The definite integrals listed above can be found in standard tables, and they remain valid if the limits are changed as long as the integration is over a complete period or over an integral number of periods.

The orthogonality property allows the derivation of explicit formulas for the constants $a_0, a_1, \ldots, b_1, b_2, \ldots$ when $f(t)$ is given. To derive a formula for a_0, integrate each term in Eq. 9.1-1 over one period. All the integrals on the right-hand side except the first will vanish, so

$$\int_0^T f(t) \; dt = a_0 \int_0^T dt$$

and

$$a_0 = \frac{1}{T} \int_0^T f(t) \; dt \tag{9.1-3}$$

which is consistent with the previous statement that a_0 is the average value of $f(t)$. Frequently, the average value can be determined by inspection, in which case Eq. 9.1-3 is not needed.

To evaluate a_1, multiply every term in Eq. 9.1-1 by $\cos \omega_0 t$ and then integrate over one period. The only nonzero terms are

$$\int_0^T f(t) \cos \omega_0 t \, dt = \int_0^T a_1 \cos^2 \omega_0 t \, dt$$

so

$$a_1 = \frac{2}{T} \int_0^T f(t) \cos \omega_0 t \, dt$$

In general,

$$a_n = \frac{2}{T} \int_0^T f(t) \cos n\omega_0 t \, dt \qquad (n \neq 0) \tag{9.1-4}$$

$$b_n = \frac{2}{T} \int_0^T f(t) \sin n\omega_0 t \, dt \tag{9.1-5}$$

Note specifically that Eq. 9.1-4 does *not* reduce to 9.1-3 if n is replaced by zero. In Eqs. 9.1-3 through 9.1-5, the integration may be carried out from $-T/2$ to $T/2$, or from $-T/4$ to $3T/4$, or over *any* convenient period. When these coefficients are inserted into Eq. 9.1-1, the series will converge to $f(t)$ at all points of continuity, and at a finite discontinuity will converge to the average of the values on either side of the discontinuity.[†]

Example 9.1-1. Find the Fourier series for each of the functions in Fig. 9.1-1.

Solution. The period for $f_1(t)$ is $T = 2\pi$, so $\omega_0 = 1$ rad per sec. Since the slope of the straight line between $t = 0$ and $t = 2\pi$ is $1/2\pi$,

$$a_0 = \frac{1}{2\pi} \int_0^{2\pi} \frac{t}{2\pi} \, dt = \frac{1}{2}$$

$$a_n = \frac{1}{\pi} \int_0^{2\pi} \frac{t}{2\pi} \cos nt \, dt = 0 \qquad (n \neq 0)$$

$$b_n = \frac{1}{\pi} \int_0^{2\pi} \frac{t}{2\pi} \sin nt \, dt = -\frac{1}{n\pi}$$

Thus,

$$f_1(t) = \frac{1}{2} - \frac{1}{\pi} \sin t - \frac{1}{2\pi} \sin 2t - \cdots$$

$$= \frac{1}{2} - \frac{1}{\pi} \sum_{n=1}^{\infty} \frac{1}{n} \sin nt$$

The detailed steps of the integrations are omitted in this chapter in order not to obscure the basic procedures. The conscientious student will no doubt carry out the details of the integration for a few of the examples, probably with the aid of integral tables. Note that the value of a_0, the d-c component of $f_1(t)$, can be written down directly from an inspection of the waveshape.

[†] A fuller discussion of the convergence of the Fourier series may be found in Sections 7.12 and 7.13 of E. A. Guillemin, *The Mathematics of Circuit Analysis* (New York: Wiley, 1949) or Chapter XIV of E. A. Guillemin, *Theory of Linear Physical Systems* (New York: Wiley, 1963).

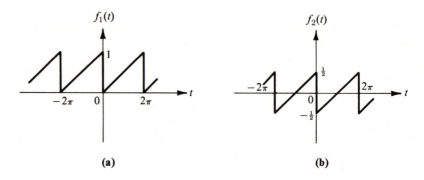

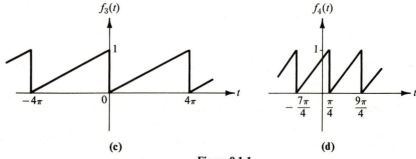

Figure 9.1-1

There is clearly some relationship between the four functions in Fig. 9.1-1, so it should not be necessary to work out each Fourier series independently. Since $f_2(t)$ is the first function shifted downward $\frac{1}{2}$ unit,

$$f_2(t) = f_1(t) - \frac{1}{2} = -\frac{1}{\pi} \sin t - \frac{1}{2\pi} \sin 2t - \cdots$$

The function $f_3(t)$ is the same as $f_1(t)$, except for a change in the time scale. Specifically,

$$f_3(t) = f_1\left(\frac{t}{2}\right) = \frac{1}{2} - \frac{1}{\pi} \sin \frac{t}{2} - \frac{1}{2\pi} \sin t - \cdots$$

Finally, $f_4(t)$ can be obtained by shifting $f_1(t)$ $\pi/4$ units in the positive t direction, so

$$f_4(t) = f_1\left(t - \frac{\pi}{4}\right) = \frac{1}{2} - \frac{1}{\pi} \sin \left(t - \frac{\pi}{4}\right) - \frac{1}{2\pi} \sin \left(2t - \frac{\pi}{2}\right)$$

$$- \frac{1}{3\pi} \sin \left(3t - \frac{3\pi}{2}\right) - \cdots$$

$$= \frac{1}{2} + \frac{1}{\sqrt{2}\,\pi} \cos t - \frac{1}{\sqrt{2}\,\pi} \sin t + \frac{1}{2\pi} \cos 2t$$

$$+ \frac{1}{3\sqrt{2}\,\pi} \cos 3t + \frac{1}{3\sqrt{2}\,\pi} \sin 3t + \cdots$$

The student should especially notice two things that are illustrated in the last example. First, a comparison of the series for $f_1(t)$ and $f_3(t)$ shows that changing the time scale *does* change the angular frequencies of the individual terms but does *not* change their coefficients. This is always true, since changing the time scale corresponds to replacing every t by at in Eq. 9.1-1 (where a is a scale constant). *For the purpose of evaluating the coefficients*, therefore, the given period may be arbitrarily changed if this seems to be convenient. In many practical problems, the value of T is such that difficult numbers and large powers of ten are introduced during the calculation of the coefficients. In such cases, it is sensible to temporarily change the period to 2π, in which case the formulas for the coefficients become

$$a_0 = \frac{1}{2\pi} \int_0^{2\pi} f_a(t)\, dt$$

$$a_n = \frac{1}{\pi} \int_0^{2\pi} f_a(t) \cos nt\, dt \qquad (n \neq 0) \tag{9.1-6}$$

$$b_n = \frac{1}{\pi} \int_0^{2\pi} f_a(t) \sin nt\, dt$$

where the subscript a denotes the function that is obtained by relabelling the time scale.

A comparison of the series for $f_1(t)$ and $f_4(t)$ in Example 9.1-1 provides one illustration of the effect of changing the time origin. In order to investigate this situation in a more general way, recall that trigonometric terms of the same frequency may be combined into a single term with a phase angle, as in Eq. 5.4-1. Then the Fourier series in Eq. 9.1-1 becomes

$$f(t) = a_0 + \sum_{n=1}^{\infty} A_n \cos(n\omega_0 t + \phi_n) \tag{9.1-7}$$

where

$$A_n = \sqrt{a_n{}^2 + b_n{}^2} \quad \text{and} \quad \phi_n = -\tan^{-1} \frac{b_n}{a_n} \tag{9.1-8}$$

When both sine and cosine terms are present in Eq. 9.1-1, the series in Eq. 9.1-7 is often preferred, because it clearly shows the magnitude A_n and the phase angle ϕ_n of the nth harmonic.

A time delay of t_0 seconds corresponds to moving the time origin back t_0 units, as in Fig. 9.1-2b, and to replacing t by $t - t_0$ in the expression for $f(t)$. Then

$$f(t - t_0) = a_0 + \sum_{n=1}^{\infty} A_n \cos(n\omega_0 t + \phi_n - n\omega_0 t_0) \tag{9.1-9}$$

The magnitudes of the harmonics are unchanged, but the phase angle is changed by the increment $-n\omega_0 t_0$, which is proportional to the order of the harmonic.

Simplifications for Symmetrical Waveshapes.   A function for which

$$f(t) = f(-t) \tag{9.1-10}$$

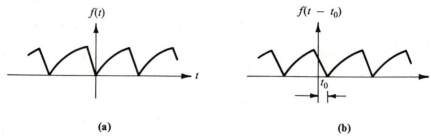

Figure 9.1-2

is symmetrical about the vertical axis, as in Fig. 9.1-3a, and is called an *even function*. The integrals $\int_0^a f(t)\, dt$ and $\int_{-a}^0 f(t)\, dt$ are represented by the two shaded areas, which are equal for an even function; hence,

$$\int_{-a}^a f(t)\, dt = 2 \int_0^a f(t)\, dt \tag{9.1-11}$$

An *odd function* is one for which

$$f(t) = -f(-t) \tag{9.1-12}$$

and it has symmetry about the origin, as in Fig. 9.1-3b. Since an area underneath the horizontal axis is regarded as negative, the net area underneath the curve from $-a$ to a is zero and

$$\int_{-a}^a f(t)\, dt = 0 \tag{9.1-13}$$

Some additional examples of odd and even functions are listed below.

Odd Functions	Even Functions
t	$|t|$
$t^3 + 6t$	$t^2 - 5$
$\sin \beta t$	$\cos \beta t$

Consider the product of two odd functions $f_1(t)$ and $f_2(t)$. By Eq. 9.1-12,

$$f(t) = f_1(t)f_2(t) = [-f_1(-t)][-f_2(-t)] = f_1(-t)f_2(-t) = f(-t)$$

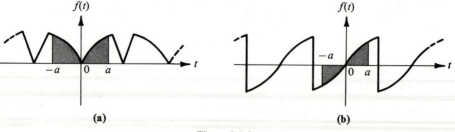

Figure 9.1-3

which by Eq. 9.1-10 must be an even function. It can be shown in a similar manner that the product of two odd or two even functions is an even function, while the product of one odd and one even function is odd.

When calculating the coefficients in the Fourier series for functions having symmetry, it is convenient to integrate from $-T/2$ to $T/2$, instead of zero to T.

$$a_0 = \frac{1}{T} \int_{-T/2}^{T/2} f(t)\, dt$$

$$a_n = \frac{2}{T} \int_{-T/2}^{T/2} f(t) \cos n\omega_0 t\, dt \qquad (n \neq 0) \qquad\qquad (9.1\text{-}14)$$

$$b_n = \frac{2}{T} \int_{-T/2}^{T/2} f(t) \sin n\omega_0 t\, dt$$

Suppose that $f(t)$ is an even function. Then $f(t) \cos n\omega_0 t$ is even but $f(t) \sin n\omega_0 t$ is odd, and by Eqs. 9.1-11 and 9.1-13 the formulas for the coefficients become

$$a_0 = \frac{2}{T} \int_0^{T/2} f(t)\, dt$$

$$a_n = \frac{4}{T} \int_0^{T/2} f(t) \cos n\omega_0 t\, dt \qquad (n \neq 0) \qquad\qquad (9.1\text{-}15)$$

$$b_n = 0$$

Similarly, if $f(t)$ is an odd function,

$$a_0 = a_n = 0$$

$$b_n = \frac{4}{T} \int_0^{T/2} f(t) \sin n\omega_0 t\, dt \qquad\qquad (9.1\text{-}16)$$

These results are consistent with what the student should expect intuitively. Since the sine terms are odd functions, they are not present in the Fourier series of an even function. The constant and the cosine terms, themselves even, do not appear in the series for an odd function. Finally, the nonzero terms are evaluated by integrating over half the period, and multiplying by two.

One example of an odd function is $f_2(t)$ in Fig. 9.1-1. Only sine terms should be present in the Fourier series, and

$$b_n = 2\left[\frac{1}{\pi} \int_0^\pi \left(-\frac{1}{2} + \frac{t}{2\pi}\right) \sin nt\, dt\right] = -\frac{1}{n\pi}$$

which is consistent with Example 9.1-1. It is sometimes desirable to temporarily shift either the horizontal or vertical axes or both in order to create an even or odd function and to make use of the simplifications for symmetrical waveshapes. The effect of moving the axes was discussed in connection with Example 9.1-1 and is illustrated again in the following example.

Example 9.1-2.   Find the Fourier series for $f_1(t)$ and $f_2(t)$ in Fig. 9.1-4.

Solution.   By inspection, the d-c component of $f_1(t)$ is $a_0 = \frac{1}{2}$. Since $f_1(t)$ is an even function, $b_n = 0$ and

$$a_n = 2\left(\frac{1}{\pi}\int_0^\pi \frac{t}{\pi}\cos nt\,dt\right) = \frac{-4}{n^2\pi^2} \qquad \text{for } n \text{ odd}$$

$$= 0 \qquad \text{for } n \text{ even}$$

Thus,

$$f_1(t) = \frac{1}{2} - \frac{4}{\pi^2}\left(\cos t + \frac{1}{9}\cos 3t + \frac{1}{25}\cos 5t + \cdots\right)$$

If the coefficients were found directly from Eqs. 9.1-6, without taking advantage of the symmetry, it would be necessary to note that $f_1(t) = t/\pi$ for $0 < t < \pi$ and is $= 2 - t/\pi$ for $\pi < t < 2\pi$. Then

$$a_n = \frac{1}{\pi}\left[\int_0^\pi \frac{t}{\pi}\cos nt\,dt + \int_\pi^{2\pi}\left(2 - \frac{t}{\pi}\right)\cos nt\,dt\right]$$

$$b_n = \frac{1}{\pi}\left[\int_0^\pi \frac{t}{\pi}\sin nt\,dt + \int_\pi^{2\pi}\left(2 - \frac{t}{\pi}\right)\sin nt\,dt\right]$$

which lead to the same result, but with considerably more computational work.

Although it is not difficult to find the Fourier series for $f_2(t)$ directly, let us try to relocate the axes in order to use the symmetry relations. While $f_2(t)$ is itself neither

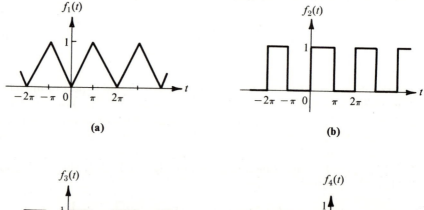

(a)

(b)

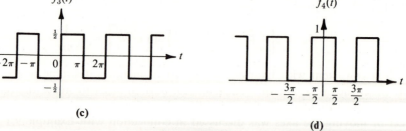

(c)

(d)

Figure 9.1-4

even nor odd, subtraction of a constant of $\frac{1}{2}$ produces the odd function $f_3(t)$, for which $a_0 = a_n = 0$ and

$$b_n = 2\left(\frac{1}{\pi}\int_0^\pi \frac{1}{2}\sin nt\, dt\right) = \frac{2}{n\pi} \qquad \text{for } n \text{ odd}$$

$$= 0 \qquad \text{for } n \text{ even}$$

Thus,

$$f_3(t) = \frac{2}{\pi}\left(\sin t + \frac{1}{3}\sin 3t + \frac{1}{5}\sin 5t + \cdots\right)$$

so

$$f_2(t) = \frac{1}{2} + \frac{2}{\pi}\left(\sin t + \frac{1}{3}\sin 3t + \frac{1}{5}\sin 5t + \cdots\right)$$

Alternatively, the vertical axis may be changed to create the even function $f_4(t)$, for which $b_n = 0$, $a_0 = \frac{1}{2}$, and

$$a_n = 2\left[\frac{1}{\pi}\int_0^{\pi/2}(1)\cos nt\, dt\right] = \frac{2}{n\pi}(-1)^{(n-1)/2} \qquad \text{for } n \text{ odd}$$

$$= 0 \qquad \text{for } n \text{ even}$$

Then

$$f_4(t) = \frac{1}{2} + \frac{2}{\pi}\left(\cos t - \frac{1}{3}\cos 3t + \frac{1}{5}\cos 5t + \cdots\right)$$

and, by Eq. 9.1-9,

$$f_2(t) = \frac{1}{2} + \frac{2}{\pi}\left[\cos\left(t - \frac{\pi}{2}\right) - \frac{1}{3}\cos\left(3t - \frac{3\pi}{2}\right) + \cdots\right]$$

$$= \frac{1}{2} + \frac{2}{\pi}\left(\sin t + \frac{1}{3}\sin 3t + \cdots\right)$$

which agrees with the previous result.

Half-wave symmetry is defined mathematically by the condition

$$f\left(t + \frac{T}{2}\right) = -f(t) \tag{9.1-17}$$

and is illustrated in Fig. 9.1-4c. The second half of each period is the same as the first half, except for a change in sign. It can be shown that for such functions $a_0 = a_2 = b_2 = a_4 = b_4 = \cdots = 0$, i.e., that all the even harmonics are missing. One way to justify this statement is to notice that the terms a_0, $a_2\cos 2\omega_0 t$, $b_2\sin 2\omega_0 t$, $\ldots$ are unchanged when t is replaced by $t + T/2$, i.e., when $\omega_0 t$ is replaced by $\omega_0 t + \pi$. In contrast, the terms $a_1\cos\omega_0 t$, $b_1\sin\omega_0 t$, $a_3\cos 3\omega_0 t$, $\ldots$ *do* suffer a sign change, as in Eq. 9.1-17. In this type of symmetry, the nonzero coefficients again may be evaluated by integrating over half the period and then multiplying by two.

All the functions in Fig. 9.1-4 have half-wave symmetry except for the addition of a constant term in some cases, so the fact that their Fourier series contain no even harmonics is consistent with the above discussion. Note that a given wave-shape may sometimes have both half-wave and even or odd symmetry.

Still other types of symmetry could be discussed, but the ones mentioned above are the most useful. Furthermore, Eqs. 9.1-3 through 9.1-5 may be evaluated by graphical or numerical integration, which is especially useful for functions which cannot be easily described analytically. Several special methods of numerical integration have been devised for the rapid calculation of the coefficients.[†]

9.2

The Response to Periodic Inputs

Suppose that a periodic input is represented by the Fourier series of Eq. 9.1-1 or 9.1-7. The forced response to each of the individual terms in the series may be found by d-c and a-c steady-state circuit theory, and then the entire forced response may be written down by superposition. The procedure is essentially the same as that used in Example 5.3-7, except that the response consists of an infinite number rather than a finite number of terms.

> **Example 9.2-1.**   For the circuit in Fig. 9.2-1a and the input waveshape in Fig. 9.2-1b, find and sketch the steady-state output voltage.
>
> **Solution.**   The given input is not really periodic, because it is zero for negative values of t. The complete response of any circuit for $t \geq 0$, however, is uniquely determined by the input for $t \geq 0$ and by the stored energy at $t = 0$. The initial stored energy affects only the size of the free response, so the forced response for $t \geq 0$ is independent of the input for $t < 0$. For the calculation of the forced (or steady-state) response, the waveshape in Fig. 9.2-1b may be replaced by the one in Fig. 9.2-1c.
>
> The Fourier series for $i(t)$ is the same as the series for $f_2(t)$ in Fig. 9.1-4b, except that the angular frequency of the first harmonic is $\omega_0 = 2\pi/2 = \pi$ rad per sec instead of 1. Thus,
>
> $$ i(t) = \frac{1}{2} + \frac{2}{\pi} \left(\sin \pi t + \frac{1}{3} \sin 3\pi t + \frac{1}{5} \sin 5\pi t + \cdots \right) $$
>
> Since the capacitance acts like an open circuit in the d-c steady state, the forced response to the constant component of $\frac{1}{2}$ is $\frac{1}{2}$. The a-c steady-state network function is the impedance
>
> $$ Z(j\omega) = \frac{1}{1/R + j\omega C} = \frac{1}{1 + j\omega} = \frac{1}{\sqrt{1 + \omega^2}} \underline{/-\tan^{-1}\omega} $$
>
> which can be evaluated for each of the angular frequencies appearing in the input.

[†] See, for example, Sections 14.16 and 14.17 of H. H. Skilling, *Electrical Engineering Circuits* (New York: Wiley, 1958).

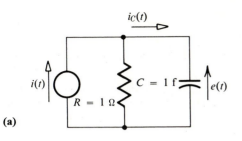

(a)

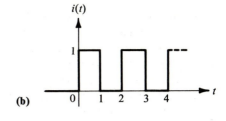

(b)

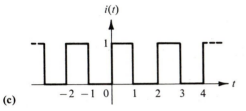

(c)

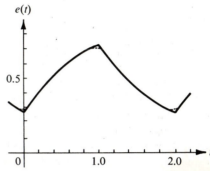

(d)

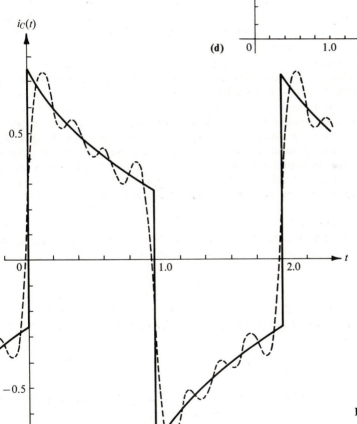

(e)

Figure 9.2-1

Since

$$Z(j\pi) = \frac{1}{\sqrt{1 + \pi^2}} \underline{/-\tan^{-1}\pi} \quad \text{and} \quad Z(j3\pi) = \frac{1}{\sqrt{1 + 9\pi^2}} \underline{/-\tan^{-1}3\pi}$$

the forced responses to $\sin \pi t$ and $\sin 3\pi t$ are $(1/\sqrt{1 + \pi^2})\sin(\pi t - \tan^{-1}\pi)$ and $(1/\sqrt{1 + 9\pi^2})\sin(3\pi t - \tan^{-1}3\pi)$, respectively. The entire forced response is

$$e(t) = \frac{1}{2} + \frac{2}{\pi}\left[\frac{1}{\sqrt{1 + \pi^2}}\sin(\pi t - \tan^{-1}\pi)\right.$$

$$\left. + \frac{1}{3\sqrt{1 + 9\pi^2}}\sin(3\pi t - \tan^{-1}3\pi) + \cdots\right]$$

Both $e(t)$ and $i_C(t)$ were needed in Example 8.1-2. The current through the capacitance can be found by considering a new network function or by differentiating the expression for $e(t)$.

$$i_C(t) = C\frac{de}{dt}$$

$$= \frac{2}{\pi}\left[\frac{\pi}{\sqrt{1 + \pi^2}}\cos(\pi t - \tan^{-1}\pi) + \frac{\pi}{\sqrt{1 + 9\pi^2}}\cos(3\pi t - \tan^{-1}3\pi) + \cdots\right]$$

The expressions for $e(t)$ and $i_C(t)$ represent the Fourier series for these quantities. The sum of the d-c component and the first four nonzero harmonics is shown by the dashed lines in Figs. 9.2-1d and 9.2-1e. The solid curves are the exact waveshapes, as the number of terms in the sum approaches infinity, and are derived in the next example. The partial sum for $e(t)$ is within 1% of the exact solution for most values of t, and the maximum deviation is about 5%, where the slope changes rapidly. Because of the discontinuities in the exact curve of $i_C(t)$, the sum of the first four terms in the series differs appreciably from the exact solution. The convergence of the Fourier series is particularly poor near a discontinuity. The height of the ripple in the approximate curve on either side of the discontinuity remains about 9% of the discontinuity, even when a larger number of terms is used.

Example 9.2-1 illustrates the three basic steps in the analysis of a circuit by the Fourier series. First, the series for the given periodic input $x(t)$ is found.

$$x(t) = a_0 + \sum_{n=1}^{\infty} A_n \cos(n\omega_0 t + \phi_n) \qquad (9.2\text{-}1)$$

Next, the series representing the output $y(t)$ has the form

$$y(t) = a_0 H(0) + \sum_{n=1}^{\infty} A_n |H(jn\omega_0)| \cos[n\omega_0 t + \phi_n + \ast H(jn\omega_0)] \qquad (9.2\text{-}2)$$

where $H(j\omega)$ is the a-c steady-state network function, which is evaluated for $\omega = 0, \omega_0, 2\omega_0, \ldots$ Lastly, a sufficient number of terms in the output series are combined to obtain the output waveshape.

Although an understanding of this procedure is most important to the rest of the chapter, the principal use of the Fourier series is *not* in solving a problem like Example 9.1-1. Two disadvantages of the above solution are that it is a slowly converging infinite series and that it contains the forced but not the free response. To maintain a proper perspective, we shall now digress to discuss one method of overcoming these disadvantages in this particular problem. (Another method is suggested in one of the problems at the end of the chapter.) The great importance of the Fourier series approach will gradually become apparent in the rest of the chapter.

Example 9.2-2. Find the complete response for the previous example in closed form by a classical solution.

Solution. The node equation describing Fig. 9.2-1a is

$$\frac{de}{dt} + e = i = \begin{cases} 1 & \text{for } 0 < t < 1 \\ 0 & \text{for } 1 < t < 2 \end{cases}$$

The classical solution, by the methods of Chapter 4, is

$$e(t) = 1 + K_1\epsilon^{-t} \qquad \text{for } 0 < t < 1$$

$$= K_2\epsilon^{-t} \qquad \text{for } 1 < t < 2$$

Since the voltage across a capacitance cannot change instantaneously unless impulses are present, these two expressions must be equal when $t = 1$.

$$1 + K_1\epsilon^{-1} = K_2\epsilon^{-1} \quad \text{or} \quad K_2 = K_1 + \epsilon$$

The forced or steady-state response to a periodic input is itself periodic, so the forced response at the start of the second cycle ($t = 2$) must be the same as at the start of the first cycle ($t = 0$).

$$1 + K_1\epsilon^{-0} = K_2\epsilon^{-2} \quad \text{or} \quad K_2 = (1 + K_1)\epsilon^2$$

Equating the two expressions for K_2, we obtain

$$K_1 = \frac{-\epsilon}{1 + \epsilon} \quad \text{and} \quad K_2 = \frac{\epsilon^2}{1 + \epsilon}$$

so

$$e(t) = 1 - \frac{\epsilon}{1 + \epsilon}\,\epsilon^{-t} \qquad \text{for } 0 < t < 1$$

$$= \frac{\epsilon^2}{1 + \epsilon}\,\epsilon^{-t} \qquad \text{for } 1 < t < 2$$

Finally,

$$i_C(t) = C\frac{de}{dt} = \frac{\epsilon}{1 + \epsilon}\,\epsilon^{-t} \qquad \text{for } 0 < t < 1$$

$$= \frac{-\epsilon^2}{1 + \epsilon}\,\epsilon^{-t} \qquad \text{for } 1 < t < 2$$

These equations describe the solid curves in Figs. 9.2-1d and 9.2-1e.

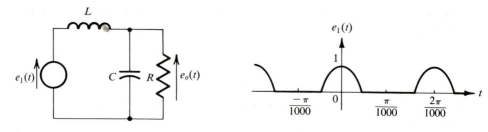

Figure 9.2-2

From the original differential equation, the free or complementary part of $e(t)$ has the form $K\epsilon^{-t}$. If there was no stored energy in the capacitance for $t < 0$, $e(0+) = 0$. Since the forced response is $\epsilon/(1 + \epsilon) = 0.731$ at $t = 0+$, the complementary solution is $-0.731\epsilon^{-t}$ for $t > 0$.

Example 9.2.3. The low-pass filter in Fig. 9.2-2 is designed to produce a nearly constant output voltage when the input is the "half-wave rectified sinusoid" shown. If $L = 1$ h and $R = 1000\ \Omega$, determine the value of C such that the peak value of the largest a-c component in the output is $\frac{1}{20}$ of the d-c component.

Solution. The Fourier series for the source voltage can be evaluated from Eqs. 9.1-15 and is found to be

$$e_1(t) = \frac{1}{\pi} + \frac{1}{2}\cos 1000t + \frac{2}{3\pi}\cos 2000t + \cdots$$

The circuit's transfer function can be determined by writing a single node equation.

$$\frac{E_o(s) - E_1(s)}{sL} + sCE_o(s) + \frac{E_o(s)}{R} = 0$$

$$H(s) = \frac{E_o(s)}{E_1(s)} = \frac{1}{s^2LC + sL/R + 1}$$

The d-c transfer function is unity, so the d-c component of the output voltage is $1/\pi$. Notice that the size of the harmonics in the input and the magnitude of

$$H(j\omega) = \frac{1}{-\omega^2LC + j\omega L/R + 1}$$

decrease as the angular frequency increases. Thus, the largest a-c component in the output will be the first harmonic, at 1000 rad per sec. With the given values of L and R,

$$H(j1000) = \frac{1}{-10^6C + j1 + 1}$$

The peak value of the first harmonic in the output is $\frac{1}{2}|H(j1000)|$, which must equal $\frac{1}{20}(1/\pi)$. Then

$$|H(j1000)| = \frac{1}{10\pi}$$

$$|1 - 10^6 C| \doteq 10\pi$$

$$10^6 C \doteq 10\pi + 1 = 32.4$$

or $C = 32.4 \ \mu\text{f}$.

The expression for $H(j\omega)$ clearly indicates that the a-c components in the output will be negligible if $\omega^2 LC \gg 1$ (or else $\omega L/R \gg 1$) at the frequency of the first harmonic. There are upper limits on the sizes of practical inductances and capacitances, however, and the load resistance is usually fixed by other considerations, so in some cases a more complicated low-pass filter may be needed.

Example 9.2-4. Find the forced response of the circuit in Fig. 9.2-3a to the input in Fig. 9.2-3b.

Solution. The square-wave input is the same as Fig. 9.1-4b, except that

$$\omega_0 = \frac{2\pi}{3\pi/5} = \frac{10}{3} \text{ rad per sec}$$

so

$$e_1(t) = \frac{1}{2} + \frac{2}{\pi}\left(\sin\frac{10}{3}t + \frac{1}{3}\sin 10t + \frac{1}{5}\sin\frac{50}{3}t + \cdots \right)$$

The circuit is the same as in Fig. 6.2-2e, and the transfer function

$$H(s) = \frac{100}{s^2 + 0.2s + 100}$$

has poles at $-0.1 \pm j10$, as shown in Fig. 6.3-4e. An inspection of the pole-zero pattern shows that the circuit is highly selective, with a resonant angular frequency of 10 rad per sec. Thus the magnitude of

$$H(j10) = -j50 = 50\underline{/-90^\circ}$$

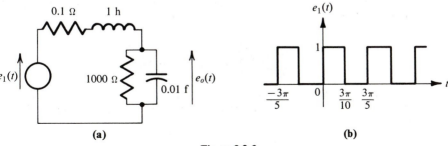

(a)   (b)

Figure 9.2-3

is much greater than that of $H(0)$, $H(j10/3)$, $H(j50/3)$, etc. All the components in the output except the third harmonic are negligible, so

$$e_o(t) \doteq \left(\frac{2}{3\pi}\right)(50)\sin(10t - 90°) = -\frac{100}{3\pi}\cos 10t$$

which is a pure sinusoidal waveshape.

The last two examples are typical of filter applications, in which the Fourier series is very useful. The filter passes certain of the components in the incoming signal, while almost entirely rejecting the other components.

Example 9.2-5. The Fourier series is useful in many other areas of electrical engineering, including modulation theory and noise theory. As one simple example, consider the "amplitude modulated signal"

$$x(t) = (1 - k\cos\beta_1 t)\cos\beta_2 t$$

where $\beta_2 \gg \beta_1$. This input might represent a musical note of angular frequency β_1 transmitted on a "carrier" frequency of β_2 radians per second, and it has the waveshape shown in Fig. 9.2-4. What frequencies must a receiving set pass in order to transmit the entire input?

Solution. Although the Fourier series for $x(t)$ may be evaluated by Eqs. 9.1-15, it is easier to use the trigonometric identity

$$\cos\beta_1 t\cos\beta_2 t = \tfrac{1}{2}[\cos(\beta_2 + \beta_1)t + \cos(\beta_2 - \beta_1)t]$$

Then

$$x(t) = \cos\beta_2 t - \frac{k}{2}\cos(\beta_2 + \beta_1)t - \frac{k}{2}\cos(\beta_2 - \beta_1)t$$

Note that in this case the Fourier series has a finite rather than an infinite number of terms. To preserve the entire input, the angular frequencies β_2, $\beta_2 + \beta_1$, and $\beta_2 - \beta_1$ must be passed by the receiving set.

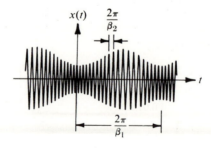

Figure 9.2-4

Power Calculations. In Section 8.1 it was shown that *if* the currents and voltages consist of several components at different frequencies, the power at each frequency may be computed individually, and the total power found by superposition. Currents and voltages that are represented by a Fourier series fall into this category. Furthermore, their rms values may be calculated from the result of Example 8.1-7.

Example 9.2-6. For the circuit in Fig. 9.2-1, the Fourier series for the current $i(t)$ and the steady-state voltage $e(t)$ are

$$i(t) = \frac{1}{2} + \frac{2}{\pi}\left(\sin \pi t + \frac{1}{3}\sin 3\pi t + \cdots\right)$$

$$e(t) = \frac{1}{2} + \frac{2}{\pi}\left[0.304 \sin (\pi t - 72.3°) + \frac{1}{3}(0.106)\sin (3\pi t - 83.9°) + \cdots\right]$$

Calculate the power delivered to the circuit and the rms values of $i(t)$ and $e(t)$.

Solution. When only the d-c components are considered,

$$P_{d-c} = (\tfrac{1}{2})(\tfrac{1}{2}) = 0.2500$$

For the first harmonic,

$$P_1 = \left(\frac{2}{\pi}\frac{1}{\sqrt{2}}\right)\left(0.304\,\frac{2}{\pi}\frac{1}{\sqrt{2}}\right)\cos 72.3° = 0.0186$$

and for the third harmonic,

$$P_3 = \left(\frac{2}{3\pi}\frac{1}{\sqrt{2}}\right)\left(0.106\,\frac{2}{3\pi}\frac{1}{\sqrt{2}}\right)\cos 83.9° = 0.0003$$

The total power delivered in the steady state is

$$P = 0.2500 + 0.0186 + 0.0003 + \cdots = 0.2689$$

which agrees with the exact answer of $1/(1 + \epsilon)$ in Example 8.1-2 to four decimal places. Also notice that

$$I_{rms} = \sqrt{\left(\frac{1}{2}\right)^2 + \left(\frac{\sqrt{2}}{\pi}\right)^2 + \left(\frac{\sqrt{2}}{3\pi}\right)^2 + \cdots} = 0.707$$

$$E_{rms} = \sqrt{\left(\frac{1}{2}\right)^2 + \left(0.304\,\frac{\sqrt{2}}{\pi}\right)^2 + \left(0.106\,\frac{\sqrt{2}}{3\pi}\right)^2 + \cdots} = 0.519$$

which agrees with Example 8.1-2.

9.3
The Complex Form of the Fourier Series

In this section, the Fourier series of Section 9.1 is rewritten as the sum of terms having the form $\epsilon^{j\omega t}$. Although the basic procedures of circuit analysis are the same no matter which form is used, the new form has certain advantages and provides a background for the subsequent development of the Fourier and Laplace transforms. The trigonometric Fourier series for a real, periodic function is

$$f(t) = a_0 + \sum_{n=1}^{\infty} (a_n \cos n\omega_0 t + b_n \sin n\omega_0 t)$$

$$= a_0 + \sum_{n=1}^{\infty} A_n \cos (n\omega_0 t + \phi_n)$$

where equations for the real quantities a_0, a_n, b_n, A_n, and ϕ_n are given in Section 9.1. To derive the complex (or exponential) form of the series, the identities

$$\cos \theta = \frac{\epsilon^{j\theta} + \epsilon^{-j\theta}}{2}, \qquad \sin \theta = \frac{\epsilon^{j\theta} - \epsilon^{-j\theta}}{j2}, \qquad \epsilon^{\pm j\theta} = \cos \theta \pm j \sin \theta \qquad (9.3\text{-}1)$$

are used. Then

$$f(t) = a_0 + \sum_{n=1}^{\infty} \left(\frac{a_n}{2} \epsilon^{jn\omega_0 t} + \frac{a_n}{2} \epsilon^{-jn\omega_0 t} - j\frac{b_n}{2} \epsilon^{jn\omega_0 t} + j\frac{b_n}{2} \epsilon^{-jn\omega_0 t} \right)$$

$$= a_0 + \sum_{n=1}^{\infty} \left(\frac{a_n - jb_n}{2} \epsilon^{jn\omega_0 t} + \frac{a_n + jb_n}{2} \epsilon^{-jn\omega_0 t} \right)$$

$$= a_0 + \sum_{n=1}^{\infty} (c_n \epsilon^{jn\omega_0 t} + c_n{}^* \epsilon^{-jn\omega_0 t}) \qquad (9.3\text{-}2)$$

where

$$c_n = \frac{a_n - jb_n}{2} = \frac{\sqrt{a_n{}^2 + b_n{}^2}}{2} \left/ -\tan^{-1} \frac{b_n}{a_n} \right.$$

$$= \frac{A_n}{2} \left/ \phi_n \right. \qquad (9.3\text{-}3)$$

and where the asterisk denotes the complex conjugate. The last step follows from Eqs. 9.1-8 and gives some physical significance to the coefficient c_n, which in contrast to a_n and b_n is a complex quantity. The magnitude of c_n is one-half the magnitude of the nth harmonic, and its angle is the phase angle of the nth harmonic with respect to a cosine wave. Thus, c_n contains the essential information about the frequency components comprising $f(t)$: their relative strength and their phase angles.

By Eqs. 9.1-14,

$$c_n = \frac{1}{2} (a_n - jb_n) = \frac{1}{T} \int_{-T/2}^{T/2} [f(t)(\cos n\omega_0 t - j \sin n\omega_0 t)] \, dt$$

so

$$c_n = \frac{1}{T} \int_{-T/2}^{T/2} f(t) \epsilon^{-jn\omega_0 t} \, dt \qquad (9.3\text{-}4)$$

Since the integration in the formulas for a_n and b_n can be over any one full period, the same freedom exists in the evaluation of c_n, but the limits of $-T/2$ to $T/2$ are usually the most convenient. Note for future reference that replacing j by $-j$ in Eq. 9.3-4 is equivalent to replacing n by $-n$, so in Eq. 9.3-2

$$c_n{}^* = [c_n]_{j \to -j} = [c_n]_{n \to -n} = c_{-n}$$

Thus,

$$f(t) = a_0 + \sum_{n=1}^{\infty} c_n \epsilon^{jn\omega_0 t} + \sum_{n=-1}^{\infty} c_n \epsilon^{jn\omega_0 t}$$

Although c_0 does not appear in the last equation, notice that when n is replaced by zero in Eq. 9.3-4,

$$c_0 = \frac{1}{T} \int_{-T/2}^{T/2} f(t)\, dt$$

which, by Eqs. 9.1-14, equals a_0. Therefore, the equation for $f(t)$ can be written as a single summation over all integral values of n.

$$f(t) = \sum_{n=-\infty}^{\infty} c_n \epsilon^{jn\omega_0 t} \tag{9.3-5}$$

Equation 9.3-5 is known as the complex form of the Fourier series, where the coefficients are given by Eq. 9.3-4.

Many of the comments made about the trigonometric form of the series carry over to the complex form. For example, c_0 is the d-c component, and often may be determined by inspection. *For the purpose of evaluating the coefficients*, the time scale may be temporarily changed so that $T = 2\pi$, in which case

$$c_n = \frac{1}{2\pi} \int_{-\pi}^{\pi} f(t) \epsilon^{-jnt}\, dt$$

If the function $f(t)$ is delayed by t_0 seconds, the resulting series

$$f(t - t_0) = \sum_{n=-\infty}^{\infty} c_n \epsilon^{jn\omega_0 (t-t_0)} = \sum_{n=-\infty}^{\infty} (c_n \epsilon^{-jn\omega_0 t_0}) \epsilon^{jn\omega_0 t}$$

has the coefficients $c_n(1 \underline{/-n\omega_0 t_0})$. The magnitudes are unchanged, but the angles are changed by the increment $-n\omega_0 t_0$.

For functions with half-wave symmetry, $c_0 = c_2 = c_4 = \cdots = 0$, and all the even harmonics are missing. For an even or an odd function, however, none of the coefficients is necessarily zero, and none may be evaluated by integrating over half the period and multiplying the result by two. This is because the factor $\epsilon^{-jn\omega_0 t}$ in the integrand of Eq. 9.3-4 is neither even nor odd. Nevertheless, the evaluation of c_n is made easier if the axes can be chosen so that $f(t)$ is even or odd. It is an interesting exercise to show that for an even function the coefficients are real and are given by

$$c_n = \frac{2}{T} \int_0^{T/2} f(t) \cos n\omega_0 t\, dt \tag{9.3-6}$$

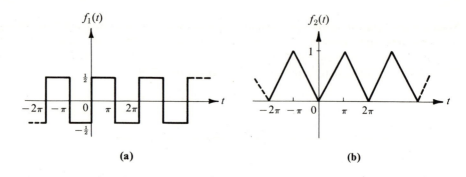

(a)   **(b)**

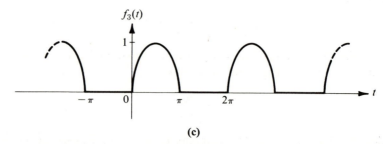

(c)

Figure 9.3-1

while for an odd function

$$c_n = -j\frac{2}{T}\int_0^{T/2} f(t)\sin n\omega_0 t\, dt \tag{9.3-7}$$

Example 9.3-1.   Find the Fourier series for each of the functions in Fig. 9.3-1. The function $f_3(t)$ is a half-wave rectified sine wave.

Solution.   In all three cases, $T = 2\pi$ and $\omega_0 = 1$ rad per sec. For the function $f_1(t)$,

$$c_n = \frac{1}{2\pi}\left[\int_{-\pi}^0 \left(-\frac{1}{2}\right)\epsilon^{-jnt}\, dt + \int_0^\pi \frac{1}{2}\epsilon^{-jnt}\, dt\right]$$

or by Eq. 9.3-7, since the function of time is odd,

$$c_n = -j\frac{1}{\pi}\int_0^\pi \frac{1}{2}\sin nt\, dt = -\frac{j}{n\pi} = \frac{1}{n\pi}\underline{/-90°} \qquad \text{for } n \text{ odd}$$

$$= 0 \qquad \text{for } n = 0 \text{ and for } n \text{ even}$$

Thus,

$$f_1(t) = -\frac{j}{\pi}\epsilon^{jt} - \frac{j}{3\pi}\epsilon^{j3t} - \frac{j}{5\pi}\epsilon^{j5t} - \cdots + \frac{j}{\pi}\epsilon^{-jt} + \frac{j}{3\pi}\epsilon^{-j3t} + \frac{j}{5\pi}\epsilon^{-j5t} - \cdots$$

where the terms for $-5 \le n \le 5$ have been explicitly written out. Notice that $c_{-1} = c_1{}^*$, $c_{-3} = c_3{}^*$, ... as expected. In order to relate this result to the series for

$f_3(t)$ in Example 9.1-2, the exponential terms may be combined in pairs by the use of Eqs. 9.3-1.

$$f_1(t) = \frac{2}{\pi}\left(\sin t + \frac{1}{3}\sin 3t + \frac{1}{5}\sin 5t + \cdots\right)$$

which agrees with Example 9.1-2. The trigonometric form of the series may also be written down directly from the expression for c_n, since the magnitude and phase angle of the nth harmonic are $A_n = 2\,|c_n|$ and $\phi_n = \angle\, c_n$. Thus,

$$f_1(t) = \frac{2}{\pi}\cos\,(t - 90°) + \frac{2}{3\pi}\cos\,(3t - 90°) + \frac{2}{5\pi}\cos\,(5t - 90°) + \cdots$$

For the even function $f_2(t)$,

$$c_n = \frac{1}{2\pi}\left[\int_{-\pi}^{0}\frac{-t}{\pi}\,\epsilon^{-jnt}\,dt + \int_{0}^{\pi}\frac{t}{\pi}\,\epsilon^{-jnt}\,dt\right]$$

or, by Eq. 9.3-6,

$$c_n = \frac{1}{\pi}\int_{0}^{\pi}\frac{t}{\pi}\cos nt\,dt = \frac{1}{2} \qquad \text{for } n = 0$$

$$= \frac{-2}{n^2\pi^2} \qquad \text{for } n \text{ odd}$$

$$= 0 \qquad \text{for } n \text{ even}$$

Hence,

$$f_2(t) = \frac{1}{2} - \frac{2}{\pi^2}\left(\epsilon^{jt} + \frac{1}{9}\epsilon^{j3t} + \cdots + \epsilon^{-jt} + \frac{1}{9}\epsilon^{-j3t} + \cdots\right)$$

$$= \frac{1}{2} - \frac{4}{\pi^2}\left(\cos t + \frac{1}{9}\cos 3t + \cdots\right)$$

which is consistent with Example 9.1-2.

For $f_3(t)$, which is zero for $-\pi < t < 0$,

$$c_n = \frac{1}{2\pi}\int_{0}^{\pi}(\sin t)\epsilon^{-jnt}\,dt = -\frac{1}{2\pi}\left(\frac{1 + \epsilon^{-jn\pi}}{n^2 - 1}\right)$$

so

$$c_0 = \frac{1}{\pi}$$

$$c_1 = -\frac{j}{4} = \frac{1}{4}\underline{/-90°}$$

$$c_2 = -\frac{1}{3\pi}$$

$$c_3 = 0$$

$$c^4 = \frac{1}{15\pi}$$

. . . .

The value of c_1 is not immediately obvious, since the general expression for c_n is an indeterminate form for $n = 1$. It can be found by evaluating the indeterminate form or by carrying out a separate integration for $n = 1$.

$$f_3(t) = \frac{1}{\pi} + \frac{1}{2}\cos(t - 90°) - \frac{2}{3\pi}\cos 2t - \frac{2}{15\pi}\cos 4t - \cdots$$

The series for $e_1(t)$ in Example 9.2-3 can be obtained from this result by changing the time scale and shifting the time origin.

The Response to Periodic Signals.   As discussed in Sections 4.6, 5.2, and 6.2, the forced response to $\epsilon^{j\omega t}$ is $H(j\omega)\epsilon^{j\omega t}$. If the input is given by the series

$$x(t) = \sum_{n=-\infty}^{\infty} c_n \epsilon^{jn\omega_0 t}$$

then, by superposition, the entire forced response is

$$y(t) = \sum_{n=-\infty}^{\infty} [c_n H(jn\omega_0)]\epsilon^{jn\omega_0 t}$$

The coefficients in the Fourier series representing the output are seen to be $c_n H(jn\omega_0)$, the product of the input coefficients and the a-c steady-state network function, evaluated at the appropriate angular frequencies. If the symbols $c_n^{(x)}$ and $c_n^{(y)}$ denote the coefficients in the input and output series, respectively,

$$c_n^{(y)} = H(jn\omega_0)c_n^{(x)} \tag{9.3-8}$$

In general, all three factors in this equation are complex quantities, so

$$|c_n^{(y)}| = |H(jn\omega_0)|\ |c_n^{(x)}|$$
$$\angle c_n^{(y)} = \angle H(jn\omega_0) + \angle c_n^{(x)}$$

In the solution of a circuit problem, the same three basic steps that were used with the trigonometric series are again present. First, the complex series for the input is calculated; second, the series representing the output is determined by Eq. 9.3-8; and third, the terms in the output are combined and plotted or else otherwise interpreted.

Example 9.3-2.   Use the complex Fourier series to find the steady-state output voltage for the circuit in Fig. 9.2-1a and the input waveshape in Fig. 9.2-1b.

Solution.   For $t > 0$, the waveshape of the input current is the same as $f_1(t)$ in Fig. 9.3-1, except for the d-c component and a change in the time scale. Changing the time scale does not affect the coefficients in the Fourier series, so

$$c_n^{(x)} = \tfrac{1}{2} \qquad \text{for } n = 0$$

$$= \frac{1}{n\pi}\ \underline{/-90°} \qquad \text{for } n \text{ odd}$$

$$= 0 \qquad \text{for other values of } n$$

The angular frequency of the first harmonic is $\omega_0 = \pi$. The a-c steady-state network function is

$$Z(j\omega) = \frac{1}{1 + j\omega} = \frac{1}{\sqrt{1 + \omega^2}} \underline{/-\tan^{-1}\omega}$$

so

$$c_n{}^{(y)} = \tfrac{1}{2} \quad \text{for } n = 0$$

$$= \frac{1}{n\pi\sqrt{1 + n^2\pi^2}} \underline{/-90° - \tan^{-1} n\pi} \qquad \text{for } n \text{ odd}$$

$$= 0 \quad \text{for other values of } n$$

Therefore, the forced response is

$$e(t) = \frac{1}{2} + \frac{2}{\pi\sqrt{1 + \pi^2}}\cos(\pi t - 90° - \tan^{-1}\pi)$$

$$+ \frac{2}{3\pi\sqrt{1 + 9\pi^2}}\cos(3\pi t - 90° - \tan^{-1} 3\pi) + \cdots$$

which agrees with the answer to Example 9.2-1.

Either the trigonometric or complex form of the Fourier series may be used in the solution of a circuit problem. Since the trigonometric form does not involve complex quantities, and since it is frequently used in undergraduate mathematics courses, it is preferred by many students. Nevertheless, the complex form is usually used in advanced courses. Its compactness in expression is evident by a comparison of Eq. 9.3-5 to Eq. 9.1-1, of Eq. 9.3-4 to Eqs. 9.1-14, and of Eq. 9.3-8 to Eq. 9.2-2. Since $c_n = \tfrac{1}{2}A_n\underline{/\phi_n}$, the magnitude and phase angle of each harmonic are given directly by the complex coefficients. Perhaps the most important advantage of the complex Fourier series is that it provides a more convenient starting point for developing techniques that apply to nonperiodic inputs.

The Frequency Spectra of Periodic Signals.   The essential information about the harmonics in a periodic signal consists of their magnitudes, phase angles, and frequencies, all of which is summarized by a knowledge of c_n and $\omega_0 = 2\pi/T$. As indicated by Eq. 9.3-8, the circuit's effect upon the harmonics is given by $H(j\omega)$. The frequency response curves of $|H(j\omega)|$ and $\measuredangle H(j\omega)$ versus ω that were used in Chapter 6 to describe the frequency discrimination of a circuit suggest that curves of $|c_n|$ and $\measuredangle c_n$ versus ω would be a useful way of graphically representing the harmonics in a periodic signal. This is indeed commonly done, but while $H(j\omega)$ is a continuous function of ω, c_n is meaningful only when n is an integer, i.e., when $\omega = 0, \pm\omega_0, \pm 2\omega_0, \ldots, \pm n\omega_0, \ldots$. The frequency spectra plots of $|c_n|$ and $\measuredangle c_n$ will have values only at these discrete angular frequencies and are therefore called *discrete spectra* or *line spectra*. Sometimes, when only the trigonometric form of the series in Eq. 9.1-1 is presented, the frequency spectrum is represented by $A_n\underline{/\phi_n}$ instead of $c_n = (A_n/2)\underline{/\phi_n}$. Since the two expressions differ only by a multiplying factor of 2 for $n > 0$, either procedure is satisfactory.

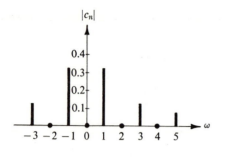

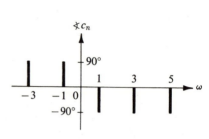

(a) Frequency spectrum for $f_1(t)$

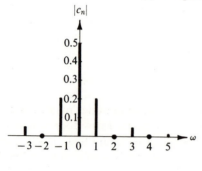

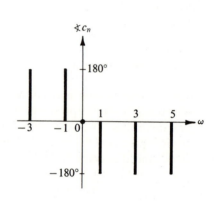

(b) Frequency spectrum for $f_2(t)$

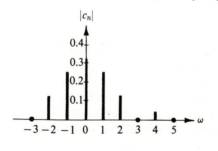

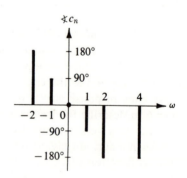

(c) Frequency spectrum for $f_3(t)$

Figure 9.3-2

Example 9.3-3.   Use the results of Example 9.3-1 to sketch the frequency spectra for the functions in Fig. 9.3-1.

Solution.   For the function $f_1(t)$, $\omega_0 = 1$ rad per sec and

$$c_n = \frac{1}{n\pi} \underline{/-90°} \qquad \text{for } n \text{ odd}$$

$$= 0 \qquad \text{for } n = 0 \text{ and for } n \text{ even}$$

In the construction of the magnitude plot in Fig. 9.3-2a, a line of height $1/\pi$ is drawn at the angular frequency of the first harmonic, a line of height $1/3\pi$ at the angular frequency of the third harmonic, etc. The angle of c_n is $-90°$ for positive, odd values of n, and (since a minus sign in front of a complex number is equivalent to a 180° change in the angle) is 90° for $n = -1, -3, \ldots$. Since $c_n = (A_n/2) \underline{/\phi_n}$ for $n > 0$, an inspection of Fig. 9.3-2a shows at a glance the frequencies of the harmonics in $f_1(t)$, their relative strength, and their phase angles with respect to a cosine wave.

Some comment should be made about the part of the graphs corresponding to negative values of ω. This half always can be constructed from the right-hand half by remembering that $c_{-n} = c_n^*$, i.e., that $|c_{-n}| = |c_n|$ and $\ast c_{-n} = - \ast c_{-n}$. Components for negative ω appear only because, in the derivation of the complex Fourier series, each trigonometric term was broken up into two terms, one with a positive exponent and one with a negative exponent. Negative frequency components do not have any physical significance and are regarded as a byproduct of the mathematical manipulations.[†] They cannot be ignored, however, since they constitute the negative half of the summation in Eq. 9.3-5. In this summation, the first harmonic includes both the c_1 and c_{-1} terms, which may intuitively explain why $|c_1|$ is only half the magnitude of the first harmonic.

The frequency spectra for the functions $f_2(t)$ and $f_3(t)$ follow from the previously derived expressions for c_n and are included in Fig. 9.3-2.

Example 9.3-4.   Sketch the frequency spectra for Example 9.3-2.

Solution.   For the waveshape of the input current, $\omega_0 = \pi$ rad per sec and

$$c_n^{(x)} = \tfrac{1}{2} \qquad \text{for } n = 0$$

$$= \frac{1}{n\pi} \underline{/-90°} \qquad \text{for } n \text{ odd}$$

which is plotted in Fig. 9.3-3a. If the output is the voltage $e(t)$,

$$H(j\omega) = \frac{1}{\sqrt{1 + \omega^2}} \underline{/-\tan^{-1} \omega}$$

$$c_n^{(y)} = \tfrac{1}{2} \qquad \text{for } n = 0$$

$$= \frac{1}{n\pi \sqrt{1 + n^2\pi^2}} \underline{/-90° - \tan^{-1} n\pi} \qquad \text{for } n \text{ odd}$$

[†] Negative values of ω correspond to the lower half of the imaginary axis in the s-plane and in some applications are not dismissed so lightly.

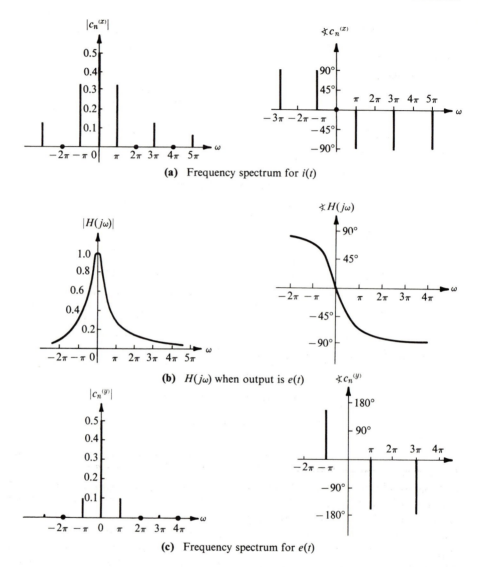

(a) Frequency spectrum for $i(t)$

(b) $H(j\omega)$ when output is $e(t)$

(c) Frequency spectrum for $e(t)$

which are shown in Figs. 9.3-3b and 9.3-3c. By Eq. 9.3-8, the last of these three plots also may be obtained graphically from the other two. If the output of interest is the current $i_C(t)$,

$$H(j\omega) = \frac{\omega}{\sqrt{1 + \omega^2}}\,\underline{/90° - \tan^{-1}\omega}$$

$$c_n^{(y)} = \frac{1}{\sqrt{1 + n^2\pi^2}}\,\underline{/-\tan^{-1}n\pi} \qquad \text{for } n \text{ odd}$$

$$= 0 \qquad \text{for other values of } n$$

which are shown in Figs. 9.3-3d and 9.3-3e.

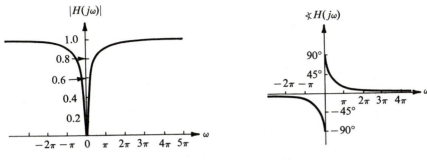

(d) $H(j\omega)$ when output is $i_C(t)$

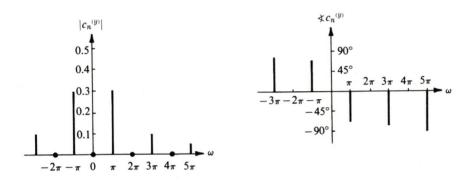

(e)   Frequency spectrum for $i_C(t)$

Figure 9.3-3

Often, the circuit's frequency response curves give some rough idea of the output to be expected. In Example 9.3-4, a glance at the sketches of $H(j\omega)$ in Fig. 9.3-3d shows that above the angular frequency $\omega_0 = \pi$ rad per sec of the first harmonic, the network function is nearly $1\underline{/0°}$. It should be anticipated, therefore, that the waveshape of $i_C(t)$ should not differ greatly from the input waveshape, except for a change in the d-c component. This is confirmed by a comparison of the curves in Figs. 9.2-1b and 9.2-1e. On the other hand, the magnitude of the network function shown in Fig. 9.3-3b is relatively small at and above π rad per sec. As shown in Fig. 9.2-1d, the waveshape of $e(t)$ has a large d-c component, but bears little resemblance to the input waveshape.

Example 9.3-5.   Sketch the frequency spectrum for the amplitude modulated signal in Example 9.2-5.

Solution.   The expression

$$x(t) = \cos \beta_2 t - \frac{k}{2} \cos (\beta_2 + \beta_1)t - \frac{k}{2} \cos (\beta_2 - \beta_1)t$$

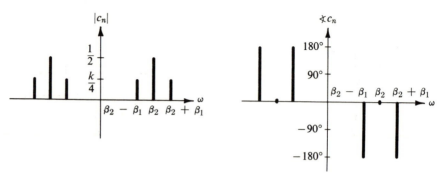

Figure 9.3-4

indicates that $c_n = \frac{1}{2}$ for $\omega = \beta_2$, and equals $-k/4 = (k/4)\underline{/-180°}$ for $\omega = \beta_2 \pm \beta_1$, as shown in Fig. 9.3-4.

Example 9.3-6. Examine the relationship between the input and output frequency spectra if $y(t) = dx/dt$.

Solution. If $x(t) = \displaystyle\sum_{n=-\infty}^{\infty} c_n^{(x)}\epsilon^{jn\omega_0 t}$, then $y(t) = \displaystyle\sum_{n=-\infty}^{\infty} (c_n^{(x)}jn\omega_0)\epsilon^{jn\omega_0 t}$, so

$$c_n^{(y)} = jn\omega_0 c_n^{(x)}$$

This conclusion agrees with Example 9.3-1, where $f_1(t) = (\pi/2)(df_2/dt)$, $\omega_0 = 1$ rad per sec, and $c_n^{(1)} = (jn\pi/2)c_n^{(2)}$.

The Effect of Increasing the Distance between Recurring Pulses. In many applications, pulses with a fixed waveshape and of fixed duration are repeated periodically, as in Fig. 9.3-5. The individual pulses may be rectangular or triangular, or they may be the short sinusoidal "burst" in part (c), which is one component of a color television signal. We shall try to separate the effects of the pulse waveshape and of the separation between pulses on the frequency spectrum. More specifically, we shall examine the effect of varying T when the pulse shape and duration is fixed.

In the Fourier series, the angular frequency of the nth harmonic is $\omega = n\omega_0$. With this substitution, Eqs. 9.3-5 and 9.3-4 become

$$f(t) = \sum_{\omega=0,\,\pm\omega_0,\,\pm 2\omega_0,\dots}^{\pm\infty} c_n \epsilon^{j\omega t}$$

(9.3-9)

$$c_n = \frac{1}{T}\int_{-T/2}^{T/2} f(t)\epsilon^{-j\omega t}\,dt$$

The expression for c_n is meaningful only if n is an integral multiple of ω_0. Since the functions in Fig. 9.3-5 are zero for $L/2 < |t| < T/2$,

$$c_n = \frac{1}{T}\left[\int_{-L/2}^{L/2} f(t)\epsilon^{-j\omega t}\,dt\right]$$

(9.3-10)

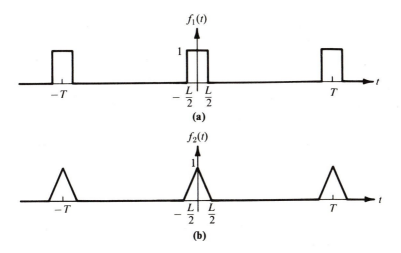

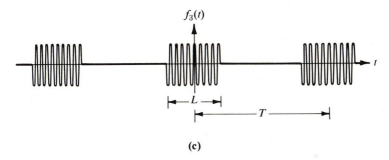

Figure 9.3-5

Note that the quantity in brackets is a function of ω that depends only upon the pulse shape and duration and not upon T.

As a particular example, the frequency spectrum for the rectangular pulse train in Fig. 9.3-5a is easily found to be

$$c_n = \frac{1}{T}\left[\frac{2\sin(\omega L/2)}{\omega}\right] = \frac{L}{T}\left[\frac{\sin(\omega L/2)}{\omega L/2}\right]$$

This expression is an indeterminate form for $\omega = 0$. If the indeterminate form is evaluated, or if ω is replaced by zero in Eq. 9.3-10, $c_0 = L/T$. This result also follows from the interpretation of c_0 as the d-c component.

In sketching the frequency spectrum for $f_1(t)$, several points should be noted. Since c_n happens to be a real quantity (a consequence of the fact that the function of time is an even function), separate plots of $|c_n|$ and $\angle c_n$ are not necessary. The quantity $(\sin u)/u$ is a frequently encountered function that is shown in Fig. 9.3-6a. For the special case of $T = 2L$, the expression for c_n describes the dashed line (or

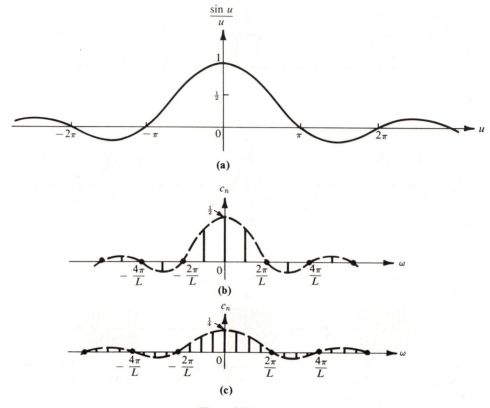

Figure 9.3-6

envelope) in Fig. 9.3-6b. Since we have emphasized that c_n is meaningful only when $\omega = n\omega_0$, the vertical lines at $\omega = 0$, $\omega = \omega_0 = \pi/L$, $\omega = 2\omega_0 = 2\pi/L$, ... indicate the harmonics in $f_1(t)$. If the period is doubled so that $T = 4L$, the envelope for c_n is only half as big, but it is otherwise unchanged. However, ω_0 equals $\pi/2L$ instead of π/L, so the vertical lines representing the harmonics are spaced closer together, as in Fig. 9.3-6c.

The above comments on the frequency spectrum of $f_1(t)$ apply more generally to *all* the functions in Fig. 9.3-5, as can be seen from Eq. 9.3-10. The shape of the envelope of the c_n plots depends only upon the shape and width of the individual pulses. Within this envelope, the harmonics are spaced ω_0 radians per second apart. As the distance between the pulses increases, the size of the envelope decreases uniformly, and since $\omega_0 = 2\pi/T$ the harmonics are crowded closer together on the frequency scale.

It is interesting to try to extend these conclusions when the period is allowed to approach infinity. The function $f(t)$ then becomes a single, nonrecurring pulse. The spacing ω_0 between the harmonics approaches zero, so that the frequency spectrum contains all possible frequencies and becomes a continuous rather than a discrete spectrum. In the next section, this train of thought will be used to develop techniques for handling nonperiodic inputs.

9.4

The Fourier Transform

Recall that the spacing between the sinusoidal components comprising a periodic signal is $\Delta\omega = \omega_0 = 2\pi/T$ radians per second. As in the figures of Section 9.3, we shall continue to regard c_n as a discrete function of ω rather than of n. Because the equation for c_n vanishes as the period T approaches infinity, we define a new quantity

$$g(\omega) = \frac{c_n}{\Delta\omega} = \frac{c_n}{\omega_0} = \frac{1}{2\pi} \int_{-T/2}^{T/2} f(t)\epsilon^{-j\omega t}\, dt$$

which is the frequency spectrum of $f(t)$ *per unit interval of angular frequency*, and which is again a function of ω.† The first of Eqs. 9.3-9 becomes

$$f(t) = \sum_{\omega=-\infty}^{\infty} g(\omega)\epsilon^{j\omega t}\, \Delta\omega$$

where the summation is over all angular frequencies in increments of $\Delta\omega = 2\pi/T$.

A nonperiodic function can be constructed from a periodic one by letting T approach infinity, as suggested at the end of the last section. In the limit,

$$g(\omega) = \frac{1}{2\pi} \int_{-\infty}^{\infty} f(t)\epsilon^{-j\omega t}\, dt \tag{9.4-1}$$

Since $\Delta\omega$ becomes infinitesimal, it is replaced by the differential $d\omega$, and the summation in the equation for $f(t)$ becomes an integration.

$$f(t) = \int_{-\infty}^{\infty} g(\omega)\epsilon^{j\omega t}\, d\omega \tag{9.4-2}$$

Equations 9.4-1 and 9.4-2 replace Eqs. 9.3-9 for nonperiodic functions. The form that is most commonly used differs slightly, however, from the above and may be obtained by the substitution $G(\omega) = 2\pi g(\omega)$.

$$G(\omega) = \int_{-\infty}^{\infty} f(t)\epsilon^{-j\omega t}\, dt \tag{9.4-3}$$

$$f(t) = \frac{1}{2\pi} \int_{-\infty}^{\infty} G(\omega)\epsilon^{j\omega t}\, d\omega \tag{9.4-4}$$

---

† This is a common mathematical procedure when preparing to change from a discrete to a continuous function, e.g., the defining of discrete and continuous probability distributions.

These two equations are called the *direct* and *inverse Fourier transforms,* respectively, and are often written symbolically as

$$G(\omega) = \mathscr{F}[f(t)]$$
$$f(t) = \mathscr{F}^{-1}[G(\omega)]$$

(9.4-5)

The second of the two expressions is also known as the *Fourier integral.*

Since $g(\omega)$ and $G(\omega)$ differ only by a multiplying constant, they have essentially the same physical significance. $G(\omega)$ presents the frequency spectrum information for a nonperiodic function and, like c_n, is in general a complex quantity. It is a *continuous* rather than a discrete function of ω, because a nonperiodic function contains components at *all* angular frequencies. The magnitude of $G(\omega)$ represents the strength of the components per unit of angular frequency, so a plot of $|G(\omega)|$ versus ω again shows the relative importance of the various components.

An intuitive appreciation of the significance of $|G(\omega)|$ as discussed above is sufficient for this book. If the student wishes to attempt to formulate a more precise analytical interpretation, he might guess that $\int_{\omega_1}^{\omega_2} |G(\omega)| \, d\omega$ would be a measure of the strength of the components in the frequency range $\omega_1 < \omega < \omega_2$. A better point of view is to consider a nonperiodic current $i(t)$ flowing into a 1 Ω resistance. The power is $|i(t)|^2(1)$, and the total energy delivered is $w = \int_{-\infty}^{\infty} |i(t)|^2 \, dt$. If $G(\omega)$ is the Fourier transform of $i(t)$ it can be proved that

$$w = \int_{-\infty}^{\infty} |i(t)|^2 \, dt = \frac{1}{2\pi} \int_{-\infty}^{\infty} |G(\omega)|^2 \, d\omega$$

so that $|G(\omega)|^2$ must be an energy density function (energy per unit of angular frequency). This important result is known as the *Fourier integral energy theorem.*

In summary, the direct Fourier transform yields $G(\omega)$ and indicates the relative strength and phase angles of the components in $f(t)$, while the inverse transform recombines the components to form $f(t)$. Note that Eq. 9.4-4 gives $f(t)$ in closed form and not as an infinite series. The conditions necessary for the existence of Eqs. 9.4-3 and 9.4-4 are similar to those for the Fourier series. The function $f(t)$ must be single valued and have a finite number of maxima, minima, and discontinuities, and the area $\int_{-\infty}^{\infty} |f(t)| \, dt$ must be finite. It can be shown that for even functions, Eq. 9.4-3 is equivalent to

$$G(\omega) = 2 \int_{0}^{\infty} f(t) \cos \omega t \, dt$$

(9.4-6)

while for odd functions

$$G(\omega) = -j2 \int_{0}^{\infty} f(t) \sin \omega t \, dt$$

(9.4-7)

Example 9.4-1.   Sketch the frequency spectrum for the rectangular pulse of unit area in Fig. 9.4-1a.

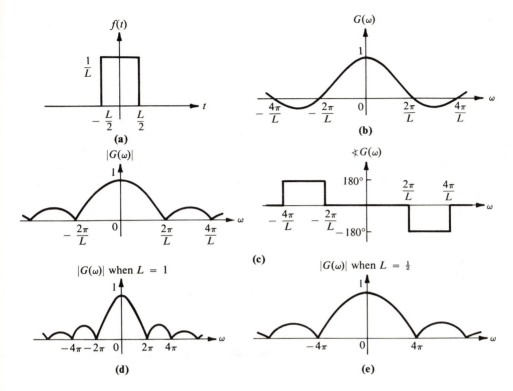

Figure 9.4-1

Solution.   By Eq. 9.4-3,

$$G(\omega) = \int_{-L/2}^{L/2} \frac{1}{L} \, \epsilon^{-j\omega t} \, dt$$

or, by Eq. 9.4-6,

$$G(\omega) = 2 \int_{0}^{L/2} \frac{1}{L} \cos \omega t \, dt = \frac{\sin (\omega L/2)}{\omega L/2}$$

which describes the curve in Fig. 9.4-1b or the equivalent set of curves in part (c). Figures 9.4-1d and 9.4-1e are drawn for the special cases of $L = 1$ and $L = \frac{1}{2}$, respectively, and show that the frequency spectrum spreads out as the pulse width is decreased.

The frequency spectra in Figs. 9.4-1d and 9.4-1e provide one illustration of an important characteristic of all pulses, regardless of their shape. The narrower the pulse width is made, the wider is the frequency spectrum, and the more important are the high-frequency components. Changing the width of a pulse corresponds to changing the time scale, i.e., to replacing t by at. Since the Fourier transform of $f(t)$ is

$$G(\omega) = \mathscr{F}[f(t)] = \int_{-\infty}^{\infty} f(t)\epsilon^{-j\omega t}\, dt$$

the transform of $f(at)$ is

$$\mathscr{F}[f(at)] = \int_{-\infty}^{\infty} f(at)\epsilon^{-j\omega t}\, dt$$

$$= \frac{1}{a}\int_{-\infty}^{\infty} f(\lambda)\epsilon^{-j(\omega/a)\lambda}\, d\lambda$$

where the substitutions $\lambda = at$ and $d\lambda = a\, dt$ have been made. The last integral is the same as $G(\omega)$, except that t has been replaced by the dummy variable λ, and ω has been replaced by ω/a. Hence,

$$\mathscr{F}[f(at)] = \frac{1}{a}\, G\!\left(\frac{\omega}{a}\right)$$

If $a > 1$, replacing t by at decreases the width of the function of time, but spreads out the frequency spectrum along the ω axis. The multiplying factor of $1/a$ on the right side of the last equation can be explained intuitively, because if the width of a pulse is decreased without changing its height, the available energy would be expected to decrease.

Another characteristic concerns pulses with about the same width but with different shapes. The frequency spectrum for a pulse with sharp corners—and especially for a pulse with discontinuities—is wider than that for a smooth waveshape. The following example illustrates these statements.

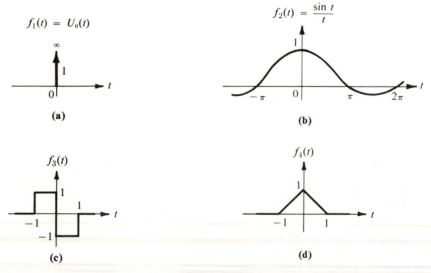

Figure 9.4-2

Example 9.4-2.   Sketch the frequency spectra for the functions in Fig. 9.4-2.

Solution.   For the unit impulse $f_1(t) = U_0(t)$,

$$G_1(\omega) = \int_{\infty}^{-\infty} U_0(t)\epsilon^{-j\omega t}\,dt = 1$$

by Eq. 3.2-8. This result also follows from the previous example by letting L approach zero. Thus, the unit impulse has the frequency spectrum in Fig. 9.4-3a and contains all frequencies in equal strength.

Since $f_2(t) = (\sin t)/t$ is an even function,

$$G_2(\omega) = 2\int_0^\infty \frac{\sin t \cos \omega t}{t}\,dt = \pi \qquad \text{for } |\omega| < 1$$

$$= 0 \qquad \text{for } |\omega| > 1$$

which is plotted in Fig. 9.4-3b. Fortunately, the last integral appears in the common tables of definite integrals, for its evaluation would otherwise be difficult. Because of the similarity of Eqs. 9.4-3 and 9.4-4, the student should expect the shape of the frequency spectrum to be similar to the function of time in Fig. 9.4-1a.

For the third function, which is odd,

$$G_3(\omega) = -j2\int_0^1 (-1)\sin \omega t\,dt = \frac{j2}{\omega}(1 - \cos \omega)$$

The final expression is an indeterminate form when $\omega = 0$, but it is easily seen that

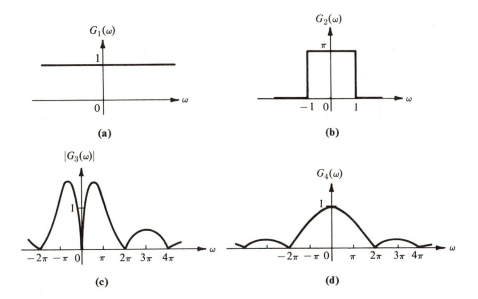

Figure 9.4-3

$G_3(0) = 0$. The curve for $|G_3(\omega)|$ is given in Fig. 9.4-3c, and $\not\prec G_3(\omega)$ is 90° for $\omega > 0$. Since $f_4(t)$ is an even function,

$$G_4(\omega) = 2 \int_0^1 (1 - t) \cos \omega t \, dt = \frac{2}{\omega^2} (1 - \cos \omega)$$

which is shown in Fig. 9.4-3d. Although the expression for $G_4(\omega)$ is again an indeterminate form when $\omega = 0$, $G_4(0) = 1$.

The results of Example 9.4-2 reinforce the general comments that were made after Example 9.4-1. Because the width of an impulse is zero, it is not surprising that the strength of the components in its frequency spectrum does not decrease as ω increases. In Fig. 9.4-2, $f_3(t)$ has discontinuities, $f_4(t)$ has sharp corners but no discontinuities, and $f_2(t)$ has the smoothest waveshape of all. Thus the high-frequency components in $G_3(\omega)$ are larger than those in $G_4(\omega)$, while $G_2(\omega)$ is zero for $|\omega| > 1$.

Several useful properties of the Fourier transform are summarized in Table 9.4-1. The proof of the first two properties is left to the student, Eq. 9.4-10 has already been derived, and Eq. 9.4-15 is discussed in Section 9.5. The rest of the table is considered now. Since

$$f(t) = \frac{1}{2\pi} \int_{-\infty}^{\infty} G(\omega) \epsilon^{j\omega t} \, d\omega = \mathscr{F}^{-1}[G(\omega)]$$

then

$$\frac{df}{dt} = \frac{1}{2\pi} \int_{-\infty}^{\infty} [j\omega G(\omega)] \epsilon^{j\omega t} \, d\omega = \mathscr{F}^{-1}[j\omega G(\omega)]$$

as in Eq. 9.4-11. Thus, differentiating a function of time corresponds to multiplying its frequency spectrum by $j\omega$. The extra factor of $j\omega$ indicates that the higher frequencies will be more important in the spectrum of the differentiated function.

TABLE 9.4-1

Property	Equation number
$\mathscr{F}[f_1(t) + f_2(t)] = \mathscr{F}[f_1(t)] + \mathscr{F}[f_2(t)] = G_1(\omega) + G_2(\omega)$	9.4-8
$\mathscr{F}[af(t)] = aG(\omega)$	9.4-9
$\mathscr{F}[f(at)] = \dfrac{1}{a} G\left(\dfrac{\omega}{a}\right)$	9.4-10
$\mathscr{F}\left[\dfrac{d}{dt} f(t)\right] = j\omega G(\omega)$	9.4-11
$\mathscr{F}\left[\displaystyle\int_{-\infty}^{t} f(\lambda) \, d\lambda\right] = \dfrac{1}{j\omega} G(\omega)$	9.4-12
$\mathscr{F}[f(t - t_0)] = \epsilon^{-j\omega t_0} G(\omega)$	9.4-13
$\mathscr{F}[f(t) \cos \omega_1 t] = \dfrac{1}{2} [G(\omega - \omega_1) + G(\omega + \omega_1)]$	9.4-14
$\mathscr{F}[f_1(t) * f_2(t)] = G_1(\omega) G_2(\omega)$	9.4-15

This is consistent with the remarks made previously, since differentiation tends to accentuate any irregularities in the function, while integration tends to smooth out the function. In Example 9.4-2, notice that $f_3(t) = df_4/dt$ and $G_3(\omega) = j\omega G_4(\omega)$. The proof and significance of Eq. 9.4-12 are similar to those of Eq. 9.4-11.

The effect on the frequency spectrum of a change in the time origin was discussed in Sections 9.1 and 9.3. Since

$$f(t) = \frac{1}{2\pi} \int_{-\infty}^{\infty} G(\omega)\epsilon^{j\omega t}\, d\omega = \mathscr{F}^{-1}[G(\omega)]$$

a time delay of t_0 seconds corresponds to the new function

$$f(t - t_0) = \frac{1}{2\pi} \int_{-\infty}^{\infty} [G(\omega)\epsilon^{-j\omega t_0}]\epsilon^{j\omega t}\, d\omega = \mathscr{F}^{-1}[G(\omega)\epsilon^{-j\omega t_0}]$$

as in Eq. 9.4-13. Since $\epsilon^{-j\omega t_0} = 1/\!\!-\omega t_0$, the angle but not the magnitude of the frequency spectrum is changed.

In the modulation stages of a communication system, two signals are sometimes multiplied together. One of the signals is usually a sinusoidal function. Therefore, we shall derive an expression for the Fourier transform of $f(t) \cos \omega_1 t$, where $f(t)$ is a nonperiodic function with a Fourier transform of $G(\omega)$. By Eqs. 9.3-1,

$$\mathscr{F}[f(t) \cos \omega_1 t] = \int_{-\infty}^{\infty} f(t)\, \frac{\epsilon^{j\omega_1 t} + \epsilon^{-j\omega_1 t}}{2}\, \epsilon^{-j\omega t}\, dt$$

$$= \tfrac{1}{2}\left[\int_{-\infty}^{\infty} f(t)\epsilon^{-j(\omega-\omega_1)t}\, dt + \int_{-\infty}^{\infty} f(t)\epsilon^{-j(\omega+\omega_1)t}\, dt \right]$$

$$= \tfrac{1}{2}[G(\omega - \omega_1) + G(\omega + \omega_1)]$$

Replacing ω by $\omega - \omega_1$ or $\omega + \omega_1$ in $G(\omega)$ shifts the frequency spectrum a distance ω_1 along the ω axis. If the frequency spectrum of $f(t)$ is the one in Fig. 9.4-4a, then

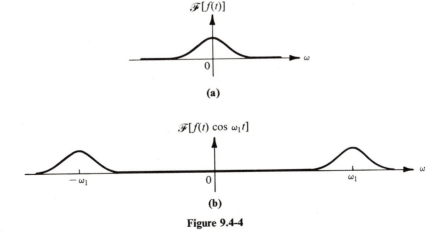

(a)

(b)

Figure 9.4-4

the frequency spectrum for $f(t) \cos \omega_1 t$ is given in Fig. 9.4-4b. This result can be regarded as a generalization of Example 9.2-5.

Example 9.4-3.   Find the Fourier transform of the unit step function.

Solution.   The function $f(t) = U_{-1}(t)$ does *not* satisfy the requirement, discussed earlier, that the area $\int_{-\infty}^{\infty} |f(t)|\, dt$ be finite. Consequently, it should be expected that a Fourier transform cannot be found. If Eq. 9.4-3 is formally used,

$$G(\omega) = \int_0^{\infty} \epsilon^{-j\omega t}\, dt = \left[\frac{\epsilon^{-j\omega t}}{-j\omega}\right]_0^{\infty}$$

But the quantity $\epsilon^{-j\omega t} = 1\underline{/-\omega t}$ radians does not approach a unique limit as t approaches infinity, since its angle continually changes. Thus, the upper limit cannot be evaluated, and the defining integral does not converge.

Many other functions, such as $f(t) = \epsilon^{at}$ and $f(t) = t$, do not satisfy the finite area requirement, and consequently do not have Fourier transforms. This is a limitation of the present method that will be overcome in the next chapter.

9.5

The Response to Nonperiodic Inputs

Let $G_x(\omega)$ and $G_y(\omega)$ denote the Fourier transforms of the input $x(t)$ and output $y(t)$, respectively. Then

$$x(t) = \frac{1}{2\pi} \int_{-\infty}^{\infty} G_x(\omega)\epsilon^{j\omega t}\, d\omega \tag{9.5-1}$$

As in Sections 4.6, 5.2, 6.2, and 9.3, the forced response to $G_x(\omega)\epsilon^{j\omega t}$ is $G_x(\omega)H(j\omega)\epsilon^{j\omega t}$, so the forced response to Eq. 9.5-1 is

$$y(t) = \frac{1}{2\pi} \int_{-\infty}^{\infty} G_y(\omega)\epsilon^{j\omega t}\, d\omega \tag{9.5-2}$$

where

$$G_y(\omega) = G_x(\omega)H(j\omega) \tag{9.5-3}$$

The three basic steps in circuit analysis by the Fourier transform are finding $G_x(\omega)$ for the given input, multiplying by the network function to obtain $G_y(\omega)$, and finally obtaining the output as a function of time by the inverse transform of Eq. 9.5-2. Although the following example is not the type of problem usually handled by Fourier methods, it demonstrates these steps for a simple circuit. The succeeding examples give some idea of the practical uses of the method.

Example 9.5-1.   Find the output of the circuit in Fig. 9.5-1a, if the capacitance is uncharged for $t < -\frac{1}{2}$.

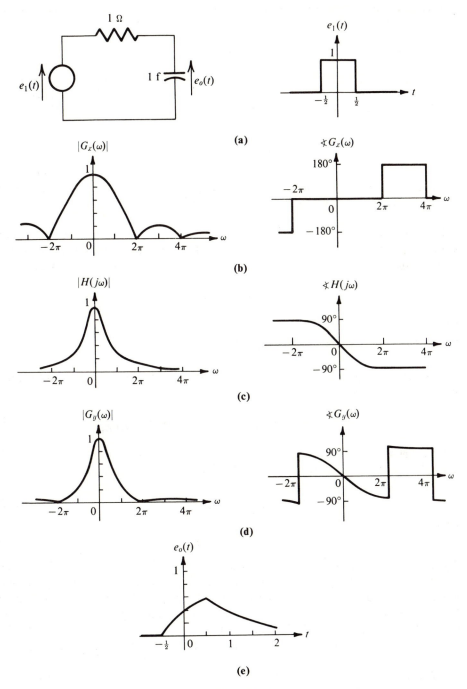

Figure 9.5-1

Solution. From Example 9.4-1,

$$G_x(\omega) = \frac{\sin(\omega/2)}{\omega/2}$$

which is sketched in Figs. 9.4-1c and 9.5-1b. The quantities

$$H(j\omega) = \frac{1}{1 + j\omega}$$

and

$$G_y(\omega) = \frac{\sin(\omega/2)}{\omega/2} \frac{1}{1 + j\omega}$$

are also sketched in the figure. Because $|H(j\omega)|$ falls off as ω increases, the high-frequency components in the output are reduced. To find the output as a function of time, Eq. 9.5-2 gives

$$e_o(t) = \frac{1}{2\pi} \int_{-\infty}^{\infty} \frac{\sin(\omega/2)}{\omega/2} \frac{\epsilon^{j\omega t}}{1 + j\omega} d\omega$$

This integral unfortunately does not appear in standard tables and is quite difficult to evaluate. It can be shown, however, that it finally reduces to

$$e_o(t) = 0 \qquad\qquad \text{for } t < -\tfrac{1}{2}$$

$$= 1 - \epsilon^{-(t+1/2)} \qquad \text{for } -\tfrac{1}{2} < t < \tfrac{1}{2}$$

$$= (\epsilon - 1)\epsilon^{-(t+1/2)} \qquad \text{for } t > \tfrac{1}{2}$$

which is plotted in Fig. 9.5-1e. Very frequently, the integration in the inverse transform is difficult, and special tables of Fourier transforms are used.[†]

It is important to notice that the solution in Example 9.5-1 is the *complete* response, whereas the Fourier series solutions in Sections 9.2 and 9.3 gave only the *forced* response. In order to appreciate the reasons for this, recall that

$$y(t) = y_P(t) + y_H(t)$$

where $y_P(t)$ is the particular or forced response, and where $y_H(t)$ is the complementary or free response. In most circuits, the free response eventually dies out as t increases, and we shall assume this condition in the ensuing discussion.

A periodic input repeats itself every T seconds over *all* values of t. Each term in its Fourier series

$$a_0 + a_1 \cos \omega_0 t + \cdots + b_1 \sin \omega_0 t + \cdots$$

is likewise defined for $-\infty < t < \infty$. Since all the components in the input start at

---

[†] See, for example, G. A. Campbell and R. M. Foster, *Fourier Integrals for Practical Applications* (Princeton, N.J.: Van Nostrand, 1948).

$t = -\infty$, the free response dies out long before $t = 0$, leaving only the forced response in the complete solution.[†]

In practice, of course, an input never starts at $t = -\infty$, and so is never really periodic. As discussed in Example 9.2-1, the forced response for $t \geq 0$ depends only on the input for $t \geq 0$ and is the same whether the input is assumed to start at $t = -\infty$ or at $t = 0$. The free response for $t \geq 0$, however, does depend upon when the input was applied and is not included in the Fourier series for the response.

The Fourier integral of Eq. 9.5-1 for a nonperiodic input has components of the form $G_x(\omega)\epsilon^{j\omega t}$ for $-\infty < t < \infty$. The free response to each such component again vanishes long before $t = 0$. Because all frequencies are present in the spectrum of a nonperiodic input, however, the summation (actually an integration) of the components gives a function that is zero as $t \to -\infty$, *even though* the components themselves are *not* zero there. Since an input that does not begin until $t = 0$ is thus correctly expressed as the summation of components beginning at $t = -\infty$, the complete output is the summation of just the forced responses to these components.

Example 9.5-2.   What properties should the transfer function $H(j\omega)$ have, if the input and output functions of time are to have the same shape?

Solution.   General expressions for $x(t)$ and $y(t)$ are given by Eqs. 9.5-1 and 9.5-2. If $H(j\omega) = K\epsilon^{-jt_0\omega}$, where K and t_0 are real constants,

$$y(t) = \frac{K}{2\pi} \int_{-\infty}^{\infty} G_x(\omega)\epsilon^{j\omega(t-t_0)}\, d\omega = Kx(t - t_0)$$

which is the input signal multiplied by K and delayed by t_0 seconds, but otherwise unchanged. For "distortionless transmission," therefore, the transfer function should have the form

$$H(j\omega) = K\underline{/-t_0\omega}$$

for all frequencies contained in the input spectrum. This result should be anticipated from Eq. 9.4-13.

Example 9.5-3.   An "ideal low-pass filter" is defined by the transfer function

$$H(j\omega) = K\underline{/-t_0\omega} \qquad \text{for } |\omega| < \omega_c$$

$$= 0 \qquad \text{for } |\omega| > \omega_c$$

as shown in Fig. 9.5-2a. It passes, with a time delay t_0, all the components below the "cut-off angular frequency" ω_c, but completely rejects the higher-frequency components. Find and sketch the impulse and step responses.

[†] The phrase $t = 0$ may be replaced by $t = t_0$, where t_0 is any finite, real time.

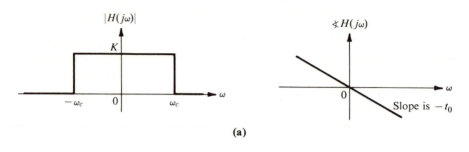

(a)

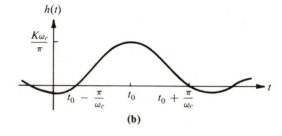

(b)

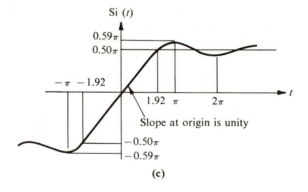

(c)

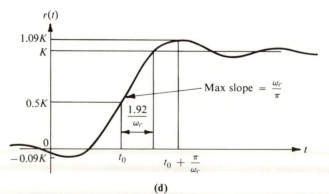

(d)

Figure 9.5-2

Solution. If the input is the unit impulse $U_0(t)$, $G_x(\omega) = 1$ by Example 9.4-2, and

$$G_y(\omega) = H(j\omega) = K\epsilon^{-jt_0\omega} \qquad \text{for } -\omega_c < \omega < \omega_c$$

$$= 0 \qquad \text{for } |\omega| > \omega_c$$

$$y(t) = h(t) = \frac{1}{2\pi} \int_{-\omega_c}^{\omega_c} K\epsilon^{-jt_0\omega} \epsilon^{j\omega t} \, d\omega$$

$$= \frac{K}{2\pi} \int_{-\omega_c}^{\omega_c} \epsilon^{j\omega(t-t_0)} \, d\omega = \frac{K}{2\pi j(t-t_0)} \left[\epsilon^{j\omega_c(t-t_0)} - \epsilon^{-j\omega_c(t-t_0)} \right]$$

$$= \frac{K}{\pi} \frac{\sin \omega_c(t-t_0)}{(t-t_0)} \qquad \text{for all } t$$

which is sketched in Fig. 9.5-2b. Note that the response is *not* zero for $t < 0$, but that it begins *before* the input is applied. This violation of the normal cause-and-effect relationship means that the ideal low-pass filter is physically unrealizable; i.e., no circuit having the postulated properties can be found. The ideal filter is still worth investigating, since circuits approximating its characteristics can exist, and since it may give some insight into the general relationship between the input, output, and cut-off frequency of a filter.

By Eq. 3.3-4, the step response is $r(t) = \int_{-\infty}^{t} h(\lambda) \, d\lambda$. To simplify the mathematics slightly, we shall temporarily assume that the time delay t_0 is zero, since this does not affect the shape of the output. Then

$$r(t) = \frac{K}{\pi} \left(\int_{-\infty}^{0} \frac{\sin \omega_c\lambda}{\lambda} \, d\lambda + \int_{0}^{t} \frac{\sin \omega_c\lambda}{\lambda} \, d\lambda \right)$$

The integration has been broken up into two parts, because the first part appears in most integral tables and equals $\pi/2$. The second integral is related to a quantity that occurs in several electrical engineering applications, which is defined by

$$\text{Si} \, (t) = \int_{0}^{t} \frac{\sin u}{u} \, du$$

where Si (t) is read "the sine integral of t." It is simply the area from zero to t underneath the curve of $(\sin u)/u$ versus u in Fig. 9.3-6a and is sketched in Fig. 9.5-2c. The integral cannot be evaluated in closed form, but it can be expressed as an infinite series, which has been accurately plotted and well tabulated. Now

$$\int_{0}^{t} \frac{\sin \omega_c\lambda}{\lambda} \, d\lambda = \int_{0}^{\omega_c t} \frac{\sin \omega_c\lambda}{\omega_c\lambda} \, d(\omega_c\lambda) = \text{Si} \, (\omega_c t)$$

and

$$r(t) = \frac{K}{\pi} \left[\frac{\pi}{2} + \text{Si} \, (\omega_c t) \right]$$

or, if t_0 is not zero,

$$r(t) = K \left\{ \frac{1}{2} + \frac{1}{\pi} \text{Si} \, [\omega_c(t - t_0)] \right\}$$

which is sketched in Fig. 9.5-2d.

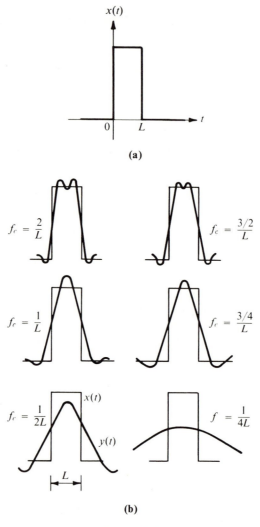

Figure 9.5-3

(b) From S. Goldman, *Frequency Analysis, Modulation and Noise* (New York: McGraw-Hill, 1948).

The results of Example 9.5-3 can be extended to the transmission of rectangular pulses through a low-pass filter. The pulse in Fig. 9.5-3a is $x(t) = U_{-1}(t) - U_{-1}(t - L)$, so the filter's response is $r(t) - r(t - L)$. This response is simply the difference of two curves (displaced by the distance L), like the one in Fig. 9.5-2d, and has been plotted for different values of ω_c in Fig. 9.5-3b. In order to concentrate on the relative *shapes* of the output and input, we assume that the time delay t_0 is zero and plot the response directly over the input pulse. In the figure, $f_c = \omega_c/2\pi$

is the cut-off frequency in cycles per second, and the light and heavy curves represent the input and output, respectively.

As expected, the output more closely resembles the input when the cut-off frequency is high. If f_c drops below $1/2L$ cycles per second, the magnitude of the response decreases rapidly, and the sharp edges of the pulse disappear. Since the leading edge of a pulse is often used to mark the occurrence of an event, or to trigger a specified reaction at a desired time, the inability to accurately locate the leading edge of the output when $f_c < 1/2L$ may be very serious.

Next consider an input consisting of the two pulses in Fig. 9.5-4a, which can be expressed as the sum of four step functions. In Fig. 9.5-4b, the response (heavy curve) and input (light curve) are compared for several values of f_c. If $f_c < 1/2L$, then an examination of the output fails to reveal whether the input contained one or two pulses.

In television, for example, black and white information can be conveyed by a train of pulses with $L = \frac{1}{8}$ μsec. Black portions of the picture are represented by pulses of maximum height and white portions by very small pulses. In order to faithfully reproduce the details of the picture, the receiver should have a cut-off frequency of $f_c > 1/2L = 4$ megacycles per second. Because the video information amplitude modulates a high-frequency carrier (as in Fig. 9.4-4), the actual bandwidth would be expected to be twice this figure. Still other considerations are involved, however, so that the width of a TV channel has been fixed at 6 megacycles

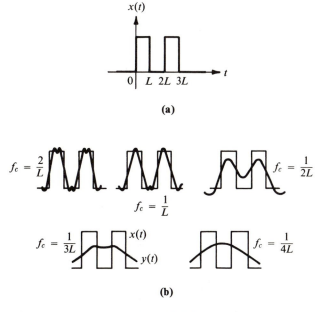

(a)

(b)

Figure 9.5-4

(b) From S. Goldman, *Frequency Analysis, Modulation and Noise* (New York: McGraw-Hill, 1948).

per second. Our crude results do serve, however, to give an order-of-magnitude answer.

In summary, Figs. 9.5-3 and 9.5-4 demonstrate that the product of the pulse width and the cut-off frequency of the circuit should be greater than $\frac{1}{2}$ for the accurate reproduction of the input. In very general terms, reproducing a rapidly changing input requires a larger bandwidth.

The following equation from Table 9.4-1 has not yet been justified:

$$\mathscr{F}[f_1(t) * f_2(t)] = G_1(\omega)G_2(\omega) \tag{9.5-4}$$

where the asterisk denotes convolution. The convolution of two functions of time occurs naturally, for example, in the scanning operations that are used in television and other recording devices.

Instead of a formal mathematical proof of Eq. 9.5-4, consider a circuit having an input $x(t)$ and output $y(t)$. Let the unit impulse response be $h(t)$ and the a-c steady-state network function be $H(j\omega)$. By Eq. 3.5-4,

$$y(t) = x(t) * h(t)$$

so

$$G_y(\omega) = \mathscr{F}[x(t) * h(t)]$$

By Eq. 9.5-3,

$$\mathscr{F}[x(t) * h(t)] = G_x(\omega)H(j\omega)$$

where $H(j\omega) = \mathscr{F}[h(t)]$, as discussed in Example 9.5-3. Except for the use of different symbols, this result is identical with Eq. 9.5-4.

9.6

Summary

The Fourier series for a function having a period T may be written in several equivalent forms.

$$f(t) = a_0 + \sum_{n=1}^{\infty} (a_n \cos n\omega_0 t + b_n \sin n\omega_0 t)$$

$$= a_0 + \sum_{n=1}^{\infty} A_n \cos (n\omega_0 t + \phi_n) = \sum_{n=-\infty}^{\infty} c_n \epsilon^{jn\omega_0 t}$$

where $\omega_0 = 2\pi/T$ is the angular frequency of the fundamental component. The coefficient c_n is given by

$$c_n = \frac{1}{T} \int_{-T/2}^{T/2} f(t)\epsilon^{-jn\omega_0 t} \, dt$$

The d-c component of $f(t)$ is $a_0 = c_0$, while the magnitude and phase angle of the nth harmonic are $A_n = 2|c_n|$ and $\phi_n = \measuredangle c_n$. A plot of the magnitude and angle of

c_n versus $\omega = n\omega_0$ (the angular frequency of the nth harmonic) shows the relative strength of the harmonics in $f(t)$, their phase angles with respect to a cosine wave, and their angular frequencies.

A three-step procedure, based upon the superposition theorem, may be used to find the steady-state response to a periodic input. First the Fourier series for the input is found, and then the series for the output is determined by a-c steady-state circuit theory. If the superscripts (x) and (y) denote the input and output, respectively, and if $H(j\omega)$ is the a-c steady-state network function,

$$c_n^{(y)} = H(jn\omega_0)c_n^{(x)}$$

Finally, the terms in the output series are plotted or otherwise interpreted.

A nonperiodic function can be expressed by the Fourier integral

$$f(t) = \frac{1}{2\pi} \int_{-\infty}^{\infty} G(\omega)\epsilon^{j\omega t}\, d\omega = \mathscr{F}^{-1}[G(\omega)]$$

where

$$G(\omega) = \int_{-\infty}^{\infty} f(t)\epsilon^{-j\omega t}\, dt = \mathscr{F}[f(t)]$$

The symbols $\mathscr{F}^{-1}$ and $\mathscr{F}$ stand for the inverse and direct Fourier transform. The first integration over all values of ω indicates that $f(t)$ now contains components at *all* angular frequencies. $G(\omega)$ is a continuous function of ω that indicates the relative strength and the phase angles of these components. The response $y(t)$ to a nonperiodic input $x(t)$ may be found by the following three-step procedure:

$$G_x(\omega) = \int_{-\infty}^{\infty} x(t)\epsilon^{-j\omega t}\, dt$$

$$G_y(\omega) = G_x(\omega)H(j\omega)$$

$$y(t) = \frac{1}{2\pi} \int_{-\infty}^{\infty} G_y(\omega)\epsilon^{j\omega t}\, d\omega$$

The method gives the complete response if the circuit contained no stored energy at some time in the past, and if all signals applied since that time are included in $x(t)$. The chief disadvantages are that $G_x(\omega)$ does not exist for certain important inputs and that the final integration may be difficult.

The Fourier series and integral place in evidence the frequency spectra of signals. Their use indicates the frequency components that a proposed circuit should pass or reject and is an important aid in circuit design, as illustrated by several of the examples in Sections 9.3 and 9.5. In order to transmit an input signal without distortion, for example, $H(j\omega)$ must have a constant magnitude and an angle proportional to ω for the important components in the input frequency spectrum. Short pulses and inputs with discontinuities have a relatively wide frequency spectrum and require a transmission system having a wide bandwidth if they are to be faithfully reproduced.

PROBLEMS

9.1 Find the trigonometric Fourier series for each of the following input voltages:

(a) $e_1(t) = \sin^2 t$. (b) $e_1(t) = \sin t \cos 2t$.

In each case find the steady-state output voltage $e_o(t)$ for the circuit in Fig. P9.1.

9.2 Find the Fourier series for each of the periodic input signals shown in Fig. P9.2.

9.3 Plot the sum of the first two nonzero terms and the sum of the first four nonzero terms in the Fourier series for $f_1(t)$ in Fig. 9.1-1a. Repeat for Fig. 9.1-4a. Compare the convergence of the series for these two cases.

9.4 Show that, for the half-wave rectified cosine wave in Fig. P9.4a,

$$f_1(t) = \frac{1}{\pi}\left(1 + \frac{\pi}{2}\cos t + \frac{2}{3}\cos 2t - \frac{2}{15}\cos 4t + \frac{2}{35}\cos 6t - \cdots\right)$$

Use *this result* to find the Fourier series for the full-wave rectified cosine wave in Fig. P9.4b.

9.5 If the Fourier series for $f(t)$ is given by Eq. 9.1-1, show that

$$f(T - t) = a_0 + \sum_{n=1}^{\infty} (a_n \cos n\omega_0 t - b_n \sin n\omega_0 t)$$

(a) Use this property and the results of Problem 9.2a to find the Fourier series for the function $f_2(t)$ in Fig. P9.5.

(b) Use the answer of part (a) to write down the Fourier series for $f_3(t)$ in Fig. P9.5. Show that the series is consistent with the series that was obtained for the function in Fig. 9.1-4a.

9.6 Find the Fourier series for the functions shown in Fig. P9.6 by temporarily relocating the axes to give symmetrical waveshapes.

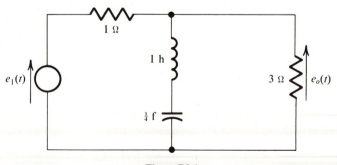

Figure P9.1

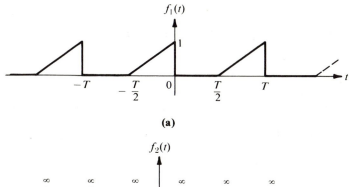

(a)

(b)

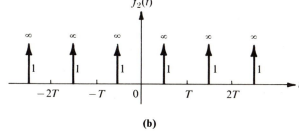

(c)

Figure P9.2

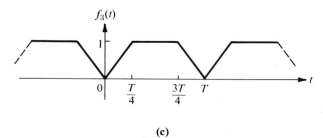

(a)   (b)

Figure P9.4

9.7 Any function of time may be expressed as the sum of an even function $f_e(t)$ and an odd function $f_o(t)$. If $f(t) = f_e(t) + f_o(t)$, show that

$$f_e(t) = \tfrac{1}{2}[f(t) + f(-t)]$$

$$f_o(t) = \tfrac{1}{2}[f(t) - f(-t)]$$

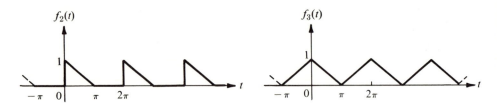

Figure P9.5

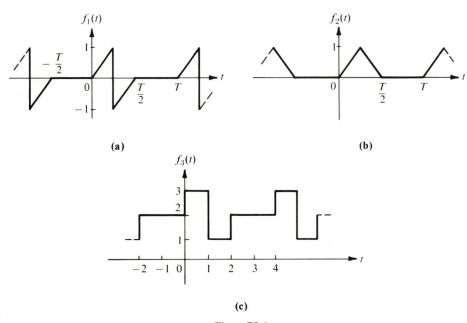

and that the coefficients in the Fourier series for $f(t)$ are

$$a_n = \frac{4}{T} \int_0^{T/2} f_e(t) \cos n\omega_0 t \, dt \qquad (n \neq 0)$$

$$b_n = \frac{4}{T} \int_0^{T/2} f_o(t) \sin n\omega_0 t \, dt$$

9.8 Formally prove that, if $f(t + T/2) = -f(t)$, $a_0 = a_2 = b_2 = a_4 = b_4 = \cdots = 0$, and the remaining coefficients may be found by integrating over half the period and then multiplying by 2.

9.9 The waveshape in Fig. 9.2-1c may be expressed as the sum of step functions, as follows:

$$i(t) = \cdots + U_{-1}(t + 2) - U_{-1}(t + 1) + U_{-1}(t) - U_{-1}(t - 1) + \cdots$$

Find the response of the circuit in Fig. 9.2-1a to the individual step functions, and show that

$$e(t) = 1 - \epsilon^{-t}(1 - \epsilon^{-1} + \epsilon^{-2} - \epsilon^{-3} + \cdots) = 1 - \frac{\epsilon}{1+\epsilon}\epsilon^{-t} \qquad \text{for } 0 < t < 1$$

$$= 1 - \frac{\epsilon}{1+\epsilon}\epsilon^{-t} - 1 + \epsilon^{-(t-2)} = \frac{\epsilon^2}{1+\epsilon}\epsilon^{-t} \qquad \text{for } 1 < t < 2$$

Compare these results to the steady-state response found in Example 9.2-2.

9.10 Find the first three nonzero terms in the Fourier series for $e_o(t)$ in Fig. P9.10.

9.11 For the circuit shown in Fig. P9.11, $R = 5 \, \Omega$ and $e_1(t) = 10|\cos 377t|$. If the peak value of the largest a-c component in the steady-state output voltage should be $\frac{1}{30}$ of the d-c component, find the value of L.

9.12 The voltage $e_1(t)$ in Fig. P9.12 is a 60 cycle per sec sine wave that is distorted by the presence of a third harmonic. Will the voltage $e_o(t)$ be more or less distorted than $e_2(t)$? If the ratio of the third to the first harmonic is $\frac{1}{20}$ for $e_1(t)$, what will this ratio be for $e_o(t)$?

9.13 For each of the circuits in Fig. P5.24, $R_2 = 100 \, \Omega$, $C_1 = 100 \, \mu f$, $L_4 = 0.1$ h, and $R_4 = 10 \, \Omega$. In each case, the values of R_1 and R_3 are chosen so that $e_o(t) = 0$ when $e_1(t)$ is a 60 cycle per sec sine wave. If the sinusoidal source is then changed to the square wave in Fig. P9.13, find the first nonzero term in the Fourier series for $e_o(t)$.

9.14 The a-c steady-state network function for a certain ideal filter is shown in Fig. P9.14. If the input is the triangular wave in Fig. 9.1-4a, find and sketch the steady-state output.

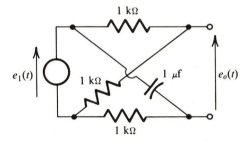

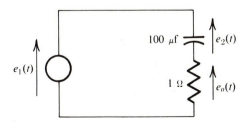

Figure P9.10

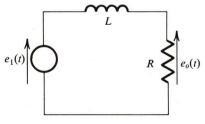

Figure P9.11

Figure P9.12

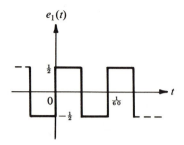

$e_1(t)$

Figure P9.13

9.15 The input current for the *RLC* circuit in Fig. 6.4-7b has the waveshape shown in Fig. P9.13. The inductance is $L = 10$ mh and the desired voltage is a 300 cycle per sec sine wave. If the peak value of the other frequencies in the output voltage should be less than $\frac{1}{20}$ of the peak value of the 300 cycle per sec component, find the values of C and R, and the selectivity Q of the circuit.

9.16 Derive Eqs. 9.3-6 and 9.3-7.

9.17 For a periodic waveshape, it is found that $c_n = 2/(1 + jn)$. If the period is $T = \frac{1}{10}$, write the first four nonzero terms in the trigonometric form of the Fourier series.

9.18 Find the complex Fourier series for the sawtooth waveshape in Fig. 9.1-1a, and sketch the frequency spectrum. Compare your answer to the trigonometric series in Example 9.1-1.

9.19 Sketch the frequency spectra for the waveshapes in Fig. P9.2.

9.20 Find the complex Fourier series for the succession of impulses in Fig. P9.20. Then use the results of Example 9.3-6 to obtain the series for a square wave.

9.21 For the circuit in Fig. P9.21a, $L = 0.1$ h, $C = 10$ μf, and $R = 10$ Ω.

  (a) If the input current $i(t)$ is the impulse train in Fig. P9.21b with $T = 2\pi/1000$, find the ratio of the first to the second harmonic in the output voltage $e(t)$.

  (b) Repeat when $i(t)$ has the waveshape in Fig. P9.21c, where $i(t) = -\sqrt{3} + 2\cos 1000t$ for $|1000t| < \pi/6$, where $i(t)$ is zero for the rest of the cycle, and where $T = 2\pi/1000$.

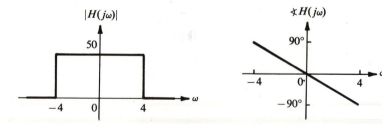

Figure P9.14

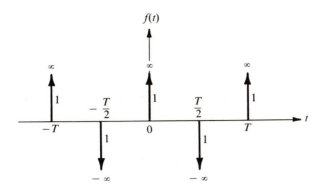

Figure P9.20

9.22 Derive Eqs. 9.4-6 and 9.4-7.

9.23 By a direct application of Eq. 9.4-3, show that the Fourier transform of the pulse in Fig. P9.23 is $(1 - \cos \omega L + j \sin \omega L)/j\omega L$. Express this answer in polar form, and show that it is equivalent to the answer obtained from Example 9.4-1 and Eq. 9.4-13.

9.24 Find and sketch $\mathscr{F}[\epsilon^{-a|t|}]$ and $\mathscr{F}[\epsilon^{-(at)^2}]$.

9.25 Let $a \to \infty$ in the expression for $\mathscr{F}[(a/2)\epsilon^{-a|t|}]$, and show that $\mathscr{F}[U_0(t)] = 1$.

9.26 Use Eq. 9.4-14 to find the Fourier transform of

$$f(t) = \cos t \quad \text{for } |t| < a$$
$$= 0 \quad \text{for } |t| > a$$

Sketch the frequency spectrum when $a = \pi$, 2π, and 10π.

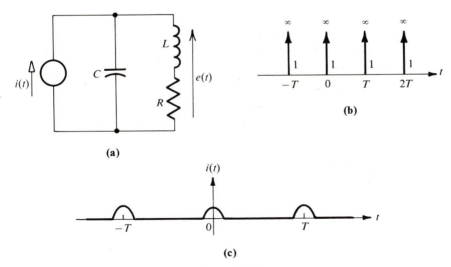

(a)

(b)

(c)

Figure P9.21

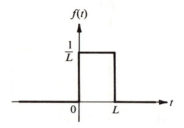

Figure P9.23

9.27 Find and discuss $\mathscr{F}^{-1}[U_0(\omega - \omega_1) + U_0(\omega + \omega_1)]$.

9.28 If $G_1(\omega)$ and $G_2(\omega)$ are real, prove that

$$\mathscr{F}[f_1(t)f_2(t)] = \frac{1}{2\pi} \int_{-\infty}^{\infty} G_1(\lambda)G_2(\omega - \lambda) \, d\lambda$$

9.29 Find $\mathscr{F}^{-1}[1/j\omega]$ by letting $\epsilon^{j\omega t} = \cos \omega t + j \sin \omega t$ and by using Eq. 9.1-11, Eq. 9.1-13, and a table of definite integrals. Discuss your answer.

9.30 Sketch the Bode diagrams corresponding to the frequency spectrum of $e_1(t) = \epsilon^{-t}U_{-1}(t)$. Also sketch the Bode diagrams corresponding to the frequency spectrum of $e_o(t)$ for each of the circuits in Fig. P9.30.

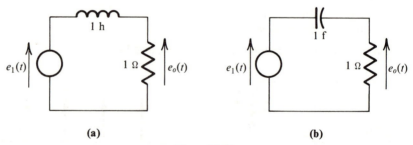

(a)   (b)

Figure P9.30

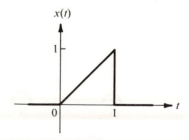

Figure P9.31

9.31 Find the frequency spectrum corresponding to the output of a circuit that is characterized by the differential equation

$$\frac{d^2y}{dt^2} + 2\frac{dy}{dt} + y = \frac{dx}{dt} + 10x$$

when the input $x(t)$ is as shown in Fig. P9.31.

9.32 If a circuit's unit impulse response is $h(t) = t\epsilon^{-t}U_{-1}(t)$ and if the input is $x(t) = \epsilon^{-t}U_{-1}(t)$, find the output frequency spectrum.

10

The Laplace Transform

The Fourier integral represents a nonperiodic function of time as the sum (or more precisely the integration) of components having the form $G(\omega)\epsilon^{j\omega t}\,d\omega$. Because the response to $\epsilon^{j\omega t}$ is easily determined, the entire output of a circuit can be found by invoking the superposition theorem. The Fourier methods also provide additional insight into how a circuit changes the waveshape of the incoming signal by discriminating against certain components. One limitation of the Fourier integral is that it is not applicable for all inputs, as illustrated by Example 9.4-3.

In searching for an extension or generalization of the Fourier method that will overcome this limitation, recall how the concepts in Chapter 5 were extended in Chapter 6. The a-c steady-state network function $H(j\omega)$ was considered as a special case of the more general function $H(s)$, where $s = \sigma + j\omega$. To understand why such a generalization should be useful here, reconsider Eq. 9.4-3,

$$G(\omega) = \int_{-\infty}^{\infty} f(t)\epsilon^{-j\omega t}\,dt$$

If $f(t)$ is the unit step function or the unit ramp function, the integrand does not vanish as t approaches infinity, and $G(\omega)$ cannot be found. Suppose that the factor $\epsilon^{-\sigma t}$ is introduced into the integrand to form

$$\int_{-\infty}^{\infty} f(t)\epsilon^{-\sigma t}\epsilon^{-j\omega t}\,dt = \int_{-\infty}^{\infty} f(t)\epsilon^{-st}\,dt$$

For a sufficiently large value of σ, the integrand now becomes zero as t approaches infinity even if $f(t)$ itself does not vanish there.

The formal implementation of this generalization, which is carried out in Section 10.1, leads to the Laplace transform. Circuit analysis with the Laplace transform can be divided into three basic steps, which are similar to those used with the Fourier transform. The first step is discussed at length in Section 10.1, and the last one is covered in Section 10.2. The complete analysis of circuits is considered in Section 10.3.

In Sections 10.4 and 10.5 the elements of complex-variable theory are introduced in order to put the Laplace transform on a firmer mathematical foundation and to prepare the student for more advanced books on transform theory. Section 10.6 contains additional applications.

10.1

The Direct Transform

The direct and inverse Fourier transforms, which are given in Eqs. 9.4-3 and 9.4-4, are

$$G(\omega) = \int_{-\infty}^{\infty} f(t)\epsilon^{-j\omega t}\, dt = \mathscr{F}[f(t)]$$

$$f(t) = \frac{1}{2\pi} \int_{-\infty}^{\infty} G(\omega)\epsilon^{j\omega t}\, d\omega = \mathscr{F}^{-1}[G(\omega)]$$

Since G is a function of ω, and since ω is a function of $j\omega$ (differing only by the complex constant $1/90°$), G also can be regarded as a function of $j\omega$. It is convenient to replace the symbol $G(\omega)$ by $F(j\omega)$ and to rewrite the above equations as

$$F(j\omega) = \int_{-\infty}^{\infty} f(t)\epsilon^{-j\omega t}\, dt$$

$$f(t) = \frac{1}{2\pi j} \int_{-j\infty}^{j\infty} F(j\omega)\epsilon^{j\omega t}\, d(j\omega)$$

The second expression has been multiplied and divided by j, and the variable of integration has been changed to $j\omega$. If $j\omega$ is replaced by $s = \sigma + j\omega$,

$$F(s) = \int_{-\infty}^{\infty} f(t)\epsilon^{-st}\, dt$$

$$f(t) = \frac{1}{2\pi j} \int_{\sigma-j\infty}^{\sigma+j\infty} F(s)\epsilon^{st}\, ds$$

In most problems, we are concerned only with a circuit's response for values of t greater than some reference time, which is usually taken as $t = 0$. As discussed in Section 4.4, the complete response for $t > 0$ depends only upon the input for $t > 0$ and upon the stored energy in the circuit at $t = 0$. It is not necessary to know the input for $t < 0$, because its effect is accounted for by the initial stored energy. As long as the initial energy in each capacitance, inductance, and transformer is known, the input and output waveshapes for $t < 0$ are not needed, and it is customary to assume that they are zero. If the function of time in the above equations can be considered to be zero for $t < 0$,

$$F(s) = \int_{0}^{\infty} f(t)\epsilon^{-st}\, dt = \mathscr{L}[f(t)] \tag{10.1-1}$$

$$f(t) = \frac{1}{2\pi j} \int_{\sigma-j\infty}^{\sigma+j\infty} F(s)\epsilon^{st}\, ds = \mathscr{L}^{-1}[F(s)] \tag{10.1-2}$$

The symbols $\mathscr{L}$ and $\mathscr{L}^{-1}$ stand for the direct and inverse Laplace transform, respectively.

The above development is not intended to be a rigorous mathematical derivation, but is instead a heuristic presentation that illustrates the relationship between the Laplace and Fourier transforms. Equation 10.1-1 could be taken as an arbitrary definition of the quantity $F(s)$, which is done in many books. The integrand contains the factor $\epsilon^{-\sigma t}$, which permits the upper limit of integration to be evaluated in nearly all cases, even if the corresponding Fourier transform does not exist.

In the inverse Laplace transform of Eq. 10.1-2, $f(t)$ is expressed as the summation (integration) of components having the form $F(s)\epsilon^{st}$. This form should be helpful in circuit analysis, because the response to ϵ^{st} is easily found. Notice, however, that the integration is with respect to the *complex* frequency $s = \sigma + j\omega$. A discussion of how such an integration should be carried out is deferred until Section 10.5.

Functions of time are normally denoted by lower-case letters, and their Laplace transforms by the corresponding capital letters. Thus, $E_3(s)$ and $I_4(s)$ would be the transforms of the voltage $e_3(t)$ and the current $i_4(t)$. The following examples illustrate the method of finding the transform of a known function of time.

Example 10.1-1.   Find the Laplace transform of the unit step function.

Solution.   Since $f(t) = 1$ for $t > 0$,

$$F(s) = \int_0^\infty \epsilon^{-st} \, dt = \left[\frac{\epsilon^{-st}}{-s} \right]_0^\infty = \frac{1}{s}$$

Note that $\lim_{t \to \infty} \epsilon^{-st} = \lim_{t \to \infty} \epsilon^{-\sigma t} \epsilon^{-j\omega t}$. The factor $\epsilon^{-j\omega t}$ has a magnitude of unity, so the value of σ, which is the real part of s, must be positive in order for the upper limit to exist. The significance of the range of σ for which the integral converges is deferred until after a discussion of complex-variable theory. For the present, the student may assume that σ may be made as large as is necessary to make the transform exist.

Example 10.1-2.   Find $F(s)$ if $f(t) = t$ for $t > 0$.

Solution

$$F(s) = \int_0^\infty t\epsilon^{-st} \, dt = \left[-\frac{t\epsilon^{-st}}{s} - \frac{\epsilon^{-st}}{s^2} \right]_0^\infty$$

The upper limit of the first term is an indeterminate form, but by l'Hôpital's rule,

$$\lim_{t \to \infty} t\epsilon^{-st} = \lim_{t \to \infty} \frac{t}{\epsilon^{st}} = \lim_{t \to \infty} \frac{1}{s\epsilon^{st}} = 0$$

so

$$F(s) = \frac{1}{s^2}$$

for $\sigma > 0$.

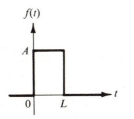

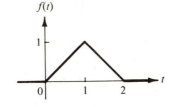

Figure 10.1-1   Figure 10.1-2

Example 10.1-3.   Find the transform of the function shown in Fig. 10.1-1.

Solution.   The definite integral in Eq. 10.1-1 may be broken into the sum of two or more integrals.

$$F(s) = \int_0^L A\epsilon^{-st}\, dt + \int_L^\infty 0\epsilon^{-st}\, dt = A\left[\frac{\epsilon^{-st}}{-s}\right]_0^L + 0 = \frac{A}{s}(-\epsilon^{-sL} + 1)$$

Notice that if $A = 1/L$, the area underneath $f(t)$ is unity and

$$F(s) = \frac{1 - \epsilon^{-sL}}{sL}$$

If L is now allowed to approach zero, $f(t)$ becomes the unit impulse function $U_0(t)$, and

$$F(s) = \lim_{L \to 0} \frac{1 - \epsilon^{-sL}}{sL} = \lim_{L \to 0} \frac{s\epsilon^{-sL}}{s} = 1$$

   The above three examples suggest that $\mathcal{L}[U_n(t)] = s^n$, where $U_n(t)$ represents the family of singularity functions discussed in Section 3.2. This generalization proves to be valid, so the subscript on a singularity function is the power of s in its Laplace transform.

Example 10.1-4.   Find the transform of the function in Fig. 10.1-2.

Solution.   Since $f(t) = t$ for $0 < t < 1$ and $f(t) = 2 - t$ for $1 < t < 2$,

$$F(s) = \int_0^1 t\epsilon^{-st}\, dt + \int_1^2 (2 - t)\epsilon^{-st}\, dt$$

$$= \left[-t\frac{\epsilon^{-st}}{s} - \frac{\epsilon^{-st}}{s^2}\right]_0^1 + \left[-2\frac{\epsilon^{-st}}{s} + t\frac{\epsilon^{-st}}{s} + \frac{\epsilon^{-st}}{s^2}\right]_1^2 = \frac{1}{s^2}(1 - 2\epsilon^{-s} + \epsilon^{-2s})$$

   If a single-valued function $f(t)$ is defined for all $t \geq 0$, with the possible exception of a number of isolated points, then the integral in Eq. 10.1-1 exists for some range of the parameter σ in all cases of practical interest.[†] Therefore, the Laplace transform of any common function may be found from the defining equation. The

---

† If $f(t) = t^t$, for example, Eq. 10.1-1 does not converge for any range of σ, but such a function does not arise in circuit analysis.

TABLE 10.1-1

$f(t)$ for $t \geq 0$	$F(s)$	Region of convergence
$U_0(t)$	1	Everywhere
$U_{-1}(t)$	$\dfrac{1}{s}$	$\sigma > 0$
t^n (n any positive integer)	$\dfrac{n!}{s^{n+1}}$	$\sigma > 0$
ϵ^{-at}	$\dfrac{1}{s + a}$	$\sigma > -a$
$\cos \beta t$	$\dfrac{s}{s^2 + \beta^2}$	$\sigma > 0$
$\sin \beta t$	$\dfrac{\beta}{s^2 + \beta^2}$	$\sigma > 0$

transforms of several functions are given in Table 10.1-1. When deriving the transforms of trigonometric functions, it is convenient to express them as the sum of two exponential terms. The regions of convergence given in the table assume that the constant a is real, but the formulas for $F(s)$ are valid for any real or complex constant.

Many reference books contain fairly lengthy tables. See, for example, C. D. Hodgman (ed.), *Mathematical Tables from the Handbook of Chemistry and Physics*, eleventh edition (Cleveland: Chemical Rubber Publishing Co., 1957). A most extensive table is found in pp. 125–302 of A. Erdélyi (ed.), *Tables of Integral Transforms*, vol. I (New York: McGraw-Hill, 1954).

These tables may be used not only to find $\mathscr{L}[f(t)]$, but also to find $f(t)$ when $F(s)$ is given, provided that a given $F(s)$ corresponds to a unique function of time. Thus, it is important to know whether or not two different functions of time can have the same Laplace transform. Figure 10.1-3 shows that this can indeed happen. The function $f_2(t)$ is the same as $f_1(t)$ except for $t < 0$, and since the defining integral extends only for $t \geq 0$, $F_2(s) = F_1(s)$. The function $f_3(t)$ differs from $f_1(t)$ only at

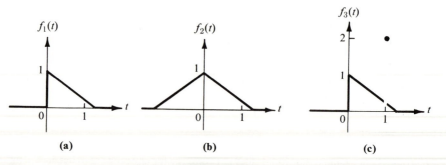

(a) (b) (c)

Figure 10.1-3

the single point $t = 1$, where its value is arbitrarily defined to be 2, and again $F_3(s) = F_1(s)$. These are rather trivial cases, however, since we shall be interested in a circuit's response only for $t \geq 0$, and since we shall not encounter functions like $f_3(t)$.

It can be shown that two functions of time have identical transforms only if they themselves are identical for all $t \geq 0$, except possibly at isolated points. This uniqueness property means that a given $F(s)$ corresponds to only one function of time for $t \geq 0$. If the function of s is given and can be found in a table, then $f(t)$ can be immediately written down.

Many of the functions of time encountered in circuit analysis, such as the one in Fig. 10.1-3a, have a discontinuity at $t = 0$. In such cases, the value of the function at the discontinuity is not uniquely defined, unless some special convention is adopted. Three of the possible conventions are $f(0-), f(0+)$, and $\frac{1}{2}[f(0-) + f(0+)]$, where $f(0-)$ and $f(0+)$ stand for $f(t)$ just before and just after $t = 0$. Any convention is satisfactory as long as it is consistently adhered to. Since we are normally interested in the response of a system for $t \geq 0$, we shall define the value of $f(t)$ at the origin to be $f(0+)$. Then Eq. 10.1-1 is interpreted as

$$F(s) = \int_{0+}^{\infty} f(t) \epsilon^{-st} \, dt$$

Properties of the Laplace Transform.   A number of very useful properties are summarized in Table 10.1-2. Some of them increase the usefulness of Table 10.1-1, while others are needed to solve circuit problems. The proofs all follow directly from Eq. 10.1-1.

TABLE 10.1-2

Property	Equation number
$\mathscr{L}[f_1(t) + f_2(t)] = \mathscr{L}[f_1(t)] + \mathscr{L}[f_2(t)] \quad F = {}_1(s) + F_2(s)$	10.1-3
$\mathscr{L}[af(t)] = a\mathscr{L}[f(t)] = aF(s)$	10.1-4
$\mathscr{L}[\epsilon^{-at}f(t)] = F(s + a)$	10.1-5
$\mathscr{L}[f(t - a) \, U_{-1}(t - a)] = \epsilon^{-as} F(s)$	10.1-6
$\mathscr{L}[f(at)] = \dfrac{1}{a} F\left(\dfrac{s}{a}\right)$	10.1-7
$\mathscr{L}\left[\dfrac{d}{dt} f(t)\right] = sF(s) - f(0+)$	10.1-8
$\mathscr{L}\left[\dfrac{d^2}{dt^2} f(t)\right] = s^2 F(s) - sf(0+) - \dfrac{df}{dt}(0+)$	10.1-9
$\mathscr{L}\left[\displaystyle\int_{0+}^{t} f(\lambda) \, d\lambda\right] = \dfrac{F(s)}{s}$	10.1-10
$\mathscr{L}\left[\displaystyle\int_{-\infty}^{t} f(\lambda) \, d\lambda\right] = \dfrac{F(s)}{s} + \dfrac{f^{-1}(0+)}{s}$	
where $f^{-1}(0+) = \displaystyle\lim_{t \to 0+} \int_{-\infty}^{t} f(\lambda) \, d\lambda$	10.1-11

Since the transform is defined in terms of an integral, the basic theorems for integrals also apply to transforms, as in Eqs. 10.1-3 and 10.1-4. To formally prove the first of these, note that

$$\mathcal{L}[f_1(t) + f_2(t)] = \int_0^\infty [f_1(t) + f_2(t)]\epsilon^{-st}\, dt$$

$$= \int_0^\infty [f_1(t)\epsilon^{-st} + f_2(t)\epsilon^{-st}]\, dt$$

$$= \int_0^\infty f_1(t)\epsilon^{-st}\, dt + \int_0^\infty f_2(t)\epsilon^{-st}\, dt$$

$$= \mathcal{L}[f_1(t)] + \mathcal{L}[f_2(t)] = F_1(s) + F_2(s)$$

Note that an analogous property is not valid for the product of two functions. In general,

$$\mathcal{L}[f_1(t)f_2(t)] \neq F_1(s)F_2(s)$$

Next we derive a formula for the transform of $\epsilon^{-at}f(t)$ when $\mathcal{L}[f(t)] = F(s)$ is known.

$$\mathcal{L}[\epsilon^{-at}f(t)] = \int_0^\infty \epsilon^{-at}f(t)\epsilon^{-st}\, dt = \int_0^\infty f(t)\epsilon^{-(s+a)t}\, dt$$

The last expression is identical with the transform of $f(t)$, except that s has been replaced by the quantity $(s + a)$. This is indicated mathematically by the symbol $F(s + a)$ in Eq. 10.1-5.

Example 10.1-5.   Find the transform of $t\epsilon^{-at}$, $\epsilon^{-at}\cos \beta t$, and $\epsilon^{-at}\sin \beta t$ by Eq. 10.1-5.

Solution.   Since the transform of t is $1/s^2$,

$$\mathcal{L}[t\epsilon^{-at}] = \frac{1}{(s + a)^2}$$

Notice that this does not equal the product of $\mathcal{L}[t]$ and $\mathcal{L}[\epsilon^{-at}]$. Table 10.1-1 indicates that the transform of $\cos \beta t$ is $s/(s^2 + \beta^2)$. Therefore,

$$\mathcal{L}[\epsilon^{-at}\cos \beta t] = \frac{s + a}{(s + a)^2 + \beta^2}$$

In a similar way,

$$\mathcal{L}[\epsilon^{-at}\sin \beta t] = \frac{\beta}{(s + a)^2 + \beta^2}$$

Equation 10.1-6 is useful when transforming functions that are zero for $t < a$. Consider the quantity

$$\epsilon^{-as}F(s) = \epsilon^{-as}\int_0^\infty f(t)\epsilon^{-st}\, dt = \int_0^\infty f(t)\epsilon^{-s(t+a)}\, dt$$

If the substitutions $\lambda = t + a$ and $d\lambda = dt$ are made,

$$\epsilon^{-as}F(s) = \int_a^\infty f(\lambda - a)\epsilon^{-s\lambda}\, d\lambda = \int_0^\infty f(\lambda - a)U_{-1}(\lambda - a)\epsilon^{-s\lambda}\, d\lambda$$

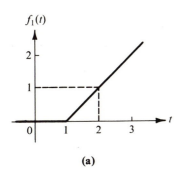

(a) (b)

Figure 10.1-4

which is recognized as the transform of $f(t - a)U_{-1}(t - a)$, and which proves Eq. 10.1-6. The fact that λ instead of t is the variable of integration in the last expression is of no significance, since it is a dummy variable which disappears after the limits are inserted. The quantity $f(t - a)$ is the function $f(t)$ shifted a units in the positive t direction. The factor $U_{-1}(t - a)$ only means that the function being transformed must be zero for $t < a$.

Example 10.1-6. Find the Laplace transform of the functions in Figs. 10.1-4a, 10.1-5a, and 10.1-6a by the use of Eq. 10.1-6.

Solution. In Fig. 10.1-4, $f_1(t) = f_2(t - 1)U_{-1}(t - 1)$, since it is the second function shifted 1 sec to the right, and since it does not begin until $t = 1$. Since $F_2(s) = 1/s^2$ by Example 10.1-2, $F_1(s) = (1/s^2)\epsilon^{-s}$.

The function $f_3(t)$ in Fig. 10.1-5a can be expressed as $f_3(t) = f_4(t) + f_5(t)$. Since $F_4(s) = A/s$ and $F_5(s) = -(A/s)\epsilon^{-sL}$, $F_3(s) = (A/s)(1 - \epsilon^{-sL})$, which agrees with Example 10.1-3.

In Fig. 10.1-6, $f_6(t) = f_7(t) + f_8(t) + f_9(t)$. The transforms of these components are $F_7(s) = 1/s^2$, $F_8(s) = -(2/s^2)\epsilon^{-s}$, and $F_9(s) = (1/s^2)\epsilon^{-2s}$. The complete answer is $F_6(s) = (1/s^2)(1 - 2\epsilon^{-s} + \epsilon^{-2s})$, which agrees with Example 10.1-4.

The proof of Eq. 10.1-7 is similar to the proof of Eq. 9.4-10. This property is discussed and applied in the next chapter.

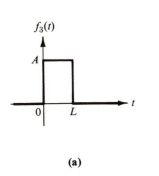

(a)

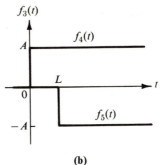

(b)

Figure 10.1-5

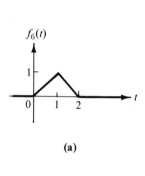

(a)

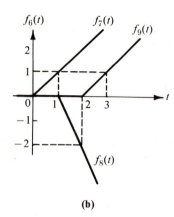

(b)

Figure 10.1-6

Since systems are often described by integral-differential equations, formulas for transforming derivatives and integrals are included in Table 10.1-2. Consider

$$\mathcal{L}\left[\frac{df}{dt}\right] = \int_0^\infty \frac{df}{dt}\,\epsilon^{-st}\,dt$$

In the standard formula for integration by parts,

$$\int_0^\infty u\,dv = \left[uv\right]_0^\infty - \int_0^\infty v\,du$$

Let $u = \epsilon^{-st}$ and $dv = (df/dt)\,dt$. Then $du = -s\epsilon^{-st}\,dt$ and $v = f(t)$, so

$$\mathcal{L}\left[\frac{df}{dt}\right] = [\epsilon^{-st}f(t)]_{0+}^\infty - \int_0^\infty f(t)(-s\epsilon^{-st})\,dt$$

$$= 0 - f(0+) + s\int_0^\infty f(t)\epsilon^{-st}\,dt = sF(s) - f(0+)$$

which proves Eq. 10.1-8. The upper limit in the uv term may be an indeterminate form in some cases, but an evaluation of the indeterminate form does give zero for all transformable functions. The lower limit of the uv term is $f(t)$ evaluated at $t = 0$. In accordance with the earlier discussion, the value at $t = 0+$ is used if $f(t)$ has a discontinuity at the origin.

Although the chief use of this property is in solving differential equations, it is occasionally used to find the transform of a given function. If, for example, it is known that $\mathcal{L}[\cos \beta t] = s/(s^2 + \beta^2)$, then

$$\mathcal{L}\left[\frac{d}{dt}\cos \beta t\right] = s\,\frac{s}{s^2 + \beta^2} - \cos 0$$

$$-\beta\mathcal{L}[\sin \beta t] = \frac{s^2}{s^2 + \beta^2} - 1 = \frac{-\beta^2}{s^2 + \beta^2}$$

$$\mathcal{L}[\sin \beta t] = \frac{\beta}{s^2 + \beta^2}$$

which agrees with Table 10.1-1.

Equation 10.1-9 may be proved by integrating by parts twice. An easier method is to apply Eq. 10.1-8 with $f(t)$ replaced by df/dt.

$$\mathscr{L}\left[\frac{d}{dt}\frac{df}{dt}\right] = s\mathscr{L}\left[\frac{df}{dt}\right] - \left[\frac{df}{dt}\right]_{t=0+}$$

$$= s[sF(s) - f(0+)] - f'(0+)$$

where the prime denotes differentiation.

The proof of Eq. 10.1-10 again follows from the integration-by-parts formula. If $u = \int_{0+}^{t} f(\lambda)\, d\lambda$ and $dv = \epsilon^{-st}\, dt$, $du = f(t)\, dt$ and $v = (-1/s)\epsilon^{-st}$. Then

$$\mathscr{L}\left[\int_{0+}^{t} f(\lambda)\, d\lambda\right] = \left[\frac{\epsilon^{-st}}{-s}\int_{0+}^{t} f(\lambda)\, d\lambda\right]_{0+}^{\infty} - \int_{0+}^{\infty}\frac{\epsilon^{-st}}{-s} f(t)\, dt$$

$$= 0 - 0 + \frac{1}{s}\int_{0+}^{\infty} f(t)\epsilon^{-st}\, dt = \frac{F(s)}{s}$$

To obtain Eq. 10.1-11, recall that

$$\int_{-\infty}^{t} f(\lambda)\, d\lambda = \int_{-\infty}^{0+} f(\lambda)\, d\lambda + \int_{0+}^{t} f(\lambda)\, d\lambda$$

The first term on the right-hand side is a constant, since neither limit is a function of t. We denote this term by the symbol $f^{-1}(0+)$, and recall that the transform of a constant is the constant divided by s. Thus,

$$\mathscr{L}\left[\int_{-\infty}^{t} f(\lambda)\, d\lambda\right] = \frac{f^{-1}(0+)}{s} + \frac{F(s)}{s}$$

The application of the last two properties in circuit analysis sometimes causes confusion. Their use arises in the two situations shown in Fig. 10.1-7. As discussed in Section 1.3, the relationship $e(t) = (1/C)\int i(t)\, dt$ for the capacitance may be interpreted in either of two ways.

$$e(t) = e(0+) + \frac{1}{C}\int_{0+}^{t} i(\lambda)\, d\lambda$$

$$e(t) = \frac{1}{C}\int_{-\infty}^{t} i(\lambda)\, d\lambda$$

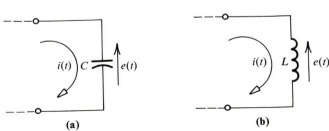

(a)　　　　　　　　　　　(b)

Figure 10.1-7

Let $E(s)$ and $I(s)$ denote the Laplace transforms of $e(t)$ and $i(t)$, respectively, and transform both sides of the first of these equations. By Eq. 10.1-10,

$$E(s) = \frac{e(0+)}{s} + \frac{I(s)}{sC} \tag{10.1-12}$$

If the alternative expression for $e(t)$ is transformed term by term,

$$E(s) = \frac{1}{C}\left[\frac{I(s)}{s} + \frac{i^{-1}(0+)}{s}\right]$$

Since $(1/C)i^{-1}(0+) = (1/C)\int_{-\infty}^{0+} i(\lambda)\,d\lambda = e(0+)$, this approach gives the same result as before. The $e(0+)$ term summarizes the effect of all the current up to the time $t = 0+$. The same result may be obtained from the equation $i = C(de/dt)$. By Eq. 10.1-8,

$$I(s) = C[sE(s) - e(0+)]$$

which can be rewritten as Eq. 10.1-12.

For the inductance in Fig. 10.1-7b,

$$i(t) = \frac{1}{L}\int_{-\infty}^{t} e(\lambda)\,d\lambda = i(0+) + \frac{1}{L}\int_{0+}^{t} e(\lambda)\,d\lambda$$

Using either Eq. 10.1-10 or 10.1-11, and noting that $(1/L)e^{-1}(0+) = i(0+)$, the student should show that

$$I(s) = \frac{E(s)}{sL} + \frac{i(0+)}{s} \tag{10.1-13}$$

The same equation may be obtained by transforming $e = L(di/dt)$.

Example 10.1-7.   If the switch K in Fig. 10.1-8 closes at $t = 0$, find $i(t)$ for $t > 0$.

Solution.   For $t > 0$, the circuit is described by the differential equation

$$e_1 = Ri + \frac{1}{C}\int i\,dt$$

The transform of the voltage source is $E_1(s) = 1/s$, and the transform of the unknown current can be denoted by $I(s)$. If $I(s)$ can be found, $i(t)$ is then indirectly specified,

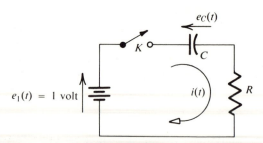

Figure 10.1-8

according to the uniqueness theorem. Transforming both sides of the differential equation, and using Eqs. 10.1-3, 10.1-4, and 10.1-12, we obtain

$$\frac{1}{s} = RI(s) + \frac{I(s)}{sC} + \frac{e_C(0+)}{s}$$

so

$$I(s) = \frac{1/s - e_C(0+)/s}{R + 1/sC} = \frac{1 - e_C(0+)}{R}\left(\frac{1}{s + 1/RC}\right)$$

From Table 10.1-1, $\mathscr{L}^{-1}[1/(s + a)] = \epsilon^{-at}$ for $t > 0$, so

$$i(t) = \mathscr{L}^{-1}[I(s)] = \frac{1 - e_C(0+)}{R}\ \epsilon^{-t/RC}\quad\text{for}\quad t > 0$$

Since the voltage across a capacitance cannot change instantaneously, $e_C(0+) = e_C(0-)$, which is the voltage just before the switch closes, and which is presumably known. If $e_C(0+) = 1$ volt, the answer indicates that no current will flow, as can also be seen by an inspection of the circuit.

Example 10.1-7 illustrates the use of Tables 10.1-1 and 10.1-2 in solving problems by means of the Laplace transform. The solution could have been carried out by using the concept of impedance and admittance, although the initial stored energy in the capacitance tends to complicate this approach. The various analysis procedures are discussed at length in Section 10.3. In each case, however, the aim is to find the transform of the desired output, and then to determine the corresponding function of time.

Example 10.1-7 is a very simple problem, whose solution can be written down by inspection or by reference to Fig. 3.3-1c. The advantages of a powerful tool such as the Laplace transform are not readily apparent in such simple problems. Before solving more complicated circuits, however, we must be able to take the inverse transform of functions which do not appear in Table 10.1-1.

10.2

Partial-Fraction Expansions

If the Laplace transform $F(s)$ is given, there are several ways of finding $f(t)$. In a few cases, $F(s)$ may correspond to an entry in Table 10.1-1, so that $f(t)$ can be written down immediately. In other cases, the student may find the properties in Table 10.1-2 helpful, or he may be able to locate the given $F(s)$ in a more exten·sive table. The function $f(t)$ may always be found by Eq. 10.1-2, but this equation requires a knowledge of complex-variable theory and is not discussed until Section 10.5.

The partial-fraction method consists of breaking $F(s)$ up into the sum of simple parts, each one of which may be found in a short table. Assume that $F(s)$ is the quotient of two polynomials with real coefficients. If the two polynomials are denoted by $P(s)$ and $Q(s)$, and if the equation $Q(s) = 0$ has distinct roots at $s_1, s_2, \ldots, s_n$, then

$$F(s) = \frac{P(s)}{Q(s)} = \frac{P(s)}{(s - s_1)(s - s_2) \cdots (s - s_n)} \tag{10.2-1}$$

In this equation, it is assumed that the coefficient of the highest power of s in $Q(s)$ has been made equal to unity, for otherwise the denominator would contain a multiplying constant. If $P(s)$ is of lower order than $Q(s)$, $F(s)$ may be expressed as the following partial-fraction expansion:

$$F(s) = \frac{K_1}{s - s_1} + \frac{K_2}{s - s_2} + \cdots + \frac{K_n}{s - s_n} \tag{10.2-2}$$

The constants K_1 through K_n always may be found so that Eq. 10.2-2 is a mathematical identity for all values of s. In order to avoid solving simultaneous equations when determining these constants, it is convenient to let s take on particular values. To find an explicit expression for K_1, multiply Eq. 10.2-2 by $s - s_1$, and then let $s = s_1$. All terms on the right-hand side except K_1 will vanish, so

$$K_1 = [(s - s_1)F(s)]_{s=s_1} \tag{10.2-3}$$

The factor $s - s_1$ should be cancelled with the corresponding factor in the denominator of $F(s)$ before substituting $s = s_1$. The constants K_2 through K_n are found in a similar way.

If $Q(s)$ has repeated factors, a modification of the partial-fraction expansion is necessary. Suppose that the root s_1 is repeated r times, so that

$$F(s) = \frac{P(s)}{(s - s_1)^r(s - s_{r+1}) \cdots (s - s_n)} \tag{10.2-4}$$

If the order of the numerator is again less than that of the denominator,

$$F(s) = \frac{K_1}{(s - s_1)} + \cdots + \frac{K_{r-1}}{(s - s_1)^{r-1}} + \frac{K_r}{(s - s_1)^r} + \frac{K_{r+1}}{s - s_{r+1}} + \cdots + \frac{K_n}{s - s_n}$$

The constants K_{r+1} through K_n may be found as before. To find K_r, multiply by $(s - s_1)^r$ to obtain

$$(s - s_1)^r F(s) = K_1(s - s_1)^{r-1} + \cdots + K_{r-1}(s - s_1) + K_r$$

$$+ \cdots + \frac{K_n(s - s_1)^r}{s - s_n} \tag{10.2-5}$$

When $s = s_1$, all but one of the terms on the right side of the equation will vanish, and

$$K_r = [(s - s_1)^r F(s)]_{s=s_1} \tag{10.2-6}$$

Differentiating Eq. 10.2-5 with respect to s, and then letting $s = s_1$, we obtain

$$K_{r-1} = \left[\frac{d}{ds} (s - s_1)^r F(s) \right]_{s=s_1} \tag{10.2-7}$$

In general,

$$K_{r-k} = \frac{1}{k!} \left[\frac{d^k}{ds^k} (s - s_1)^r F(s) \right]_{s=s_1} \tag{10.2-8}$$

In the use of these formulas, $(s - s_1)^r$ should be cancelled with the corresponding factor in the denominator of $F(s)$ before differentiating.

Once the partial-fraction expansion of $F(s)$ has been determined, the inverse transform of each term can be found by Table 10.2-1. The entries in this table follow directly from Table 10.1-1 and Eq. 10.1-5. The quantity 0! is defined to be unity.

Example 10.2-1.   Find the inverse transform of $F(s) = \dfrac{5s - 1}{(s + 1)^2(s - 2)}$.

Solution.   The form of the partial-fraction expansion is

$$F(s) = \frac{K_1}{s + 1} + \frac{K_2}{(s + 1)^2} + \frac{K_3}{s - 2}$$

where

$$K_2 = \left[\frac{5s - 1}{s - 2} \right]_{s=-1} = 2$$

$$K_1 = \left[\frac{d}{ds} \frac{5s - 1}{s - 2} \right]_{s=-1} = \left[\frac{-9}{(s - 2)^2} \right]_{s=-1} = -1$$

$$K_3 = \left[\frac{5s - 1}{(s + 1)^2} \right]_{s=2} = 1$$

$$F(s) = \frac{-1}{s + 1} + \frac{2}{(s + 1)^2} + \frac{1}{s - 2}$$

The validity of this result may be checked by recombining the individual terms over a common denominator to reconstruct the original function of s. By Table 10.2-1,

$$f(t) = -\epsilon^{-t} + 2t\epsilon^{-t} + \epsilon^{2t} \quad \text{for} \quad t \geq 0$$

TABLE 10.2-1

$f(t)$ for $t \geq 0$	$F(s)$
$U_n(t)$	s^n
$\dfrac{t^{n-1}}{(n - 1)!}$ for $n = 1, 2, 3, \ldots$	$\dfrac{1}{s^n}$
ϵ^{-at}	$\dfrac{1}{s + a}$
$\dfrac{t^{n-1}}{(n - 1)!} \epsilon^{-at}$ for $n = 1, 2, 3, \ldots$	$\dfrac{1}{(s + a)^n}$

Example 10.2-2. Find $\mathscr{L}^{-1}\left[\dfrac{4s^2 + 9s + 1}{2s^2 + 4s}\right]$.

Solution. Since the numerator and denominator have the same order, $F(s)$ cannot be correctly represented by the expansion of Eq. 10.2-2. However, a preliminary step of long division gives

$$F(s) = \frac{4s^2 + 9s + 1}{2s(s + 2)} = 2 + \frac{s + 1}{2s(s + 2)}$$

In the second of these two terms, the numerator is of lower order than the denominator, so

$$F(s) = 2 + \frac{K_1}{s} + \frac{K_2}{s + 2}$$

Notice that when s becomes large, the original function of s approaches 2, while the terms K_1/s and $K_2/(s + 2)$ approach zero. Thus, the last equation, which must be valid for all values of s, would certainly not be correct without the 2 on the right-hand side. One or more preliminary steps of long division are always necessary when the order of the numerator equals or exceeds that of the denominator.

$$K_1 = \left[\frac{s + 1}{2(s + 2)}\right]_{s=0} = \frac{1}{4}$$

$$K_2 = \left[\frac{s + 1}{2s}\right]_{s=-2} = \frac{1}{4}$$

$$f(t) = 2U_0(t) + \tfrac{1}{4}(1 + \epsilon^{-2t}) \qquad \text{for } t \geq 0$$

If the polynomial $Q(s)$ has complex factors, Eqs. 10.2-1 through 10.2-8 are still valid, but some of the quantities $s_1, s_2, \ldots, s_n, K_1, K_2, \ldots, K_n$ will be complex. If $F(s)$ is the quotient of two polynomials with real coefficients, $f(t)$ is a real function of time, so complex quantities in the final answer should be avoided. Such quantities always can be eliminated by using the identities

$$\cos \theta = \frac{\epsilon^{j\theta} + \epsilon^{-j\theta}}{2}, \qquad \sin \theta = \frac{\epsilon^{j\theta} - \epsilon^{-j\theta}}{j2} \tag{10.2-9}$$

It is helpful to know that complex roots must occur in complex-conjugate pairs and that the partial-fraction coefficients for complex-conjugate roots are themselves complex conjugates. In a partial-fraction expansion, typical terms corresponding to a pair of distinct complex roots are

$$\frac{|K|\epsilon^{j\theta}}{s + \alpha - j\beta} + \frac{|K|\epsilon^{-j\theta}}{s + \alpha + j\beta}$$

where the complex coefficients in the numerator have been expressed in polar form. The inverse transform of these two terms is

$$|K|\epsilon^{j\theta}\epsilon^{-(\alpha - j\beta)t} + |K|\epsilon^{-j\theta}\epsilon^{-(\alpha + j\beta)t} = 2|K|\epsilon^{-\alpha t}\frac{\epsilon^{j(\beta t + \theta)} + \epsilon^{-j(\beta t + \theta)}}{2}$$

$$= 2|K|\epsilon^{-\alpha t}\cos(\beta t + \theta) \tag{10.2-10}$$

Example 10.2-3. Find the inverse transform of $F(s) = \dfrac{s + 15}{s^2(s^2 + 2s + 5)}$.

Solution

$$F(s) = \frac{s + 15}{s^2(s + 1 - j2)(s + 1 + j2)} = \frac{K_1}{s} + \frac{K_2}{s^2} + \frac{K_3}{s + 1 - j2} + \frac{K_4}{s + 1 + j2}$$

$$K_2 = \left[\frac{s + 15}{s^2 + 2s + 5} \right]_{s=0} = 3$$

$$K_1 = \left[\frac{d}{ds} \frac{s + 15}{s^2 + 2s + 5} \right]_{s=0} = -1$$

$$K_3 = \left[\frac{s + 15}{s^2(s + 1 + j2)} \right]_{s=-1+j2}$$

$$= \left(\frac{7 + j1}{8 - j6} \right) \left(\frac{8 + j6}{8 + j6} \right) = \frac{50 + j50}{100} = \frac{1}{2} + j\frac{1}{2} = \frac{\sqrt{2}}{2} \underline{/45°}$$

$$K_4 = K_3{}^* = \frac{1}{2} - j\frac{1}{2} = \frac{\sqrt{2}}{2} \underline{/-45°}$$

$$f(t) = -1 + 3t + \sqrt{2}\, \epsilon^{-t} \cos (2t + 45°) \quad \text{for} \quad t \geq 0$$

Although the last term follows from Eq. 10.2-10, it can be obtained almost as easily by using Eqs. 10.2-9.

Example 10.2-4. Find $\mathcal{L}^{-1} \left[\dfrac{s}{(s^2 + 1)^2} \right]$.

Solution

$$F(s) = \frac{s}{(s - j)^2(s + j)^2} = \frac{K_1}{s - j} + \frac{K_2}{(s - j)^2} + \frac{K_3}{s + j} + \frac{K_4}{(s + j)^2}$$

$$K_2 = \left[\frac{s}{(s + j)^2} \right]_{s=j} = -\frac{j}{4}$$

$$K_1 = \left[\frac{d}{ds} \frac{s}{(s + j)^2} \right]_{s=j} = 0$$

$$K_4 = K_2{}^* = \frac{j}{4} \quad \text{and} \quad K_3 = K_1{}^* = 0$$

$$f(t) = -\frac{j}{4} t\epsilon^{jt} + \frac{j}{4} t\epsilon^{-jt} = \frac{t}{2} \left(\frac{\epsilon^{jt} - \epsilon^{-jt}}{j2} \right) = \frac{1}{2} t \sin t \quad \text{for} \quad t \geq 0$$

A partial-fraction expansion can be used only when $F(s)$ is the quotient of two polynomials. However, this method, together with Eq. 10.1-6, is sufficient for most common functions of s, including all those encountered in the next section. An even more general method of finding the inverse transform is discussed in Section 10.5.

10.3

The Complete Solution of Circuits

There are two basic methods of solving fixed, linear circuits by the Laplace transform. In the first of these, the differential equations describing the circuit are transformed term by term, while the second approach is based upon the network function $H(s)$. Since Section 4.6 showed how $H(s)$ may be obtained from the differential equations, the two methods are closely related. The first method may be summarized as follows:

1. Write and immediately transform the integral-differential equations describing the circuit for $t > 0$. Evaluate all the initial condition terms.
2. Solve these equations for the transform of the output.
3. Evaluate the inverse transform to obtain the output as a function of time.

This procedure is applicable to any electrical or nonelectrical system that can be described by a set of integral-differential equations with constant coefficients. It is illustrated by Example 10.1-7 and by the following examples.

> **Example 10.3-1.** The circuit in Fig. 10.3-1 is operating in the steady state with the switch K closed. If the switch suddenly opens at $t = 0$, find an expression for the voltage across the capacitance for all $t > 0$. The classical solution of this problem is given in Example 4.5-2.
>
> **Solution.**   For $t > 0$, with the switch open, the circuit is described by the following two node equations:
>
> $$\frac{1}{2}(e_C - 16) + \frac{1}{8}\frac{de_C}{dt} + \frac{1}{2}(e_C - e_B) = 0$$
>
> $$\frac{1}{2}(e_B - e_C) + 2\int e_B\,dt = 0$$
>
> Let $E_B(s)$ and $E_C(s)$ denote the Laplace transforms of $e_B(t)$ and $e_C(t)$, and transform these equations term by term. The derivative and integral are handled by the properties in Table 10.1-2.
>
> $$\frac{1}{2}E_C(s) - \frac{8}{s} + \frac{1}{8}[sE_C(s) - e_C(0+)] + \frac{1}{2}[E_C(s) - E_B(s)] = 0$$
>
> $$\frac{1}{2}[E_B(s) - E_C(s)] + \left[\frac{2E_B(s)}{s} + \frac{i_L(0+)}{s}\right] = 0$$

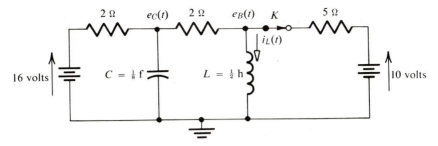

Figure 10.3-1

Since $i_L(t)$ and $e_C(t)$ are the current through the inductance and the voltage across the capacitance, respectively, their values do not change instantaneously, so $i_L(0+) = i_L(0-)$ and $e_C(0+) = e_C(0-)$. From d-c steady-state theory, with K closed and with the inductance and capacitance acting like short and open circuits, respectively, $i_L(0+) = i_L(0-) = 6$ amp and $e_C(0+) = e_C(0-) = 8$ volts. Inserting these values into the transformed equations and collecting like terms, we obtain

$$\left(\frac{1}{8}s + 1\right)E_C(s) - \frac{1}{2}E_B(s) = \frac{8}{s} + 1$$

$$-\frac{1}{2}E_C(s) + \left(\frac{1}{2} + \frac{2}{s}\right)E_B(s) = -\frac{6}{s}$$

Next we solve these equations simultaneously for the transform of the desired voltage.

$$E_C(s) = \frac{8(s^2 + 6s + 32)}{s(s^2 + 8s + 32)}$$

The inverse transform of this expression, which may be found by the methods of Section 10.2, is

$$e_C(t) = \mathcal{L}^{-1}[E_C(s)] = 8 - 4\epsilon^{-4t}\sin 4t \qquad \text{for } t > 0$$

which agrees with the answer to Example 4.5-2.

Example 10.3-1 shows how the Laplace transform converts a set of *integral-differential* equations in the independent variable t into a set of *algebraic* equations in the independent variable s. Then the algebraic equations may be solved by any convenient method for the transformed output. The inverse transform includes both the steady-state and transient components, and unlike the classical approach of Chapter 4 contains no arbitrary constants that must still be evaluated.

The values of the arbitrary constants in a classical solution depend upon the circuit, the input, and any initial stored energy. In Example 10.3-1, the initial energy is completely specified by $i_L(0+)$ and $e_C(0+)$, which are terms in the transformed equations. Thus, all of the required information is automatically introduced into the solution at an early stage. If the integral-differential equations describing a circuit are transformed immediately, without any preliminary steps of differentiation or simplification, the only initial conditions that are ever needed are the currents through the inductances and the voltages across the capacitances. Since these quantities cannot normally change instantaneously, their values at $t = 0+$ are easily found. This is one of the advantages of the transform method compared to a classical solution. In a classical solution, the required initial conditions frequently involve quantities that change instantaneously at $t = 0$, and their evaluation may require considerable thought.

Suppose that the second node equation in Example 10.3-1 were differentiated term by term to remove the integral sign. Then

$$\frac{1}{2}\left(\frac{de_B}{dt} - \frac{de_C}{dt}\right) + 2e_B = 0$$

and

$$\tfrac{1}{2}[sE_B(s) - sE_C(s)] - \tfrac{1}{2}[e_B(0+) - e_C(0+)] + 2E_B(s) = 0$$

At $t = 0+$, the current through the middle resistance is 6 amp, so $e_B(0+) - e_C(0+) = -12$ volts and

$$-\tfrac{1}{2}sE_C(s) + (\tfrac{1}{2}s + 2)E_B(s) = -6$$

which is equivalent to the second transformed equation in Example 10.3-1. The preliminary step of differentiation, however, made it necessary to find the initial value of a current which changed instantaneously.

Suppose that the original node equations in Example 10.3-1 were solved simultaneously to give

$$\frac{d^2e_C}{dt^2} + 8\frac{de_C}{dt} + 32e_C = 256$$

The transformed equation is

$$s^2E_C(s) - se_C(0+) - \frac{de_C}{dt}(0+) + 8sE_C(s) - 8e_C(0+) + 32E_C(s) = \frac{256}{s}$$

As discussed in Example 4.5-2,

$$e_C(0+) = 8 \quad \text{and} \quad \frac{de_C}{dt}(0+) = -16$$

so

$$E_C(s) = \frac{8(s^2 + 6s + 32)}{s(s^2 + 8s + 32)}$$

In some problems, failure to immediately transform the original integral-differential equations may require initial conditions that are not found so easily.

Example 10.3-2.   If the circuit in Fig. 10.3-2 contains no initial stored energy, find $e_o(t)$ for $t > 0$.

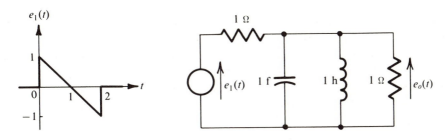

Figure 10.3-2

Solution.   Transforming the node equation

$$e_o - e_1 + \frac{de_o}{dt} + \int e_o \, dt + e_o = 0$$

we find that

$$E_o(s) - E_1(s) + sE_o(s) - e_o(0+) + \frac{E_o(s)}{s} + \frac{i_L(0+)}{s} + E_o(s) = 0$$

where

$$E_1(s) = \mathcal{L}[e_1(t)] = \frac{1}{s} - \frac{1}{s^2} + \frac{\epsilon^{-2s}}{s} + \frac{\epsilon^{-2s}}{s^2}$$

Since there is no initial energy stored in the inductance or capacitance, $i_L(0+) = i_L(0-) = 0$ and $e_o(0+) = e_o(0-) = 0$. Thus,

$$E_o(s) = \frac{sE_1(s)}{s^2 + 2s + 1} = \frac{1 - 1/s + \epsilon^{-2s} + \epsilon^{-2s}/s}{(s + 1)^2}$$

Now,

$$\mathcal{L}^{-1}\left[\frac{1}{(s + 1)^2}\right] = t\epsilon^{-t}$$

$$\mathcal{L}^{-1}\left[\frac{1}{s(s + 1)^2}\right] = \mathcal{L}^{-1}\left[\frac{1}{s} - \frac{1}{(s + 1)^2} - \frac{1}{s + 1}\right] = 1 - (t + 1)\epsilon^{-t}$$

The inverse transforms of $\epsilon^{-2s}/(s + 1)^2$ and $\epsilon^{-2s}/s(s + 1)^2$ cannot be found directly from a partial-fraction expansion, because the functions are not the quotients of two polynomials. However, Eq. 10.1-6 may be used to find these inverse transforms.

$$e_o(t) = \mathcal{L}^{-1}[E_o(s)] = t\epsilon^{-t} - 1 + (t + 1)\epsilon^{-t}$$

$$+ [(t - 2)\epsilon^{-(t-2)} + 1 - (t - 1)\epsilon^{-(t-2)}]U_{-1}(t - 2)$$

$$= -1 + (2t + 1)\epsilon^{-t} + [1 - \epsilon^{-(t-2)}]U_{-1}(t - 2)$$

Since $U_{-1}(t - 2)$ is zero until $t = 2$,

$$e_o(t) = -1 + (2t + 1)\epsilon^{-t} \qquad \text{for } 0 < t < 2$$

$$= (2t + 1 - \epsilon^2)\epsilon^{-t} \qquad \text{for } t > 2$$

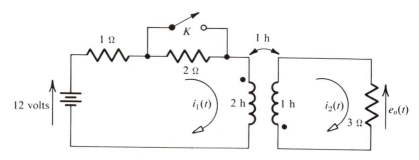

Figure 10.3-3

The classical solution of Example 10.3-2, which was required in Problem 4.12, is similar to but more complicated than the solution of Example 4.3-5. In order to evaluate the arbitrary constants of integration, the values of $e_o(t)$ and de_o/dt at $t = 0+$ and at $t = 2+$ would be needed. In contrast, the Laplace transform approach yields the answer for all $t > 0$ when $e_o(0+)$ and $i_L(0+)$ are known.

Example 10.3-3.   With the circuit in Fig. 10.3-3 operating in the steady state, the switch K suddenly closes at $t = 0$. Find $e_o(t)$ for $t > 0$.

Solution.   The loop equations for $t > 0$ are

$$\left(2\frac{di_1}{dt} + \frac{di_2}{dt}\right) + i_1 = 12$$

$$3i_2 + \left(\frac{di_2}{dt} + \frac{di_1}{dt}\right) = 0$$

The corresponding transformed equations are

$$2sI_1(s) - 2i_1(0+) + sI_2(s) - i_2(0+) + I_1(s) = \frac{12}{s}$$

$$3I_2(s) + sI_2(s) - i_2(0+) + sI_1(s) - i_1(0+) = 0$$

Since $i_1(0+) = i_1(0-) = 4$ amp and $i_2(0+) = i_2(0-) = 0$,

$$(2s + 1)I_1(s) + sI_2(s) = 8 + \frac{12}{s}$$

$$sI_1(s) + (s + 3)I_2(s) = 4$$

If these equations are solved simultaneously for $I_2(s)$,

$$E_o(s) = 3I_2(s) = \frac{-24}{s^2 + 7s + 3} = \frac{-24}{(s + 0.46)(s + 6.54)}$$

$$e_o(t) = \mathcal{L}^{-1}[E_o(s)] = -3.95(\epsilon^{-0.46t} - \epsilon^{-6.54t})$$

for $t > 0$, where the details of the inverse transform are left to the student. This expression agrees with the answer to Problem 7.14b.

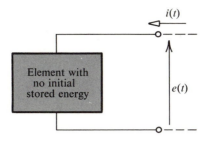

Figure 10.3-4

Impedance and Admittance for Circuits with No Initial Stored Energy. The second method of circuit analysis is associated with the concepts of impedance, admittance, and the network function. For the typical element shown in Fig. 10.3-4, let $E(s) = \mathscr{L}[e(t)]$ and $I(s) = \mathscr{L}[i(t)]$. If the element is a resistance, $e(t) = Ri(t)$, $E(s) = RI(s)$, and

$$Z_R(s) = \frac{E(s)}{I(s)} = R \tag{10.3-1}$$

For a capacitance, $i = C(de/dt)$, so $I(s) = sCE(s) - Ce(0+)$. If there is no initial stored energy, $e(0+) = 0$ and

$$Z_C(s) = \frac{E(s)}{I(s)} = \frac{1}{sC} \tag{10.3-2}$$

Finally, for an inductance, $e = L(di/dt)$ and $E(s) = sLI(s) - Li(0+)$. If there is no initial stored energy, $i(0+) = 0$ and

$$Z_L(s) = \frac{E(s)}{I(s)} = sL \tag{10.3-3}$$

In each case the impedance $Z(s)$, which is now defined as the ratio of the transformed voltage to the transformed current when there is no initial stored energy, depends only upon the complex frequency s and the value of the element. The admittance is again defined as $Y(s) = 1/Z(s)$.

Equations 10.3-1 through 10.3-3 are identical with the entries in Table 6.2-1, where $Z(s)$ was defined as the ratio of the voltage and current phasors. Thus, the methods used in Chapter 6 to find the relationship between the input and output phasors may all be applied to find the relationship between the transformed input and the transformed output. Passive elements are characterized by their impedances, and voltages and currents by their Laplace transforms. For an input $x(t)$ and output $y(t)$, and for a circuit whose network function is $H(s)$,

$$Y(s) = X(s)H(s) \tag{10.3-4}$$

This discussion leads to the following alternative method for solving circuits that contain no initial stored energy.

1. Find $X(s) = \mathcal{L}[x(t)]$.
2. Find $Y(s)$ from the complex-frequency-domain circuit, where the passive elements are characterized by their impedances and the voltages and currents by their transforms, or from Eq. 10.3-4.
3. Find $y(t) = \mathcal{L}^{-1}[Y(s)]$.

Example 10.3-4.   Find the unit step response of the circuit in Fig. 10.3-5a when $R_1 = R_2 = 1\,\Omega$, $L = 1$ h, and $C = 1$ f.

Solution.   The Laplace transform of $e_1(t) = U_{-1}(t)$ is $E_1(s) = 1/s$. The circuit is relabelled in part (b) of the figure with impedances and transformed quantities. The node equation at node 2 is

$$\frac{E_2(s) - E_1(s)}{1/sC} + \frac{E_2(s)}{R_1} + \frac{E_2(s)}{sL + R_2} = 0$$

which gives

$$E_2(s) = \frac{sC}{sC + 1/R_1 + 1/(sL + R_2)} E_1(s)$$

By the voltage divider rule,

$$E_o(s) = \frac{R_2}{sL + R_2} E_2(s) = \left[\frac{sCR_2}{(sL + R_2)(sC + 1/R_1) + 1} \right] E_1(s)$$

The quantity in the brackets is the transfer function $H(s)$. When the known values of R_1, R_2, L, C, and $E_1(s)$ are inserted,

$$E_o(s) = \frac{1}{(s + 1)^2 + 1}$$

Although the inverse Laplace transform could be found by the partial-fraction method, an easier procedure is to recall that $\mathcal{L}^{-1}[1/(s^2 + 1)] = \sin t$. By Eq. 10.1-5,

$$e_o(t) = \mathcal{L}^{-1}[E_o(s)] = \epsilon^{-t} \sin t \quad \text{for} \quad t > 0$$

which agrees with Example 4.3-4.

In practice, when there is no initial stored energy, the student is able to find $H(s)$ without explicitly redrawing and relabelling the circuit. If the circuit contains a transformer, it may be represented by the self-impedances sL_1 and sL_2 and the mutual impedance sM, as in Fig. 7.2-2b.

Circuits with Initial Stored Energy.   Although circuits with initial stored energy always may be solved by writing and transforming the integral-differential equations, we wish to see whether or not a complex-frequency-domain circuit still can be used. This might seem impossible, since only those passive elements which do not contain initial energy may be completely characterized by an impedance $Z(s)$. When calculating the output for $t > 0$, however, the initial energy may be represented by an added independent source, as in Figs. 3.3-7 and 3.3-8.

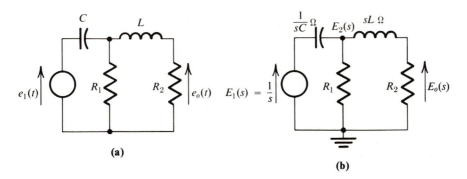

Figure 10.3-5

The equivalent circuits that were developed for an initially charged capacitance are repeated in Figs. 10.3-6b and 10.3-6c. Since the uncharged capacitance may be represented by the impedance $1/sC$, and since $\mathscr{L}[e(0+)U_{-1}(t)] = e(0+)/s$ and $\mathscr{L}[Ce(0+)U_0(t)] = Ce(0+)$, Figs. 10.3-6d and 10.3-6e adequately represent the charged capacitance in a complex-frequency-domain circuit. These figures also may be justified by writing and transforming the equations for an initially charged capacitance.

$$e(t) = e(0+) + \frac{1}{C} \int_{0+}^{t} i(\lambda)\, d\lambda$$

$$i(t) = C \frac{de}{dt}$$

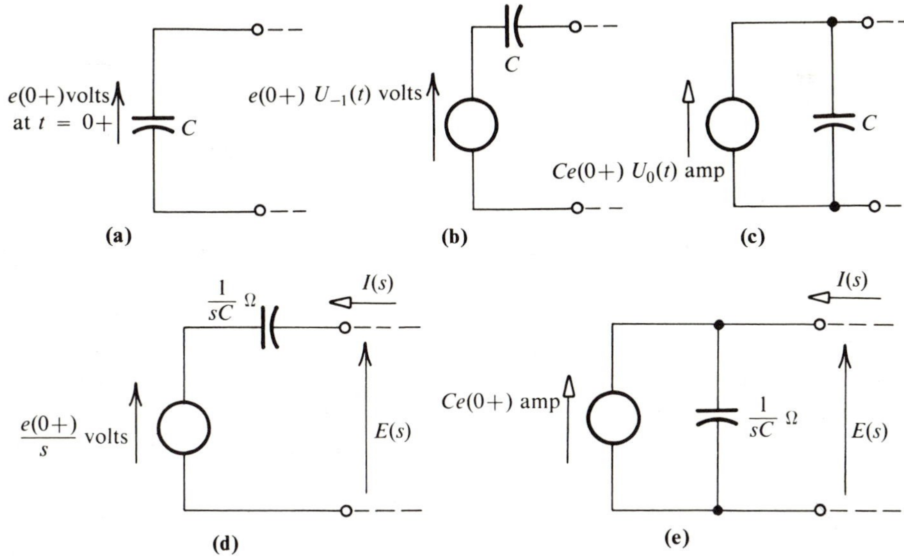

Figure 10.3-6

The transformed equations

$$E(s) = \frac{e(0+)}{s} + \frac{1}{sC} I(s)$$

$$I(s) = sCE(s) - Ce(0+)$$

describe Figs. 10.3-6d and 10.3-6e, respectively.

The equivalent circuits for an inductance with an initial current $i(0+)$ are repeated in Figs. 10.3-7b and 10.3-7c. The corresponding complex-frequency-domain circuits, labelled with impedances and transformed quantities, are shown in Figs. 10.3-7d and 10.3-7e. An alternative way of obtaining the last two circuits is to transform the equations

$$i(t) = i(0+) + \frac{1}{L} \int_{0+}^{t} e(\lambda) \, d\lambda$$

$$e(t) = L \frac{di}{dt}$$

to obtain

$$I(s) = \frac{i(0+)}{s} + \frac{1}{sL} E(s)$$

$$E(s) = sLI(s) - Li(0+)$$

The following examples show how the transformed output may be found from the complex-frequency-domain circuit, when the initial stored energy is represented

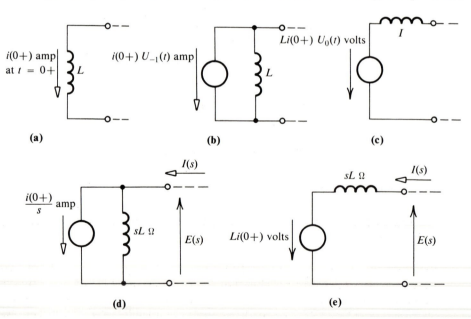

Figure 10.3-7

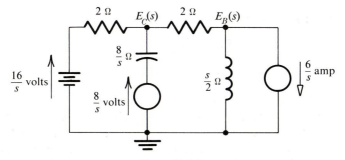

Figure 10.3-8

by added sources. It should be emphasized that Eq. 10.3-4 assumes that $x(t)$ is the only independent source and is not applicable to circuits with initial stored energy.

Example 10.3-5. Solve Example 10.3-1 by representing the initial stored energy by added sources.

Solution. The complex-frequency-domain circuit for Fig. 10.3-1 is shown in Fig. 10.3-8. The initial current through L is represented by a current step of value $i_L(0+) = 6$ amp, and the initial voltage across C is represented by a voltage step of value $e_C(0+) = 8$ volts. The two node equations are

$$\frac{1}{2}\left[E_C(s) - \frac{16}{s}\right] + \frac{s}{8}\left[E_C(s) - \frac{8}{s}\right] + \frac{1}{2}[E_C(s) - E_B(s)] = 0$$

$$\frac{1}{2}[E_B(s) - E_C(s)] + \frac{2}{s}E_B(s) + \frac{6}{s} = 0$$

or

$$\left(\frac{s}{8} + 1\right)E_C(s) - \frac{1}{2}E_B(s) = \frac{8}{s} + 1$$

$$-\frac{1}{2}E_C(s) + \left(\frac{1}{2} + \frac{2}{s}\right)E_B(s) = -\frac{6}{s}$$

which agree with Example 10.3-1.

Example 10.3-6. Solve Example 10.3-3 by representing the initial stored energy by added sources.

Solution. The transformer in Fig. 10.3-3 may not be represented by Fig. 7.2-2b, since it does contain some initial energy. Since this energy is stored in a magnetic field that links *both* windings, it is not immediately clear how it can be represented by added sources. One approach is to replace the transformer by the T equivalent circuit of Fig. 7.4-1c, which yields Fig. 10.3-9a. The initial current of 4 amp gives rise to the

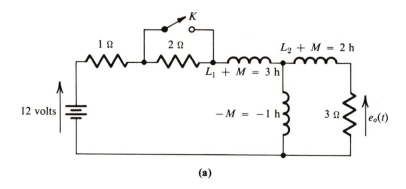

(a)

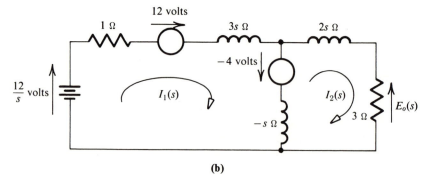

(b)

Figure 10.3-9

two added voltage sources shown in the complex-frequency-domain circuit of Fig.
10.3-9b. The loop equations are

$$(2s + 1)I_1(s) + sI_2(s) = 8 + \frac{12}{s}$$

$$sI_1(s) + (s + 3)I_2(s) = 4$$

which agree with Example 10.3-3.

Example 10.3-7.   The switch K in Fig. 10.3-10a, after being in position ① for a long
period of time, switches instantaneously to position ② at $t = 0$. Find $e_o(t)$ for $t > 0$.

Solution.   We must find the voltage across the capacitance just before $t = 0$. Since
any transient component created when the switch was originally moved to position ①
is assumed to have disappeared, and since only the a-c source contributes to $e_C(t)$ in
this position, a-c steady-state circuit theory may be used. With capital boldface letters
representing phasors (and *not* transformed quantities), the voltage divider rule gives

$$\mathbf{E}_C = \mathbf{E}_1 \frac{12/j\omega}{6 + 12/j\omega}$$

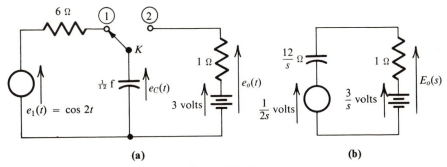

Figure 10.3-10

where $E_1 = 1 \underline{/0°}$ and $\omega = 2$ rad per sec.

$$Ec = \frac{-j6}{6 - j6} = \frac{1}{\sqrt{2}} \underline{/-45°}$$

$$e_C(t) = \frac{1}{\sqrt{2}} \cos (2t - 45°)$$

$$e_C(0-) = \frac{1}{\sqrt{2}} \cos (-45°) = \frac{1}{2} = e_C(0+)$$

For $t > 0$, with the switch in position ②, the complex-frequency-domain circuit in Fig. 10.3-10b is applicable. The capital letters now represent transformed quantities, and the initial stored energy in the capacitance has been represented by a voltage source. The node equation is

$$\frac{E_o(s) - 1/2s}{12/s} + \frac{E_o(s) - 3/s}{1} = 0$$

which gives

$$E_o(s) = \frac{s + 72}{2s(s + 12)} = \frac{3}{s} - \frac{5/2}{s + 12}$$

$$e_o(t) = 3 - \tfrac{5}{2} \epsilon^{-12t} \quad \text{for} \quad t > 0$$

10.4
Review of Complex-Variable Theory

The Laplace transform $F(s)$ is a function of the complex variable s, and the use of Eq. 10.1-2,

$$f(t) = \frac{1}{2\pi j} \int_{\sigma-j\infty}^{\sigma+j\infty} F(s)\epsilon^{st} \, ds$$

involves an integration in the complex plane. Although the circuits in this book can be solved without a detailed knowledge of complex-variable theory, the theory does help to place the Laplace transform on a firmer mathematical foundation and is essential to understanding more advanced books on transform methods.

This section presents, without detailed proofs, only the barest skeleton of that part of complex-variable theory needed for transform techniques.[†] Since the theory is to be applied to the Laplace transform, the independent variable is taken to be

$$s = \sigma + j\omega \tag{10.4-1}$$

where σ and ω are the real and imaginary parts of s, as indicated in Fig. 10.4-1. The dependent variable is denoted as

$$F(s) = R + jX \tag{10.4-2}$$

where R and X are the real and imaginary parts of $F(s)$ and are real functions of the two real variables σ and ω. Many mathematics books use $z = x + jy$ and $w = u + jv$, respectively, as the independent and dependent variables.

Analytic Functions.   The derivative of $F(s)$ is defined as

$$\frac{dF(s)}{ds} = \lim_{\Delta s \to 0} \frac{\Delta F(s)}{\Delta s} = \lim_{\Delta s \to 0} \frac{F(s + \Delta s) - F(s)}{\Delta s}$$

where Δs represents a small distance in the s-plane and may be taken in any direction. The usefulness of the derivative concept is greatly increased if the derivative

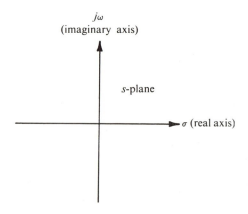

Figure 10.4-1

[†] Except for some modifications and additions, this section is the same as Section 3.6 of P. M. DeRusso, R. J. Roy, and C. M. Close, *State Variables for Engineers* (New York: Wiley, 1965) and is included here with the permission of the publisher. For a more complete treatment, see, for example, R. V. Churchill, *Complex Variables and Applications*, 2nd ed. (New York: McGraw-Hill, 1960).

turns out to be independent of the direction of Δs. The above expression is independent of the direction of Δs at a point s_0 if and only if

$$\frac{\partial R}{\partial \sigma} = \frac{\partial X}{\partial \omega} \quad \text{and} \quad \frac{\partial R}{\partial \omega} = -\frac{\partial X}{\partial \sigma} \tag{10.4-3}$$

and if $F(s)$ and these partial derivatives are continuous in the neighborhood of s_0. Equations 10.4-3 are known as the Cauchy-Riemann conditions for a unique derivative.

A function $F(s)$ is said to be *analytic* at a point s_0 in the s-plane if and only if it is single valued and has a finite and unique derivative at and in the neighborhood of s_0.[†] The vast majority of commonly encountered functions are analytic except at a finite number of points. The quotient of two polynomials in s, for example, is analytic for all finite values of s except where the denominator becomes zero. All of the usual rules of differentiation, such as those given below, hold for analytic functions.

$$\frac{d}{ds}(ks^n) = kns^{n-1}$$

$$\frac{d}{ds}[F_1(s)F_2(s)] = F_1(s)\frac{dF_2(s)}{ds} + F_2(s)\frac{dF_1(s)}{ds}$$

Integration in the Complex Plane. Integration along a contour C in the s-plane can be interpreted in terms of real integrals by writing

$$\int_C F(s)\,ds = \int_C (R + jX)(d\sigma + j\,d\omega)$$

$$= \int_C (R\,d\sigma - X\,d\omega) + j\int_C (X\,d\sigma + R\,d\omega)$$

The values of σ and ω are related by the contour along which the integration is carried out.

Example 10.4-1. Find $\int (1/s)\,ds$ from A to B along contours C_1, C_2, and C_3 in Fig. 10.4-2.

Solution. Consider first the contour C_1. Along AD, $s = \sigma - j1$, $ds = d\sigma$, and $1/s = (\sigma + j)/(\sigma^2 + 1)$, so

$$\int_A^D \frac{ds}{s} = \int_{-1}^1 \left(\frac{\sigma}{\sigma^2 + 1}\right) d\sigma + j\int_{-1}^1 \left(\frac{1}{\sigma^2 + 1}\right) d\sigma = \frac{j\pi}{2}$$

Along path DB, $s = 1 + j\omega$, $ds = j\,d\omega$, and $1/s = (1 - j\omega)/(1 + \omega^2)$, so

$$\int_D^B \frac{ds}{s} = j\int_{-1}^1 \left(\frac{1}{1 + \omega^2}\right) d\omega + \int_{-1}^1 \left(\frac{\omega}{1 + \omega^2}\right) d\omega = \frac{j\pi}{2}$$

[†] The phrase "analytic at and in the neighborhood of s_0" is more correct, but is also more cumbersome.

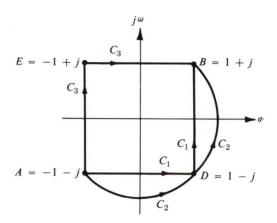

Figure 10.4-2

Thus,

$$\int_{C_1} \frac{1}{s}\, ds = \frac{j\pi}{2} + \frac{j\pi}{2} = j\pi$$

Next, along the semicircular contour C_2, $s = \sqrt{2}\, \epsilon^{j\theta}$ where $-3\pi/4 < \theta < \pi/4$, and $ds = j\sqrt{2}\, \epsilon^{j\theta}\, d\theta$.

$$\int_{C_2} \frac{1}{s}\, ds = j \int_{-3\pi/4}^{\pi/4} d\theta = j\pi$$

For contour C_3, note that along AE,

$$s = -1 + j\omega, \qquad ds = j\, d\omega, \qquad \frac{1}{s} = \frac{-1 - j\omega}{1 + \omega^2}$$

and

$$\int_A^E \frac{ds}{s} = -j \int_{-1}^{1} \left(\frac{1}{1 + \omega^2} \right) d\omega + \int_{-1}^{1} \left(\frac{\omega}{1 + \omega^2} \right) d\omega = -\frac{j\pi}{2}$$

while along EB, $s = \sigma + j1$, $ds = d\sigma$, $1/s = (\sigma - j)/(\sigma^2 + 1)$, and

$$\int_E^B \frac{ds}{s} = \int_{-1}^{1} \left(\frac{\sigma}{\sigma^2 + 1} \right) d\sigma - j \int_{-1}^{1} \left(\frac{1}{\sigma^2 + 1} \right) d\sigma = -\frac{j\pi}{2}$$

so

$$\int_{C_3} \frac{1}{s}\, ds = -j\pi$$

In this example, notice that $\int_{C_1} F(s)\, ds = \int_{C_2} F(s)\, ds \neq \int_{C_3} F(s)\, ds$, so a contour integral between two fixed points is not, in general, independent of the path taken.

If the path of integration is a closed contour, as in Fig. 10.4-3a, this fact is indicated by a circle superimposed upon the integral sign. An arrow is sometimes shown on the circle to indicate that the integration is to be carried out in a specific

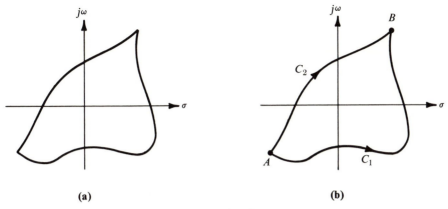

(a)   (b)

Figure 10.4-3

direction, either clockwise or counterclockwise. Since reversing the direction of integration along a path reverses the sign of the integral,

$$\oint F(s)\, ds = -\oint F(s)\, ds$$

In Fig. 10.4-3b,

$$\oint F(s)\, ds = \int_{C_1} F(s)\, ds - \int_{C_2} F(s)\, ds$$

Since the integrals along contours C_1 and C_2 are not necessarily equal, the integral around a closed path is not necessarily zero.

According to an important theorem attributed to Cauchy, if $F(s)$ is analytic on and inside a closed contour C in the s-plane,

$$\oint F(s)\, ds = 0 \qquad\qquad (10.4\text{-}4)$$

A corollary to this theorem is that

$$\int_{C_1} F(s)\, ds = \int_{C_2} F(s)\, ds \qquad\qquad (10.4\text{-}5)$$

in Fig. 10.4-3b if $F(s)$ is analytic on and between the two paths. The function $F(s) = 1/s$ in Example 10.4-1 is analytic except at the origin. The theorem therefore indicates that the integrals along paths C_1 and C_2 in Fig. 10.4-2 are equal but are not necessarily equal to the integral along C_3. These observations agree with the results obtained in the example.

Another corollary applies to a multiply connected region, which is a region that is bounded by more than one closed curve. In Fig. 10.4-4a, $F(s)$ is assumed to be analytic on the closed contours C, C_1, and C_2, and in the shaded region R. It is not, however, necessarily analytic inside C_1 and C_2. In Fig. 10.4-4b, cuts labelled C_3 through C_6 are constructed so as to give a simply connected region. The distance

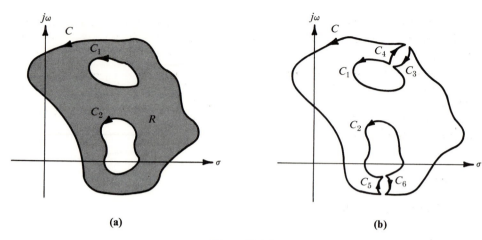

Figure 10.4-4

between corresponding ends of C_3 and C_4, and also C_5 and C_6, is assumed to be infinitesimal. By Cauchy's theorem,

$$\int_C F(s)\, ds + \int_{C_3} F(s)\, ds - \int_{C_1} F(s)\, ds$$

$$+ \int_{C_4} F(s)\, ds + \int_{C_5} F(s)\, ds - \int_{C_2} F(s)\, ds + \int_{C_6} F(s)\, ds = 0$$

since $F(s)$ is analytic on and inside the closed contour formed by the sum of these individual contours. By Eq. 10.4-5, the integration along C_3 and C_4, and also along C_5 and C_6, cancel, so

$$\oint_C F(s)\, ds = \oint_{C_1} F(s)\, ds + \oint_{C_2} F(s)\, ds \qquad (10.4\text{-}6)$$

Taylor's Series.   If $F(s)$ is analytic at a point s_0, then it can be represented by the following infinite power series in the vicinity of s_0.

$$F(s) = b_0 + b_1(s - s_0) + b_2(s - s_0)^2 + \cdots + b_n(s - s_0)^n + \cdots \qquad (10.4\text{-}7)$$

where

$$b_0 = F(s_0)$$

$$b_n = \frac{1}{n!}\left[\frac{d^n F(s)}{ds^n}\right]_{s=s_0} \qquad (10.4\text{-}8)$$

The form of Eq. 10.4-7 is seen to be the same as the Taylor series for a function of a real variable. If s_0 happens to be zero, the series is called *Maclaurin's series*.

Taylor's series converges to $F(s)$ for all points in the s-plane inside a *circle of convergence*. This is the largest circle that can be drawn about s_0 without enclosing any points at which $F(s)$ is not analytic. For all points outside the circle of convergence, the series diverges.

The coefficients in Taylor's series always can be evaluated by Eqs. 10.4-8, since all the derivatives of an analytic function exist. If $F(s)$ and the first $k - 1$ derivatives are zero at $s = s_0$, the first k terms in Eq. 10.4-7 are missing, and $F(s)$ is said to have a *zero* of order k at s_0.

A uniqueness theorem states that if two power series represent $F(s)$ in the neighborhood of s_0, then they must be identical. The coefficients may therefore be found in any convenient way and not necessarily from Eq. 10.4-8. In the following examples, long division is used.

Example 10.4-2.   Expand $F(s) = 1/(s + 1)$ in a power series about $s = 1$, i.e., in powers of $s - 1$.

Solution.   By long division,

$$\frac{1}{s + 1} = \frac{1}{2 + (s - 1)} = \frac{1}{2} - \frac{1}{4}(s - 1) + \frac{1}{8}(s - 1)^2 - \frac{1}{16}(s - 1)^3 + \cdots$$

The same series results if Eqs. 10.4-7 and 10.4-8 are used with $s_0 = 1$. The function $F(s)$ is analytic except at the point $s = -1$, shown by the cross in Fig. 10.4-5. The circle of convergence indicates that the series converges for $|s - 1| < 2$.

Example 10.4-3.   Expand $F(s) = s/(s - 1)(s - 3)$ about $s = 1$.

Solution.   Although $F(s)$ is not analytic at $s = 1$, the function $s/(s - 3)$ is analytic and may be expanded in a Taylor series. By long division,

$$\frac{s}{s - 3} = \frac{1 + (s - 1)}{-2 + (s - 1)} = -\frac{1}{2} - \frac{3}{4}(s - 1) - \frac{3}{8}(s - 1)^2 - \cdots$$

which converges for $|s - 1| < 2$. When each term of the series is divided by $s - 1$,

$$\frac{s}{(s - 1)(s - 3)} = \frac{-1/2}{s - 1} - \frac{3}{4} - \frac{3}{8}(s - 1) - \cdots$$

which converges for $0 < |s - 1| < 2$.

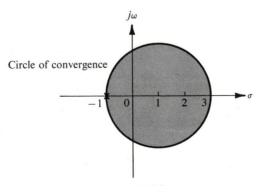

Figure 10.4-5

Example 10.4-4.   Expand $\epsilon^{st}/(s + 1)^2$, where t is to be treated as an independent parameter, about the point $s = -1$.

Solution.   The factor ϵ^{st} is analytic everywhere in the finite s-plane. By Eqs. 10.4-7 and 10.4-8 with $s_0 = -1$,

$$\epsilon^{st} = \epsilon^{-t} + t\epsilon^{-t}(s + 1) + \frac{1}{2}\,t^2\epsilon^{-t}(s + 1)^2 + \frac{1}{6}\,t^3\epsilon^{-t}(s + 1)^3 + \cdots$$

which converges for all finite values of s. Then

$$\frac{\epsilon^{st}}{(s + 1)^2} = \frac{\epsilon^{-t}}{(s + 1)^2} + \frac{t\epsilon^{-t}}{s + 1} + \frac{1}{2}\,t^2\epsilon^{-t} + \frac{1}{6}\,t^3\epsilon^{-t}\,(s + 1) + \cdots$$

for $s \neq -1$.

Principle of Analytic Continuation.   The uniqueness theorem associated with the Taylor series implies that if a series representation of $F(s)$ is valid in any small region, then it is the unique representation wherever it converges. Suppose that a function $F(s)$ is known in some small region R of the s-plane. A Taylor series written about a point s_0 in this region then represents $F(s)$ at all points inside the circle of convergence drawn about s_0. The region inside this circle is denoted by R_0 in Fig. 10.4-6.

The function is now known throughout the union of R and R_0. If a Taylor series is written about a new point s_1 in R_0, the series representation of $F(s)$ in R_1 is found. By continuing this process, which is called *analytic continuation*, the entire finite s-plane can be covered. In summary, defining $F(s)$ within an arbitrarily small region of the s-plane is sufficient to determine it uniquely throughout the entire finite s-plane, except at points where it is not analytic.

Laurent's Series.   Even if $F(s)$ is not analytic at the point s_0, it can be represented

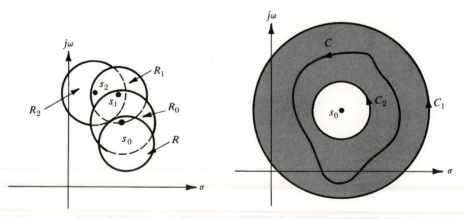

Figure 10.4-6   Figure 10.4-7

by an infinite series in powers of $s - s_0$. In this case, however, the series contains both positive and negative powers of $s - s_0$.

$$F(s) = b_0 + b_1(s - s_0) + b_2(s - s_0)^2 + \cdots$$
$$+ b_{-1}(s - s_0)^{-1} + b_{-2}(s - s_0)^{-2} + \cdots \qquad (10.4\text{-}9)$$

where

$$b_k = \frac{1}{2\pi j} \oint_C \frac{F(s)\, ds}{(s - s_0)^{k+1}} \qquad \text{for all } k \qquad (10.4\text{-}10)$$

The first part of the series is known as the ascending part; the part with the negative exponents is the *principal* or *descending* part. The series converges between two circles of convergence, both centered at s_0. $F(s)$ must be analytic between these circles, which are labelled C_1 and C_2 in Fig. 10.4-7. If there is an "isolated singularity" of $F(s)$ at s_0, the circle C_2 may shrink to infinitesimal size, and the Laurent series then represents $F(s)$ in the vicinity of s_0. As discussed later in this section, the coefficient b_{-1}, called the *residue* of $F(s)$ at s_0, is particularly important.

In the use of Eq. 10.4-10, the contour C may be any closed contour between C_1 and C_2, shown in Fig. 10.4-7. Since this equation is difficult to evaluate, however, the coefficients in Laurent's series are normally found by some other means. Any convenient method may be used, because the representation of $F(s)$ by a series in powers of $s - s_0$ in any given region is unique. Example 10.4-3 develops the Laurent series for $F(s) = s/(s - 1)(s - 3)$ about the point $s = 1$. The residue of $F(s)$ as $s = 1$ is seen to be $-\frac{1}{2}$. The Laurent series for $\epsilon^{st}/(s + 1)^2$ in Example 10.4-4 indicates that the residue at $s = -1$ is $t\epsilon^{-t}$.

Classification of Singularities. Singularities of $F(s)$, which are also called *singular points*, are points in the s-plane at which $F(s)$ is not analytic. If nonoverlapping circles, no matter how small, can be drawn around each singular point, the points are called *isolated singularities*. The function $F(s) = (s + 1)/s^3(s^2 + 1)$ has isolated singularities at $s = 0$, $+j$, and $-j$. The function $F(s) = 1/\sin(\pi/s)$ has isolated singularities at $s = 1, \frac{1}{2}, \frac{1}{3}, \ldots$, but a nonisolated singularity at the origin. Fortunately, the commonly encountered functions have only isolated singularities. The student should remember that $F(s)$ can be represented by a Laurent series in the vicinity of every isolated singularity.

An isolated singularity is classified further by examining the Laurent series written about it. If the principal or descending part of the series has an infinite number of terms, the singularity is called an *essential singularity*. Otherwise, the singularity is called a *pole*. The order of the pole is equal to the largest negative exponent of $(s - s_0)$ in the principal part of the series. Thus, $s/(s - 1)(s - 3)$ has first-order poles at $s = 1$ and $s = 3$, while $\epsilon^{st}/(s + 1)^2$ has a second-order pole at $s = -1$. If $F(s)$ has a pole of order n at s_0, then there are no terms in the principal part of the power series for $(s - s_0)^n F(s)$, so that $(s - s_0)^n F(s)$ is analytic at s_0.

The Residue Theorem. When $k = -1$ in Eq. 10.4-10,

$$\oint_C F(s)\, ds = 2\pi j b_{-1} \qquad (10.4\text{-}11)$$

where b_{-1} is the residue of $F(s)$ at s_0. The contour of integration, which is shown in Fig. 10.4-7, is understood to enclose no singularities other than an isolated singularity at s_0. If s_0 is not a singular point, then b_{-1} is zero, and this equation reduces to Eq. 10.4-4.

In Example 10.4-3, if C is a circle of radius 2 about the origin,

$$\oint_C F(s)\, ds = 2\pi j(-\tfrac{1}{2}) = -\pi j$$

In Example 10.4-1, the residue of $F(s) = 1/s$ at the origin is unity, so

$$\int_{C_1} \frac{1}{s}\, ds - \int_{C_3} \frac{1}{s}\, ds = \oint \frac{1}{s}\, ds = 2\pi j$$

which agrees with the results previously obtained.

Suppose the contour of integration encloses several isolated singularities. Since nonoverlapping circles can be drawn about each of them, the integral along contour C can be replaced by the sum of the integrals around the individual circles, as in Eq. 10.4-6 and Fig. 10.4-4. Applying Eq. 10.4-11 to each individual circle, we obtain

$$\oint_C F(s)\, ds = 2\pi j \sum \text{residues} \qquad (10.4\text{-}12)$$

where the summation includes the residues at all the singularities inside C.

The Evaluation of Residues. Although the residues always can be found by writing a Laurent series about each singularity and selecting the coefficient of the $(s - s_0)^{-1}$ term, several special formulas prove to be useful. If $F(s)$ has a first-order pole at $s = s_0$, the function $g(s) = (s - s_0)F(s)$ is analytic at s_0 and may be expanded in a Taylor series about s_0.

$$g(s) = g(s_0) + g'(s_0)(s - s_0) + \tfrac{1}{2}g''(s_0)(s - s_0)^2 + \cdots$$

where the primes denote differentiation with respect to s. Then

$$F(s) = \frac{g(s_0)}{s - s_0} + g'(s_0) + \frac{1}{2}g''(s_0)(s - s_0) + \cdots$$

so the residue of $F(s)$ at $s = s_0$ is $g(s_0)$, i.e.,

$$b_{-1} = [(s - s_0)F(s)]_{s=s_0} \qquad (10.4\text{-}13)$$

For the more general case where $F(s)$ has a pole of order n at $s = s_0$, the function $g(s) = (s - s_0)^n F(s)$ may be expanded in a Taylor series about s_0.

$$g(s) = g(s_0) + \cdots + \left[\frac{g^{(n-1)}(s_0)}{(n-1)!} \right](s - s_0)^{n-1} + \left[\frac{g^{(n)}(s_0)}{n!} \right](s - s_0)^n + \cdots$$

where the superscripts in parentheses denote differentiation with respect to s. Then the Laurent series for $F(s)$ is

$$F(s) = \frac{g(s)}{(s - s_0)^n} = \frac{g(s_0)}{(s - s_0)^n} + \cdots + \left[\frac{g^{(n-1)}(s_0)}{(n-1)!} \right] \frac{1}{s - s_0} + \cdots$$

and the residue as $s = s_0$ is

$$b_{-1} = \frac{1}{(n-1)!} \left[\frac{d^{n-1}}{ds^{n-1}} (s - s_0)^n F(s) \right]_{s=s_0} \tag{10.4-14}$$

As an example of the use of this formula, the residue of $\epsilon^{st}/(s + 1)^2$ at the second-order pole at $s = -1$ is

$$\left[\frac{d}{ds} \epsilon^{st} \right]_{s=-1} = \left[t\epsilon^{st} \right]_{s=-1} = t\epsilon^{-t}$$

which agrees with Example 10.4-4.

The following approach is valid only when $F(s)$ can be expressed as

$$F(s) = \frac{P(s)}{Q(s)} \tag{10.4-15}$$

where both $P(s)$ and $Q(s)$ are analytic at s_0, and where $P(s_0) \neq 0$. If $F(s)$ has a first-order pole at s_0, then $Q(s)$ must have a first-order zero, so $Q(s_0) = 0$ but $Q'(s_0) \neq 0$. We write a Taylor series for both $P(s)$ and $Q(s)$ and carry out the indicated long division.

$$F(s) = \frac{P(s_0) + P'(s_0)(s - s_0) + \cdots}{Q'(s_0)(s - s_0) + \frac{1}{2}Q''(s_0)(s - s_0)^2 + \cdots} = \frac{P(s_0)}{Q'(s_0)} (s - s_0)^{-1} + \cdots$$

The residue at s_0 is seen to be

$$b_{-1} = \frac{P(s_0)}{Q'(s_0)} = \left[\frac{P(s)}{Q'(s)} \right]_{s=s_0} \tag{10.4-16}$$

In Example 10.4-3, $F(s) = s/(s - 1)(s - 3) = s/(s^2 - 4s + 3)$, so the residue at $s = 1$ is

$$\left[\frac{s}{\dfrac{d}{ds}(s^2 - 4s + 3)} \right]_{s=1} = \left[\frac{s}{2s - 4} \right]_{s=1} = -\frac{1}{2}$$

This approach is extended to second-order poles in one of the problems at the end of this chapter.

Example 10.4-5. Find the singularities of $1/\sinh s$, and determine the residues of $\epsilon^{st}/\sinh s$ at these singularities.

Solution. The factor $\sinh s = \sinh (\sigma + j\omega) = \sinh \sigma \cos \omega + j \cosh \sigma \sin \omega$ is zero when both its real and imaginary parts are zero. The imaginary part is zero only when $\sin \omega = 0$, i.e., when $\omega = \pm n\pi$. Since $\cos \omega$ is not zero for these values of ω, $\sinh \sigma$ and hence σ must be zero. Thus, the only zeros of $\sinh s$ are at $s = \pm jn\pi$. These are first-order zeros, since (d/ds) sinh $s =$ cosh s is not zero at these points. The residues at the first-order poles of $\epsilon^{st}/\sinh s$ at $s = \pm jn\pi$ are

$$\left[\frac{\epsilon^{st}}{\cosh s}\right]_{s=\pm jn\pi} = \frac{\epsilon^{\pm jn\pi t}}{\cos n\pi} = (-1)^n \epsilon^{\pm jn\pi t}$$

In the next section, Eqs. 10.4-12 through 10.4-16 are used in the evaluation of the inverse Laplace transform.

10.5

The Inversion Integral

As illustrated in Section 10.1, the defining equation for the direct transform

$$F(s) = \int_0^\infty f(t)\epsilon^{-st}\,dt \qquad (10.5\text{-}1)$$

exists for all functions of practical interest when σ, the real part of s, is made sufficiently large. Even though this equation may converge only for a certain region in the s-plane, this is sufficient, according to the principle of analytic continuation, to define $F(s)$ throughout the entire s-plane, except at singular points, where the function does not exist.

The region of convergence for a few common functions of time is indicated in Table 10.1-1. Notice that this region corresponds to the part of the s-plane that lies to the right of all the poles of $F(s)$. This result happens to be always true and can be partially explained by the following plausibility argument.

If, as t approaches infinity, $f(t)$ remains finite or approaches infinity more slowly than an exponential function, then the entire integrand approaches zero if, in the factor $\epsilon^{-(\sigma+j\omega)t}$, σ is positive. Next, recall the derivation of Eq. 10.1-5, and notice that

$$\mathscr{L}[\epsilon^{at}f(t)] = \int_0^\infty f(t)\epsilon^{-(s-a)t}\,dt$$

$$= \int_0^\infty f(t)\epsilon^{-(\sigma-a)t}\epsilon^{-j\omega t}\,dt = F(s-a)$$

Again assume that $\lim_{t\to\infty} f(t)$ is a nonzero constant or that $f(t)$ approaches infinity more slowly than an exponential. The multiplication by ϵ^{at} requires $\sigma > a$ in order for the integral to converge. When s is replaced by $s - a$ in $F(s)$, the pole-zero plot is shifted a units to the right in the s-plane. In summary, moving the poles a units to the right moves the region of convergence a units to the right.

Equation 10.1-2, which is rewritten below, is known as the *inversion integral* and always can be used to find the inverse transform of $F(s)$.

$$f(t) = \frac{1}{2\pi j} \lim_{R\to\infty} \int_{\sigma-jR}^{\sigma+jR} F(s)\epsilon^{st}\,ds \qquad (10.5\text{-}2)$$

The path of integration is restricted to values of σ for which Eq. 10.5-1 converges and therefore must pass to the right of all the poles of $F(s)$. This path is usually taken as the vertical line ABD shown in Fig. 10.5-1.[†]

Equations 10.5-1 and 10.5-2 were derived heuristically at the beginning of this chapter. A more rigorous approach is to take Eq. 10.5-1 as the formal definition of the Laplace transform, and then to prove the validity of Eq. 10.5-2 by showing that the successive application of these two equations to any $f(t)$ yields the original function of time. Although the proof is not included here, it can be shown that Eq. 10.5-2, with the above restriction on the path of integration, yields the original $f(t)$ at all points of continuity for $t > 0$. The result is zero for $t < 0$ and $\frac{1}{2}f(0+)$ for $t = 0$.

Since the evaluation of an integral between two points in the complex plane can be difficult, the student may wonder why the inversion integral is ever used. If $F(s)$ is the quotient of two polynomials, the inverse transform may be found by a partial-fraction expansion, and properties such as Eq. 10.1-6 can be used for certain other cases. In fact, the circuits in this book may be completely solved by the methods already discussed.

Some problems, however, cannot be adequately handled by the previous methods, and familiarity with the inversion integral is essential for an understanding of more advanced transform techniques. Furthermore, the evaluation of Eq. 10.5-2 is not nearly as difficult as one might expect, because the path ABD in Fig. 10.5-1 usually can be changed to a closed contour without affecting the final result. If this is done, the integral can be easily evaluated by the residue theorem of Eq. 10.4-12.

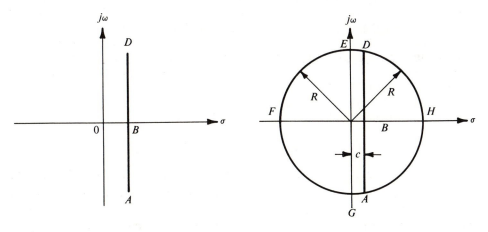

Figure 10.5-1   **Figure 10.5-2**

[†] It is sometimes convenient to regard the s-plane as mapped on the surface of an infinite sphere, with the point at infinity diametrically opposite the origin. Such a concept is not particularly convenient here, however, since ϵ^{st} has an essential singularity at infinity. We shall consider a finite s-plane, and apply the limiting process of $R \to \infty$ after the integration has been carried out.

We now try to add to the contour ABD a return path from D to A such that $\int F(s)\epsilon^{st}\, ds$ vanishes along the added path when R approaches infinity. Two paths that might be considered are $DEFGA$ and DHA in Fig. 10.5-2. In evaluating the integral along these paths, two relationships from complex-variable theory are needed.

$$|F_1(s)F_2(s)| = |F_1(s)|\, |F_2(s)|$$

$$\left|\int_C F(s)\, ds\right| \le \int_C |F(s)|\, |ds| \le |F(s)|_{\text{max on } C} \int_C |ds|$$

where $\int_C |ds|$ is simply the length of the contour C.

On the semicircle EFG, $s = R\epsilon^{j\theta} = R\cos\theta + jR\sin\theta$, where R is treated as a constant since all points are the same distance from the origin. Then $ds = jR\epsilon^{j\theta}\, d\theta$ and

$$\left|\int_{EFG} F(s)\epsilon^{st}\, ds\right| \le \int_{EFG} |F(s)|\, |\epsilon^{R(\cos\theta)t}|\, |\epsilon^{jR(\sin\theta)t}|\, |j|R|\epsilon^{j\theta}|\, |d\theta|$$

The quantities j and $\epsilon^{j\theta}$ have a magnitude of unity, so

$$\left|\int_{EFG} F(s)\epsilon^{st}\, ds\right| \le \int_{EFG} R|F(s)|\epsilon^{R(\cos\theta)t}|d\theta|$$

Because the angle θ is in the second and third quadrant for the path EFG, $\cos\theta$ is negative, and for $t > 0$ the exponent $R(\cos\theta)t$ is negative. We still need to know the nature of $F(s)$, however, before we can be sure that the last integral will vanish. Suppose that on the semicircle, $|F(s)|$ approaches $|N/s^k| = N/R^k$ when the radius becomes large, where N and k are constants, and where $k \ge 1$. Then as R approaches infinity, $R|F(s)|$ remains finite and the integral does vanish for $t > 0$.

For the semicircle DHA, $|\int F(s)\epsilon^{st}\, ds|$ is no greater than

$$\int_{DHA} R|F(s)|\epsilon^{R(\cos\theta)t}|d\theta|$$

Since $\cos\theta$ is positive in the first and fourth quadrants, this integral vanishes for $k \ge 1$ and $t < 0$, when R approaches infinity.

For the horizontal line DE, $s = \sigma + jR$ and $ds = d\sigma$.

$$\left|\int_{DE} F(s)\epsilon^{st}\, ds\right| \le \int_{DE} |F(s)|\, |\epsilon^{\sigma t}|\, |\epsilon^{jRt}|\, |d\sigma|$$

If R approaches infinity and $|F(s)|$ approaches N/R^k,

$$\left|\int_{DE} F(s)\epsilon^{st}\, ds\right| \le \int_{DE} \frac{N}{R^k}\epsilon^{\sigma t}|d\sigma| \le \left[\frac{N}{R^k}\epsilon^{ct}\right][c] = 0$$

for $k > 0$. In a similar way, as R approaches infinity, the integral along path GA is zero for $k > 0$.

In summary, for $k \ge 1$,

$$\lim_{R \to \infty} \int_{DEFGA} F(s)\epsilon^{st}\, ds = 0$$

for $t > 0$, and

$$\lim_{R \to \infty} \int_{DHA} F(s)\epsilon^{st}\, ds = 0$$

for $t < 0$. Then the path ABD may be changed to $ABDEFGA$ for $t > 0$ and to $ABDHA$ for $t < 0$. By the residue theorem of Eq. 10.4-12,

$$f(t) = \frac{1}{2\pi j} \lim_{R \to \infty} \oint_{ABDEFGA} F(s)\epsilon^{st}\, ds$$

$$= \sum [\text{residues of } F(s)\epsilon^{st} \text{ at the singularities}$$
$$\text{to the left of } ABD] \qquad \text{for } t > 0 \qquad (10.5\text{-}3)$$

$$f(t) = \frac{1}{2\pi j} \lim_{R \to \infty} \oint_{ABDHA} F(s)\epsilon^{st}\, ds$$

$$= -\sum [\text{residues of } F(s)\epsilon^{st} \text{ at the singularities}$$
$$\text{to the right of } ABD] \qquad \text{for } t < 0 \qquad (10.5\text{-}4)$$

Equations 10.5-3 and 10.5-4 always may be used if, when s is replaced by the real quantity R and R approaches infinity, $F(R)$ approaches zero at least as fast as N/R. Notice that this condition is always satisfied if $F(s)$ is the quotient of two polynomials, and if the order of the numerator is less than that of the denominator.

The conditions in the previous paragraph for the application of Eqs. 10.5-3 and 10.5-4, although sufficient, are really unnecessarily restrictive. If there is any question about the validity of these equations, the contribution along the paths $DEFGA$ and DHA may be examined for the particular $F(s)$ under consideration. Alternatively, the equations may be used to find $f(t)$ even without such an examination, provided that Eq. 10.5-1 is then applied to the resulting $f(t)$. If the transform of the answer is found to be the original $F(s)$, the function of time must be correct, because of the uniqueness theorem.

The factor ϵ^{st} in Eqs. 10.5-2 through 10.5-4 is analytic throughout the entire finite s-plane. Since the path ABD must pass to the right of all of the singularities of $F(s)$, Eq. 10.5-4 always yields zero for $t < 0$. In using Eq. 10.5-3, the residue formulas developed in Section 10.4 prove useful. For a first-order pole at s_0, the residue of $F(s)\epsilon^{st}$ is given by

$$[(s - s_0)F(s)\epsilon^{st}]_{s=s_0} \qquad (10.5\text{-}5)$$

or

$$\left[\frac{P(s)\epsilon^{st}}{\dfrac{d}{ds}\, Q(s)} \right]_{s=s_0} \qquad (10.5\text{-}6)$$

where $F(s) = P(s)/Q(s)$, and where $P(s)$ and $Q(s)$ are analytic at s_0, and $P(s_0) \neq 0$. For a pole of order n at s_0, the residue of $F(s)\epsilon^{st}$ is

$$\frac{1}{(n-1)!} \left[\frac{d^{n-1}}{ds^{n-1}} (s - s_0)^n F(s)\epsilon^{st} \right]_{s=s_0} \qquad (10.5\text{-}7)$$

Example 10.5-1.   Evaluate $\mathcal{L}^{-1}[1/(s + 1)^2]$.

Solution.   From Example 10.4-4, the residue of $\epsilon^{st}/(s + 1)^2$ at the second-order pole at $s = -1$ is $t\epsilon^{-t}$, so $f(t) = t\epsilon^{-t}$ for $t > 0$.

Example 10.5-2.   Find $f(t)$ if

$$F(s) = \frac{s + 3}{s^2 + 4s + 5} = \frac{s + 3}{(s + 2 - j)(s + 2 + j)}$$

Solution. $F(s)$ has first-order poles at $s = -2 \pm j$. By Eq. 10.5-6, the residue of $F(s)\epsilon^{st}$ at $s = -2 + j$ is

$$\left[\frac{(s + 3)\epsilon^{st}}{2s + 4}\right]_{s=-2+j} = \frac{(1 + j)\epsilon^{(-2+j)t}}{j2} = \frac{1 - j}{2}\epsilon^{(-2+j)t} = \frac{\sqrt{2}}{2}\epsilon^{-j45°}\epsilon^{(-2+j)t}$$

while the residue at $-2 - j$ is

$$\left[\frac{(s + 3)\epsilon^{st}}{2s + 4}\right]_{s=-2-j} = \frac{1 + j}{2}\epsilon^{(-2-j)t} = \frac{\sqrt{2}}{2}\epsilon^{j45°}\epsilon^{(-2-j)t}$$

If the rectangular form of the above coefficients is used in Eq. 10.5-3,

$$f(t) = \epsilon^{-2t}[(\tfrac{1}{2} - j\tfrac{1}{2})\epsilon^{jt} + (\tfrac{1}{2} + j\tfrac{1}{2})\epsilon^{-jt}] = \epsilon^{-2t}(\cos t + \sin t)$$

for $t > 0$. The polar form of the coefficients gives the equivalent answer of

$$f(t) = \frac{\sqrt{2}}{2}\epsilon^{-2t}[\epsilon^{j(t-45°)} + \epsilon^{-j(t-45°)}] = \sqrt{2}\,\epsilon^{-2t}\cos(t - 45°)$$

Whenever $F(s)$ is the transform of a real function of time, the residues of $F(s)\epsilon^{st}$ at complex-conjugate poles are themselves complex conjugates. This fact can be used either as a check on the work or to save time in the calculations.

Example 10.5-3.   Find $\mathcal{L}^{-1}\left[\dfrac{s^2 + 2}{s(s + 1)^2(s + 2)}\right]$.

Solution. $F(s)$ has first-order poles at $s = 0$ and -2, and a second-order pole at $s = -1$. By Eqs. 10.5-5 and 10.5-7 the residue of $F(s)\epsilon^{st}$ at $s = 0$ is

$$\left[\frac{(s^2 + 2)\epsilon^{st}}{(s + 1)^2(s + 2)}\right]_{s=0} = 1$$

and at $s = -2$ is

$$\left[\frac{(s^2 + 2)\epsilon^{st}}{s(s + 1)^2}\right]_{s=-2} = -3\epsilon^{-2t}$$

and at $s = -1$ is

$$\left[\frac{d}{ds}\frac{(s^2 + 2)\epsilon^{st}}{s(s + 2)}\right]_{s=-1}$$

$$= \left[\frac{(s^2 + 2s)(2s\epsilon^{st} + s^2t\epsilon^{st} + 2t\epsilon^{st}) - (s^2 + 2)\epsilon^{st}(2s + 2)}{(s^2 + 2s)^2}\right]_{s=-1}$$

$$= 2\epsilon^{-t} - 3t\epsilon^{-t}$$

Hence, for $t > 0$,

$$f(t) = 1 - 3\epsilon^{-2t} + (2 - 3t)\epsilon^{-t}$$

Notice that there are two terms in the expression for the residue at the second-order pole. In general, n terms may be expected for an nth-order pole.

Example 10.5-4. Find $\mathcal{L}^{-1}\left[\epsilon^{-2s}/(s + 1)^2\right]$.

Solution. The result of Example 10.5-1, together with Eq. 10.1-6, immediately yields the answer of

$$(t - 2)\epsilon^{-(t-2)}U_{-1}(t - 2)$$

Although this method is by far the simplest, Eq. 10.5-2 always may be used for the given $F(s)$.

$$F(s)\epsilon^{st} = \frac{\epsilon^{s(t-2)}}{(s + 1)^2}$$

By analogy to the proof of Eqs. 10.5-3 and 10.5-4, $\int F(s)\epsilon^{st} \, ds$ vanishes along contour *DEFGA* in Fig. 10.5-2 only for $t - 2 > 0$ and vanishes along *DHA* for $t - 2 < 0$. Thus, the answer is zero for $t < 2$, while for $t > 2$

$$f(t) = \text{residue of } F(s)\epsilon^{st} \quad \text{at} \quad s = -1$$

$$= \left[\frac{d}{ds} \, \epsilon^{s(t-2)}\right]_{s=-1} = (t - 2)\epsilon^{-(t-2)}$$

When $F(s)$ is the quotient of two polynomials, the residue and partial-fraction methods are essentially identical. Suppose that

$$F(s) = \frac{P(s)}{Q(s)} = \frac{P(s)}{(s - s_1)(s - s_2)^2}$$

where $P(s)$ is of lower order than $Q(s)$. The partial-fraction expansion is

$$F(s) = \frac{K_1}{s - s_1} + \frac{K_2}{s - s_2} + \frac{K_3}{(s - s_2)^2}$$

where

$$K_1 = [(s - s_1)F(s)]_{s=s_1}$$

$$K_3 = [(s - s_2)^2 F(s)]_{s=s_2}$$

$$K_2 = \left[\frac{d}{ds} (s - s_2)^2 F(s)\right]_{s=s_2}$$

The function of time is

$$f(t) = K_1\epsilon^{s_1 t} + K_2\epsilon^{s_2 t} + K_3 t\epsilon^{s_2 t} \qquad \text{for } t > 0$$

The residue of $F(s)\epsilon^{st}$ at $s = s_1$ is, by Eq. 10.5-5,

$$[(s - s_1)F(s)\epsilon^{st}]_{s=s_1} = K_1\epsilon^{s_1 t}$$

The residue at $s = s_2$ is, by Eq. 10.5-7,

$$\left[\frac{d}{ds}(s - s_2)^2 F(s)\epsilon^{st}\right]_{s=s_2} = \left[\epsilon^{st}\frac{d}{ds}(s - s_2)^2 F(s)\right]_{s=s_2} + \left[(s - s_2)^2 F(s)t\epsilon^{st}\right]_{s=s_2}$$

$$= K_2\epsilon^{s_2 t} + K_3 t\epsilon^{s_2 t}$$

which agrees with the partial-fraction method.

Although the partial-fraction method is restricted to the quotient of two polynomials in s, the more powerful residue method is not. Several examples involving irrational functions are considered in P. M. DeRusso, R. J. Roy, and C. M. Close, *State Variables for Engineers* (New York: Wiley, 1965). One simple example is given below.

Example 10.5-5. Find $\mathscr{L}^{-1}[1/\sinh s]$.

Solution. From Example 10.4-5, $\epsilon^{st}/\sinh s$ has first-order poles at $s = jn\pi$, with a residue of $(-1)^n \epsilon^{jn\pi t}$, for all integral values of n. The fact that $F(s)\epsilon^{st}$ has an infinite number of poles on the imaginary axis of the s-plane, as shown in Fig. 10.5-3a, might raise some doubts about the validity of using Eq. 10.5-3. If the equation is used despite this fact,

$$f(t) = \sum_{n=-\infty}^{\infty}(-1)^n \epsilon^{jn\pi t} = 1 + \sum_{n=1}^{\infty}(-1)^n(\epsilon^{jn\pi t} + \epsilon^{-jn\pi t})$$

$$= 1 + \sum_{n=1}^{\infty} 2(-1)^n \cos n\pi t$$

which, from Chapter 9, is the Fourier series for the impulse train in Fig. 10.5-3b.

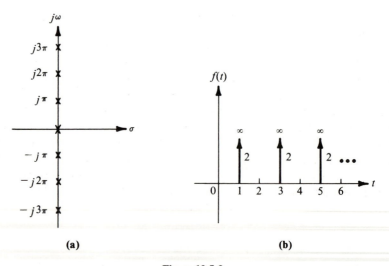

(a)   (b)

Figure 10.5-3

To resolve any doubts about the validity of the answer we have formally obtained, we now take the direct transform of the impulse train. Since $\mathcal{L}[U_0(t)] = 1$, $\mathcal{L}[U_0(t-n)] = \epsilon^{-ns}$ and

$$F(s) = 2\epsilon^{-s} + 2\epsilon^{-3s} + 2\epsilon^{-5s} + \cdots$$

$$= 2\epsilon^{-s}(1 + \epsilon^{-2s} + \epsilon^{-4s} + \cdots)$$

The quantity in parentheses is similar to the Maclaurin series $1/(1-z) = 1 + z + z^2 + \cdots$, which converges for $|z| < 1$.

$$F(s) = \frac{2\epsilon^{-s}}{1 - \epsilon^{-2s}} = \frac{2}{\epsilon^s - \epsilon^{-s}} = \frac{1}{\sinh s}$$

with a region of convergence given by $\sigma > 0$, so the impulse train is indeed the correct answer.

10.6
The Significance of the Laplace Transform

The purpose of this section is to relate the Laplace transform to the methods of earlier chapters and to emphasize the physical and mathematical insight into circuit analysis that the student should be acquiring. First, we consider the underlying importance of the concept of superposition in the analysis of linear circuits. Second, we look at the relationship between the classical and transform solutions, with particular emphasis on finding the initial values of a response and its derivatives. Third, we examine more carefully the calculation of the step and impulse responses by the Laplace transform and their relationships to the material in earlier chapters. Last, we consider the significance of the location of the poles and zeros in the transformed output.

Consider a circuit with no initial stored energy and with an input $x(t)$ and an output $y(t)$. There are certain inputs for which the response is often known or else easily found. The responses to the unit step $U_{-1}(t - \lambda)$ and to the unit impulse $U_0(t - \lambda)$ were denoted in Chapter 3 by the symbols $r(t - \lambda)$ and $h(t - \lambda)$, respectively, while the forced response to ϵ^{st} was shown in Chapter 4 to be $H(s)\epsilon^{st}$.

The superposition principle provides, among other things, a means of determining the response to an arbitrary input once the response to one of the inputs of the previous paragraph is known. Figure 3.5-1 showed how an arbitrary input can be approximately decomposed into a series of step functions starting at different times and how this approximation becomes exact as the number of step functions that are used increases without limit. In the equation

$$x(t) = \int_{-\infty}^{\infty} \frac{dx(\lambda)}{d\lambda} U_{-1}(t - \lambda) \, d\lambda$$

$U_{-1}(t - \lambda)$ is a step function beginning at $t = \lambda$, and $[dx(\lambda)/d\lambda]d\lambda$ is the size of this particular component in the input. The integral sign serves to sum up all such components in order to reconstruct the function $x(t)$. Since the response to $[dx(\lambda)/d\lambda] \, d\lambda \, U_{-1}(t - \lambda)$ is $[dx(\lambda)/d\lambda] \, d\lambda \, r(t - \lambda)$, and since the total response is the sum of the responses to the individual components,

$$y(t) = \int_{-\infty}^{\infty} \frac{dx(\lambda)}{d\lambda} r(t - \lambda) \, d\lambda$$

which is Eq. 3.5-6.[†] Equation 3.5-5 is obtained in a similar way when $x(t)$ is decomposed into impulses instead of step functions.

It is possible to regard the Laplace transform in much the same way. Equation 10.1-2,

$$x(t) = \frac{1}{2\pi j} \int X(s) \epsilon^{st} \, ds$$

suggests that $x(t)$ has been decomposed into components of the form $X(s) \epsilon^{st} \, ds$ for all t. These individual components all start at the same time, but their shape depends upon the value of the parameter s. The coefficient $X(s)$ is again a measure of the strength of the particular component, and the integral sign again serves to sum up all the components. Since the response to ϵ^{st} is $H(s) \epsilon^{st}$,

$$y(t) = \frac{1}{2\pi j} \int X(s)H(s) \epsilon^{st} \, ds$$

$$= \frac{1}{2\pi j} Y(s) \epsilon^{st} \, ds = \mathscr{L}^{-1}\left[Y(s)\right]$$

where

$$Y(s) = X(s)H(s) \tag{10.6-1}$$

represents the strength of the various components in the output. The last equation is the same as Eq. 10.3-4.

Other transformations, including the Fourier transform, can be interpreted in a similar way and involve the same three steps. First, the direct transform of the source indicates the relative strength of the components into which it is decomposed (the complex-frequency spectrum in the case of the Laplace transform). Second, the output spectrum is found, as in Eq. 10.6-1. Last, the inverse transform recombines the components in the output to yield the output as a function of time. Since this process is based upon the superposition concept, the Fourier and Laplace transforms are not useful for nonlinear circuits.

Any stored energy that may be initially present can be represented as extra sources. In many circuits, it produces terms that are added directly to the transform of the external source.

The Relationship between the Classical and the Transform Solution of Circuits. Recall that there is a one-to-one correspondence between a circuit's differential

---

[†] As explained in Section 3.5, the limits of integration may be changed to zero and t in most problems.

equation and its network function. The network function for a circuit that is described by the differential equation

$$a_n \frac{d^n y}{dt^n} + \cdots + a_1 \frac{dy}{dt} + a_0 y = b_m \frac{d^m x}{dt^m} + \cdots + b_1 \frac{dx}{dt} + b_0 x \qquad (10.6\text{-}2)$$

is

$$H(s) = \frac{b_m s^m + \cdots + b_1 s + b_0}{a_n s^n + \cdots + a_1 s + a_0} \qquad (10.6\text{-}3)$$

as in Eq. 4.6-3. In the classical solution, not only $x(t)$ but also the values of $y(0+)$, $y'(0+)$, $\ldots$, $y^{(n-1)}(0+)$ were needed. According to Eq. 10.6-1, however, $x(t)$ and the statement that there is no initial stored energy should suffice for a complete solution. Quite clearly, the n initial conditions required for the classical solution, which are normally found from the circuit with considerable effort, should be obtainable directly from the differential equation and the input, without reference to the circuit. Equivalently, we should be able to obtain them from $Y(s)$ without having to evaluate the inverse transform.

A method of determining the initial values of the output and its derivatives when there is no initial energy storage can be derived independently of the Laplace transform. We choose, however, to use an additional transform property, known as the *initial value theorem*.

$$\lim_{t \to 0+} f(t) = \lim_{s \to \infty} sF(s) \qquad \text{when the limit exists} \qquad (10.6\text{-}4)$$

To justify this theorem, consider Eq. 10.1-8.

$$\lim_{s \to \infty} \int_0^\infty \frac{df}{dt} \epsilon^{-st} \, dt = \lim_{s \to \infty} [sF(s) - f(0+)]$$

If the limiting and integration processes can be interchanged, the left-hand side is zero. Since $\lim_{s \to \infty} f(0+) = f(0+) = \lim_{t \to 0+} f(t)$,

$$0 = \lim_{s \to \infty} sF(s) - \lim_{t \to 0+} f(t)$$

which gives Eq. 10.6-4.

To find the initial value of the successive derivatives of a function whose transform is known, replace $f(t)$ in Eq. 10.6-4 by $f'(t), f''(t), \ldots$. Thus,

$$\lim_{t \to 0+} f'(t) = \lim_{s \to \infty} s\mathscr{L}[f'(t)] = \lim_{s \to \infty} s[sF(s) - f(0+)]$$

$$= \lim_{s \to \infty} [s^2 F(s) - sf(0+)]$$

Similarly,

$$\lim_{t \to 0+} f''(t) = \lim_{s \to \infty} s\mathscr{L}[f''(t)] = \lim_{s \to \infty} [s^3 F(s) - s^2 f(0+) - sf'(0+)]$$

Example 10.6-1. If $R_1 = R_2 = 1\ \Omega$, $L = 1$ h, $C = 1$ f, and $e_1(t) = U_{-1}(t)$ for the circuit in Fig. 10.3-5a, the transformed output is

$$E_o(s) = \frac{1}{s^2 + 2s + 2}$$

Find the values of $e_o(t)$, de_o/dt, and d^2e_o/dt^2 at $t = 0+$.

Solution. By Eq. 10.6-4,

$$e_o(0+) = \lim_{s \to \infty} \frac{s}{s^2 + 2s + 2} = 0$$

Also,

$$e_o'(0+) = \lim_{s \to \infty} \left(\frac{s^2}{s^2 + 2s + 2} - 0 \right) = 1$$

$$e_o''(0+) = \lim_{s \to \infty} \left(\frac{s^3}{s^2 + 2s + 2} - 0 - s \right) = \lim_{s \to \infty} \left(\frac{-2s^2 - 2}{s^2 + 2s + 2} \right) = -2$$

These answers are consistent with the expression for $e_o(t)$ in Example 10.3-4.

When calculating the initial value of several derivatives, perhaps the easiest procedure is to express $F(s)$, by long division, as[†]

$$F(s) = \frac{a_1}{s} + \frac{a_2}{s^2} + \frac{a_3}{s^3} + \cdots$$

The series may be obtained by long division, or by finding the Maclaurin series

$$G(s) = F\left(\frac{1}{s}\right) = a_1 s + a_2 s^2 + a_3 s^3 + \cdots$$

and then replacing s by $1/s$. By Eq. 10.6-4 and the equations preceding Example 10.6-1,

$$f(0+) = \lim_{s \to \infty} \left(a_1 + \frac{a_2}{s} + \frac{a_3}{s^2} + \cdots \right) = a_1$$

$$f'(0+) = \lim_{s \to \infty} \left(a_1 s + a_2 + \frac{a_3}{s} + \cdots - a_1 s \right) = a_2$$

$$f''(0+) = \lim_{s \to \infty} [a_1 s^2 + a_2 s + a_3 + \cdots - a_1 s^2 - a_2 s] = a_3$$

In Example 10.6-1, $E_o(s)$ may be written as

$$E_o(s) = \frac{1}{s^2} - \frac{2}{s^3} + \frac{6}{s^4} - \cdots$$

so

$$e_o(0+) = 0, \qquad e_o'(0+) = 1, \qquad e_o''(0+) = -2, \qquad e_o'''(0+) = 6, \ldots$$

[†] The form of the series implies that $F(s)$ has a zero at infinity. If it does not, Eq. 10.6-4 cannot be used, since the limit would not exist.

The initial value theorem of Eq. 10.6-4 enables the initial behavior of $f(t)$ to be determined from $F(s)$ *without* the evaluation of the entire inverse transform. In many cases, the steady-state behavior of $f(t)$ also can be obtained by an inspection of $F(s)$. The *final value theorem*, whose proof and application is left to the student, says that

$$\lim_{t\to\infty} f(t) = \lim_{s\to 0} sF(s) \qquad \text{provided that } sF(s) \text{ is analytic on and to the right of the imaginary axis of the } s\text{-plane} \qquad (10.6\text{-}5)$$

The Step and Impulse Response.   The unit impulse response $h(t)$ can be found from Eq. 10.6-1, with $x(t) = U_0(t)$ and $X(s) = 1$. The output is then

$$h(t) = \mathscr{L}^{-1}[H(s)] \qquad (10.6\text{-}6)$$

which is one reason for denoting the unit impulse response by $h(t)$. For the unit step response, let $y(t) = r(t)$, $x(t) = U_{-1}(t)$, and $X(s) = 1/s$. Then

$$r(t) = \mathscr{L}^{-1}\left[\frac{H(s)}{s}\right] \qquad (10.6\text{-}7)$$

At the risk of temporary confusion, we shall examine Eq. 10.6-6 more closely. It is based upon the fact that $\mathscr{L}[U_0(t)] = 1$, as in Example 10.1-3 and Table 10.1-1. If, however, the function $f(t) = U_{-1}(t)$ is inserted into Eq. 10.1-8,

$$\mathscr{L}[U_0(t)] = s\mathscr{L}[U_{-1}(t)] - [U_{-1}(t)]_{t=0+} = s\frac{1}{s} - 1 = 0$$

which does not agree with the previously derived answer.

The unit impulse was originally introduced as the limit of a narrow, high pulse of unit area. In order to resolve the above inconsistency, consider the functions $f_1(t)$ through $f_3(t)$ in Fig. 10.6-1. Their respective transforms are $F_1(s) = (1 - \epsilon^{-sL})/sL$, $F_2(s) = (1 - \epsilon^{-sL/2})/sL$, and $F_3(s) = 0$. In the limit as L approaches zero, all three functions of time become the unit impulse $U_0(t)$, but the limits of the transformed quantities are $F_1(s) = 1$, $F_2(s) = \frac{1}{2}$, and $F_3(s) = 0$. This suggests that we should somehow differentiate between an impulse occurring immediately before $t = 0$, as derived from $f_3(t)$, and one immediately after $t = 0$, as derived from $f_1(t)$. This distinction is illustrated symbolically by Figs. 10.6-1d and 10.6-1e. If Eq. 10.1-8 is used with $f(t) = f_5(t)$ as L approaches zero,

$$\mathscr{L}[U_0(t)] = s\frac{1}{s} - 0 = 1$$

but with $f(t) = f_7(t)$,

$$\mathscr{L}[U_0(t)] = s\frac{1}{s} - 1 = 0$$

Once again, the transforms of a unit impulse immediately before and after $t = 0$ are seen to be zero and unity, respectively.

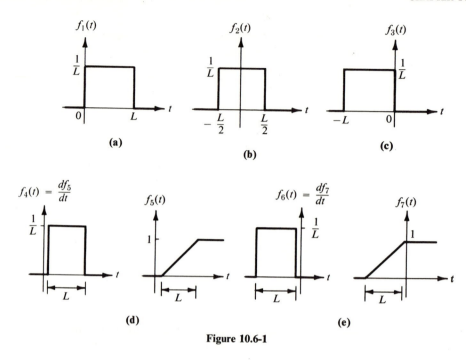

Figure 10.6-1

The student should keep in mind that when dealing with impulses the objective is to find a circuit's impulse response $h(t)$. Perhaps an input pulse whose width is small compared to the time constants of the circuit is to be approximated by an impulse of equal area, or perhaps $h(t)$ is needed for the convolution formulas. In any case, the expression for $h(t)$ should certainly *not* depend on whether $x(t) = U_0(t)$ is interpreted as an impulse occurring just before or just after $t = 0$, as long as a consistent interpretation is used. Remember that impulses can and usually do change a circuit's stored energy instantaneously. If $x(t) = U_0(t)$ is assumed to occur immediately before $t = 0$ so that $X(s) = 0$, the initial conditions used should include the stored energy inserted by the impulse. If, on the other hand, the impulse is regarded as occurring just after $t = 0$ so that $X(s) = 1$, the initial conditions used should not include the effect of the impulse. These two approaches are illustrated by the following example.

Example 10.6-2. Find $h(t)$ for the circuit in Fig. 10.6-2a, which is identical with Fig. 10.3-5a, when $R_1 = R_2 = 1\ \Omega$, $L = 1$ h, and $C = 1$ f.

Solution. First, assume that $e_1(t)$ is a unit impulse occurring just before $t = 0$, so that $E_1(s) = 0$. Since $e_C(t)$ and $i_L(t)$ must remain finite, the impulse appears across the resistance R_1 and across the inductance. Because of the unit impulse of voltage across L, $i_L(0+) = 1/L$. The unit impulse of voltage across R_1 results in a current impulse of value $1/R_1$ through the capacitance, so $e_C(0+) = 1/R_1C$. The initial

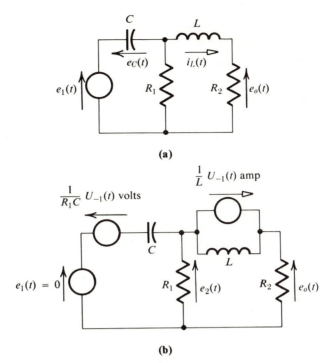

(a)

(b)

Figure 10.6-2

stored energy may be represented by the added sources in Fig. 10.6-2b. The transformed node equations are

$$sC\left[E_2(s) + \frac{1}{sR_1C}\right] + \frac{E_2(s)}{R_1} + \frac{E_2(s) - E_o(s)}{sL} + \frac{1}{sL} = 0$$

$$\frac{E_o(s) - E_2(s)}{sL} - \frac{1}{sL} + \frac{E_o(s)}{R_2} = 0$$

When these equations are solved simultaneously,

$$E_o(s) = \frac{sCR_2}{(sL + R_2)(sC + 1/R_1) + 1} = \frac{s}{s^2 + 2s + 2}$$

so

$$h(t) = \mathcal{L}^{-1}[E_o(s)] = \epsilon^{-t}(\cos t - \sin t)$$

An alternative solution is to show, as in Example 10.3-4, that the circuit's network function is

$$H(s) = \frac{E_o(s)}{E_1(s)} = \frac{sCR_2}{(sL + R_2)(sC + 1/R_1) + 1} = \frac{s}{s^2 + 2s + 2}$$

If $e_1(t)$ is a unit impulse occurring just after $t = 0$, $E_1(s) = 1$ and there is no initial stored energy. The response is

$$h(t) = \mathscr{L}^{-1}[H(s)] = \epsilon^{-t}(\cos t - \sin t)$$

Notice that this answer may also be obtained by differentiating the step response that was found in Example 10.3-4.

Since the evaluation of the stored energy that is inserted into a circuit by an impulse may be tricky, the second method in Example 10.6-2 is the one usually used. Any impulse at the time origin is assumed to occur immediately after $t = 0$.

The impulse and step responses were used in the convolution formulas in Section 3.5. These formulas also can be derived from the following property in Laplace transform theory.[†]

$$\mathscr{L}^{-1}[F_1(s)F_2(s)] = \int_0^t f_1(\lambda)f_2(t - \lambda)\,d\lambda$$

$$= \int_0^t f_1(t - \lambda)f_2(\lambda)\,d\lambda \tag{10.6-8}$$

For a circuit with no initial stored energy, by Eq. 10.6-1,

$$y(t) = \mathscr{L}^{-1}[X(s)H(s)] = \int_0^t x(\lambda)h(t - \lambda)\,d\lambda = \int_0^t x(t - \lambda)h(\lambda)\,d\lambda$$

which agrees with Eq. 3.5-2 for a circuit that is unexcited for $t < 0$.

The Significance of the Poles and Zeros.   When the inverse transform is evaluated by Eq. 10.5-3 or by the partial-fraction method of Section 10.2, each pole in $F(s)$ is seen to give rise to one or more terms in $f(t)$. The *form* of each term depends only upon the location and order of the corresponding pole. An nth-order pole at $s = s_1$, for example, produces the terms $K_1\epsilon^{s_1t} + K_2t\epsilon^{s_1t} + \cdots + K_nt^{n-1}\epsilon^{s_1t}$ in $f(t)$. If a pair of complex-conjugate poles of order n are at $s = \alpha \pm j\beta$, $f(t)$ contains the terms $\epsilon^{\alpha t}[K_1 \cos(\beta t + \phi_1) + K_2 \cos(\beta t + \phi_2) + \cdots + K_nt^{n-1} \cos(\beta t + \phi_n)]$.

The responses corresponding to several typical pole patterns are tabulated in Table 10.6-1. Poles in the open right half plane or higher-order poles on the imaginary axis result in terms that increase without limit as t approaches infinity. The distance from the imaginary axis of first-order poles in the left half plane is the reciprocal of the time constant of the corresponding terms in $f(t)$. The distance of a pair of complex-conjugate poles from the real axis is the angular frequency of oscillation of the terms in $f(t)$.

Table 10.6-1 closely resembles Fig. 4.4-1, and the present discussion is indeed closely related to Sections 4.4 and 6.3. Since a unit impulse merely serves to place some energy into the circuit instantaneously, the impulse response, $h(t) = \mathscr{L}^{-1}[H(s)]$, indicates the form of the free response. If $H(s)$ has poles in the open right half plane or higher-order poles on the imaginary axis, the free response increases without

---

[†] For the proof of this property, see, for example, pp. 228–231 of M. F. Gardner and J. L. Barnes, *Transients in Linear Systems* (New York: Wiley, 1942).

TABLE 10.6-1

Poles of F(s) in the s-plane	*Corresponding terms in f(t)*	*Description of terms in f(t)*
	K	Constant
	$K\epsilon^{at}$	Growing exponential
	$K\epsilon^{-at}$	Decaying exponential
	$K\epsilon^{at}\cos(\beta t + \phi)$	Growing oscillation
	$K\cos(\beta t + \phi)$	Constant-amplitude oscillation
	$K\epsilon^{-at}\cos(\beta t + \phi)$	Decaying oscillation
	$K_1\cos(\beta t + \phi_1)$ $+ K_2 t\cos(\beta t + \phi_2)$	Growing oscillation

limit, and the circuit is unstable. As discussed in Section 6.3, the poles of $H(s)$ are identical with the roots of the characteristic equation in the classical solution. The statements made here are consistent with those in Section 4.4.

Example 10.6-3. Circuits with controlled sources, as in Figs. 4.4-5a and 7.2-4a, may be unstable under certain conditions. A more complicated example is the Colpitts oscillator in Fig. 10.6-3. Under what conditions will the circuit be unstable? Under what conditions will the free response contain an oscillation of constant amplitude, and what will be the angular frequency of this oscillation?

Solution. In order to find the network function, we write two transformed node equations.

$$sC_1E_2(s) - g_mE_g(s) + \left(sC_2 + \frac{1}{r_p}\right)[E_2(s) - E_o(s)] = 0$$

$$\left(sC_2 + \frac{1}{r_p}\right)[E_o(s) - E_2(s)] + \frac{E_o(s)}{sL} + g_mE_g(s) = 0$$

Replacing $E_g(s)$ by $E_1(s) - E_2(s)$, solving simultaneously for $E_o(s)$, and noting that $g_m = \mu/r_p$, we obtain

$$H(s) = \frac{E_o(s)}{E_1(s)} = \frac{-g_ms^2/C_2}{s^3 + (1/r_pC_2)s^2 + (1/LC_1 + 1/LC_2)s + (\mu + 1)/Lr_pC_1C_2}$$

It is necessary to factor the cubic in the denominator of $H(s)$ to determine the position of the poles of $H(s)$. Locating the roots of a high-order polynomial is a difficulty frequently encountered in stability analysis, and special methods for doing this are usually developed in advanced courses. It happens that the cubic $a_3s^3 + a_2s^2 + a_1s + a_0$, where the constants are real and positive, contains right half plane zeros if and only if $a_2a_1 < a_3a_0$. For instability, therefore,

$$\left(\frac{1}{r_pC_2}\right)\left(\frac{C_1 + C_2}{LC_1C_2}\right) < \frac{\mu + 1}{Lr_pC_1C_2}$$

or $\mu > C_1/C_2$.

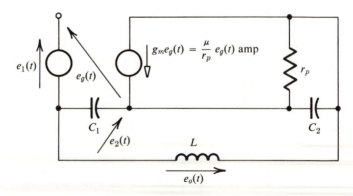

$$g_me_g(t) = \frac{\mu}{r_p}e_g(t) \text{ amp}$$

Figure 10.6-3

The free response will contain a constant-amplitude oscillation only if $H(s)$ has a pair of poles on the $j\omega$ axis. The cubic in the denominator must then vanish for some value of β when s is replaced by $j\beta$.

$$-j\beta^3 - \frac{\beta^2}{r_pC_2} + j\left(\frac{C_1 + C_2}{LC_1C_2}\right)\beta + \frac{\mu + 1}{Lr_pC_1C_2} = 0$$

The complex quantity on the left is zero only if its real and imaginary parts are zero.

$$-\frac{\beta^2}{r_pC_2} + \frac{\mu + 1}{Lr_pC_1C_2} = 0 \quad \text{so} \quad \beta^2 = \frac{\mu + 1}{LC_1}$$

and

$$-\beta^3 + \frac{C_1 + C_2}{LC_1C_2}\beta = 0 \quad \text{so} \quad \beta^2 = \frac{C_1 + C_2}{LC_1C_2}$$

Then the condition for constant-amplitude oscillations is

$$\frac{\mu + 1}{LC_1} = \frac{C_1 + C_2}{LC_1C_2} \quad \text{or} \quad \mu = \frac{C_1}{C_2}$$

and the angular frequency of this oscillation is

$$\sqrt{\frac{\mu + 1}{LC_1}} \text{ rad per sec}$$

The output of a circuit with no initial stored energy is $y(t) = \mathcal{L}^{-1}[X(s)H(s)]$. We have already discussed how the poles of $H(s)$ determine the form of the free response. The poles of $X(s)$ depend only upon the input signal and determine the form of the forced response.

For the circuit in Fig. 10.6-4, for example,

$$E_1(s) = \frac{1}{s}$$

$$H(s) = \frac{s + 1}{s^2 + 3s + 1} = \frac{s + 1}{(s + 2.618)(s + 0.382)}$$

$$E_0(s) = \frac{s + 1}{s(s + 2.618)(s + 0.382)}$$

The forced response, which is the particular solution of Chapter 4, is the residue of $E_0(s)\epsilon^{st}$ at $s = 0$, the pole contributed by the transformed input.

$$(e_0)_P = \left[\frac{s + 1}{(s + 2.618)(s + 0.382)}\epsilon^{st}\right]_{s=0} = 1 \quad \text{for} \quad t > 0$$

The free response, or the complementary solution, is the sum of the residues at $s = -2.618$ and -0.382, the poles of $H(s)$.

$$(e_0)_H = \left[\frac{s + 1}{s(s + 0.382)}\epsilon^{st}\right]_{s=-2.618} + \left[\frac{s + 1}{s(s + 2.618)}\epsilon^{st}\right]_{s=-0.382}$$

$$= -0.277\epsilon^{-2.618t} - 0.723\epsilon^{-0.382t} \quad \text{for} \quad t > 0$$

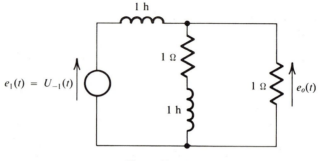

Figure 10.6-4

The *form* of the forced and free response can be seen immediately by an inspection of the pole-zero pattern in Fig. 10.6-5a. To see how the *size* of the various terms might be found directly from the pole-zero pattern, consider

$$E_o(s) = \frac{s - z}{(s - p_1)(s - p_2)(s - p_3)}$$

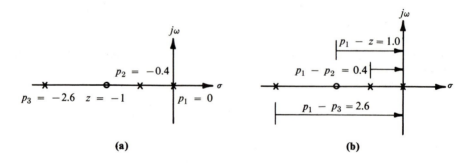

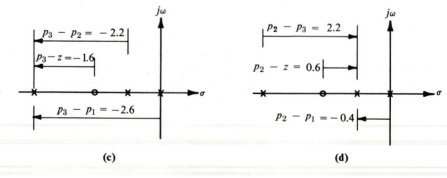

Figure 10.6-5

The residue of $E_o(s)\epsilon^{st}$ at $s = p_1$ is

$$\left[\frac{s-z}{(s-p_2)(s-p_3)}\,\epsilon^{st}\right]_{s=p_1} = \left[\frac{p_1-z}{(p_1-p_2)(p_1-p_3)}\right]\epsilon^{p_1t}$$

Each of the factors in the last bracket may be represented graphically, as in Fig. 10.6-5b. This coefficient is therefore $1/(0.4)(2.6) = 1$.

In summary, the numerator factors are represented by vectors from the zeros to the pole under consideration, and the denominator factors are represented by vectors drawn from the other poles. From Fig. 10.6-5c, the coefficient of the $\epsilon^{-2.6t}$ term is $-1.6/(-2.2)(-2.6) = -0.28$. With a little practice, the sign and approximate size of each term in $f(t)$ can be quickly estimated. Notice particularly that when two poles are close together, as are p_1 and p_2, the size of the corresponding terms in the function of time is relatively large. A zero close to the pole being considered reduces the size of the term in $f(t)$. Thus, the coefficient of the ϵ^{p_2t} term is somewhat smaller than that of the ϵ^{p_1t} term.

This graphical approach is easily extended to problems that have complex poles and zeros, as in the following example. It also can be extended to poles of higher order, but the method then loses its simplicity.

Example 10.6-4.   Find the steady-state response of the circuit in Fig. 10.6-6a, which is similar to Fig. 6.2-2e.

Solution.   The transfer function is easily found to be $H(s) = 100/(s^2 + 0.2s + 100)$, which has poles at $-0.1 \pm j10$. Since $E_1(s) = s/(s^2 + 100)$,

$$E_o(s) = \frac{100s}{(s^2 + 100)(s^2 + 0.2s + 100)}$$

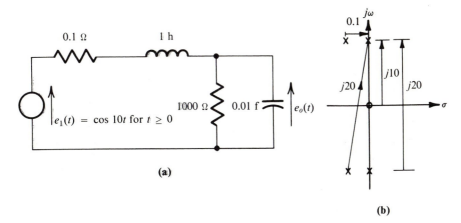

(a)

(b)

Figure 10.6-6

From the pole-zero pattern in Fig. 10.6-6b, the residue of $E_o(s)\epsilon^{st}$ at $s = j10$ is

$$\frac{(100)(j10)}{(0.1)(j20)(j20)} \epsilon^{j10t} = -25j\epsilon^{j10t}$$

and at $s = -j10$ is $25j\epsilon^{-j10t}$. The forced or steady-state response is

$$(e_o)_P = -25j(\epsilon^{j10t} - \epsilon^{-j10t}) = 50 \sin 10t$$

The student may have already noticed the similarity of this example with the methods of Section 6.4. In fact, this problem normally would be handled by the procedure of Chapter 6, but either viewpoint leads to essentially the same method of solution.

10.7

Summary

The direct Laplace transform,

$$F(s) = \mathscr{L}[f(t)] = \int_0^\infty f(t)\epsilon^{-st}\, dt$$

transforms a function of time into a function of complex frequency s. The most important properties are summarized in Table 10.1-2. If $F(s)$ is given as the quotient of two polynomials, with the numerator of lower order than the denominator, the inverse transform can be found by a partial-fraction expansion, as discussed in Section 10.2. An even more general method is to use the inversion integral

$$f(t) = \mathscr{L}^{-1}[F(s)] = \frac{1}{2\pi j}\int_C F(s)\epsilon^{st}\, ds$$

where the contour of integration C is described in Section 10.5. In practice, the integration may be carried out by evaluating the residues of $F(s)\epsilon^{st}$.

One method of circuit analysis is to transform the *integral-differential* equations that describe the circuit into a set of *algebraic* equations, which may be solved for the transform of the output. The output as a function of time is found by taking the inverse transform. Another method is to draw a complex-frequency-domain circuit, where the passive elements are characterized by their impedances and the voltages and currents by their transforms. Any initial stored energy is represented by added sources. If there is no initial stored energy, $Y(s) = H(s)X(s)$.

The transform methods give the complete response of a circuit. The effect of any initial stored energy is introduced into the solution at an early stage, and there are no arbitrary constants of integration that must be evaluated at the end of the problem. The advantages of the Laplace transform are most apparent in the solution of complicated circuits. In very simple circuits, where the forced and free response can be written down by inspection, the transform method would not be used.

The significance of the Laplace transform is examined in Section 10.6. The effect of the location of poles and zeros in the transformed output is discussed, and the transform method is related to the material in earlier chapters. In the next chapter, the methods of circuit analysis are extended to nonelectrical systems.

PROBLEMS

10.1 Solve Problem 3.1 by writing and transforming a differential equation that involves the current $i(t)$ for $t > 0$. After finding $I(s)$, use Table 10.1-1 and the fact that $K/s(s + a) = (K/a)/s - (K/a)/(s + a)$ in order to find $i(t) = \mathscr{L}^{-1}[I(s)]$.

10.2 Derive the Laplace transform of each of the following functions of time.

(a) $6t^2\epsilon^{-t} + 4$.

(b) $\cosh \beta t$.

(c) $\cos (\beta t + \phi)$.

(d) $t \sin \beta t$.

(e) $f(t) = \epsilon^{-t}$ for $0 < t < 1$ and 0 for $t > 1$.

(f) $f(t) = \epsilon^{-at}$ for $0 < t < 1/a$ and 0 for $t > 1/a$.

10.3 Find the Laplace transform of each of the functions shown in Fig. P10.3. The functions in parts (d) through (f) are portions of sinusoidal waves. The functions in parts (c) and (f) are repetitive for $t > \pi$.

10.4 Prove that $\mathscr{L}[t^n] = n!/s^{n+1}$ for all positive, integral values of n.

(a) Derive the expression by performing n steps of integration by parts.

(b) The given expression was shown to be valid for the special case of $n = 1$ in Example 10.1-2. Verify the general expression by showing that *if* it is true for n, *then* it is also true for $n + 1$. (This is a standard "induction proof.")

10.5 Prove that $\mathscr{L}[tf(t)] = -dF(s)/ds$. Use this property to find the transform of $t^2\epsilon^{at}$, $t \cos \beta t$, and t^n.

10.6 Show that $\mathscr{L}[f(t)/t] = \int_s^\infty F(\lambda)\, d\lambda$, and use this property to find $\mathscr{L}[(1 - \epsilon^{-t})/t]$.

10.7 Consider the periodic function $f(t)$ in Fig. P10.7a. If $F_1(s)$ is the transform of the first cycle, which is labelled $f_1(t)$ in Fig. P10.7b, show that

$$F(s) = F_1(s)[1 + \epsilon^{-Ts} + \cdots] = \frac{F_1(s)}{1 - \epsilon^{-Ts}}$$

Find the transform of the full-wave rectified sine wave in Fig. P10.7c.

10.8 Write a node equation describing the circuit in Fig. P10.8 for $t > 0$. If the capacitance was uncharged for $t < 0$, find $E_o(s) = \mathscr{L}[e_o(t)]$. Take the inverse transform by the use of partial fractions or Eq. 10.1-10. (In this simple problem, you should be able to check the final answer by an inspection of the circuit.)

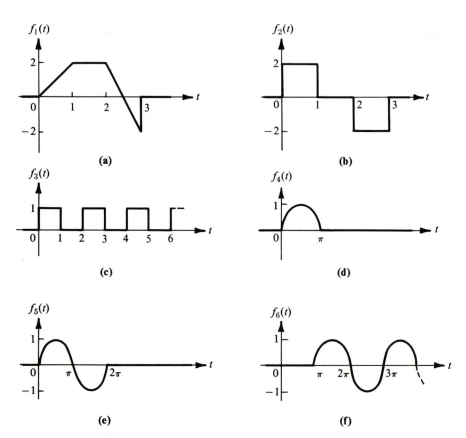

Figure P10.3

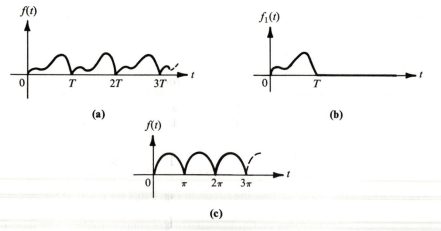

Figure P10.7

10.9 Transform the following differential equations term by term, and find $y(t)$ for all $t > 0$. (Other differential equations that may be solved by this method are given in Problem 4.1.)

(a) $\dfrac{d^2y}{dt^2} + 4\dfrac{dy}{dt} + 3y = 0,$ $y(0+) = 1, \dfrac{dy}{dt}(0+) = -1$

(b) $\dfrac{d^2y}{dt^2} + 4\dfrac{dy}{dt} + 3y = 3,$ $y(0+) = 1, \dfrac{dy}{dt}(0+) = 3$

(c) $\dfrac{d^2y}{dt^2} + 4y = 2,$ $y(0+) = \dfrac{dy}{dt}(0+) = 0$

10.10 Find the inverse transform of $F_1(s) = 1/(s + 1)(s + 2)$. Then, by Table 10.1-2, find the inverse transform of each of the following functions.

$$F_2(s) = \frac{s}{(s + 1)(s + 2)}$$

$$F_3(s) = \frac{s^2}{(s + 1)(s + 2)}$$

$$F_4(s) = \frac{10}{s(s + 1)(s + 2)}$$

$$F_5(s) = \frac{1 - \epsilon^{-3s}}{(s + 1)(s + 2)}$$

10.11 Use a partial-fraction expansion to find the inverse transform of each of the following functions.

(a) $\dfrac{s^2 - s + 1}{s^2(s + 1)}$

(b) $\dfrac{s + 4}{2s^2 + 5s + 3}$

(c) $\dfrac{s(s + 2)}{s^2 + 2s + 2}$

(d) $\dfrac{8s^2 + 8s + 1}{4s^2 + 6s + 2}$

(e) $\dfrac{\beta}{s(s^2 + \beta^2)}$

(f) $\dfrac{10}{s^2(s^2 + 4)}$

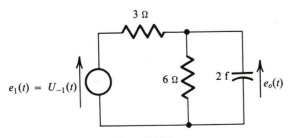

Figure P10.8

(g) $\dfrac{s + 2}{s^3(s^2 + 1)}$ (h) $\dfrac{s^2 - 4}{(s^2 + 4)^2}$ (i) $\dfrac{8(s^2 + 6s + 32)}{s(s^2 + 8s + 32)}$

(j) $\dfrac{7s^2 - 16s + 4}{s(s^2 + 4)}$ (k) $\dfrac{s}{(s + 1)^2(s^2 + 2s + 10)}$

10.12 Find the inverse transform of

$$F(s) = \frac{2s - 6}{s^2 + 2s + 5} = \frac{2s}{(s + 1)^2 + (2)^2} - \frac{6}{(s + 1)^2 + (2)^2}$$

by using the results of Example 10.1-5.

10.13 Problem 10.12 indicates one method for determining the inverse transform of $(As + B)/(s^2 + 2\alpha s + \omega_0^2)$ when the denominator has complex factors. An alternative form for a partial-fraction expansion, when $P(s)$ is of lower order than $Q(s)$, is indicated below.

$$F(s) = \frac{P(s)}{Q(s)} = \frac{P(s)}{(s - s_1) \cdots (s^2 + 2\alpha s + \omega_0^2)} = \frac{K_1}{s - s_1} + \cdots$$

$$+ \frac{As + B}{s^2 + 2\alpha s + \omega_0^2} + \cdots$$

Although this method does not involve complex numbers in the intermediate steps, explicit general formulas for the constants A and B are not available. Use such an expansion to solve Problems 10.11i, 10.11j, and 10.11k.

10.14 Show that Eq. 10.2-3 may be rewritten as

$$K_1 = \left[\frac{P(s)}{dQ(s)/ds} \right]_{s = s_1}$$

Use this result to solve Problem 10.11b.

10.15 Sometimes a partial-fraction expansion may be used to find a circuit that has a given input impedance. Find the partial-fraction expansion for each of the functions listed below. Terms in the expansion that correspond to a pair of complex-conjugate poles should be combined into a single term. Draw the circuit that is described by each of the expansions. (As indicated by Problems 6.7 through 6.10, this procedure might be an intermediate step in finding a circuit with a prescribed transfer function.)

(a) $Z(s) = \dfrac{s^2 + 9s + 10}{(s + 1)(s + 2)}$ (b) $Z(s) = \dfrac{3s(s^2 + \frac{5}{3})}{(s^2 + 1)(s^2 + 2)}$

(c) $\dfrac{1}{Z(s)} = \dfrac{(s^2 + 1)(s^2 + 2)}{3s(s^2 + \frac{5}{3})}$

NOTE: In addition to the following circuit problems, many of the problems in other chapters can be solved conveniently by the Laplace transform. These include Examples 3.5-1, 4.3-1 through 4.3-3, 4.3-5, 4.3-6, 7.2-6, and 7.2-7, and Problems 4.13 through 4.16, 4.18 through 4.20, 6.16 through 6.18, and 7.14.

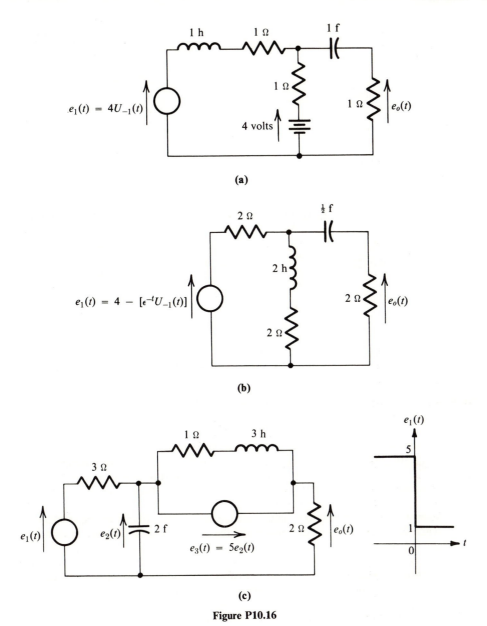

(a)

(b)

(c)

Figure P10.16

10.16 For the circuits in Fig. P10.16, find $e_o(t)$ for $t > 0$. Assume that steady-state conditions exist at $t = 0-$.

10.17 The circuits in Fig. P10.17 are operating in the steady state for $t < 0$, with the switch open. If the switch closes at $t = 0$, find $e_o(t)$ or $i_o(t)$ for $t > 0$.

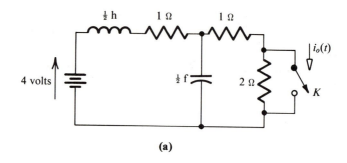

(a)

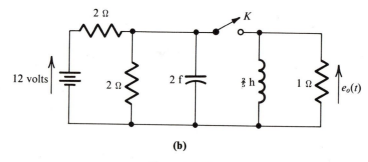

(b)

Figure P10.17

10.18 After steady-state conditions have been reached for the circuits in Fig. P10.18, the switch K opens at $t = 0$. Find $e_o(t)$ or $i_o(t)$ for $t > 0$.

10.19 Use the Laplace transform and not Norton's theorem to find $e_o(t)$ in Fig. 3.3-5a. Notice that an impulse of current flows and that the voltages across the capacitances change instantaneously. Is it necessary to calculate this change in voltage? (The discussion associated with Example 10.6-2 may be helpful.)

10.20 After steady-state conditions have been reached in the circuits in Fig. P10.20, the switch K opens. Find an expression for the voltage across the open switch as a function of time.

10.21 In Problem 5.22, use the Laplace transform to find $e_o(t)$ for all $t > 0$. Use the initial value theorem to check the values of $e_o(t)$ and de_o/dt at $t = 0+$.

10.22 For what values of s are the following functions analytic?

(a) ϵ^{st} (b) s^* (c) $|s|^2$ (d) $\tan s$

10.23 Write a Laurent series about each of the singularities in the following functions. Indicate the region of convergence of the series, and the residue at each singularity.

(a) $\dfrac{s+1}{s(s-2)^2}$ (b) $\dfrac{\epsilon^{st}}{s(s+1)}$ (c) $\dfrac{1-\epsilon^{-s}}{s}$ (d) $\dfrac{\sinh s}{s^3}$

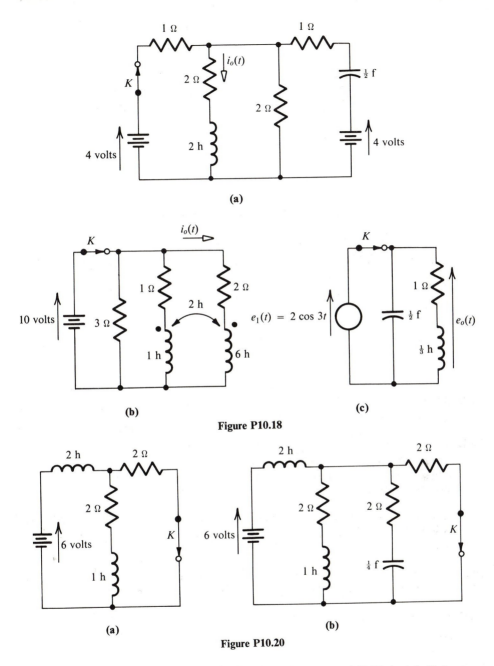

(a)

(b)

(c)

Figure P10.18

(a)

(b)

Figure P10.20

10.24 If $F(s)$ in Eq. 10.4-15 has a second-order pole at s_o, and if $P(s_o) \neq 0$, find an expression for the residue at this pole.

10.25 If $F(s)$ has a first-order pole at s_o, and if $G(s) = 1/F(s)$, show that the residue of $F(s)$ is $1/G'(s_o)$, where the prime denotes differentiation with respect to s.

10.26 Find the inverse transforms of the functions in Problem 10.11 by the methods in Section 10.5.

10.27 Find the inverse transforms of the following functions of s.

(a) $\dfrac{1}{s \cosh s}$ (b) $\dfrac{1}{s \sinh s}$

10.28 Find the unit impulse response and the unit step response for the circuits in Fig. P10.28.

10.29 The transformed output voltage for the circuit in Fig. 10.6-4 is

$$E_o(s) = \frac{s + 1}{s(s^2 + 3s + 1)}$$

Use the initial value theorem to find the value of $e_o(t)$ and its first three derivatives at $t = 0+$.

10.30 Find the initial and final values of the functions of time that correspond to the following functions of s.

(a) $\dfrac{7s - 2}{s^2 + 2s + 3}$ (b) $\dfrac{s^2 + 3}{s(s + 2)(s + 4)}$ (c) $\dfrac{1}{s(s^2 - 4s + 3)}$

(d) $\dfrac{3s^2 + 6s - 2}{s(s^2 + 4)}$ (e) $\dfrac{s^2 - 2s + 2}{s(s^2 + 2s + 2)}$

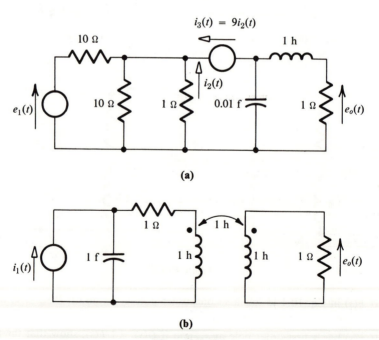

(a)

(b)

Figure P10.28

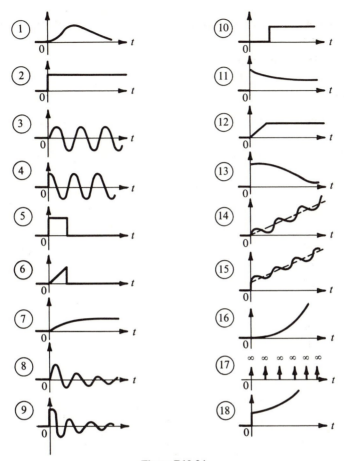

Figure P10.34

10.31 Use the convolution theorem of Eq. 10.6-8 to find $\mathscr{L}^{-1}[1/(s + a)^2]$.

10.32 Use Eq. 10.6-8 to prove Eqs. 10.1-6 and 10.1-10.

10.33 From Eq. 10.6-8, derive an equation similar to Eq. 3.5-1. For the case when $x(0+) \neq 0$, resolve any apparent difference between the two equations.

10.34 A number of functions of time are shown in Fig. P10.34. *Without any written calculations*, list the function of time that corresponds to each of the following functions of s.

(a) $\dfrac{2s^2 + 2}{s^2(s^2 + 2)}$

(b) $\dfrac{1}{s(s + 2)}$

(c) $\dfrac{1 - \epsilon^{-2s}}{s}$

(d) $\dfrac{1 - \epsilon^{-2s}}{s^2}$

(e) $\dfrac{1}{1 - \epsilon^{-2s}}$

(f) $\dfrac{1}{(s + 2)^2}$

(g) $\dfrac{s + 3}{(s + 2)^2}$

(h) $\dfrac{7s + 10}{(s + 1)(s + 2)(s - 2)}$

(i) $\dfrac{s^2 + 10s + 16}{(s + 1)(s + 2)(s - 2)}$

(j) $\dfrac{s + 2}{s^2 + 4s + 8}$

11

Scaling and Analogs

In this chapter we indicate how the results of the examples in earlier chapters may be applied to other circuits and to nonelectrical systems. In order to simplify the mathematics, the values of the passive elements in most of the previous problems have been in the vicinity of 1 Ω, 1 f, or 1 h, and the frequency of any oscillations in the response often has been only a few cycles per second. In actual practice, however, values of capacitance are usually in the microfarad or picofarad range, while resistances and inductances may be given in kilohms and millihenries. Sinusoidal oscillations may have frequencies of several kilocycles or megacycles per second. In Section 11.1, we show how the previous circuits may be converted into ones that have typical responses and elements of practical size.

In Section 2.7 we pointed out that two different circuits may be described by integral-differential equations of exactly the same form. More generally, many nonelectrical and electrical systems may be described by the same set of equations, in which case their behavior also will be the same. Thus, the methods of circuit analysis can be extended to nonelectrical systems. Furthermore, we can find a circuit that has the same behavior as a given nonelectrical system. In Sections 11.2 and 11.3 we restrict the discussion to simple mechanical and electrical systems, because they are familiar to most students. Other types of systems are discussed briefly later in the chapter.

11.1
Magnitude and Frequency Scaling

Consider a circuit with no initial stored energy, with an input $x(t)$ and an output $y(t)$, and with a transfer function $H(s) = Y(s)/X(s)$. The impedances of the passive elements in the circuit are given by $Z(s) = R$, sL, and $1/sC$, as in Table 11.1-1. *Magnitude scaling* consists of multiplying every impedance by the same constant A. For a given voltage source, all the currents will be divided by A, but all voltages

561

TABLE 11.1-1

Element	Resistance	Inductance	Capacitance
$Z(s)$	R	sL	$\dfrac{1}{sC}$
$AZ(s)$	AR	$s(AL)$	$\dfrac{1}{s(C/A)}$
$Z\left(\dfrac{s}{a}\right)$	R	$s\left(\dfrac{L}{a}\right)$	$\dfrac{1}{s(C/a)}$

will remain unchanged. Input and transfer impedances will be multiplied by A; input and transfer admittances will be divided by A; and transfer functions with units of volts per volt or amp per amp will be unaffected. By Table 11.1-1, magnitude scaling can be accomplished by multiplying the value of every resistance and inductance by A, and by dividing the value of every capacitance by A.

Magnitude scaling does not affect the pole-zero pattern of the network function. It has no effect on a circuit's free or forced response, except that it may affect the size of the response. In contrast, *frequency scaling* affects all of these things. Every s in the expression for $H(s)$ is replaced by s/a, so the network function

$$H(s) = \frac{K(s - z_1)(s - z_2)\cdots}{(s - p_1)(s - p_2)\cdots}$$

becomes

$$H\left(\frac{s}{a}\right) = \frac{K(s/a - z_1)(s/a - z_2)\cdots}{(s/a - p_1)(s/a - p_2)\cdots} = \frac{K'(s - az_1)(s - az_2)\cdots}{(s - ap_1)(s - ap_2)\cdots}$$

Thus, a pole or zero that was previously located at $s = s_k$ is now at as_k. The *relative* positions of the poles and zeros will be unchanged, but they all will be moved away from the origin of the s-plane by a multiplying factor of a. Since the poles of the transfer function determine the form of the free response, the free response will occur a times faster than before, i.e., the time constants of the circuit will be reduced.

The frequency scaling process derives its name from its effect on the a-c steady-state response, where s is replaced by $j\omega$. If the subscripts 1 and 2 refer to the original and scaled circuit, respectively, $H_2(j\omega) = H_1(j\omega/a)$. If $x_1(t) = \cos \omega_k t$ and $x_2(t) = \cos a\omega_k t$, the network functions and therefore the steady-state responses for the two circuits will have the same magnitude and phase angle. Thus, the frequency response curves of $|H(j\omega)|$ and $\measuredangle H(j\omega)$ versus ω in Chapter 6 will be stretched out along the horizontal axis by a multiplying factor of a for the scaled circuit. In the case of a filter, the bandwidth and the resonant frequency will be increased by the multiplying factor of a. These conclusions also follow from the discussion in Section 6.4 and the fact that the pole-zero configuration is moved away from the origin of the s-plane.

Replacing s by s/a in the network function is equivalent to making the same change in the impedance of every passive circuit element, as in the last row of Table 11.1-1. To frequency scale a circuit, the value of every resistance is left unchanged, and the value of every inductance and capacitance is divided by a.

Example 11.1-1. Consider the low-pass filter shown in Fig. 11.1-1a, which is similar to Fig. 6.4-5a. Its transfer function

$$H_1(s) = \frac{E_o(s)}{E_1(s)} = \frac{2}{s^2 + 2s + 2}$$

has poles at $s = -1 \pm j1$. The frequency response curve of Fig. 11.1-1c is identical with Fig. 6.4-6c, with a bandwidth of $\omega_0 = \sqrt{2}$ rad per sec. Raise the impedance level of the circuit by a factor of 1000, and then increase the bandwidth by a factor of 10^4.

Solution. To increase the impedance level of the circuit, multiply R and L and divide C by 1000, as shown in Fig. 11.1-1d. This reduces the current but leaves the transfer function unchanged.

$$H_2(s) = \frac{2}{s^2 + 2s + 2}$$

To increase the bandwidth to $\sqrt{2} \times 10^4$ rad per sec, divide L and C by 10^4. The new circuit in Fig. 11.1-1e has the transfer function

$$H_3(s) = \frac{2}{(s/10^4)^2 + 2(s/10^4) + 2} = \frac{2 \times 10^8}{s^2 + 2 \times 10^4 s + 2 \times 10^8}$$

The pole-zero pattern and the curve of $|H_3(j\omega)|$ are shown in Figs. 11.1-1f and 11.1-1g.

It is interesting to compare the impulse response for the circuits in Figs. 11.1-1a and 11.1-1e. For the first circuit,

$$h_1(t) = \mathcal{L}^{-1}[H_1(s)] = 2\epsilon^{-t} \sin t$$

while

$$h_3(t) = \mathcal{L}^{-1}[H_3(s)] = 2 \times 10^4 \epsilon^{-10^4 t} \sin 10^4 t$$

The last circuit has a shorter time constant and responds faster; the shape of its output waveshape more closely resembles the impulsive input; and the magnitude of its output is increased. These things should be expected, since by Example 9.4-2 the spectrum of an impulse contains all frequencies in equal strength. More of these frequency components "get through" the circuit that has a broad frequency response curve. The observations also agree with the fact that an impulse only inserts some energy into the circuit instantaneously, so that the impulse response indicates the form of the free response, which will occur faster for the scaled circuit.

Equation 10.1-7, which is repeated below, can contribute to an understanding of frequency scaling.

$$\mathcal{L}[f(at)] = \frac{1}{a} F\left(\frac{s}{a}\right)$$

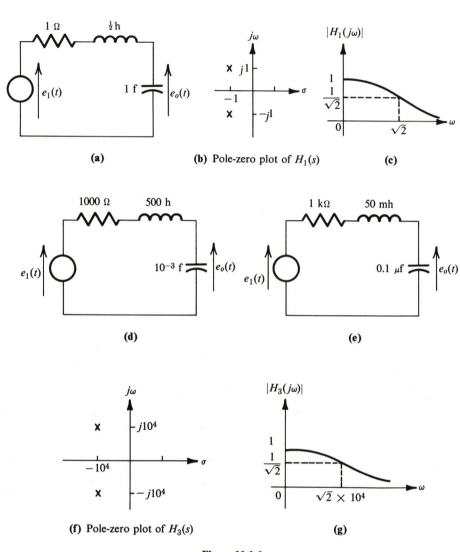

(a)

(b) Pole-zero plot of $H_1(s)$

(c)

(d)

(e)

(f) Pole-zero plot of $H_3(s)$

(g)

Figure 11.1-1

For a circuit with no initial stored energy and with an input $x_1(t)$ and an output $y_1(t)$,

$$Y_1(s) = X_1(s)H_1(s)$$

Suppose that we wish to find a second circuit for which the response is $y_2(t) = y_1(at)$. The transform of the new response is

$$Y_2(s) = \frac{1}{a} Y_1\left(\frac{s}{a}\right) = \left[\frac{1}{a} X_1\left(\frac{s}{a}\right)\right]\left[H_1\left(\frac{s}{a}\right)\right]$$

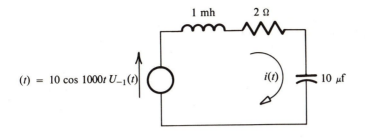

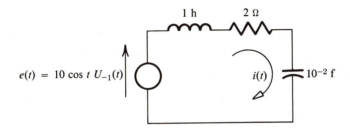

Figure 11.1-2

The quantity in the first bracket is the transform of the input $x_2(t) = x_1(at)$, while $H_2(s) = H_1(s/a)$ is the network function for a frequency scaled circuit. If $a > 1$, $y_2(t)$ and $x_2(t)$ have the same waveshapes as $y_1(t)$ and $x_1(t)$, except that they occur a times as fast. The network function $H_2(s) = H_1(s/a)$ describes a circuit whose free response also occurs a times as fast. As an example, the current $i(t)$ will be the same for the two circuits in Fig. 11.1-2, except for a change in the time scale.

Example 11.1-2. The circuit in Fig. 11.1-3 simulates the behavior of a certain non-electrical system. Modify the circuit, without changing the shape of the output waveshape, so that the largest capacitance in the final circuit is 1 μf, and so that the output occurs 1000 times faster.

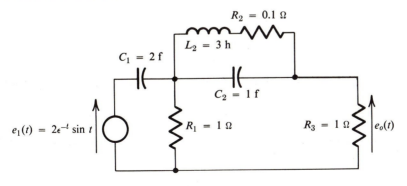

Figure 11.1-3

Solution. Because frequency scaling affects both the time scale and the size of the elements, we shall use magnitude scaling at the *end* of the problem to meet the requirement on the size of the capacitances. To make the response 1000 times faster, change the input to $e_1(t) = 2\epsilon^{-1000t} \sin 1000t$, and divide every L and C by 1000.

$$L_2 = 3 \times 10^{-3}\,\text{h}, \qquad C_1 = 2 \times 10^{-3}\,\text{f}, \qquad C_2 = 10^{-3}\,\text{f}$$

We can now multiply every R and L and divide every C by the constant A without changing the output voltage. To satisfy the problem statement, $A = 2000$ and the final element values are

$$e_1(t) = 2\epsilon^{-1000t} \sin 1000t$$

$$R_1 = R_3 = 2\,\text{k}\Omega, \qquad R_2 = 200\,\Omega$$

$$L_2 = 6\,\text{h}, \qquad C_1 = 1\,\mu\text{f}, \qquad C_2 = 0.5\,\mu\text{f}$$

The Low-Pass to Band-Pass Transformation. Another means of extending the solution of some of the earlier problems to additional circuits is by use of the low-pass to band-pass transformation. A pole-zero pattern and a frequency response curve for the network function $H_1(s)$ of a typical low-pass filter are shown in Fig. 11.1-4a. If s is replaced by $s - j\omega_0$ to form the new network function $H_2(s) = H_1(s - j\omega_0)$, the pole-zero pattern is moved vertically upward in the s-plane, as in Fig. 11.1-4b. The frequency response curve is displaced a distance ω_0 along the ω-axis. This can be seen from the discussion in Section 6.4 or by realizing that every ω has been replaced by $\omega - \omega_0$ in the a-c steady-state network function.

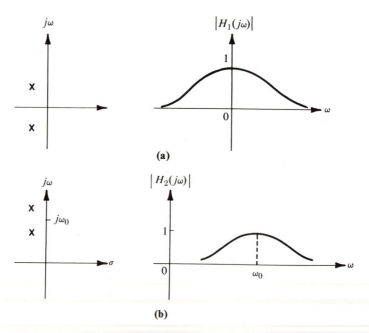

(a)

(b)

Figure 11.1-4

TABLE 11.1-2

	R	L	C
Low-pass element	⏤〰〰〰⏤	⏤◠◠◠⏤	⏤⊦⊦⏤
$Z(s)$	R	sL	$\dfrac{1}{sC}$
$Z\left(\dfrac{s^2 + \omega_0{}^2}{2s}\right)$	R	$\dfrac{sL}{2} + \dfrac{\omega_0{}^2 L}{2s}$	$\dfrac{1}{sC/2 + \omega_0{}^2 C/2s}$
Band-pass elements	R ⏤〰〰〰⏤	$\dfrac{L}{2}$ h $\quad \dfrac{2}{\omega_0{}^2 L}$ f ⏤◠◠◠⊦⊦⏤	$\dfrac{2}{\omega_0{}^2 C}$ h $\quad \dfrac{C}{2}$ f

Notice that the low-pass characteristics appear to have been converted into the characteristics of a band-pass filter. However, any complex poles in the network function of a circuit with real elements must occur in conjugate pairs. Hence, we are not able to find a circuit that has the pole-zero pattern in Fig. 11.1-4b.

In a high Q, band-pass filter, we are normally interested in the steady-state response only in the vicinity of the angular frequency ω_0. Since the factors $s + j\omega_0$ and $2s$ are approximately equal when $s \doteq j\omega_0$, let us replace every s in $H_1(s)$ by $(s - j\omega_0)(s + j\omega_0)/2s$ to form

$$H_2(s) = H_1\left(\frac{s^2 + \omega_0{}^2}{2s}\right)$$

To see how the band-pass circuit may be constructed directly from its low-pass counterpart, consider Table 11.1-2. The impedance $Z(s)$ for each of the three types of passive two-terminal elements is given in the second row. In the next row, s has

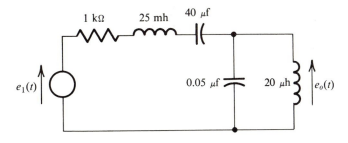

Figure 11.1-5

been replaced by $(s^2 + \omega_0^2)/2s$. From these expressions, we can list the corresponding band-pass components in the last row. Notice that the inductance and capacitance in the low-pass circuit act like short and open circuits, respectively, when $\omega = 0$. Their counterparts in a band-pass circuit act like short and open circuits when $\omega = \omega_0$.

Example 11.1-3. Convert the circuit in Fig. 11.1-1e to a band-pass circuit with a resonant angular frequency of 1 megaradian per second.

Solution. By the use of Table 11.1-2, with $\omega_0 = 10^6$, we obtain the circuit in Fig. 11.1-5. The bandwidth is $2\sqrt{2} \times 10^4$ rad per sec.

11.2
Mechanical Systems

The initial discussion of nonelectrical systems is devoted to translational mechanical systems. For simplicity, we shall assume that all the elements are constrained to move only in the horizontal (or only in the vertical) direction. Just as the behavior of circuits is expressed in terms of voltage and current, the behavior of mechanical systems is also characterized by two basic kinds of measurable variables. One of these variables is *force*, which is denoted by the symbol $f(t)$. The other basic variable may be taken to be the *displacement* $x(t)$, or the *velocity* $v(t) = dx/dt$, or the *acceleration* $a(t) = dv/dt$. In general, the variables are functions of time, even if the t in parentheses is omitted.

The passive mechanical elements that we shall consider are mass, viscous friction, and stiffness. If a force $f(t)$ is applied to a *mass M*, as in Fig. 11.2-1a,

$$f(t) = M \frac{d^2x}{dt^2} = M \frac{dv}{dt} \tag{11.2-1}$$

where $v(t)$ is the velocity of the mass relative to the earth's surface. We assume that all points on the mass move at the same velocity, that the velocity is small compared to the speed of light, and that the mass M has a constant value. If the assumed positive direction of the velocity $v(t)$ in Fig. 11.2-1 were reversed, then a minus sign would be needed in Eqs. 11.2-1 through 11.2-3.

The element *viscous friction* is used in approximating the behavior of two metal bodies that are separated by a thin film of oil, as in Fig. 11.2-1b. For simplicity, the mass of the movable block in the figure is assumed to be negligible. For laminar flow, it is found experimentally that

$$f(t) = \frac{\eta A}{d} v(t)$$

where A is the area of the mass that rests on the oil film and where d is the thickness

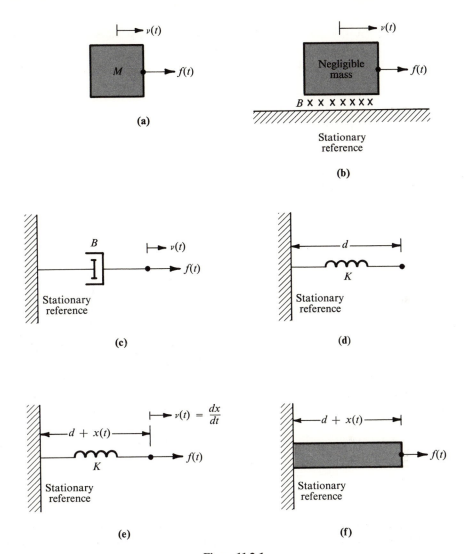

Figure 11.2-1

of the film. The symbol η denotes the viscosity of the oil that is used. These three constants are combined into the viscous friction B, so that the previous equation becomes

$$f(t) = Bv(t) \qquad (11.2\text{-}2)$$

Viscous friction also may be represented by the dashpot shown in Fig. 11.2-1c. The inner part of the symbol suggests a piston that can move in and out of a cylinder that is filled with oil.

If neither of the two masses in Fig. 11.2-1b (or neither side of the dashpot in

Fig. 11.2-1c) is stationary, then $v(t)$ in Eq. 11.2-2 is the *difference* between the velocities of the two parts. Finally, we should emphasize that viscous friction is quite different from dry friction, where two masses are in direct contact with each other.

The *stiffness* (or *spring constant*) K, which is symbolized by the spring shown in Figs. 11.2-1d and 11.2-1e, is defined by

$$f(t) = Kx(t) = K \int v \, dt \tag{11.2-3}$$

where $x(t)$ is the displacement from the equilibrium position. As in Section 1.3, the integration may be interpreted as $x(t) = \int_{-\infty}^{t} v(\lambda) \, d\lambda = x(0+) + \int_{0+}^{t} v(\lambda) \, d\lambda$. Notice for future use that, if the Laplace transform is applied to Eq. 11.2-3,

$$F(s) = K \left[\frac{N(s)}{s} + \frac{x(0+)}{s} \right]$$

where $F(s)$ and $N(s)$ are the transforms of $f(t)$ and $v(t)$, respectively. Again, if one side of the spring is not connected to a fixed reference, $v(t)$ is the difference between the velocities of the two ends.

The stiffness K may be associated with other components, such as the metallic shaft shown in Fig. 11.2-1f. Let E, A, and d denote Young's modulus (a characteristic of the metal), the cross-sectional area, and the length when no force is applied. If $x(t) = \Delta d =$ the elongation that is produced by the force $f(t)$, then

$$E = \frac{\text{stress}}{\text{strain}} = \frac{f(t)/A}{x(t)/d}$$

or

$$f(t) = \frac{EA}{d} x(t)$$

so $K = EA/d$.

A list of the mks units and of the English gravitational units that are sometimes used is given in Table 11.2-1. Notice that slugs are equivalent to lb-sec² per ft.

As in Section 1.3, the elements that we have defined are ideal and only approximate the behavior of their physical counterparts. The behavior of a metallic shaft, for example, was shown in more detail in Fig. 1.1-3. In fact, it may be difficult to obtain linear dashpots and springs.

TABLE 11.2-1

	$f(t)$	$v(t)$	M	B	K
mks units	newton	meter/sec	kilogram (kg)	newton-sec/meter	newton/meter
English units	pound (lb)	ft/sec	slug	lb-sec/ft	lb/ft

Power can be dissipated by but not stored in the element viscous friction. The power, which is dissipated as heat, is given by

$$p(t) = f(t)v(t) = Bv^2(t) = \frac{1}{B}f^2(t) \qquad (11.2\text{-}4)$$

Mass and stiffness can store energy and can later return it to the rest of the system. The kinetic energy of a moving mass is

$$w(t) = \tfrac{1}{2}Mv^2(t) \qquad (11.2\text{-}5)$$

while the potential energy stored in a spring is

$$w(t) = \tfrac{1}{2}Kx^2(t) = \frac{1}{2K}f^2(t) \qquad (11.2\text{-}6)$$

Since the stored energy of a component cannot change instantaneously unless impulses are present, the velocity of a mass and the displacement of a spring do not normally change instantaneously.

A mechanical system may be excited by either force or velocity sources (or inputs). A *force source* applies a specified force, which is usually a function of time, regardless of what the details of the system might be, while a *velocity source* moves one part of the system with a predetermined velocity. Sources are again ideal elements.

The interconnection of circuit elements was described mathematically by Kirchhoff's voltage and current laws. In a mechanical system, Newton's law says that the algebraic sum of the external forces applied to a mass is[†]

$$\sum f_{\text{external}} = M\frac{dv}{dt}$$

or

$$\sum f_{\text{external}} - M\frac{dv}{dt} = 0$$

or

$$\sum f(t) = 0 \qquad (11.2\text{-}7)$$

In the last form of the equation, $M(dv/dt)$ is regarded as an additional, retarding force that is included in the summation and that accounts for the inertia of the mass. Equation 11.2-7, which is usually called *D'Alembert's principle*, says that the algebraic sum of the forces on any body is zero. The law may be applied at the junction of several components, even if the mass of the junction is zero.

The second basic law is that the algebraic sum of the displacements around any closed path is zero. This law is equivalent to the statement that the net displacement

---

† In general, a vector sum is needed, but we are considering only systems where the motion is restricted to a straight line. This equation also assumes that the mass M has a constant value.

TABLE 11.2-2

Mechanical laws		Electrical laws	
$f = M\dfrac{dv}{dt}$	$v = \dfrac{1}{K}\dfrac{df}{dt}$	$i = C\dfrac{de}{dt}$	$e = L\dfrac{di}{dt}$
$f = Bv$	$v = \dfrac{1}{B}f$	$i = \dfrac{1}{R}e$	$e = Ri$
$f = K\displaystyle\int v\, dt$	$v = \dfrac{1}{M}\displaystyle\int f\, dt$	$i = \dfrac{1}{L}\displaystyle\int e\, dt$	$e = \dfrac{1}{C}\displaystyle\int i\, dt$
	$\sum f = 0$		$\sum i = 0$
	$\sum v = 0$		$\sum e = 0$

between any two points in the system can be uniquely defined. It can be indicated symbolically by the equation $\sum x(t) = 0$. Since $v(t) = dx/dt$,

$$\sum v(t) = 0 \tag{11.2-8}$$

around any closed path.

The element and interconnection laws for mechanical and electrical systems are summarized in Table 11.2-2. The mathematical similarities between the two types of systems will be exploited in Section 11.3.

Example 11.2-1. Write the integral-differential equation describing the system in Fig. 11.2-2. If, in a consistent set of units, $M = 1$, $B = 2$, $K = 2$, and $f(t) = U_{-1}(t)$, find $v(t)$ for all $t > 0$. Assume that the mass has no initial velocity, but that it has been displaced two units to the right of its equilibrium position before the force source is applied.

Solution. An assumed positive direction for the velocity $v(t)$ of the mass (analogous to a voltage or current reference arrow) is shown on the diagram. Setting the algebraic sum of the forces on the mass equal to zero, and expressing the forces in terms of the velocity $v(t)$, we obtain

$$f(t) - K\int v\, dt - Bv(t) - M\frac{dv}{dt} = 0$$

The first term is the applied force and the next three terms represent the retarding effects of the passive components. Notice that Eq. 11.2-8 was automatically satisfied when the same velocity symbol was used in all three terms.

The rearranged equation

$$M\frac{dv}{dt} + Bv(t) + K\int v\, dt = f(t)$$

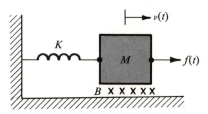

Figure 11.2-2

can be solved by either the classical or transform method. If N(s) and F(s) denote the Laplace transforms of $v(t)$ and $f(t)$, respectively, then

$$M[sN(s) - v(0+)] + BN(s) + K\left[\frac{N(s)}{s} + \frac{x(0+)}{s}\right] = F(s)$$

so

$$N(s) = \frac{sMv(0+) - Kx(0+) + sF(s)}{s^2M + sB + K}$$

For the values given in the problem statement, $v(0+) = 0$, $x(0+) = 2$, $F(s) = 1/s$, and

$$N(s) = \frac{-3}{s^2 + 2s + 2}$$

$$v(t) = \mathcal{L}^{-1}[N(s)] = -3\epsilon^{-t}\sin t \quad \text{for} \quad t > 0$$

The displacement of the mass is

$$x(t) = x(0+) + \int_{0+}^{t} v(\lambda)\, d\lambda = 2 - 3\int_{0+}^{t} \epsilon^{-\lambda} \sin \lambda\, d\lambda$$

$$= \tfrac{1}{2}[1 + 3\epsilon^{-t}(\sin t + \cos t)] \quad \text{for} \quad t > 0$$

Notice that the initial and final values of the displacement are 2 and $\tfrac{1}{2}$, as expected.

Because of the student's familiarity with circuit diagrams, it may be helpful to be able to draw a mechanical equivalent circuit. Springs and dashpots are natural two-terminal elements, since the two ends may move with two different velocities. Mass is not a natural two-terminal component, because all points on the mass move with the same velocity. However, the equation $f(t) = M(dv/dt)$ may be written as

$$f(t) = M\frac{d}{dt}(v - 0)$$

where the zero may be regarded as the velocity of the fixed reference. Thus, in the

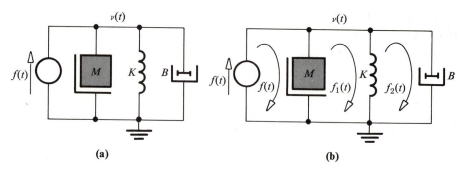

(a) **(b)**

Figure 11.2-3

construction of an equivalent mechanical circuit, one end of every mass is connected to the reference node.

The equivalent circuit for the system in Fig. 11.2-2 is shown in Fig. 11.2-3a. The fixed wall and floor are taken as the reference node. Points at the junction of M, B, and K move with an unknown velocity $v(t)$ and constitute another node. The elements K and B are connected between the two nodes, and one end of the mass is connected to the reference node, as discussed in the previous paragraph. The symbol that is used for the mass in the equivalent circuit is intended to suggest that we are regarding it as a two-terminal element. Summation of the forces for the elements connected to the top node yields the same equation that was obtained in Example 11.2-1. In fact, the proof of the validity of the equivalent circuit is that it does give the same equation as the original system.

In the solution of Example 11.2-1, we equated to zero the algebraic sum of the forces at each junction whose velocity was unknown. A junction, which usually but not always includes a mass, consists of points that move with a common velocity. It is analogous to a node in circuit diagrams, which is a collection of points at the same potential. The method of solution has obvious similarities to the node equations developed in Section 2.5.

A procedure that is analogous to the loop equation method in Section 2.6 is to apply Eq. 11.2-8 to the mechanical equivalent circuit in Fig. 11.2-3b. Although the procedure is rarely used, we include it here to emphasize the similarities between the analysis of mechanical and electrical systems. Instead of loop currents, loop forces are drawn around the three meshes. The actual force on any element is the algebraic sum of the loop forces passing through it. The drawing of these forces automatically satisfies Eq. 11.2-7 at every junction. A velocity-law equation is written around every loop that corresponds to an unknown force.

$$\frac{1}{K}\frac{d}{dt}\left(f_1 - f_2\right) + \frac{1}{M}\int \left(f_1 - f\right) dt = 0$$

$$\frac{1}{B}f_2 + \frac{1}{K}\frac{d}{dt}\left(f_2 - f_1\right) = 0$$

These equations can be transformed term by term and solved for $F_2(s)$. Notice that $(1/K)[f_1(0+) - f(0+)] = x(0+)$ and that

$$\mathscr{L}\left[\frac{1}{M}\int_{-\infty}^{t}[f(\lambda) - f_1(\lambda)]\,d\lambda\right] = \frac{1}{sM}[F(s) - F_1(s)] + \frac{1}{sM}\int_{-\infty}^{0+}[f(\lambda) - f_1(\lambda)]\,d\lambda$$

$$= \frac{1}{sM}[F(s) - F_1(s)] + \frac{v(0+)}{s}$$

The two transformed equations are

$$\frac{1}{K}[sF_1(s) - sF_2(s)] + x(0+) + \frac{1}{sM}[F_1(s) - F(s)] - \frac{v(0+)}{s} = 0$$

$$\frac{1}{B}F_2(s) + \frac{1}{K}[sF_2(s) - sF_1(s)] - x(0+) = 0$$

When these equations are solved simultaneously for $F_2(s)$, there results

$$F_2(s) = \frac{sMBv(0+) - KBx(0+) + sBF(s)}{s^2M + sB + K}$$

$$N(s) = \frac{1}{B}F_2(s) = \frac{sMv(0+) - Kx(0+) + sF(s)}{s^2M + sB + K}$$

which agrees with the expression in Example 11.2-1.

    For a system with no initial stored energy, the ratio of the transformed output to the transformed input is defined as the system function $H(s)$.[†] If, for example, $x(0+) = v(0+) = 0$ in Fig. 11.2-2, then

$$H(s) = \frac{N(s)}{F(s)} = \frac{s}{s^2M + sB + K}$$

Consider the three types of passive mechanical elements. For viscous friction, $f(t) = Bv(t)$ and

$$\frac{F(s)}{N(s)} = B$$

For a mass, $f(t) = M(dv/dt)$ and, if $v(0+) = 0$,

$$\frac{F(s)}{N(s)} = sM$$

For the stiffness element, $f(t) = K\int_{-\infty}^{t}v(\lambda)\,d\lambda$, and, if $x(0+) = K\int_{-\infty}^{0+}v(\lambda)\,d\lambda = 0$, then

$$\frac{F(s)}{N(s)} = \frac{K}{s}$$

Using these results, which are analogous to the expressions for the admittance of a

---

[†] When the output and input measurements are made at different places in the system, $H(s)$ is usually called the *transfer function*.

circuit element, we can find the system function without having to write down and formally transform the integral-differential equations. For the mechanical equivalent circuit in Fig. 11.2-3a,

$$\frac{F(s)}{N(s)} = sM + B + \frac{K}{s}$$

Example 11.2-2. Write the integral-differential equations and draw the equivalent circuit for the mechanical system shown in Fig. 11.2-4a.† The top of the spring K_1 moves with a predetermined velocity $v_1(t)$. If, in a consistent set of units, $K_1 = K_2 = K_3 = B = M_2 = M_3 = 1$, find the transfer function $N_3(s)/N_1(s)$.

Solution. There are three places in the system at which different velocity measurements can be made: at the mass M_2, at the mass M_3, and at the top of the spring K_1. The assumed positive directions for the velocities are shown in the figure. An equation is required for each junction whose velocity is unknown. At masses M_2 and M_3, respectively,

$$K_1 \int (v_2 - v_1)\, dt + M_2 \frac{dv_2}{dt} + B(v_2 - v_3) + K_2 \int (v_2 - v_3)\, dt = 0$$

$$B(v_3 - v_2) + K_2 \int (v_3 - v_2)\, dt + M_3 \frac{dv_3}{dt} + K_3 \int v_3\, dt = 0$$

When drawing the mechanical equivalent circuit, we need a reference node and a node for each part of the system that moves with a different velocity. The location of the dashpot and springs in the diagram corresponds to the original system, and one end of each of the masses is connected to the reference node. The previous two integral-differential equations may be written directly from the mechanical equivalent circuit in Fig. 11.2-4b.

The transfer function may be found by solving the following two equations simultaneously.

$$\frac{K_1}{s}[N_2(s) - N_1(s)] + sM_2N_2(s) + B[N_2(s) - N_3(s)] + \frac{K_2}{s}[N_2(s) - N_3(s)] = 0$$

$$B[N_3(s) - N_2(s)] + \frac{K_2}{s}[N_3(s) - N_2(s)] + sM_3N_3(s) + \frac{K_3}{s}N_3(s) = 0$$

When the given numerical values are inserted,

$$\left(s + 1 + \frac{2}{s}\right)N_2(s) - \left(1 + \frac{1}{s}\right)N_3(s) = \frac{1}{s}N_1(s)$$

$$-\left(1 + \frac{1}{s}\right)N_2(s) + \left(s + 1 + \frac{2}{s}\right)N_3(s) = 0$$

and

$$\frac{N_3(s)}{N_1(s)} = \frac{s(s + 1)}{s^4 + 2s^3 + 4s^2 + 2s + 3}$$

† In all the examples, the force of gravity on the masses is neglected. This force simply causes a constant displacement, and the total displacement always can be found by superposition.

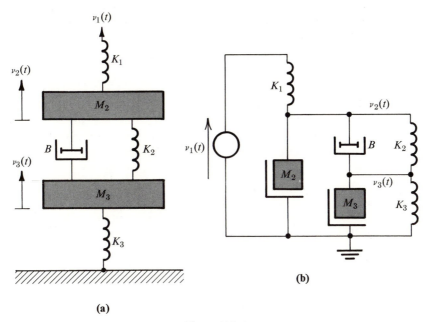

Figure 11.2-4

Rotational Mechanical Systems.   The discussion of rotational systems is parallel to that for translational systems and is therefore somewhat more concise. One of the two basic types of measurable variables in a rotational system is the torque $\tau(t)$. The other may be taken to be the angular displacement $\theta(t)$, or the angular velocity $\omega_r(t) = d\theta/dt$, or the angular acceleration $\alpha(t) = d\omega_r/dt$. A body rotating about a fixed axis is shown in Fig. 11.2-5a. If $v(t)$ is the velocity of a differential mass dm that is a distance r from the axis of rotation, $v(t) = r\omega_r(t)$.

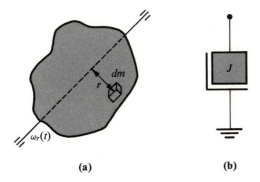

Figure 11.2-5

If the only external force on the rotating body in Fig. 11.2-5a is an applied torque $\tau(t)$, then

$$\tau(t) = J \frac{d^2\theta}{dt^2} = J \frac{d\omega_r}{dt} \tag{11.2-9}$$

In one of the problems at the end of this chapter, the student is asked to show that the *moment of inertia J* is given by

$$J = \int r^2 \, dm$$

where the integration is over the entire body. In a mechanical equivalent circuit, the moment of inertia is represented by the symbol in Fig. 11.2-5b. If J is to be regarded as a two-terminal element in an equivalent circuit, one of the two terminals must be connected to the stationary reference.

Rotational viscous friction B_r arises when two bodies are separated by a film of oil or air, as in Fig. 11.2-6a. If the moment of inertia of the rotating body is negligible, then

$$\tau(t) = B_r \frac{d\theta}{dt} = B_r \omega_r(t) \tag{11.2-10}$$

If both bodies are in motion, $\omega_r(t)$ is the *difference* between the two angular velocities. Rotational viscous friction also might approximate the effect of the damping vanes shown in Fig. 11.2-6b. In a mechanical equivalent circuit, it is represented by the symbol in Fig. 11.2-6c.

The *rotational stiffness K_r* is usually associated with a thin metal shaft. One end of the circular shaft in Fig. 11.2-7a is fixed, and the top end is subjected to a torque $\tau(t)$, which causes an angular displacement $\theta(t)$.

$$\tau(t) = K_r \theta(t) = K_r \int \omega_r \, dt \tag{11.2-11}$$

In one of the problems at the end of the chapter, the student is asked to show that

$$K_r = \frac{\pi E_s R^4}{2d}$$

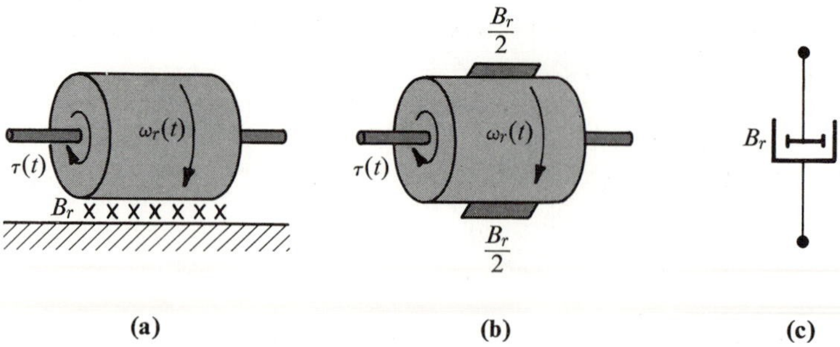

(a)   (b)   (c)

Figure 11.2-6

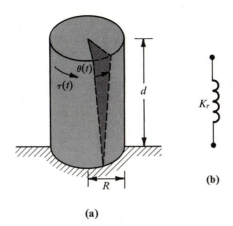

Figure 11.2-7

where E_s is the shear modulus for the material used. If both ends of the shaft are allowed to move, $\omega_r(t)$ is the difference between the angular velocities of the two ends. The symbol used in a mechanical equivalent circuit is shown in Fig. 11.2-7b.

The integral in Eq. 11.2-11 is interpreted as

$$\tau(t) = K_r \int_{-\infty}^{t} \omega_r(\lambda)\, d\lambda = K_r\left[\theta(0+) + \int_{0+}^{t} \omega_r(\lambda)\, d\lambda\right]$$

If $T(s)$ and $\Omega(s)$ denote the Laplace transforms of $\tau(t)$ and $\omega_r(t)$, respectively,

$$T(s) = \frac{K_r}{s}\left[\theta(0+) + \Omega(s)\right]$$

Energy can be dissipated as heat in the rotational viscous friction, and it can be stored in a body exhibiting moment of inertia or rotational stiffness. The kinetic energy associated with the moment of inertia is $w(t) = \frac{1}{2}J\omega_r^2(t)$, while the potential energy associated with the stiffness is $w(t) = \frac{1}{2}K_r\theta^2(t)$. If no impulses are present, therefore, the angular velocity of a rotating mass and the angular displacement of a shaft cannot change instantaneously.

The input to a rotational system may be a torque or an angular velocity source. Once again, there are two interconnection laws. The algebraic sum of the torques acting on a rotating mass (or applied to any junction) is zero. Symbolically,

$$\sum \tau(t) = 0 \qquad\qquad (11.2\text{-}12)$$

In this equation, the quantity $J(d\omega_r/dt)$ is regarded as an additional, retarding force that must be included in the summation. The second interconnection law says that the net angular displacement between any two points can be uniquely defined, i.e., that the algebraic sum of the angular displacements around a closed loop is zero. Symbolically, $\sum \theta(t) = 0$, or, equivalently,

$$\sum \omega_r(t) = 0 \qquad (11.2\text{-}13)$$

By the use of Eqs. 11.2-9 through 11.2-13, we can write a set of integral-differential equations that describe a given system.

Example 11.2-3.   Write the integral-differential equation describing the system in Fig. 11.2-8a, and draw the mechanical equivalent circuit. The rod connecting the disk to the fixed reference is assumed to have negligible mass. If $J = 1$ slug-ft^2 and $K_r = 100$ lb-ft, what is the smallest value of B_r that will prevent oscillations in the free response?

Solution.   Summation of the torques acting on the disk gives

$$J \frac{d\omega_r}{dt} + B_r \omega_r + K_r \int \omega_r \, dt = \tau$$

or

$$J \frac{d^2\theta}{dt^2} + B_r \frac{d\theta}{dt} + K_r \theta = \tau$$

The mechanical equivalent circuit is shown in Fig. 11.2-8b. The form of the free response is determined by the roots of the characteristic equation $r^2 + (B_r/J)r + (K_r/J) = 0$ or, equivalently, by the poles of the system function $H(s)$. As discussed in connection with Fig. 4.4-2b, the roots must be real if the free response is to contain no oscillations. In the notation of Eq. 4.4-2, $\omega_0^2 = K_r/J$ and $2\zeta\omega_0 = B_r/J$. The damping ratio $\zeta = B_r/2\sqrt{JK_r}$ must be at least equal to unity, so for the given numerical values

$$B_r \geq 2 \sqrt{JK_r} = 20 \text{ lb-ft per sec}$$

Example 11.2-4.   Draw the mechanical equivalent circuit and write the integral-differential equations for the rotational system in Fig. 11.2-9a. For simplicity, the

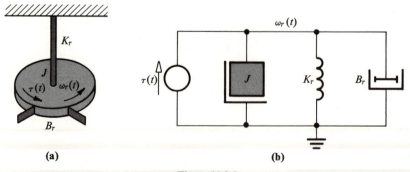

(a)

(b)

Figure 11.2-8

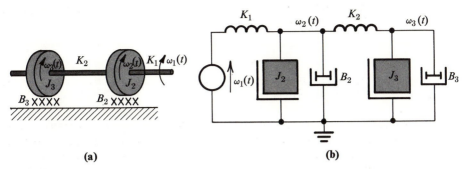

Figure 11.2-9

subscript r for the elements and for the angular velocities has been dropped. The right end of the system is moved with a predetermined angular velocity $\omega_1(t)$. If, in a consistent set of units, $K_1 = K_2 = B_2 = B_3 = J_2 = J_3 = 1$, find the system function $H(s) = \Omega_3(s)/\Omega_1(s)$, where $\Omega(s)$ is the Laplace transform of $\omega(t)$.

Solution. The mechanical equivalent circuit is shown in Fig. 11.2-9b. Summing the torques at every junction whose angular velocity is unknown, we obtain

$$K_1 \int (\omega_2 - \omega_1)\, dt + J_2 \frac{d\omega_2}{dt} + B_2 \omega_2 + K_2 \int (\omega_2 - \omega_3)\, dt = 0$$

$$K_2 \int (\omega_3 - \omega_2)\, dt + J_3 \frac{d\omega_3}{dt} + B_3 \omega_3 = 0$$

When the system contains no initial stored energy, and when the given numerical values are inserted, the transformed equations are

$$\left(s + 1 + \frac{2}{s}\right)\Omega_2(s) = \frac{1}{s}\Omega_3(s) = \frac{1}{s}\Omega_1(s)$$

$$-\frac{1}{s}\Omega_2(s) + \left(s + 1 + \frac{1}{s}\right)\Omega_3(s) = 0$$

which yield

$$H(s) = \frac{\Omega_3(s)}{\Omega_1(s)} = \frac{1}{s^4 + 2s^3 + 4s^2 + 3s + 1}$$

11.3

Electrical-Mechanical Analogs

The similarity between the basic laws for electrical and for translational systems was pointed out in Table 11.2-2. The similarity also applies to rotational systems, as can be seen from Eqs. 11.2-9 through 11.2-13.

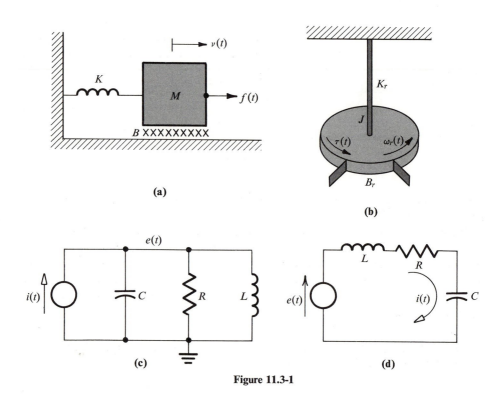

Figure 11.3-1

To make the analogy still clearer, reconsider the four systems shown in Fig. 11.3-1. The equations that describe them are, respectively,

$$M\frac{dv}{dt} + Bv + K\int v\,dt = f, \qquad J\frac{d\omega_r}{dt} + B_r\omega_r + K_r\int \omega_r\,dt = \tau$$

$$C\frac{de}{dt} + Ge + \frac{1}{L}\int e\,dt = i, \qquad L\frac{di}{dt} + Ri + \frac{1}{C}\int i\,dt = e$$

Since the equations are identical, except for the use of different symbols, the systems are called *analogs*. If the systems have the same input and the same initial stored energy, their responses also will be the same. As soon as one system has been solved, the other solutions may be written down immediately. Two systems of the same type that are described by similar equations, such as the two circuits in Fig. 11.3-1, are called *duals*.

Frequently, an electrical analog is desired for a given mechanical system. Perhaps the system is to be studied experimentally when some of its components are changed, and it is much easier to use variable circuit elements in the analog than to physically change the mechanical components. Measurements made on the electrical analog can be converted into the corresponding mechanical quantities.

TABLE 11.3-1

Translational-electrical analogs		Rotational-electrical analogs	
$f \sim i$	$f \sim e$	$\tau \sim i$	$\tau \sim e$
$v \sim e$	$v \sim i$	$\omega_r \sim e$	$\omega_r \sim i$
$M \sim C$	$M \sim L$	$J \sim C$	$J \sim L$
$B \sim G$	$B \sim R$	$B_r \sim G$	$B_r \sim R$
$K \sim \dfrac{1}{L}$	$K \sim \dfrac{1}{C}$	$K_r \sim \dfrac{1}{L}$	$K_r \sim \dfrac{1}{C}$

The study of analogs also provides an introduction to the general area of simulation and analog computers.

The equations for the systems in Fig. 11.3-1 lead to the four lists of analogous quantities in Table 11.3-1. We may find an electrical analog for a translational system by writing the equations describing the mechanical system, and then replacing the mechanical symbols by the analogous electrical symbols, using either the force-current or the force-voltage analog. Finally, the circuit described by the new equations is drawn.

Example 11.3-1.   Find two electrical analogs for the system in Fig. 11.2-4a, which is described by the following equations:

$$K_1 \int (v_2 - v_1)\, dt + M_2 \frac{dv_2}{dt} + B(v_2 - v_3) + K_2 \int (v_2 - v_3)\, dt = 0$$

$$B(v_3 - v_2) + K_2 \int (v_3 - v_2)\, dt + M_3 \frac{dv_3}{dt} + K_3 \int v_3\, dt = 0$$

Solution.   If the force-current analog in the first column of Table 11.3-1 is used, the equations become

$$\frac{1}{L_1} \int (e_2 - e_1)\, dt + C_2 \frac{de_2}{dt} + G(e_2 - e_3) + \frac{1}{L_2} \int (e_2 - e_3)\, dt = 0$$

$$G(e_3 - e_2) + \frac{1}{L_2} \int (e_3 - e_2)\, dt + C_3 \frac{de_3}{dt} + \frac{1}{L_3} \int e_3\, dt = 0$$

These equations resemble a set of node equations, since every term has the units of amperes. Because the equations contain three different voltages, the circuit must have three nodes plus a reference node. According to the first equation, L_1 is connected between nodes 2 and 1, a resistance R and an inductance L_2 between nodes 2 and 3, and C_2 between node 2 and the reference node. The second equation is interpreted in a similar way. The complete circuit is shown in Fig. 11.3-2a.

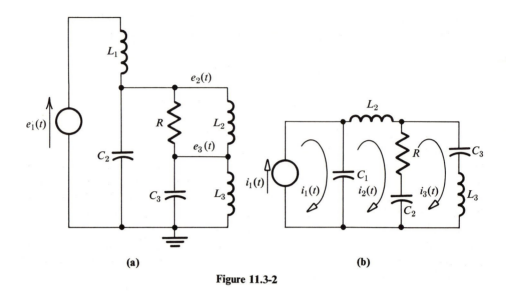

(a)   **(b)**

Figure 11.3-2

If the force-voltage analog is used instead, the original equations become

$$\frac{1}{C_1} \int (i_2 - i_1)\, dt + L_2 \frac{di_2}{dt} + R(i_2 - i_3) + \frac{1}{C_2} \int (i_2 - i_3)\, dt = 0$$

$$R(i_3 - i_2) + \frac{1}{C_2} \int (i_3 - i_2)\, dt + L_3 \frac{di_3}{dt} + \frac{1}{C_3} \int i_3\, dt = 0$$

The symbols i_1, i_2, and i_3 represent three loop currents, and the equations describe the circuit in Fig. 11.3-2b.

The two circuits in Fig. 11.3-2 are duals, and one may be derived from the other by the methods in Section 2.7. Notice that the force-current analog is the same as the mechanical equivalent circuit in Fig. 11.2-4b, except that the mechanical elements have been replaced by the analogous electrical elements. This is always true, so we may draw the force-current analog without writing any equations. The use of the force-voltage analog gives a circuit that is not similar to the mechanical equivalent circuit. One way to obtain it without any written equations is to draw the dual of the force-current analog.

Although historically the force-voltage analog appears to have been the most popular, many contemporary authors prefer the force-current analog. As has already been pointed out, the latter may be easier to find. Furthermore, the velocity used in the equations for a dashpot or a spring is the difference between the velocities at the two ends, just as the voltage for a circuit element is the difference between the potentials of the two terminals. It seems natural, therefore, to regard velocity as a "potential" or "across" variable, and force as a "flow" or "through" variable.[†] This approach is reinforced by comparison of the equations

$$\sum f(t) = 0 \quad \text{at a junction}$$

$$\sum v(t) = 0 \quad \text{around a closed path}$$

to Kirchhoff's current and voltage laws, respectively.

After an electrical analog for a mechanical system has been obtained, it is necessary to relate the currents and voltages in the analog to the desired displacements, velocities, or accelerations in the original system. Then measurements made on the analog can be converted into the corresponding mechanical quantities. This is easily done if the values of the electrical elements are numerically equal to their mechanical counterparts. In practice, however, such a procedure gives unreasonable sizes for the electrical elements. Furthermore, the operation of mechanical systems is usually much slower than that of most circuits. Hence the scaling processes described in Section 11.1 must be used to obtain practical circuits.

Example 11.3-2. Consider the force-current analog in Fig. 11.3-2a for the mechanical system in Fig. 11.2-4a. Assume that $M_2 = 0.2$ slugs, $M_3 = 0.3$ slugs, $K_1 = 1$ lb per ft, $K_2 = 2$ lb per ft, $K_3 = 3$ lb per ft, $B = 1$ lb-sec per ft, and $v_1(t) = \cos 2t$ ft per sec. Modify the electrical analog so that the largest capacitance is 6 μf and so that a 20 volt rms, 1000 rad per sec source may be used. How can the displacement, velocity,

TABLE 11.3-2

Elements	Original values	After frequency scaling	After magnitude scaling	Final values
C_2	0.2 f	400 μf	4 μf	4 μf
C_3	0.3 f	600 μf	6 μf	6 μf
R	1 Ω	1 Ω	100 Ω	100 Ω
L_1	1 h	0.002 h	0.2 h	0.2 h
L_2	$\frac{1}{2}$ h	0.001 h	0.1 h	0.1 h
L_3	$\frac{1}{3}$ h	0.67 mh	0.067 h	0.067 h
$e_1(t)$	$\cos 2t$	$\cos 1000t$	$\cos 1000t$	$20\sqrt{2} \cos 1000t$

[†] The adjectives "across" and "through" are intended to suggest how the variables might be measured.

and acceleration of the mass M_3 be determined from measurements made on the electrical analog?

Solution. If $e_1(t) = v_1(t)$, $C_2 = M_2$, $G = B$, $1/L_1 = K_1$, etc., as in the second column of Table 11.3-2, then $v_3(t) = e_3(t)$. However, the angular frequency of the electrical input must be 500 times that of the mechanical input. Therefore, the circuit must act 500 times as fast, so the inductances and capacitances are divided by 500 in the third column of Table 11.3-2.

Since the largest capacitance is now 600 μf instead of 6 μf, we must increase the impedance level of the circuit by multiplying the resistance and inductances by 100 and by dividing the capacitances by 100. This step, which is carried out in the fourth column of the table, does not change the relationship between the input and output voltages. Except for a change in the time scale, $v_3 = e_3$. Finally, if a 20 volt rms source is used, as in the last column, all voltages are increased proportionately, and $v_3 = (1/20\sqrt{2})e_3$.

Care must be exercised when measured voltages and currents in the analog are converted into the corresponding mechanical quantities. Since the time scale of the circuit has been changed, let t_m and t_e represent time in the mechanical and electrical systems, respectively. In this example, the electrical system responds 500 times faster and $t_m = 500t_e$; i.e., what happens at $t = 1$ sec in the circuit occurs at $t = 500$ sec in the mechanical system.

As previously discussed, the velocity of the mass M_3 is given by $v_3 = (1/20\sqrt{2})e_3$. Although the displacement and acceleration could be obtained by integration and

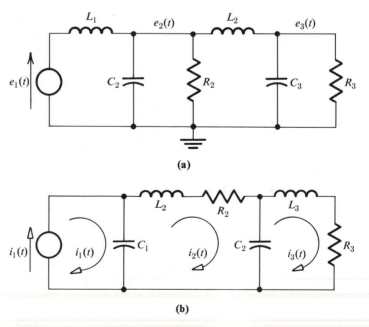

(a)

(b)

Figure 11.3-3

differentiation of the velocity waveshape, they also can be obtained directly from current measurements.

$$x_3 = \int v_3 \, dt_m = \int \left(\frac{e_3}{20\sqrt{2}}\right)(500 \, dt_e) = \left(\frac{25}{\sqrt{2}} L_3\right)\frac{1}{L_3}\int e_3 \, dt_e = 1.18 i_{L_3}$$

$$a_3 = \frac{dv_3}{dt_m} = \frac{1}{\sqrt{2}\times 10^4}\frac{de_3}{dt_e} = \frac{1}{\sqrt{2}\times 10^4 C_3} C_3 \frac{de_3}{dt} = 11.8 i_{c_3}$$

Thus, the displacement and the acceleration of the mass M_3 are proportional to the currents through L_3 and C_3, respectively.

The procedures for obtaining an electrical analog for a rotational mechanical system are similar to those developed for translational systems. The last two columns in Table 11.3-1 list the two sets of analogous quantities. The torque-current and torque-voltage analogs for the system in Fig. 11.2-9a are shown in Fig. 11.3-3. The first of these may be drawn directly from the mechanical equivalent circuit, and the student should justify the second one.

11.4

Other Types of Systems

Among other types of systems that often may be approximated by the inter-connection of idealized, lumped, linear elements are hydraulic, pneumatic, acousti-cal, and thermal systems.[†] In this section we present the laws describing some of these systems and discuss the calculation of the system function and the drawing of an electrical analog. In some applications, these systems exhibit appreciable nonlinear and distributed effects. It may be more difficult to approximate them by lumped linear models than it is for electrical and mechanical systems.

Hydraulic Systems. In addition to problems involving water flow, hydraulic systems are frequently used in control systems and in machines for highway and building construction. For a given horsepower rating, hydraulic motors are usually smaller, more reliable, and have a faster speed of response than electromechanical motors.

One of the two basic, measurable variables in a hydraulic system is the *rate of flow*, which is denoted by $q(t)$ and which has units of cubic meters per second or

---

[†] Several books, including some on feedback control systems, consider systems having many kinds of components. See, for example, J. E. Gibson and F. B. Tuteur, *Control System Com-ponents* (New York: McGraw-Hill, 1958); H. Chestnut and R. W. Mayer, *Servomechanisms and Regulating System Design*, vol. 1, 2nd ed. (New York: Wiley, 1959); G. J. Thaler and R. G. Brown, *Analysis and Design of Feedback Control Systems*, 2nd ed. (New York: McGraw-Hill, 1960); G. F. Paskusz and B. Bussell, *Linear Circuit Analysis* (Englewood Cliffs, N.J.: Prentice-Hall, 1963); and R. S. Sanford, *Physical Networks* (Englewood Cliffs, N.J.: Prentice-Hall, 1965).

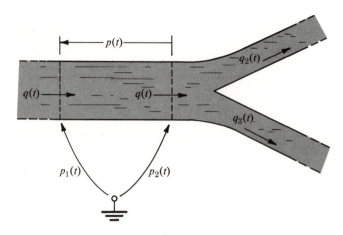

Figure 11.4-1

cubic feet per second. For the fluid-filled pipes shown in Fig. 11.4-1, $q(t)$ is the rate at which the fluid enters at the left. If, at some instant of time, the direction of fluid flow is opposite to that of the reference arrow, then $q(t)$ is negative. If the fluid is incompressible, the algebraic sum of the flow rates at any junction must be zero.

$$\sum q(t) = 0 \tag{11.4-1}$$

This law was implicitly invoked when the same symbol $q(t)$ was used in two places in Fig. 11.4-1. Also in this figure, $q(t) = q_2(t) + q_3(t)$. Notice the similarity of the flow rate to current, and the similarity of Eq. 11.4-1 to Kirchhoff's current law.

The second basic variable is the *pressure difference* $p(t)$, which is measured in newtons per square meter or pounds per square foot. In Fig. 11.4-1, $p(t)$ represents the difference in pressure between the two surfaces that are denoted by the dashed lines. Frequently, the pressure of the atmosphere or the pressure of a vacuum is assigned a value of zero, and all other pressures are expressed with respect to this reference.[†] A point of zero pressure is indicated in Fig. 11.4-1 by the ground symbol, and $p_1(t)$ and $p_2(t)$ denote pressures with respect to this reference. The algebraic sum of the pressure differences around any closed path is zero.

$$\sum p(t) = 0 \tag{11.4-2}$$

Thus, in Fig. 11.4-1, $p(t) = p_1(t) - p_2(t)$. Pressure and Eq. 11.4-2 are analogous to voltage and Kirchhoff's voltage law. The power consumed in the pipe between the two dashed lines in Fig. 11.4-1 is $p(t)q(t)$.

Fluids flowing through pipes or around obstacles exhibit frictional effects, which

---

[†] "Absolute pressure" is measured with respect to the pressure of a vacuum, while "gage pressure" is measured with respect to the pressure of the atmosphere.

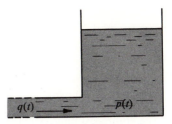

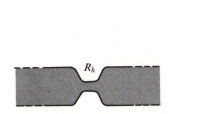

Figure 11.4-2   Figure 11.4-3

can be approximated by a lumped element, *hydraulic resistance* R_h. For the notation in Fig. 11.4-1,

$$p(t) = R_h q(t) \qquad (11.4\text{-}3)$$

It is found experimentally that the hydraulic resistance is directly proportional to the length of a pipe and decreases as the cross-sectional area increases. Although the resistance may depend to some extent upon the flow rate (and is therefore a nonlinear element), we shall assume that it is a constant, independent of $p(t)$ or $q(t)$. The symbol for hydraulic resistance is shown in Fig. 11.4-2. Any frictional effects in a pipe are usually lumped into one hydraulic resistance, so, unless otherwise stated, ordinary pipes in the figures are assumed to be frictionless (which is analogous to assuming that connecting wires in circuit diagrams have no resistance). The power dissipated in the hydraulic resistance is

$$q(t)p(t) = R_h q^2(t) = \frac{1}{R_h} p^2(t)$$

Consider the open storage tank shown in Fig. 11.4-3. If the atmospheric pressure is taken as the reference, the pressure $p(t)$ at the inlet is $\rho g h(t)$, where ρ is the density of the fluid, g is the acceleration of gravity, and $h(t)$ is the height of the fluid. If the tank has a constant cross-sectional area A, the flow rate is

$$q(t) = \frac{d}{dt}\left[Ah(t) \right] = \frac{d}{dt}\left[\frac{A}{\rho g} p(t) \right] = \frac{A}{\rho g} \frac{dp}{dt}$$

Thus,

$$q(t) = C_h \frac{dp}{dt} \quad \text{and} \quad p(t) = \frac{1}{C_h} \int q \, dt \qquad (11.4\text{-}4)$$

where the *hydraulic capacitance* is $C_h = A/\rho g$. As in Section 1.3, the integral may be interpreted as

$$p(t) = \frac{1}{C_h} \int_{-\infty}^{t} q(\lambda) \, d\lambda = p(0+) + \frac{1}{C_h} \int_{0+}^{t} q(\lambda) \, d\lambda$$

The potential energy associated with the hydraulic capacitance is $\frac{1}{2} C_h p^2(t)$.

A hydraulic capacitance also may represent the fact that all fluids are to some extent compressible. (Also, it may account for the fact that under extreme conditions the diameter of a pipe may increase slightly as the pressure increases.) The *bulk modulus B* of a fluid is defined by the equation

$$dV = \frac{V}{B} dp$$

where V is the volume of fluid in a container and where dV is the differential decrease in volume resulting from a differential increase in the pressure.

For the pipe shown in Fig. 11.4-4a, $q_1(t) = q_2(t)$ only for incompressible fluids. Otherwise, if $q(t) = q_1(t) - q_2(t)$,

$$q(t) = \frac{V}{B} \frac{dp}{dt} = C_h \frac{dp}{dt}$$

In order to maintain the validity of Eq. 11.4-1, the compressibility is accounted for by the hydraulic capacitance in Fig. 11.4-4b. In many hydraulic systems, the compressibility of the fluid can be neglected. It becomes a very important factor, however, if the techniques of this section are extended to acoustic and pneumatic systems.

A third passive element accounts for the fact that Newton's laws of motion apply to any moving mass. The mass of the fluid in a frictionless pipe of length d and cross-sectional area A is $\rho A d$. The velocity of the fluid is $v(t) = q(t)/A$, so the force required to move the fluid is

$$f(t) = \rho A d \frac{d}{dt} \frac{q(t)}{A}$$

The corresponding pressure difference $p(t)$ equals $f(t)/A$, so

$$p(t) = \frac{\rho d}{A} \frac{dq}{dt}$$

or

$$p(t) = I \frac{dq}{dt} \quad \text{and} \quad q(t) = \frac{1}{I} \int p \, dt \tag{11.4-5}$$

The *inertance I*, which equals $\rho d/A$, is often indicated by the same symbol that is used for hydraulic resistance, except that the letter R_h is replaced by I. The kinetic energy associated with the inertance is $\frac{1}{2} I q^2(t)$. In many, but not all, hydraulic problems, the inertance is negligible.

The two types of hydraulic sources are pressure sources and velocity sources. A pressure source may represent a mechanically applied force or perhaps the outlet of a water tank in which the water level is varied in a predetermined way. On the other hand, many hydraulic pumps maintain a fluid flow rate that is proportional to the speed of the pump's shaft. (If the shaft speed is held constant, the flow rate may be varied by moving a control vane.)

The element and interconnection laws in Eqs. 11.4-1 through 11.4-5 provide the basis for writing the integral-differential equations to describe a given system. The

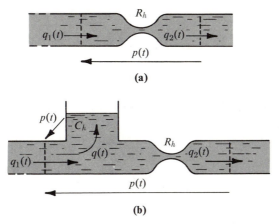

Figure 11.4-4

necessary initial conditions may be obtained by noting that the pressure across a hydraulic capacitance and the flow rate for an inertance cannot normally change instantaneously, since the stored energy of any component does not change instantaneously unless impulses are present. When an hydraulic equivalent circuit is drawn, the symbols used for the hydraulic resistance, capacitance, and inertance are similar to those for electrical resistance, capacitance, and inductance. A pressure-voltage analog may be obtained from either the original equations or from the hydraulic equivalent circuit. In most cases a pressure-current analog, which is the dual of the first analog, also can be found.

In the following examples of hydraulic systems, the subscript h is omitted when this does not cause confusion. It is assumed that the effects of any compressibility and inertance are negligible. A problem involving both hydraulic and mechanical components is solved in Section 11.5.

> **Example 11.4-1.**   A pump delivers $q_1(t)$ cubic meters per sec to the left end of the pipe shown in Fig. 11.4-5a. The right end of the pipe is open to the atmosphere, and $f_3(t)$ is a mechanically applied force at the top of a tank of cross-sectional area A. Write the differential equation relating $q_1(t)$ and $q_0(t)$. If $f_3(t) = 0$ and if there is no stored energy in the tank at $t = 0+$, find the transfer function $H(s) = Q_0(s)/Q_1(s)$. Draw the hydraulic equivalent circuit and two electrical analogs.
>
> **Solution.**   The atmospheric pressure $p_0(t)$ is assigned a value of zero, and the other pressures indicated on the figure are expressed with respect to this reference. Since the algebraic sum of the flow rates at any junction is zero,
>
> $$q_1(t) = q_2(t) + q_0(t) = C\frac{d}{dt}[p_2(t) - p_3(t)] + \frac{1}{R_2}p_2(t)$$
>
> where $p_3(t) = f_3(t)/A$. After $p_2(t)$ has been obtained, $p_1(t)$ may be found from the equation
>
> $$p_1(t) - p_2(t) = R_1 q_1(t)$$

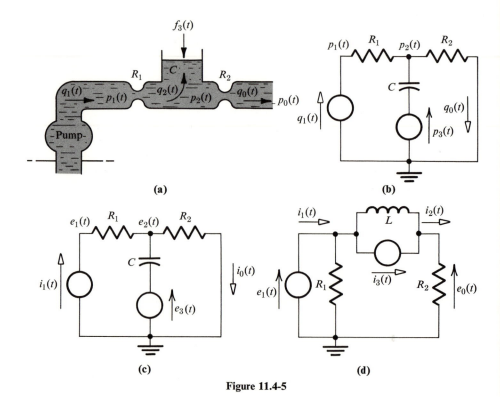

Figure 11.4-5

If $f_3(t) = 0$ and if $p_2(0+) = 0$, the Laplace transform of the first equation is

$$Q_1(s) = sCP_2(s) + \frac{1}{R_2} P_2(s)$$

so

$$H(s) = \frac{Q_0(s)}{Q_1(s)} = \frac{1}{R_2} \frac{P_2(s)}{Q_1(s)} = \frac{1}{sCR_2 + 1}$$

In the hydraulic equivalent circuit, there must be one node corresponding to each different pressure, including the reference pressure $p_0(t) = 0$. A complete equivalent circuit is shown in Fig. 11.4-5b. The pressure-voltage and pressure-current analogs are given in parts (c) and (d) of the figure. One way to obtain the latter analog is to replace $p(t)$ by $i(t)$, $q(t)$ by $e(t)$, C_h by L, R_h by G, and I by C in the equations for the hydraulic system. The resulting equations

$$e_1(t) = L \frac{d}{dt} [i_2(t) - i_3(t)] + R_2 i_2(t)$$

$$i_1(t) - i_2(t) = G_1 e_1(t)$$

do describe the circuit in Fig. 11.4-5d.

Example 11.4-2.   In the hydraulic system shown in Fig. 11.4-6a, the value is closed for $t < 0$, but opens at $t = 0$. The two horizontal pipes are assumed to be on the same level, although this is difficult to show in the figure. The pump produces a constant flow rate $q_1(t) = 1$ cubic meter per sec for all values of t. In mks units, assume that $R_1 = 2$, $R_2 = R_3 = 1$, and $C_1 = C_2 = 1$. If steady-state conditions exist at $t = 0-$, find $q_0(t)$ for all $t > 0$.

Solution.   At $t = 0-$, all pressures and flow rates are constant, so $p_b(0-) = p_0(0-) = 0$. Otherwise, there would be fluid flow through R_2, and the height of the fluid in the right-hand tank and hence $p_b(t)$ would be changing. The entire output of the pump flows through R_1, so $p_a(0-) = R_1 q_1(0-) = 2$. Since the pressure at the bottom of a tank cannot change instantaneously, $p_a(0+) = 2$ and $p_b(0+) = 0$.

For $t > 0$, with the valve open,

$$\frac{1}{R_1} p_a(t) + C_1 \frac{dp_a}{dt} + \frac{1}{R_3} [p_a(t) - p_b(t)] = q_1(t)$$

$$\frac{1}{R_3} [p_b(t) - p_a(t)] + C_2 \frac{dp_b}{dt} + \frac{1}{R_2} p_b(t) = 0$$

The transformed equations are

$$\frac{1}{R_1} P_a(s) + C_1[sP_a(s) - p_a(0+)] + \frac{1}{R_3} [P_a(s) - P_b(s)] = Q_1(s)$$

$$\frac{1}{R_3} [P_b(s) - P_a(s)] + C_2[sP_b(s) - p_b(0+)] + \frac{1}{R_2} P_b(s) = 0$$

where $Q_1(s) = \mathcal{L}[q_1(t)] = 1/s$. When the numerical values are inserted, the equations become

$$\left(s + \frac{3}{2}\right) P_a(s) - P_b(s) = 2 + \frac{1}{s}$$

$$-P_a(s) + (s + 2)P_b(s) = 0$$

which yield

$$P_a(s) = \frac{2s^2 + 5s + 2}{s(s^2 + \tfrac{7}{2}s + 2)}$$

$$P_b(s) = \frac{2s + 1}{s(s^2 + \tfrac{7}{2}s + 2)}$$

Finally, for $t > 0$,

$$p_a(t) = \mathcal{L}^{-1}[P_a(s)] = 1 + 0.38\epsilon^{-0.72t} + 0.62\epsilon^{-2.78t}$$

$$p_b(t) = \mathcal{L}^{-1}[P_b(s)] = \tfrac{1}{2} + 0.30\epsilon^{-0.72t} - 0.80\epsilon^{-2.78t}$$

$$q_0(t) = \tfrac{1}{2}p_a(t) + p_b(t) = 1 + 0.49\epsilon^{-0.72t} - 0.49\epsilon^{-2.78t}$$

Notice that as t increases from zero, $p_a(t)$ changes from 2 to 1 and $p_b(t)$ from 0 to $\tfrac{1}{2}$. When the valve is first opened, the flow from the left-hand tank is greater than that

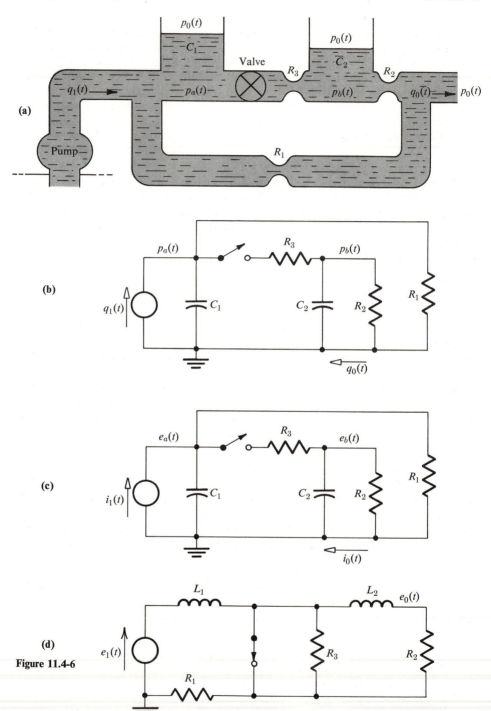

(a)

(b)

(c)

(d)

Figure 11.4-6

into the right-hand tank, and the output flow rate increases temporarily.[†] As t approaches infinity, steady-state conditions are re-established. Since $R_1 = R_2 + R_3$, the input flow rate $q_1(t)$ is divided equally between the two pipes for large values of t.

The similarity between this problem and a circuit that is excited by a d-c source can be emphasized by examining the hydraulic equivalent circuit in Fig. 11.4-6b. Since a closed valve interrupts the fluid flow, it is equivalent to an open switch in the hydraulic equivalent circuit. The pressure-voltage and pressure-current analogs are shown in Figs. 11.4-6c and 11.4-6d, respectively. The positions of all the switches are shown for t < 0.

Thermal Systems.   Among the thermal systems that are frequently encountered are the thermometer and the boiler. Simplified models for these devices are discussed in Examples 11.4-3 and 11.4-4. In addition to being an essential part of many energy-conversion systems, thermal components are often needed to maintain reasonable ambient temperatures for electrical and mechanical systems. Overheating can be a common cause of failure for such systems.

One of the two basic, measurable variables in a thermal system is the *temperature difference* $\theta(t)$, which is usually taken as the potential variable.[††] Temperatures are commonly expressed with respect to some temperature reference. The freezing point of water is used as a temperature reference for the centigrade scale, and the temperature at which molecular motion ceases is used for the Kelvin scale. In many problems, the reference temperature is chosen to be the ambient, or room temperature.

Heat is a form of energy that is related to the molecular activity in the body under consideration. Although the units of joules may be used, heat is also expressed in kilocalories (kcal) and British thermal units (Btu).[†††] The flow variable in thermal systems is the *rate of heat flow* $q(t)$, which may be expressed in joules per sec (watts). In contrast to the systems previously discussed, the flow variable has units of power.

If a given amount of heat is supplied to a body, the increase in its temperature will depend upon the size of the body and the substance of which it is made. The *specific heat* c_p is defined as the heat required to increase the temperature of one unit of mass one degree. The heat required to produce a temperature change $\theta(t)$ is $c_p M \theta(t)$, where M is the mass of the body. The rate of heat flow into the body is

$$q(t) = C_T \frac{d\theta}{dt} \qquad (11.4\text{-}6)$$

where $C_T = c_p M$ is called the *thermal capacitance* of the mass.

[†] If the element values are chosen so that $R_2 C_2 = R_1 C_1$, all the fluid from the left-hand tank enters the right-hand tank, and $q_0(t) = 1$ for all $t > 0$.

[††] The symbol θ is used for temperature to avoid confusion with the time t.

[†††] One kilocalorie is the amount of heat required to increase the temperature of 1 kg of water by 1°C, and it equals 4186 joules. One Btu is the heat required to increase the temperature of 1 lb of water by 1°F.

Equation 11.4-6 is meaningful only when the temperature is uniform throughout the body. If the temperature is nonuniform, and if the effect of energy storage within the body cannot be neglected, the body can be treated as a distributed system, and described by partial rather than ordinary differential equations. Fortunately, the behavior of many thermal systems can be approximated fairly well by the interconnection of lumped elements.

Heat may not only be stored within a body, but may be transferred through a body if there is a temperature difference between opposite surfaces. Consider the bar of length d and of uniform cross-sectional area A, shown in Fig. 11.4-7. The sides of the bar are insulated, so that heat can enter and leave only at the ends. The thermal capacity of the bar is assumed to be negligible, and the temperatures of the two ends are $\theta_1(t)$ and $\theta_2(t)$. If $\theta(t) = \theta_1(t) - \theta_2(t)$, the rate of heat flow between the two ends is $q(t) = (kA/d)\theta(t)$ or

$$q(t) = \frac{1}{R_T}\,\theta(t) \tag{11.4-7}$$

where $R_T = d/kA$ is the *thermal resistance*. The constant k is the coefficient of thermal conductivity for the material in the bar.

The transfer of heat through the solid bar in Fig. 11.4-7 is an example of thermal conduction. Heat can be transferred between a solid body and a fluid (usually air or water) in contact with it by convection. In the latter case, the rate of heat flow is also proportional to the difference between the body and fluid temperatures, so Eq. 11.4-7 still can be used. When heat is transferred between two bodies by radiation, the rate of heat flow depends upon the fourth power of the absolute temperatures of the bodies. Except as an approximation for a very small temperature difference, Eq. 11.4-7 cannot be used.

In thermal systems, there is only one kind of energy-storing element. In contrast to other systems, there is no element for which $q(t)$ is proportional to the integral of $\theta(t)$.

There are two interconnection laws, which are analogous to Kirchhoff's current and voltage laws. The algebraic sum of the heat flow rates at any junction must be zero. (Any storage of heat at the junction is accounted for by a thermal capacitance.) Symbolically,

$$\sum q(t) = 0 \tag{11.4-8}$$

The algebraic sum of the temperature differences around any closed path must also

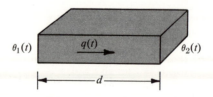

Figure 11.4-7

be zero. Otherwise, the temperature difference between two points could not be uniquely defined.

$$\sum \theta(t) = 0 \tag{11.4-9}$$

When a thermal equivalent circuit is drawn, the symbols for thermal resistance and capacitance are similar to those for electrical resistance and capacitance. A temperature-voltage or a temperature-current analog may be drawn for a given thermal system.

Example 11.4-3. A thermometer is frequently used to measure the temperature of a fluid in contact with it. A column of mercury enclosed in a glass envelope and immersed in a fluid is shown in Fig. 11.4-8a. Heat may be transferred from the fluid to the mercury by both convection at the surface of and conduction through the envelope, and

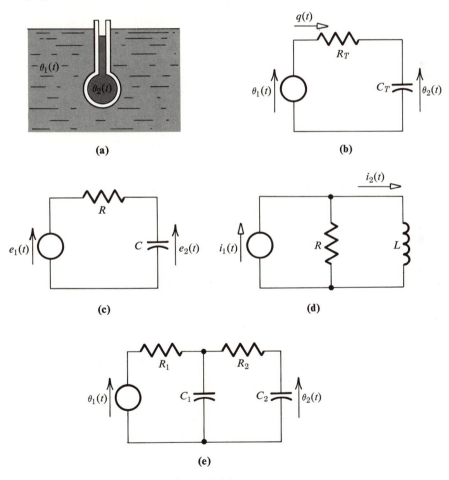

(a)

(b)

(c)

(d)

(e)

Figure 11.4-8

the rate of heat transfer is proportional to the temperature difference $\theta_1(t) - \theta_2(t)$. Write an equation for the system, and draw the equivalent circuit and the electrical analogs.

Solution. The heat transferred to the mercury is given by $q(t) = (1/R_T)[\theta_1(t) - \theta_2(t)]$ and by $q(t) = C_T(d\theta_2/dt)$, where R_T is the thermal resistance associated with the envelope and where C_T is the thermal capacitance of the mercury. Thus,

$$C_T \frac{d\theta_2}{dt} + \frac{1}{R_T} [\theta_2(t) - \theta_1(t)] = 0$$

The quantity $\theta_1(t)$ acts as a temperature source, and $\theta_2(t)$ is the desired response. The equation has the form of a node equation based upon Eq. 11.4-8. An equivalent equation, which corresponds to a loop equation, can be obtained by writing

$$\theta_1(t) = [\theta_1(t) - \theta_2(t)] + \theta_2(t) = R_T q(t) + \frac{1}{C_T} \int q \, dt$$

The thermal equivalent circuit is shown in Fig. 11.4-8b.

The equations for the temperature-voltage and temperature-current analogs are

$$C \frac{de_2}{dt} + \frac{1}{R} [e_2(t) - e_1(t)] = 0$$

and

$$L \frac{di_2}{dt} + \frac{1}{G} [i_2(t) - i_1(t)] = 0$$

The corresponding circuits are given in Figs. 11.4-8c and 11.4-8d. As expected, the temperature-voltage analog can be drawn directly from the thermal equivalent circuit, and the temperature-current analog is the dual of the first analog.

If, in Example 11.4-3, the thermometer is suddenly immersed in the fluid—corresponding to a step function for $\theta_1(t)$—the temperature of the mercury cannot change instantaneously because of the thermal capacitance C_T. Instead, $\theta_2(t)$ rises exponentially toward its final value with a time constant equal to $R_T C_T$. This behavior is also immediately evident from an inspection of the electrical analogs.

A somewhat more precise equivalent circuit for the thermometer might result if the thermal capacitance of the envelope were considered (despite the fact that the envelope is probably not at a uniform temperature). Let R_1 account for the transfer of heat from the fluid to the envelope by convection and R_2 the heat transfer to the mercury by conduction. If C_1 and C_2 represent the thermal capacitances for the envelope and the mercury, the thermal equivalent circuit takes the form shown in Fig. 11.4-8e. Because of the *two* capacitances, the initial response to a change in $\theta_1(t)$ is even slower than the first equivalent circuit would indicate.

Example 11.4-4. The system shown in Fig. 11.4-9a is used to heat the water that passes through the tank. The tank is insulated so that no heat can be lost through its walls, and it is heated by an externally controlled heater. Assume that the heat from the heater is transmitted directly to the water in the tank and that the motion of the

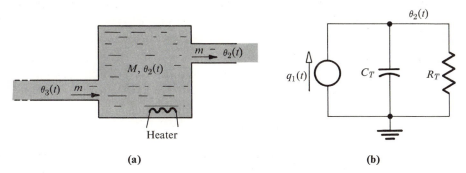

Figure 11.4-9

water is sufficient to keep the entire tank at a substantially uniform temperature $\theta_2(t)$. Let c_p, M, and m denote, respectively, the specific heat of water, the mass of water in the tank, and the mass leaving (and also entering) the tank per unit time. Write a differential equation describing the system, and draw the thermal equivalent circuit.

Solution. Let $q_1(t)$ denote the rate of heat flow from the heater. The rate of change of the heat stored in the tank is

$$q_T(t) = c_p M \frac{d\theta_2}{dt}$$

The difference in the rates of heat flow for the water leaving and entering the tank is

$$q_2(t) - q_3(t) = c_p m[\theta_2(t) - \theta_3(t)]$$

Recall that m has the units of mass per unit time, while M has units of mass.
Since heat must be conserved, as indicated by Eq. 11.4-8,

$$q_1(t) = c_p M \frac{d\theta_2}{dt} + c_p m[\theta_2(t) - \theta_3(t)] = C_T \frac{d\theta_2}{dt} + \frac{1}{R_T}[\theta_2(t) - \theta_3(t)]$$

or, if $\theta_3(t)$ is taken as the reference temperature,

$$q_1(t) = C_T \frac{d\theta_2}{dt} + \frac{1}{R_T} \theta_2(t)$$

The last equation describes the equivalent circuit in Fig. 11.4-9b. An electrical analog could be obtained by replacing $q_1(t)$ and $\theta_2(t)$ by $i_1(t)$ and $e_2(t)$, respectively. Another satisfactory analog would be the dual of the first one and would correspond to the equation

$$e_1(t) = L \frac{di_2}{dt} + \frac{1}{G} i_2(t)$$

The heat removed by the circulating water in Example 11.4-4 was accounted for

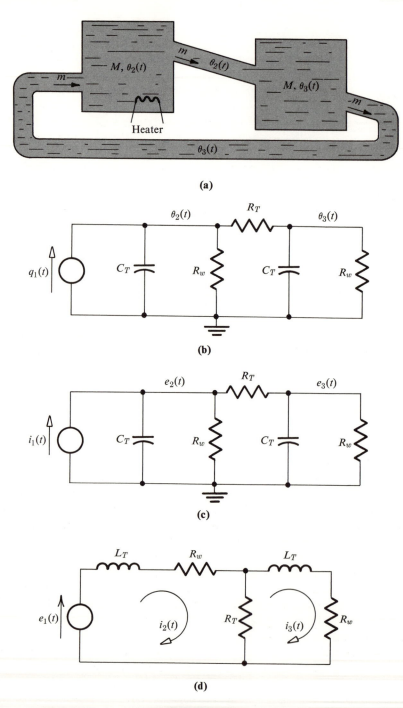

(a)

(b)

(c)

(d)

Figure 11.4-10

by a thermal resistance. The next example considers the effect of heat leakage through the walls of the tank, and includes a complete path for the circulating water.

> **Example 11.4-5.** The mass of water in each tank shown in Fig. 11.4-10a is M, while the mass of water circulating through the pipes per unit time is m. The walls of the tanks have a thermal resistance R_w and negligible thermal capacitance. The pipes are insulated to prevent any loss of heat through their walls. The water temperatures in each tank are substantially uniform and are denoted by $\theta_2(t)$ and $\theta_3(t)$. Write the differential equations describing the system, and draw two electrical analogs.

> **Solution.**   Let the ambient temperature $\theta_0(t)$ be taken as a reference, so that $\theta_0(t) = 0$. The rate of heat flow from the heater is again denoted by $q_1(t)$. The application of Eq. 11.4-8 to each of the tanks gives

$$q_1(t) = c_p M \frac{d\theta_2}{dt} + c_p m[\theta_2(t) - \theta_3(t)] + \frac{1}{R_w} \theta_2(t)$$

$$0 = c_p M \frac{d\theta_3}{dt} + c_p m[\theta_3(t) - \theta_2(t)] + \frac{1}{R_w} \theta_3(t)$$

> If $C_T = c_p M$ and $R_T = 1/c_p m$, the equations describe the equivalent circuit in Fig. 11.4-10b. The two electrical analogs are given in parts (c) and (d) of the figure.

11.5

Mixed Systems

One part of a system may consist of mechanical components, while other parts may be composed of hydraulic, electrical, or other kinds of components. In addition to the element and interconnection laws for the individual parts of the system, a mathematical description of the interaction of the parts is needed. Once the necessary equations have been obtained, they can be solved for the system function or for the response to a given input. An electrical analog, whose behavior is similar to that of the original system, also may be found.

Perhaps the most frequently encountered mixed systems are electromechanical ones, which are examined first. Since the behavior of many mixed systems may be approximated by a second-order differential equation, some general properties of such systems are discussed briefly. Finally, we shall consider an example of a hydraulic-mechanical system.

Electromechanical Systems.   Examples of electromechanical systems include microphones, loudspeakers, motors, generators, and measuring instruments. The electrical and mechanical parts interact through a magnetic field. The interaction laws are discussed in elementary physics books and are reviewed next.

A current-carrying wire that is perpendicular to a uniform magnetic field is

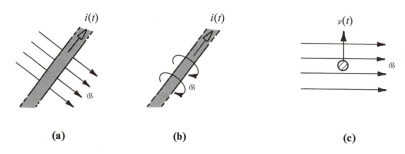

Figure 11.5-1

shown in Fig. 11.5-1a. If the current $i(t)$ is expressed in amperes, the length of the wire l in meters, and the magnetic flux density $\mathscr{B}$ in webers per square meter, then the force exerted on the wire is

$$f(t) = \mathscr{B}li(t) \qquad (11.5\text{-}1)$$

newtons.[†] The direction of the force is perpendicular to the wire and to the direction of the magnetic field. If the bent fingers of the right hand are pointed from the positive direction of the current toward the positive direction of the field (through the 90° angle), the thumb will point in the positive direction of the force. In Fig. 11.5-1a, for example, the force is into the paper.

Another method of determining the direction of the force is based upon the fact that an isolated current-carrying wire produces a magnetic field surrounding it, as shown in Fig. 11.5-1b. If the thumb of the closed right hand points in the direction of the current, the fingers will point in the direction of the field. If the wire is now placed in the external magnetic field shown in Fig. 11.5-1a, the two fields will add in front of the conductor but will subtract in back of the conductor. Then the force on the wire is directed into the paper (from the strengthened toward the weakened field).

An end view of a wire moving with a velocity $v(t)$ perpendicularly to a uniform magnetic field $\mathscr{B}$ is shown in Fig. 11.5-1c. As a result of the motion, there is induced in the wire a voltage

$$e_c(t) = \mathscr{B}lv(t) \qquad (11.5\text{-}2)$$

where l is the length of the wire that is moving through the field.[††] This induced voltage has such a polarity as to tend to cause a current to flow as follows. The current's direction is indicated by the thumb of the right hand when the bent fingers point from the positive direction of the velocity toward the positive direction

---

[†] Although this form of the equation implies that $\mathscr{B}$ and l are constants, this is not necessary. Also, an equation that is valid even if the wire is not perpendicular to the field and that automatically gives the direction of the force is $\mathbf{f}(t) = i(t)\mathbf{l} \times \mathscr{B}$, where $\mathbf{l} \times \mathscr{B}$ is the cross product of the vectors $\mathbf{l}$ and $\mathscr{B}$.

[††] A general expression for the induced voltage in terms of the cross and dot product of vector mathematics is $e_c(t) = [\mathbf{v}(t) \times \mathscr{B}] \cdot \mathbf{l}$.

of the magnetic field (through the 90° angle). For the sketch in Fig. 11.5-1c, the induced voltage tends to produce a current into the paper.

Another method of determining the polarity of the induced voltage is based upon *Lenz*'s *law*. The law states that the induced voltage tends to produce a current that in turn will tend to oppose the original mechanical motion. The voltage induced in the wire shown in Fig. 11.5-1c will tend to produce a current into the paper, because such a current would cause a downward force on the wire.

Example 11.5-1. Most students probably are familiar with the galvanometer, which should ideally produce a deflection that is proportional to an electrical input. A simplified sketch of the top view of a galvanometer is shown in Fig. 11.5-2a. The stationary pole pieces marked N and S (for North and South) represent a permanent magnet or a separately excited electromagnet that produces a constant, uniform magnetic flux density $\mathscr{B}$. A coil of wire is wound on a movable, cylindrical iron rotor. The current $i(t)$ flows into the top of the conductors marked with a cross and out of the top of those marked with a dot. The coil's resistance is represented by the lumped element R, and the inductance of the coil is neglected. A needle attached to the rotor measures the angular displacement $\theta(t)$ from the equilibrium position. Let J and B_r denote the moment of inertia of the rotor and the viscous damping that results from air friction. In addition, there is a spring (not shown) attached to the rotor that may be represented by the rotational stiffness K_r. The total length of the conductors in the magnetic field is l, the radius of the rotor is a, and the constant D is defined as $\mathscr{B}la$. Write the necessary equations for the system, draw an electrical analog, and find a differential equation relating $\theta(t)$ and $e_1(t)$.

Solution. The equations for the mechanical and electrical parts of the system are

$$J\frac{d\omega_r}{dt} + B_r\omega_r(t) + K_r\int \omega_r\,dt = \tau(t)$$

$$Ri(t) + e_c(t) = e_1(t)$$

where $\omega_r(t)$ is the angular velocity of the rotor. For a positive current $i(t)$, the force exerted on the conductors (and hence on the rotor) as a result of the magnetic field is $f(t) = \mathscr{B}li(t)$. The force on the left-hand conductors is upward and on the right-hand conductors is downward. Thus, the torque tends to move the rotor in the assumed positive direction (clockwise) and is

$$\tau(t) = \mathscr{B}lai(t) = Di(t)$$

If the rotor is displaced in the assumed positive direction, the voltage induced in the coils,

$$e_c(t) = \mathscr{B}la\omega_r(t) = D\omega_r(t)$$

tends to produce a current in opposition to $i(t)$. This corresponds to a negative term on the right side of the electrical equation or a positive term on the left side.

The four equations above suggest the equivalent circuit shown in Fig. 11.5-2b, where the electrical and mechanical parts are interconnected by the action of two controlled sources. Except for the use of different symbols, this combination of controlled sources is identical with the model for an ideal transformer in Fig. 7.3-4c. Thus, the equivalent circuit for the system can be redrawn as in Fig. 11.5-2c, where

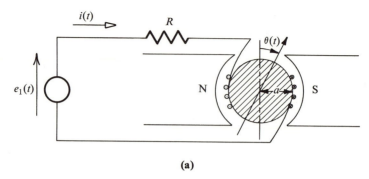

(a)

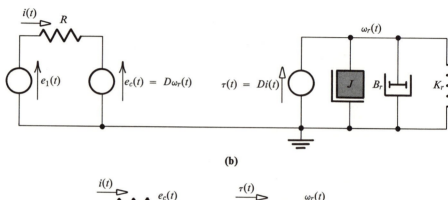

(b)

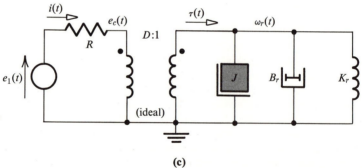

(c)

$D = \tau(t)/i(t) = e_c(t)/\omega_r(t)$. The transformer-like device, which represents the interconnection of the electrical and mechanical components, is treated exactly like the ideal transformer defined in Section 7.3, except for the fact that the turns ratio is no longer a dimensionless quantity. Notice that a transformer may be used only because both controlled sources had the same constant of proportionality.

If, in the mechanical equation, $\tau(t)$ and $\omega_r(t)$ are replaced by $Di(t)$ and $e_c(t)/D$, respectively, and if every term is divided by D, the equations for the system become

$$\frac{J}{D^2}\frac{de_c}{dt} + \frac{B_r}{D^2}e_c(t) + \frac{K_r}{D^2}\int e_c\, dt = i(t)$$

$$Ri(t) + e_c(t) = e_1(t)$$

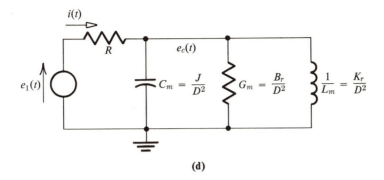

(d)

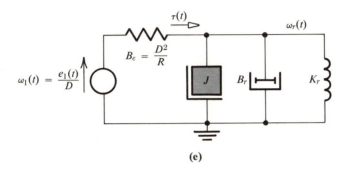

(e)

Figure 11.5-2

In the first of these two equations, every term has the units of amperes. The equations describe the electrical analog in Fig. 11.5-2d. This analog also may be obtained by moving all the elements in Fig. 11.5-2c to the left-hand side of the transformer. By analogy to the discussion following Example 7.3-1, this is done by multiplying angular velocities by D, dividing torques by D, and dividing J, B_r, and K_r by D^2.

Another suitable electrical analog may be obtained by finding the dual of Fig. 11.5-2d. The final analog may be scaled, but the portions corresponding to the electrical and mechanical elements cannot be scaled independently of each other.

If, in the original electrical equation, $e_c(t)$ and $i(t)$ are replaced by $D\omega_r(t)$ and $\tau(t)/D$, respectively, and if every term is divided by D,

$$J\frac{d\omega_r}{dt} + B_r\omega_r(t) + K_r \int \omega_r \, dt = \tau(t)$$

$$\frac{R}{D^2}\tau(t) + \omega_r(t) = \frac{e_1(t)}{D}$$

which describe the mechanical analog in Fig. 11.5-2e. The same result may be obtained by moving all the elements in Fig. 11.5-2c to the right-hand side of the transformer.

An equation relating $\theta(t)$ and $e_1(t)$ may be found from the mechanical analog, or by combining the equations

$$J\frac{d^2\theta}{dt^2} + B_r\frac{d\theta}{dt} + K_r\theta(t) = Di(t)$$

$$Ri(t) + D\frac{d\theta}{dt} = e_1(t)$$

to give

$$J\frac{d^2\theta}{dt^2} + \left(B_r + \frac{D^2}{R}\right)\frac{d\theta}{dt} + K_r\theta(t) = \frac{D}{R}e_1(t)$$

If there is no initial stored energy, $\theta(0+) = \omega_r(0+) = 0$ and the system function is

$$\frac{\Theta(s)}{E_1(s)} = \frac{D/R}{Js^2 + (B_r + D^2/R)s + K_r}$$

Second-Order Systems. Second-order systems are systems that can be described by a second-order differential equation. Even some fairly complicated systems may be approximated by such an equation. If the input $x(t)$ and the output $y(t)$ are related by

$$\frac{d^2y}{dt^2} + 2\zeta\omega_0\frac{dy}{dt} + \omega_0^2 y(t) = Ax(t) \tag{11.5-3}$$

and if $y(t)$ and dy/dt are zero at $t = 0+$,

$$\frac{Y(s)}{X(s)} = \frac{A}{s^2 + 2\zeta\omega_0 s + \omega_0^2} \tag{11.5-4}$$

The damping ratio ζ and undamped natural angular frequency ω_0 were first discussed in Example 4.4-1. If ζ is less than unity, the poles of the network function are complex, and the free response is oscillatory. In Example 11.5-1, $\omega_0 = \sqrt{K_r/J}$, $2\zeta\omega_0 = B_r/J + D^2/RJ$, and

$$\zeta = \frac{B_r}{2\sqrt{K_rJ}} + \frac{D^2}{2R\sqrt{K_rJ}}$$

The first of the two terms on the right side of the equals sign is the damping ratio of an isolated rotational system, while the second term is the contribution that results from the interaction of the electrical and mechanical parts.

If the input to a system described by Eq. 11.5-4 is the unit step function, $X(s) = 1/s$ and

$$Y(s) = \frac{A}{s(s^2 + 2\zeta\omega_0 s + \omega_0^2)}$$

The inverse transform of this expression can be found and plotted for different values of ζ to obtain curves like those in Fig. 11.5-3a. If $\zeta > 1$ the response is sluggish, while for $\zeta < 1$ the response is oscillatory. It can be shown that for $\zeta < 1$ (the "underdamped" case)

$$y(t) = \frac{A}{\omega_0^2}\left[1 - \frac{\epsilon^{-\zeta\omega_0 t}}{\sqrt{1 - \zeta^2}}\sin(\omega_d t + \phi)\right]$$

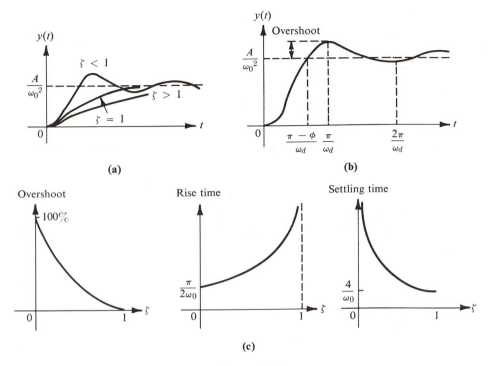

Figure 11.5-3

where $\omega_d = \omega_0\sqrt{1 - \zeta^2}$ and $\phi = \tan^{-1}(\sqrt{1 - \zeta^2}/\zeta) = \cos^{-1}\zeta$. Some additional quantities of interest for this case are shown in Fig. 11.5-3b. In one of the problems at the end of this chapter the student is asked to show that the *overshoot* is $(A/\omega_0^2)\epsilon^{-\pi\zeta/\sqrt{1-\zeta^2}}$ or, expressed as a percentage of the steady-state response,

$$\epsilon^{-\pi\zeta/\sqrt{1-\zeta^2}} \times 100\%$$

The time to reach the final value of the response first is $(\pi - \phi)/\omega_d$ and is sometimes called the *rise time*. (Another common definition of rise time is the time required to go from 10% to 90% of the final value.) The time constant of the response is $1/\zeta\omega_0$. The *stabilization time* or *settling time* is often defined as four time constants, since the oscillations are reduced to less than 2% of the steady-state response by that time. Curves of percent overshoot, rise time, and settling time are given in Fig. 11.5-3c.

It might be desirable to have a system with a low overshoot, a short rise time, and a short settling time, but the curves indicate that these are contradictory objectives. The choice of ζ depends upon the application, but a value between 0.6 and 0.8 might be suitable for an ammeter or voltmeter. For a ballistic galvanometer, ζ should be small, while for galvanometers used in cardiographs an overshoot cannot be tolerated, and ζ should not be less than 1.

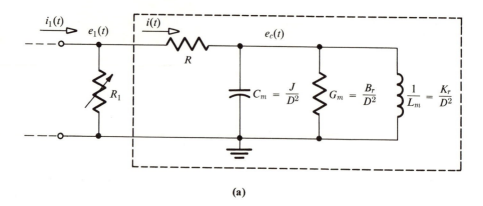

(a)

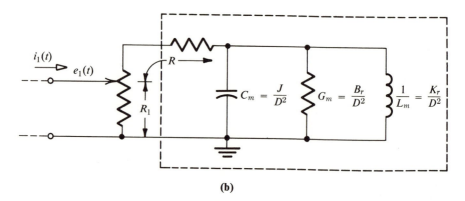

(b)

Figure 11.5-4

Example 11.5-2.   The electrical analog developed in Example 11.5-1 for the galvanometer is shown within the dashed lines in Fig. 11.5-4a. In order to use the galvanometer with higher currents, a shunt resistance R_1 is added. Find and discuss the transfer function $\Theta(s)/I_1(s)$.

The sensitivity of the instrument can be changed by varying the value of R_1 and thereby changing the portion of $i_1(t)$ that passes through the meter movement. Unfortunately, this also changes the damping ratio ζ. Modify the system so that the sensitivity can be varied without changing ζ.

Solution.   The electrical analog is described by the following transformed node equations, when there is no initial stored energy.

$$\left(\frac{1}{R_1} + \frac{1}{R}\right) E_1(s) - \frac{1}{R} E_c(s) = I_1(s)$$

$$-\frac{1}{R} E_1(s) + \left(\frac{J}{D^2} s + \frac{B_r}{D^2} + \frac{K_r}{D^2 s}\right) E_c(s) = 0$$

Solving these equations simultaneously for $E_c(s)$, and noting that $E_c(s) = sD\Theta(s)$, we obtain

$$\frac{\Theta(s)}{I_1(s)} = \frac{R_1 D/(R + R_1)}{Js^2 + [B_r + D^2/(R + R_1)]s + K_r}$$

By a comparison of this expression with Eq. 11.5-4, $\omega_0 = \sqrt{K_r/J}$, $2\zeta\omega_0 = B_r/J + D^2/(R + R_1)J$, and

$$\zeta = \frac{B_r}{2\sqrt{K_r J}} + \frac{D^2}{2(R + R_1)\sqrt{K_r J}}$$

Notice that the damping ratio does depend upon the value of R_1.

In order for ζ to remain constant, the resistance R should be increased whenever R_1 is decreased. Although R was defined as the resistance of the galvanometer coils, it could include an external series resistance. One way to keep the factor $R + R_1$ constant is to replace the shunt resistance by a potentiometer with a movable tap, as in Fig. 11.5-4b. Such an arrangement is called an *Ayrton shunt*.

Hydraulic-Mechanical Systems.   The individual parts of a hydraulic-rotational system are usually interconnected by a hydraulic pump or motor. As stated in Section 11.4, a hydraulic pump produces a fluid flow rate $q(t)$ that is proportional to the angular velocity of its shaft and to the position of a control vane. If the position of the control vane is fixed,

$$q(t) = k_p\omega_r(t) \tag{11.5-5}$$

where k_p is a constant of proportionality. For the pump shown in Fig. 11.5-5, the power imparted to the fluid is $p(t)q(t)$. If there are no losses, this must equal the mechanical power $\tau(t)\omega_r(t)$ supplied to the pump. Then, if $q(t)$ is given by the previous equation,

$$p(t) = \frac{1}{k_p}\tau(t) \tag{11.5-6}$$

Equations 11.5-5 and 11.5-6 also apply to a hydraulic motor. The only essential

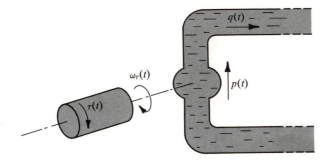

Figure 11.5-5

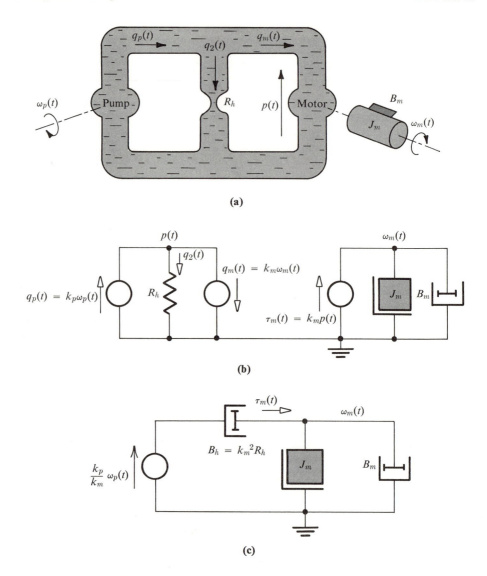

(a)

(b)

(c)

difference in the behavior of a pump and a motor is that in the latter the fluid is used to supply the power to turn the motor's shaft.

Example 11.5-3. The pump and motor constants for the system shown in Fig. 11.5-6a are defined by the equations $q_p(t) = k_p \omega_p(t)$ and $q_m(t) = k_m \omega_m(t)$. The symbols J_m and B_m represent the moment of inertia and the rotational viscous friction associated with the motor. The hydraulic resistance R_h accounts for the effect of the leakage paths around the motor. If the fluid is incompressible and has no inertance, find the transfer function $\Omega_m(s)/\Omega_p(s)$. Although in practice the pump speed could not be changed instantaneously, find $\omega_m(t)$ when $\omega_p(t) = U_{-1}(t)$.

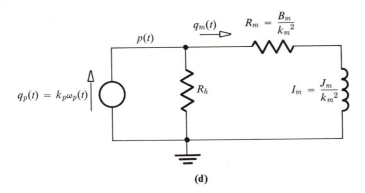

(d)

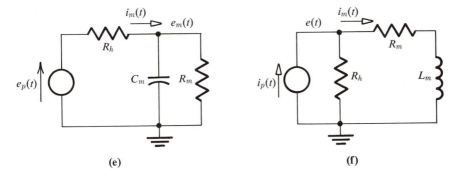

(e)   (f)

Figure 11.5-6

Solution.   The equation describing the hydraulic portion of the system is $q_m(t) + q_2(t) = q_p(t)$ or

$$q_m(t) + \frac{1}{R_h} p(t) = k_p \omega_p(t)$$

where $p(t)$ is the pressure across the pump and across the motor. If $\tau_m(t)$ denotes the torque applied by the fluid to the motor shaft,

$$J_m \frac{d\omega_m}{dt} + B_m \omega_m(t) = \tau_m(t)$$

The interaction of the hydraulic and mechanical components is given by the equations

$$q_m(t) = k_m \omega_m(t)$$

$$\tau_m(t) = k_m p(t)$$

Thus, we may draw the equivalent circuit in Fig. 11.5-6b. This time, the two controlled sources cannot be replaced by a transformer even though they have the same constant of proportionality k_m. Each controlled source is analogous to a voltage-controlled current source and not to a voltage-controlled voltage source or a current-controlled current source. The transformer equivalent circuits in Fig. 7.3-4 are therefore not directly applicable.[†]

We may still obtain a mechanical or hydraulic analog by manipulating the equations. If, in the hydraulic equation, $q_m(t)$ is replaced by $k_m \omega_m(t)$ and $p(t)$ by $\tau_m(t)/k_m$, and if every term is divided by k_m, we obtain

$$\omega_m(t) + \frac{\tau_m(t)}{k_m^2 R_h} = \frac{k_p}{k_m} \omega_p(t)$$

$$J_m \frac{d\omega_m}{dt} + B_m \omega_m(t) = \tau_m(t)$$

These equations describe the mechanical analog in Fig. 11.5-6c. If, in the mechanical equation, $\omega_m(t)$ is replaced by $q_m(t)/k_m$ and $\tau_m(t)$ by $k_m p(t)$, and if every term is again divided by k_m,

$$q_m(t) + \frac{1}{R_h} p(t) = k_p \omega_p(t)$$

$$\frac{J_m}{k_m^2} \frac{dq_m}{dt} + \frac{B_m}{k_m^2} q_m(t) = p(t)$$

which describe the hydraulic analog in Fig. 11.5-6d. Two electrical analogs are given in parts (e) and (f) of the figure.

The equations for the system have been manipulated in such a way as to suggest equivalent circuits. If the only goal were to obtain the transfer function, perhaps the easiest procedure would have been to set the algebraic sum of the pressures around the right-hand mesh in the original sketch equal to zero.

$$R_h[q_p(t) - q_m(t)] - \frac{1}{k_m} \tau_m(t) = 0$$

or

$$R_h[k_p \omega_p(t) - k_m \omega_m(t)] = \frac{1}{k_m} \left[J_m \frac{d\omega_m}{dt} + B_m \omega_m(t) \right]$$

If the motor is initially at rest so that $\omega_m(0+) = 0$, the transformed equation is

$$R_h[k_p \Omega_p(s) - k_m \Omega_m(s)] = \frac{1}{k_m} [sJ_m + B_m]\Omega_m(s)$$

Then

$$H(s) = \frac{\Omega_m(s)}{\Omega_p(s)} = \frac{k_p k_m R_h}{sJ_m + (B_m + k_m^2 R_h)}$$

If $\omega_p(t) = U_{-1}(t)$,

$$\Omega_m(s) = \frac{k_p k_m R_h / J_m}{s[s + (B_m + k_m^2 R_h)/J_m]} = \frac{a}{s(s + b)}$$

$$\omega_m(t) = \mathcal{L}^{-1}[\Omega_m(s)] = \frac{a}{b}(1 - \epsilon^{-bt})$$

[†] The two controlled sources could be replaced by a gyrator, which was examined in one of the problems at the end of Chapter 7. A method for modifying Fig. 11.5-6b to include an ideal transformer is suggested in one of the problems at the end of this chapter.

For small values of t, $\omega_m(t)$ and $q_m(t)$ are small, so $q_2(t)$ and $p(t)$ are large. Thus, a relatively large starting torque is applied to the motor.

If the compressibility of the fluid in Example 11.5-3 is considered, the equivalent circuit in Fig. 11.5-6b will contain a hydraulic capacitance C_h connected in parallel with the hydraulic resistance R_h. In the transfer function, R_h is replaced by $(R_h/sC_h)/(R_h + 1/sC_h)$, and

$$H(s) = \frac{k_p k_m R_h}{C_h R_h J_m s^2 + (C_h R_h B_m + J_m)s + (B_m + k_m{}^2 R_h)}$$

Notice that this expression has the same form as the transfer function for a second-order system given in Eq. 11.5-4. If C_h is large and B_m is small, the damping ratio ζ may be less than unity, in which case there will be oscillations in the free response. From Section 11.4, C_h is proportional to the volume of fluid in the system.

11.6

Summary

Magnitude scaling (multiplying every R, L, and $1/C$ by a constant A) changes the impedance level of a circuit but does not affect the pole-zero pattern of the network function $H(s)$. Frequency scaling (dividing every L and C by a constant a) increases the speed of a circuit's free response and stretches out the frequency response curves along the ω axis by the multiplying factor a. In order to change the time scale of a response, the time scale of the input must be changed and the circuit elements must be frequency scaled.

The mathematical description of mechanical, hydraulic, and thermal systems is remarkably similar to that for electrical circuits, as is emphasized by Table 11.6-1. In the first column of the table, the generalized flow variable is denoted by $q(t)$ and the generalized potential variable by $p(t)$. The passive elements defined in the last three rows of the table are ones for which the flow is proportional to the potential variable, or to its derivative, or to its integral. Notice that for thermal systems there are only two kinds of passive elements and that the usual expression for power does not apply.

For every system the two interconnection laws, in terms of the generalized variables, are

$$\sum q(t) = 0 \qquad \text{at a junction}$$

$$\sum p(t) = 0 \qquad \text{around a closed path}$$

Unless impulses are present, the potential across a capacitance-type element or the flow through an inductance-type element cannot change instantaneously.

Table 11.6-1 also constitutes, in effect, a list of analogous quantities that may be

TABLE 11.6-1

Generalized quantities	Electrical quantities	Mechanical translational quantities	Mechanical rotational quantities	Hydraulic quantities	Thermal quantities
Flow variable $q(t)$	Current $i(t)$	Force $f(t)$	Torque $\tau(t)$	Fluid flow rate $q(t)$	Heat flow rate $q(t)$
Potential variable $p(t)$	Voltage $e(t)$	Velocity $v(t)$	Angular velocity $\omega_r(t)$	Pressure $p(t)$	Temperature $\theta(t)$
Power	$e(t)i(t)$	$f(t)v(t)$	$\tau(t)\omega_r(t)$	$q(t)p(t)$	$q(t)$
$q(t) = Gp(t)$	$i(t) = \dfrac{1}{R}e(t)$	$f(t) = Bv(t)$	$\tau(t) = B_r\omega_r(t)$	$q(t) = \dfrac{1}{R_h}p(t)$	$q(t) = \dfrac{1}{R_T}\theta(t)$
$q(t) = C\dfrac{dp}{dt}$	$i(t) = C\dfrac{de}{dt}$	$f(t) = M\dfrac{dv}{dt}$	$\tau(t) = J\dfrac{d\omega_r}{dt}$	$q(t) = C_h\dfrac{dp}{dt}$	$q(t) = C_T\dfrac{d\theta}{dt}$
$q(t) = K\displaystyle\int p\, dt$	$i(t) = \dfrac{1}{L}\displaystyle\int e\, dt$	$f(t) = K\displaystyle\int v\, dt$	$\tau(t) = K_r\displaystyle\int \omega_r\, dt$	$q(t) = \dfrac{1}{I}\displaystyle\int p\, dt$	

used to obtain an electrical analog for a given nonelectrical system. The non-electrical symbols in the system's integral-differential equations can be replaced by their electrical counterparts, and the analog can be found by drawing the circuit that is described by the new equations. A simpler procedure is to draw the analog directly from the system's equivalent circuit. A second electrical analog may be found by constructing the dual of the first one. If the analog is to be used in experimental work, the scaling techniques of Section 11.1 are usually needed to give circuit elements of practical size.

When mechanical, hydraulic, and electrical parts are contained in the same system, additional laws are needed to describe the interaction of the individual parts. Examples of electromechanical and hydraulic-mechanical systems are given in Section 11.5.

PROBLEMS

11.1 Change the values of the elements in Fig. 4.3-7a so that $e_o(t) = 10^3(\epsilon^{-10^3 t} - \epsilon^{-10^4 t})$ and so that the largest capacitance is 1 μf.

11.2 Show that the transfer function for the circuit in Fig. P11.2 is

$$H(s) = \frac{I_2(s)}{I_1(s)} = \frac{10(100s + 50)}{(100s + 1)(s + 10.6)}$$

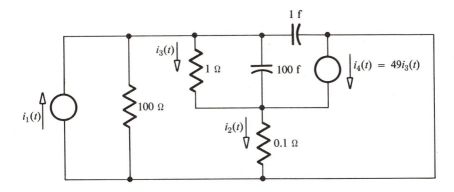

Figure P11.2

Sketch the Bode diagram of the decibel gain versus ω in the a-c steady state. If all resistances are multiplied by 10^3 and all capacitances are divided by 10^{12} (so that the circuit might be a simple model for a grounded collector transistor amplifier), draw the new Bode diagram.

11.3 Show that the transfer function for the circuit in Fig. P11.3 is

$$H(s) = \frac{E_o(s)}{E_1(s)} = \frac{1}{2(s^3 + 2s^2 + 2s + 1)} = \frac{1}{2(s + 1)(s^2 + s + 1)}$$

and that $|H(j\omega)|^2 = (\frac{1}{4})/(1 + \omega^6)$. The circuit is a third-order Butterworth filter and can be compared to the second-order Butterworth filter in Example 6.4-6. Convert the circuit into a band-pass filter having a bandwidth of 2 kiloradians per second and an angular resonant frequency of 100 kiloradians per second. The largest capacitance in the final circuit should be 10 μf.

11.4 Suppose that the bandwidth of a low-pass filter extends from $\omega = 0$ to $\omega = \omega_0$. How can the circuit be converted into a high-pass filter with a bandwidth from $\omega = \omega_0$ to $\omega \to \infty$? Use this low-pass to high-pass transformation to obtain a high-pass filter from the circuit in Fig. 11.1-1e.

11.5 Write a set of integral-differential equations to describe each of the mechanical systems shown in Fig. P11.5. Draw the mechanical equivalent circuit. An optional

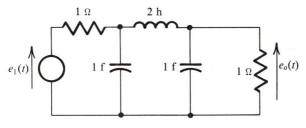

Figure P11.3

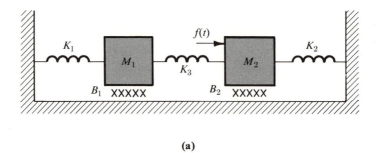

(a)

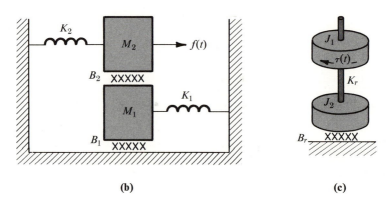

(b)   **(c)**

Figure P11.5

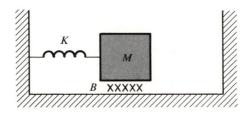

Figure P11.7

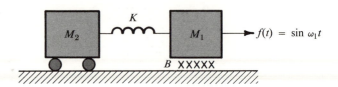

Figure P11.9

question is to find the system function that relates the transforms of the force source and the velocity of M_2 or the transforms of the torque source and the angular velocity of J_2, when there is no initial stored energy.

11.6   Derive Eqs. 11.2-5 and 11.2-6.

11.7   At $t = 0$, the mass shown in Fig. P11.7 is in its equilibrium position but is moving to the right with a velocity of 10 ft per sec. If $M = 2$ slugs and $K = 50$ lb per ft, what is the smallest value of B that will prevent oscillations in the response? For this value of B, what is the maximum displacement of the mass from its equilibrium position for $t > 0$?

11.8   Redraw Fig. 11.2-4a with the velocity source $v_1(t)$ replaced by an equivalent force source. Discuss in general terms the interchange of velocity and force sources.

11.9   The mass M_2 shown in Fig. P11.9 moves on frictionless bearings.

(a) Show that, by the proper choice of M_2, the mass M_1 can be made to stand still in the steady state.

(b) Draw the force-current and force-voltage analogs for the system.

11.10   Potential energy can be stored in the spring shown in Fig. P11.10 by compressing it and then locking it permanently in its compressed state. In order to do this, a force $f(t) = U_{-1}(t)$ pounds is applied to the mass M. Before the force is applied, the mass is at rest and is $\frac{1}{4}$ ft from the end of the spring. Let $M = 1$ slug, $B = 2$ lb-sec per ft, and $K = 9$ lb per ft.

(a) Find an expression for the displacement of the mass up to the time when it hits the spring.

(b) What is the velocity of the mass when it first hits the spring?

(c) Find an expression for the displacement of the mass when it is in contact with the spring. (The force source is still applied to the mass.)

(d) At what time should the spring be permanently locked into place in order to retain the maximum potential energy? What is the value of this maximum potential energy?

(e) Draw an electrical analog for the mechanical system.

11.11   Derive the equation $J = \int r^2 \, dm$ for Fig. 11.2-5a and the equation $K_r = \pi E_s R^4 / 2d$ for Fig. 11.2-7a. The quantity E_s is the shear modulus for the material in the shaft.

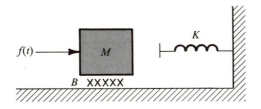

Figure P11.10

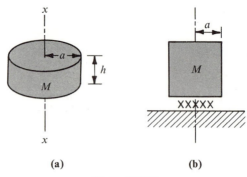

(a)   (b)

Figure P11.12

11.12 A homogeneous cylindrical body of mass M, radius a, and height h rotates about the central axis x-x, as shown in Fig. P11.12a.

(a) Calculate the moment of inertia J of the body.

(b) If the rotating cylinder is separated from a stationary reference by an oil film, as in Fig. P11.12b, calculate the rotational viscous friction B_r. Let η and d denote the viscosity and thickness of the oil film.

11.13 An irregularly shaped motor armature is suspended from the ceiling by a thin steel wire in order to determine the armature's moment of inertia. When the armature is given an initial angular displacement, it rotates with a period of 12 sec. When an object whose moment of inertia is known to be 0.1 slug-ft^2 is attached to the same wire, the period of rotation is 3 sec. What is the moment of inertia for the armature?

11.14 A simplified model of a fluid drive system is shown in Fig. P11.14. The rotational viscous friction B_{r1} represents the fluid drive coupling, and J and B_{r2} represent the load. If $\omega_1(t) = U_{-1}(t)$, find an expression for the angular velocity $\omega_2(t)$.

11.15 Find two electrical analogs for each of the systems shown in Fig. P11.5.

11.16 Repeat Example 11.3-2 for the force-voltage analog, with a 2 amp rms, 1000 rad per sec current source.

11.17 A pendulum, moving in a vertical plane and consisting of a mass attached to an ideal string that cannot be stretched, is shown in Fig. P11.17. Show that if $\theta(t)$ is

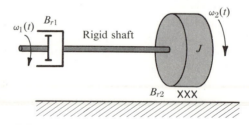

Figure P11.14

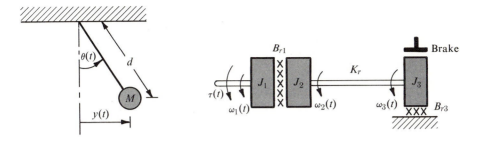

Figure P11.17   Figure P11.18

small the motion is described by the equation $d^2y/dt^2 + (g/d)y(t) = 0$, where g is the acceleration of gravity. Draw electrical, translational, and rotational systems that are analogous to the given one.

11.18 The rotational system shown in Fig. P11.18 is operating in the steady state at $t = 0-$. The torque source $\tau(t)$ has a constant value of A, and J_3 is held motionless by the brake. At $t = 0$, the brake is removed.

(a) Write the equations describing the motion for $t > 0$. What are the values of $\omega_1(t)$, $\omega_2(t)$, $\omega_3(t)$, and $d\omega_3/dt$ at $t = 0+$?

(b) Draw the torque-current analog. What operation in the analog corresponds to removing the brake in the mechanical system?

11.19 Draw two electrical analogs for the mechanical system shown in Fig. P11.19. In a consistent set of units, $M_1 = 1$, $M_2 = 2$, $M_3 = 5$, $B_1 = 4$, $B_2 = 2$, $K_1 = 10$, and $K_2 = 20$, and $f(t) = \cos t \, U_{-1}(t)$. Scale the force-voltage analog so that the largest

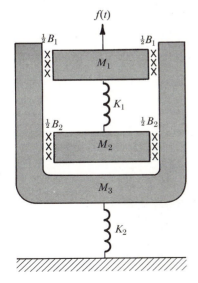

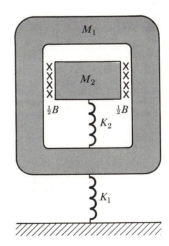

Figure P11.19   Figure P11.20

capacitance is 1 μf and so that a voltage source $e(t) = 10 \cos 500t\; U_{-1}(t)$ can be used. Find an expression that gives the displacement of the mass M_3 as a constant times an electrical quantity.

11.20 The mass M_2 shown in Fig. P11.20 is moved with a known velocity $v_2(t)$. The element values are $M_1 = M_2 = 1$ slug, $B = 2$ lb-sec per ft, $K_1 = K_2 = 2$ lb per ft. Draw and label with numerical values a force-voltage analog in which the largest capacitance is 10^{-3} f. The electrical circuit should act 100 times as fast as the mechanical system. What electrical quantities are directly proportional to the displacement, velocity, and acceleration of the mass M_1?

11.21 Draw mechanical translational and mechanical rotational systems that are analogous to the low-pass filter in Fig. P11.3.

11.22 Is it always possible to find a mechanical analog for a given electrical circuit? Explain your answer.

11.23 What mechanical and hydraulic devices (not discussed in the text) are analogous to an ideal transformer in an electrical system?

11.24 In the hydraulic system shown in Fig. P11.24, $q_1(t)$ is an externally controlled flow rate into the left-hand storage tank. Each tank has a constant cross-sectional area A, and the density of the incompressible fluid is ρ. If the system contains no initial stored energy, find the transfer function $H(s)$ that relates the transforms of $q_1(t)$ and $h_2(t)$, the height of the fluid in the right-hand tank.

11.25 Current through the copper conductor shown in Fig. P11.25 produces an $i^2(t)R$ loss that must be dissipated as heat. Assume that the thermal capacitance of the insulation is negligible and that the temperatures throughout the copper and throughout the iron are uniform. Draw and label the thermal equivalent circuit and two electrical analogs. If the current is constant, give expressions for the steady-state temperatures of the copper and of the iron.

11.26 Two water-filled tanks are separated by a thin wall, as shown in Fig. P11.26. An uninsulated pipe carrying steam at a known temperature of $\theta_1(t)$ passes through the left-hand tank. Circulating water leaves the right-hand tank and returns at a known temperature $\theta_4(t)$. The walls of the tanks have thermal resistance but negligible

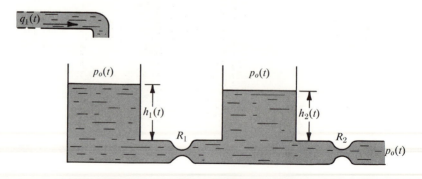

Figure P11.24

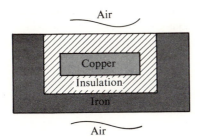

Figure P11.25

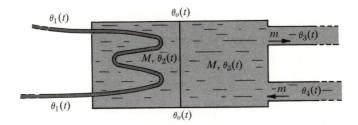

Figure P11.26

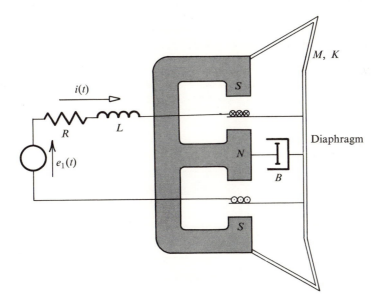

Figure P11.27

thermal capacitance, and the temperature throughout each tank is substantially uniform. Write the equations describing the system, and draw an electrical analog. Express all temperatures with respect to the ambient temperature $\theta_0(t)$, and define all the symbols that are used.

11.27 A cross-sectional view of a loudspeaker is shown in Fig. P11.27. A coil is rigidly connected to a movable diaphragm, so that an electrical signal will result in the movement of the diaphragm and the production of sound waves. The diaphragm movement is retarded by its mass M and by the stiffness K and viscous friction B of the structure that connects it to the stationary permanent magnet. The acoustical properties of the air can be accounted for by modifying the values of M, B, and K. If l denotes the total length of the conductors in a uniform magnetic field with a flux density $\mathscr{B}$, write the integral-differential equations that describe the system. Draw an electrical and a mechanical analog.

11.28 A simplified sketch of a moving coil microphone is shown in Fig. P11.28. Sound waves impinging on the diaphragm are represented by the force source $f(t)$. The diaphragm assembly has a mass M, a viscous friction B, and a stiffness K. The symbol l denotes the total length of conductors in a uniform magnetic field of flux density $\mathscr{B}$.

(a) Write the equations that describe the system, and draw an electrical analog.

(b) If the inductance L is negligible, find the transfer function $H(s) = I(s)/F(s)$. What is the damping ratio ζ?

11.29 The permanent magnet in Fig. P11.29a moves with a velocity $v_1(t) = \sin t\, U_{-1}(t)$, while the electrical conductors are fixed in space.

(a) If the total length of conductors in the magnetic field is l, and if the flux density is $\mathscr{B}$, write the equation(s) that describe the system. Find $i(t)$ for $t > 0$ if there is no initial stored energy and if $M = \mathscr{B} = l = L = R$ in a consistent set of units.

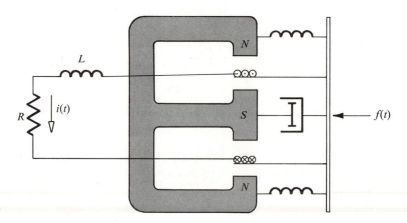

Figure P11.28

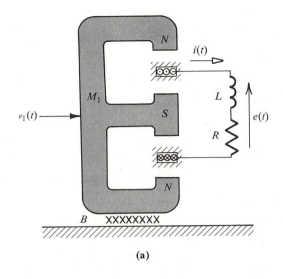

(a)

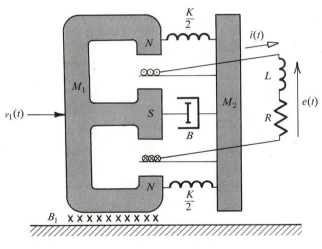

(b)

Figure P11.29

(b) In Fig. P11.29b, the coils are attached to the moving magnet by springs and a dashpot. Write the equation(s) that describe the system, and draw the electrical analog.

11.30 For the system shown in Fig. P11.30, let l denote the total length of the conductors in a uniform magnetic field with a flux density $\mathscr{B}$. The external electrical circuit is connected to the front of the bottom conductor on the left and to the front of the top conductor on the right.

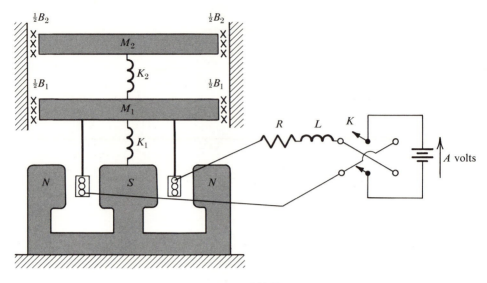

Figure **P11.30**

(a) Which way should the switch K be thrown in order to raise the two masses?

(b) What is the steady-state displacement of the mass M_1 after the switch is thrown?

(c) Draw and completely label an electrical analog for the system.

11.31 The armature of a generator has a moment of inertia J_2 and is connected by a shaft to another body, which is subjected to a torque source as shown in Fig. P11.31. Let $\mathscr{B}$ denote the flux density of the magnetic field that is cut by the conductors on the armature, and let l denote the total length of the conductors in the field. Assume that any viscous damping can be neglected. Write the integral-differential equations that describe the system. Draw and fully label an electrical analog.

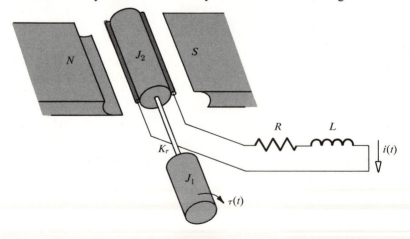

Figure **P11.31**

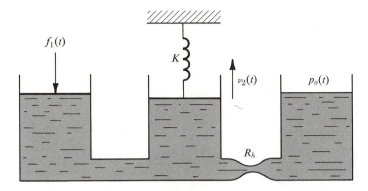

Figure P11.35

11.32 Derive the expressions given in Section 11.5 for the percent overshoot and rise time of a second-order system.

11.33 Replace the controlled sources in Fig. 11.5-6b by a gyrator, which was examined in Problem 7.32. Then obtain the analogs in Figs. 11.5-6c and 11.5-6d by using the transformation properties of the gyrator.

11.34 Redraw the mechanical portion of Fig. 11.5-6b with the angular velocity treated as the flow variable and torque as the potential variable. Then replace the two controlled sources by an ideal transformer and obtain Fig. 11.5-6d.

11.35 The mechanical parts of the system shown in Fig. P11.35 are assumed to have negligible mass, and the fluid is incompressible. The cross-sectional area of all three tanks is A. The atmospheric pressure is taken as a reference, so $p_0(t) = 0$.

(a) Find an equation relating the force source $f_1(t)$ and the velocity $v_2(t)$.

(b) Draw mechanical and electrical analogs for the system.

(c) Modify the analogs to account for any compressibility of the fluid and any viscous friction experienced by the mechanical components.

11.36 The shaft of a hydraulic pump is driven by a torque source $\tau(t)$, as shown in Fig. P11.36. Let J and B_r denote the moment of inertia and the rotational viscous friction associated with the pump. Assume that $q(t) = k_p\omega_r(t)$, as in Eq. 11.5-5, and let A be the cross-sectional area of the storage tank and ρ be the density of the fluid.

(a) Find an equation relating $\tau(t)$ and $h(t)$, the height of the fluid in the tank.

(b) If $\tau(t) = U_{-1}(t)$, what is the steady-state value of $h(t)$?

(c) Repeat the problem when the pump is used in the hydraulic system shown in Fig. 11.4-5a, if $f_3(t) = 0$.

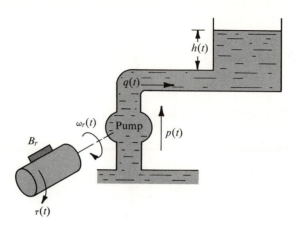

Figure P11.36

12

Linear Electronic
Circuits

In Chapter 11 we extended the methods of circuit analysis to nonelectrical systems. In the present chapter another important subject that relies heavily upon the techniques discussed in the first ten chapters is introduced.

The external characteristics of transistors and vacuum tubes and the graphical analysis of simple electronic circuits is considered in Section 12.1. In Section 12.2, circuit models that may be used to find the response to a time-varying input signal are developed. These models contain one or more controlled sources and justify the attention given to controlled sources in earlier chapters. As soon as the electronic device has been replaced by an appropriate model, all the analysis techniques of the previous chapters may be used.

In order to provide some depth of coverage in one of the important classes of electronic circuits, Section 12.3 is devoted to transistor and vacuum-tube amplifiers that are designed to amplify equally well all signals over a wide frequency range in the a-c steady state. Many other applications can be found in textbooks devoted solely to electronic circuits, e.g., A. J. Cote, Jr., and J. B. Oakes, *Linear Vacuum-tube and Transistor Circuits* (New York: McGraw-Hill, 1961) or E. J. Angelo, Jr., *Electronic Circuits*, 2nd ed. (New York: McGraw-Hill, 1964). The more general problem of how to characterize any linear network that has an input pair of terminals and an output pair of terminals is considered in Section 12.4. Since some of the results can be expressed most conveniently in terms of matrix algebra, a very brief introduction to this mathematical technique is included.

12.1
Characteristics of Transistors and Vacuum Tubes

Transistors and vacuum tubes are among the most important examples of three-terminal devices. As discussed in Section 7.4, and as shown in Fig. 12.1-1, the external behavior of any component with three external terminals can be completely

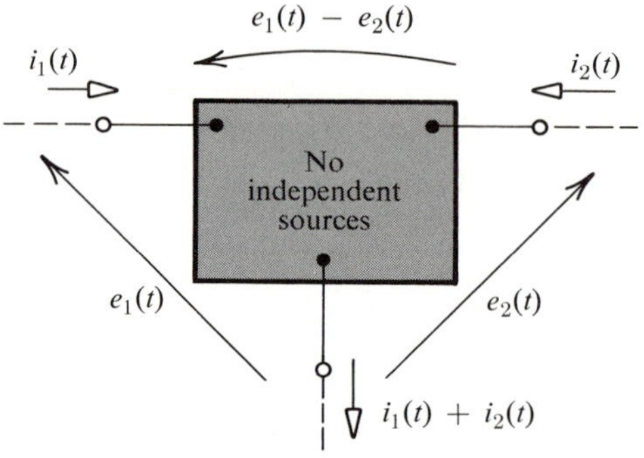

Figure 12.1-1

described by the two currents $i_1(t)$ and $i_2(t)$ and the two voltages $e_1(t)$ and $e_2(t)$. If the component contains no independent sources, two of these four variables may be expressed as functions of the other two.

The output waveshapes of a linear circuit need not be similar to the input wave-shape, so in general $i_1(t)$, $i_2(t)$, $e_1(t)$, and $e_2(t)$ cannot be directly related to each other by a pair of *algebraic* equations. If, however, a circuit contains only resist-ances, then the output function of time is directly proportional to the input, as discussed in Chapter 2. From the discussion associated with Table 4.5-1, all circuits reduce to resistive circuits in the d-c steady state. If the voltages and currents for a three-terminal circuit are constant, or if the circuit contains only resistances,

$$i_1(t) = k_{11}e_1(t) + k_{12}e_2(t)$$
$$i_2(t) = k_{21}e_1(t) + k_{22}e_2(t)$$

where the k_{ij}'s are constants of proportionality. One way to present the first of these two equations graphically is to draw a family of curves of $i_1(t)$ versus $e_1(t)$ for different values of $e_2(t)$, as in Fig. 12.1-2a. Each curve is a straight line with a slope k_{11} and with a vertical intercept $k_{12}e_2(t)$. If the curves are drawn for equal incre-ments of $e_2(t)$, they will be equally spaced.

A family of curves of $i_2(t)$ versus $e_2(t)$ for different values of $e_1(t)$ is shown in Fig. 12.1-2b. The curves are again equally spaced, straight lines. The slope is k_{22}, and the vertical intercept is $k_{21}e_1(t)$. The current $i_2(t)$ could just as well be expressed as a function of $e_2(t)$ and $i_1(t)$, which is done in Fig. 12.1-2c.

The typical curves in Fig. 12.1-2 are valid whether the external inputs are current or voltage sources. The only assumptions are that the three-terminal component does not contain independent sources and that it acts like a resistive network. It is interesting to draw the characteristic curves that correspond to a given circuit model. For the circuit in Fig. 12.1-3a, $i_1(t) = e_1(t)/R_1$ regardless of the value of $e_2(t)$, so there is only one straight line in the i_1-e_1 plot. The slope of this line is $1/R_1$, and as R_1 approaches infinity the line coincides with the horizontal axis.

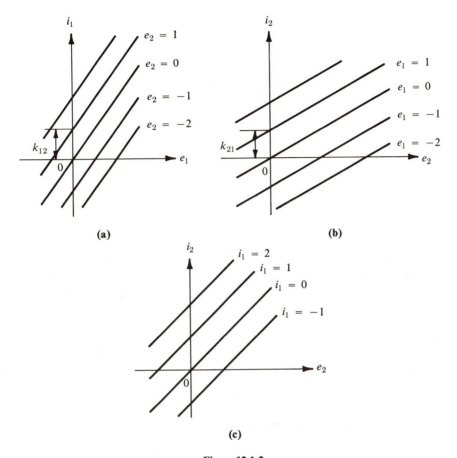

Figure 12.1-2

Since $i_2(t) = e_2(t)/R_2 - Ae_1(t)/R_2$, the slope of the lines in the i_2-e_2 plot is $1/R_2$. Notice that $i_2(t) = 0$ when $e_2(t) = Ae_1(t)$, so the horizontal spacing between lines is A. If R_2 approaches zero, the curves become vertical lines. This limiting case corresponds to an ideal voltage amplifier with an amplification A and with an infinite input resistance.

The curves for another circuit model are shown in Fig. 12.1-3b. Once again, the line in the i_1-e_1 plot coincides with the horizontal axis if R_1 approaches infinity. Since $i_2(t) = e_2(t)/R_2 + De_1(t)$, the lines in the i_2-e_2 plot have a slope of $1/R_2$ and a vertical spacing of D. These curves become horizontal lines if R_2 approaches infinity.

For the circuit in Fig. 12.1-3c, the curve in the i_1-e_1 plot becomes a vertical line if R_1 approaches zero. The family of curves in the i_2-e_2 plot have a slope of $1/R_2$ and a vertical spacing of B. These lines become horizontal for an ideal current amplifier with $R_1 = 0$ and $R_2 \to \infty$.

The Transistor. Characteristic curves for the transistor and vacuum tube can

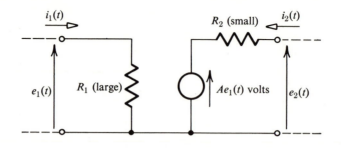

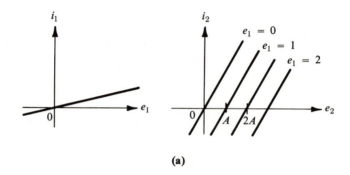

(a)

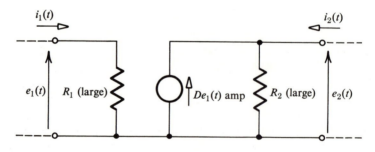

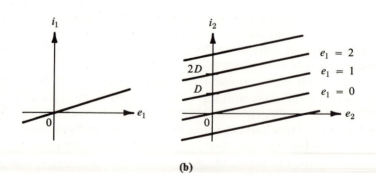

(b)

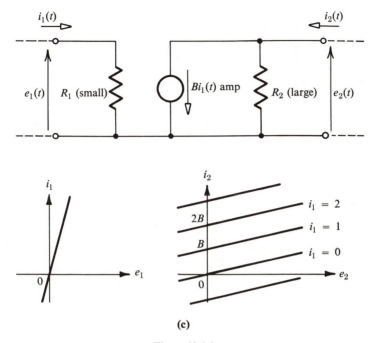

(c)

Figure 12.1-3

be found experimentally and can be compared to those in Fig. 12.1-2. The transistor consists of a germanium or silicon crystal to which certain impurities have been added in small amounts and to which metal contacts have been attached. If an impurity whose molecules have five valence electrons (such as arsenic) is added to a crystal, it produces an *N*-type region. If the molecules of the impurity have only three valence electrons (such as indium), a *P*-type region results. The transistor has three regions, each with an externally accessible terminal: the emitter (*e*), the base (*b*), and the collector (*c*). Although the emitter and collector regions contain the same type of impurity, they are not identical and in general cannot be interchanged.

The symbols for an *NPN* and *PNP* transistor are shown in Figs. 12.1-4a and 12.1-4b, respectively. In circuit diagrams, the symbols with the circle are normally used. The arrow on the emitter distinguishes it from the collector, and the direction of the arrow is the direction in which the total current flows when the transistor is used in standard circuits. The notation used for the voltages and currents is fairly standard. If the transistor is compared with the general three-terminal component in Fig. 12.1-1, $i_B(t) = i_1(t)$, $e_{BE}(t) = e_1(t)$, $i_C(t) = i_2(t)$, and $e_{CE}(t) = e_2(t)$.

Typical experimental curves for an *NPN* transistor are shown in Fig. 12.1-4c.[†]

[†] The curves for a *PNP* transistor are similar to those in the figure if all voltage and current symbols are preceded by a minus sign. In the transistor amplifier to be discussed shortly, the polarity of the bias supplies are reversed for a *PNP* transistor. Since the basic analysis procedures are the same for either type, only the *NPN* transistor is discussed in this section.

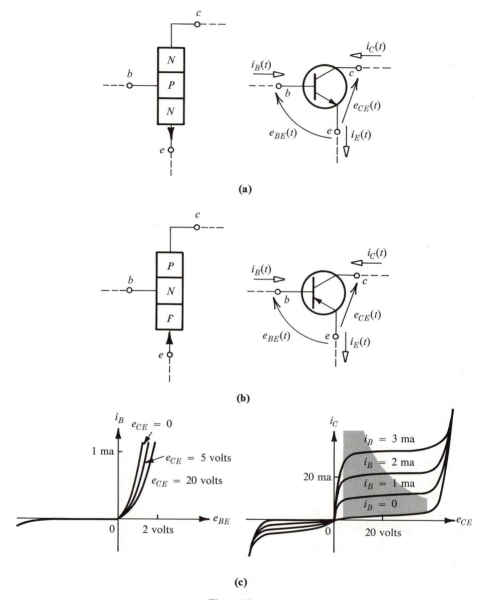

Figure 12.1-4

There are striking differences between these curves and those in Figs. 12.1-2a and 12.1-2c. The transistor curves are not symmetrical about the origin, and they can be approximated by a family of equally spaced, straight lines only in a limited region. Unless the operation of the device is confined to such a region, it cannot be treated as a linear component.

The behavior of the *NPN* transistor when $e_{BE}(t)$ and $e_{CE}(t)$ are both positive—which they are in normal amplifier operation—can be determined by examining the first quadrant of the curves in Figs. 12.1-4c. Notice that all the curves in the i_B-e_{BE} plot could be approximated by a single curve for all positive values of $e_{CE}(t)$, and that the typical values for $i_B(t)$ are small compared to those for $i_C(t)$. Since $i_E(t) = i_B(t) + i_C(t)$, the emitter and collector currents are approximately equal.

The sudden increase in the magnitude of the collector current at the ends of the curves represents an electrical breakdown of the device as a result of excessive voltage. The usual linear operating region is indicated by the shaded area in the i_C-e_{CE} plot, where the curves can be approximated by equally-spaced, straight lines. Very small collector voltages would result in severe nonlinearities. There are usually recommended values for the maximum allowable collector current and voltage, which lead to a horizontal and vertical line at the top and right side of the operating region. The hyperbola that completes the boundary of this region is based upon a consideration of the power supplied to the transistor, which is $p(t) = e_{CE}(t)i_C(t) + e_{BE}(t)i_B(t)$. Because the base voltage and current are relatively small, $p(t) \doteq e_{CE}(t)i_C(t)$. This power must be dissipated as heat, and there is a maximum power-dissipation rating for each transistor. If $p(t)$ is set equal to this maximum power, the equation describes a hyperbola in the i_C-e_{CE} plane. Except for brief periods, the operating region must lie below the hyperbola.

A simple *NPN* transistor amplifier is shown in Fig. 12.1-5a. It is customary in electronic circuits to use capital letters for d-c voltages and currents, and these must not be confused with phasors (which are in boldface type) or with transformed quantities (which are followed by an *s* in parentheses). The input signal is represented by the current source $i_s(t)$. The d-c sources, which are sometimes called *bias supplies,* are used to establish a quiescent point within the shaded region in Fig. 12.1-4c when $i_s(t) = 0$.

The series combination of R_1 and E_1 usually represents the Thévenin equivalent circuit for a more complicated combination, such as the one in Fig. 12.1-5b. As discussed in Example 2.3-1, $E_1 = R_b E_2/(R_a + R_b)$ and $R_1 = R_a R_b/(R_a + R_b)$. The original circuit can be redrawn as in Fig. 12.1-5c so that the use of two d-c sources is avoided.

When $i_s(t) = 0$, the quiescent operating point can be found from the characteristic curves for the transistor and from the following equations for the external circuit. Capital letters are used to indicate that the voltages and currents are d-c steady-state quantities.

$$I_B = \frac{E_1 - E_{BE}}{R_1}, \qquad I_C = \frac{E_2 - E_{CE}}{R_2}$$

These equations describe the two straight lines (sometimes called *load lines*) that are plotted on the characteristic curves in Figs. 12.1-5d and 12.1-5e. Their slopes are $-1/R_1$ and $-1/R_2$. The values of I_B and E_{BE} must correspond to a point on the characteristic curve in Fig. 12.1-5d and also to a point on the load line for the

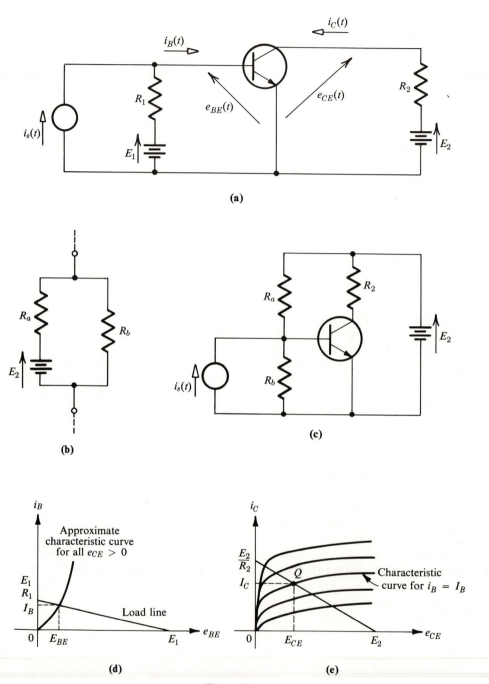

Figure 12.1-5

external circuit. Thus, the intersection of the two curves will determine the quiescent values I_B and E_{BE}. In practice, E_{BE} is small compared to E_1, so

$$I_B \doteq \frac{E_1}{R_1}$$

In Fig. 12.1-5e, the intersection of the load line with the curve corresponding to the above value of I_B is labelled Q and gives the quiescent collector current I_C and collector voltage E_{CE}.

After the quiescent point has been located on the i_C-e_{CE} characteristics, suppose that an external input $i_s(t) = A \sin \omega t$ is applied. The circuit equations are

$$i_B(t) = \frac{E_1 - e_{BE}(t)}{R_1} + i_s(t)$$

$$\doteq \frac{E_1}{R_1} + i_s(t) = I_B + i_s(t) \quad \text{if} \quad e_{BE}(t) \ll E_1$$

$$i_C(t) = \frac{E_2 - e_{CE}(t)}{R_2}$$

According to the first of these equations, $i_s(t)$ is added to the quiescent base current I_B. The second equation describes the load line already shown in Fig. 12.1-5e. Since the operating point must always lie on the load line in order to satisfy the equation for the external circuit, the path of operation is the heavy line in Fig. 12.1-6a.

Waveshapes for the source and collector currents are sketched in Fig. 12.1-6b. The time-varying component of the collector current is approximately proportional to but much larger than $i_s(t)$, so the incoming signal is amplified.[†] The amplification is increased somewhat as the load line becomes more vertical. Because the characteristic curves may not be equally spaced, the dimensions B and C may not be quite equal. If they are not equal, the nonlinearity of the transistor has caused distortion in the shape of the output sinusoid.

If the peak value of the input signal is increased so that the operating path extends almost to the ends of the load line, very serious distortion will occur in the waveshape of the collector current. A careful examination of the characteristics in Fig. 12.1-5e shows that for very small values of $e_{CE}(t)$ the curves for different values of $i_B(t)$ nearly coincide. Thus, after a certain point, further increases in the instantaneous values of $i_s(t)$ and $i_B(t)$ will have little additional effect on the value of the collector current, which will not exceed E_2/R_2 (the "collector saturation current").

It is very difficult to make the base current of an *NPN* transistor negative, even if $i_s(t)$ has a large negative value. The assumption that $e_{BE}(t)$ is approximately zero is not necessarily valid when there is no base current in the assumed positive

---

[†] This amplification, which may be between 20 and 100, is not apparent from the figure, because different scales are used for the base and collector currents.

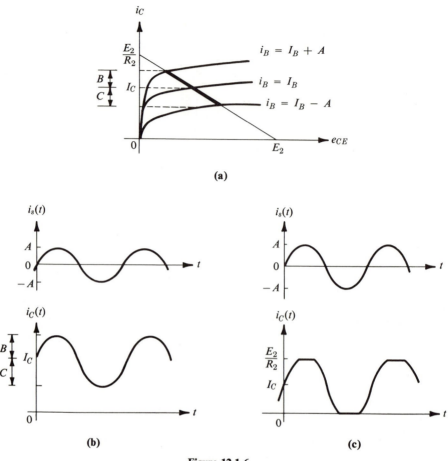

(a)

(b) **(c)**

Figure 12.1-6

direction. If the base current is zero, $e_{BE}(t) = E_1 + R_1 i_s(t)$, and as $i_s(t)$ becomes more negative than $-E_1/R_1$ the voltage $e_{BE}(t)$ becomes more and more negative. By the i_B-e_{BE} curves in Fig. 12.1-4c, the base current remains zero for negative values of base voltage (unless electrical breakdown of the transistor occurs). As seen from the i_C-e_{CE} curves, the collector current is then approximately zero ("collector cut-off"). In summary, if the peak value of $i_s(t)$ is sufficiently large, the collector current will have the waveshape shown in Fig. 12.1-6c.

Example 12.1-1. For the circuit in Fig. 12.1-5a, $E_1 = 10$ volts, $E_2 = 30$ volts, and the quiescent collector voltage and current should be $E_{CE} = 10$ volts and $I_C = 20$ ma. The characteristic curves for the transistor are given in Fig. 12.1-7, and $e_{BE}(t)$ is negligibly small. Find the values of R_1 and R_2. Calculate the average power delivered to R_2 and to the transistor under quiescent conditions; also calculate the average power delivered to R_2 and to the transistor when $i_s(t) = 0.3 \sin \omega t$ ma.

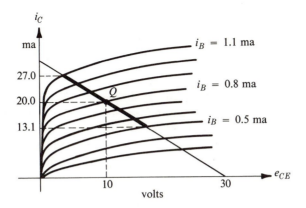

Figure 12.1-7

Solution. By plotting the quiescent point Q on the characteristic curves, we see that $I_B = 0.8$ ma. Then

$$R_1 = \frac{E_1}{I_B} = 12.5 \text{ k}\Omega, \qquad R_2 = \frac{E_2 - E_{CE}}{I_C} = 1 \text{ k}\Omega$$

Under quiescent conditions, the power supplied to R_2 is

$$P_{R_2} = I_C{}^2 R_2 = 0.4 \text{ watt}$$

and the power to the transistor is

$$P_T = E_{BE}I_B + E_{CE}I_C \doteq E_{CE}I_C = 0.2 \text{ watt}$$

The power supplied by the source in the collector circuit must be the sum of these two quantities and is $E_2 I_C = 0.6$ watt.

The load line for the collector circuit and the operating path for the given input signal $i_s(t)$ are shown in Fig. 12.1-7. The time-varying component of $i_C(t)$ is approximately sinusoidal and has a peak value of about 7 ma (which corresponds to a current amplification of 23), so

$$i_C(t) \doteq 20 + 7 \sin \omega t \text{ ma}$$

By the discussion at the end of Section 8.1, the average power delivered to R_2 is

$$P_{R_2} = (0.02)^2(10^3) + \left(\frac{7 \times 10^{-3}}{\sqrt{2}}\right)^2 (10^3) = 0.4 + 0.025 = 0.425 \text{ watt}$$

Notice that since $e_{BE}(t) \doteq 0$ the power drawn from the current source is negligible, and all the power from E_1 is consumed in R_1. The average power from E_2 is

$$[E_2 i_C(t)]_{\text{avg}} = (30)(0.02) = 0.6 \text{ watt}$$

and is not changed by the presence of the current source. The average power dissipated in the transistor is

$$P_T = 0.6 - 0.425 = 0.175 \text{ watt}$$

The collector-circuit efficiency η_c is defined as the power delivered to the load resistance R_2 by the time-varying signal divided by the power supplied by the d-c source in the collector circuit. In this example, $\eta_c = 0.025/0.6 = 4\%$. Ways to improve the efficiency are explored in one of the problems at the end of this chapter.

The most serious disadvantage of the simple amplifier in Fig. 12.1-5a is the fact that its operation is very sensitive to temperature changes. With no external signal, it is found that an increase in temperature increases the collector current, so the quiescent point moves to a higher position on the load line. Not only may this cause the path of operation to be in a less linear portion of the i_C-e_{CE} plot, but it may even result in "thermal runaway."

Suppose that because of an increase in temperature the quiescent point moves from Q to Q_1 in Fig. 12.1-8. A typical load line and lines of constant transistor power dissipation are shown in the figure. The characteristic curves are not shown because they move upward when the temperature is increased. For the load line in part (a), the power dissipation is greater at Q_1 than at Q. This increased power will tend to raise the temperature still further, which in turn will increase the collector current still further. This cumulative effect may result in the electrical destruction of the transistor.

If the slope of the load line is more nearly horizontal, as in Fig. 12.1-8b, the temperature problem is somewhat less serious. The power dissipation at Q_1 is less than at Q, so the cumulative effect described in the preceding paragraph does not occur. Recall, however, that for a given input signal the current amplification is less for the load line in part (b).

One way to limit the effects of temperature variation is to add a resistance in series with the emitter, as in Fig. 12.1-9a. The purpose of the resistance can be explained qualitatively by noting that, with no external current source and for a negligibly small base-to-emitter voltage, $I_B = (E_1 - I_E R_3)/R_1$. Since the emitter current I_E is approximately equal to I_C, any increase in I_C will cause a decrease in I_B, which will tend to move the quiescent point back down the load line.

In order to locate the quiescent point when R_3 is present, consider the loop equations

$$(I_B + I_C)R_3 + E_{BE} + I_B R_1 - E_1 = 0$$

$$(I_B + I_C)R_3 + E_{BC} + I_C R_2 - E_2 = 0$$

If $E_{BE} \ll E_1$ in the first equation, and if $I_B \ll I_C$ in the second equation,

$$I_C \doteq \frac{E_1 - I_B(R_1 + R_3)}{R_3}$$

$$I_C \doteq -\frac{E_{CE}}{R_2 + R_3} + \frac{E_2}{R_2 + R_3}$$

The first of these equations can be plotted on the i_C-e_{CE} characteristics (not necessarily as a straight line), and the second equation describes a load line with a

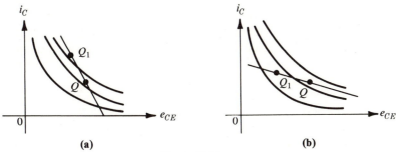

Figure 12.1-8

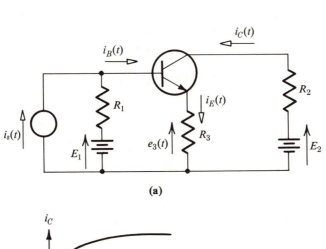

(a)

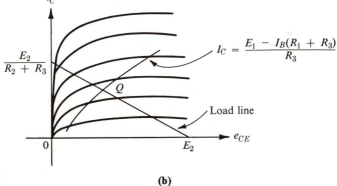

$$I_C = \frac{E_1 - I_B(R_1 + R_3)}{R_3}$$

Load line

(b)

Figure 12.1-9

slope $-1/(R_2 + R_3)$. The intersection of these two curves determines the quiescent point Q in Fig. 12.1-9b.

Although the presence of R_3 makes the circuit less susceptible to changes in temperature, it has a deleterious effect on the amplification of a time-varying input signal $i_s(t)$. An increase in $i_s(t)$ would be expected to increase $i_B(t)$ and therefore to increase $i_C(t)$. But the increase in the collector current increases the voltage $e_3(t)$, which in turn tends to *oppose* the increase in $i_B(t)$. Since the change in the collector current of a transistor is normally much larger than the change in the base current, the effect is much greater than it would be if only $i_B(t)$ flowed through R_3. The current source sees a large resistance looking into the base, so a substantial part of $i_s(t)$ flows instead through R_1.[†]

The harmful effect of R_3 on the amplification of the time-varying input signal can be avoided by adding a large capacitance in parallel with R_3, as in Fig. 12.1-10a. If $i_s(t) = A \sin \omega t$, the impedance of the capacitance in the a-c steady state is $1/j\omega C_3$. For a sufficiently large value of ωC_3, the capacitance will act like a short circuit for the sinusoidal input. Nearly all of the current $i_s(t)$ will flow into the base of the transistor, and

$$i_B(t) = I_B + i_s(t)$$

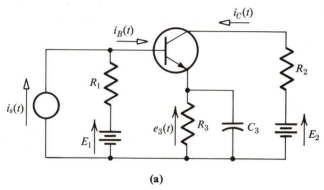

(a)

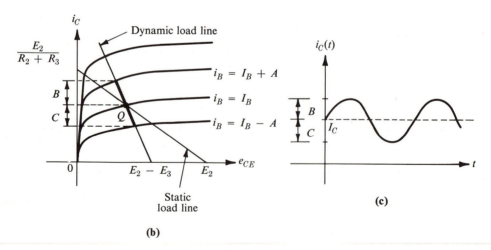

(b)

(c)

Figure 12.1-10

[†] A similar situation was examined analytically in Example 5.3-3 and will be considered again later in this chapter.

Because of the capacitance, the time-varying component of the emitter current will not affect the voltage $e_3(t)$, which will remain equal to the d-c voltage $E_3 = (I_B + I_C)R_3$. The equation for the collector circuit is

$$i_C(t) = \frac{-e_{CE}(t)}{R_2} + \frac{E_2 - E_3}{R_2}$$

which describes a load line with a slope $-1/R_2$. This new line, which is called the *dynamic* load line in order to distinguish it from the *static* load line that is used for d-c quantities, passes through the quiescent point and is shown in Fig. 12.1-10b. It is used to determine the response to a time-varying signal *after* the quiescent point has been located. The operating path for the input $i_s(t) = A \sin \omega t$ is the heavy part of the dynamic load line in the figure. The waveshape of the steady-state collector current is given in Fig. 12.1-10c. Other circuits for which the dynamic load line does not coincide with the static load line are found in the problems at the end of this chapter and in the following example.

Example 12.1-2. For the circuit in Fig. 12.1-11a,

$$R_3 = 2 \text{ k}\Omega \quad \text{and} \quad R_1 = \frac{R_a R_b}{R_a + R_b} = 8 \text{ k}\Omega$$

(The choice of these values makes the transistor sufficiently insensitive to changes in temperature.) Choose R_a and R_b so that the quiescent base current is $I_B = 20 \ \mu$a, for the characteristic curves in Fig. 12.1-11b. Then find the voltage $e_o(t)$ when $i_s(t) = 12 \sin 10^5 t \ \mu$a.

Solution. In the d-c steady state, the inductance and capacitances may be replaced by short and open circuits, respectively. For the determination of the quiescent point, the circuit reduces to the one in Fig. 12.1-9a, where $E_2 = 10$ volts, $R_2 = 3$ kΩ, $R_1 = R_a R_b/(R_a + R_b)$, $E_1 = E_2 R_b/(R_a + R_b) = E_2 R_1/R_a$, and $R_3 = 2$ kΩ. The static load line drawn in Fig. 12.1-11b is described by the equation

$$I_C \doteq \frac{-E_{CE}}{R_2 + R_3} + \frac{E_2}{R_2 + R_3}$$

and has a slope $-1/(R_2 + R_3) = -\frac{1}{5}$ ma per volt. At the quiescent point Q, $E_{CE} = 5$ volts, $I_C = 1$ ma, and $I_B = 20 \ \mu$a. If the base to emitter voltage is negligibly small,

$$E_1 = (I_B + I_C)R_3 + I_B R_1 = (1.02)(2) + (0.02)(8) = 2.2 \text{ volts}$$

$$R_a = \frac{R_1 E_2}{E_1} = 36 \text{ k}\Omega$$

$$R_b = \frac{(36)(8)}{36 - 8} = 10.3 \text{ k}\Omega$$

At the angular frequency $\omega = 10^5$ rad per sec, $\omega L = 10^5 \ \Omega$ and $1/\omega C = 1 \ \Omega$, so the inductance and capacitance can be treated as open and short circuits, respectively, for the time-varying component of the voltages and currents. The dynamic load line

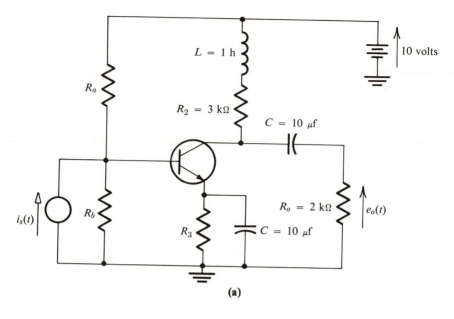

(a)

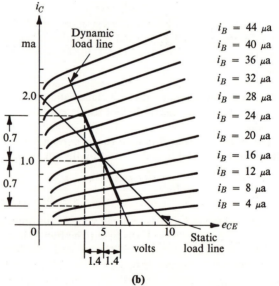

(b)

Figure 12.1-11

has a slope of $-1/R_o = -\frac{1}{2}$ ma per volt, and is shown in Fig. 12.1-11b. The operating path is indicated by the heavy line, and

$$i_C(t) \doteq 1.0 + 0.7 \sin \omega t \text{ ma}$$

$$e_{CE}(t) \doteq 5.0 - 1.4 \sin \omega t \text{ volts}$$

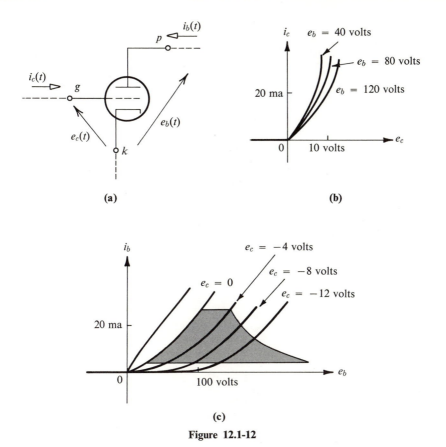

(a)

(b)

(c)

Figure 12.1-12

The time-varying component but not the d-c component of $i_C(t)$ flows through the capacitances, so

$$e_o(t) = -(0.7 \times 10^{-3})(2 \times 10^3) \sin \omega t = -1.4 \sin \omega t \text{ volts}$$

The Triode. The symbol for a vacuum-tube triode is shown in Fig. 12.1-12a. The cathode (k), the control grid (g), and the plate (p) are three metal electrodes that are placed in an evacuated envelope. Not shown is a heater that maintains the cathode at an elevated temperature so that electrons can easily leave the cathode surface. The voltage applied to the grid, which can be a wire helix or a wire mesh, exerts a strong influence on the number of electrons that can pass from the cathode to the plate. The standard notation for the instantaneous voltages and currents is shown in the figure. In terms of the general three-terminal component in Fig. 12.1-1, $i_c(t) = i_1(t)$, $e_c(t) = e_1(t)$, $i_b(t) = i_2(t)$, and $e_b(t) = e_2(t)$.

Typical experimental curves for the triode are given in Figs. 12.1-12b and 12.1-12c. Once again, the characteristic curves are not symmetrical about the origin and are not straight lines, but they can be approximated by a family of equally spaced, straight lines in a limited region. The i_c-e_c characteristics may be approximated by a single curve, but in most $e_c(t)$ is negative, in which case the grid current $i_c(t)$ is negligibly small. The usual region of linear operation is indicated by the shaded area in Fig. 12.1-12c. Very small plate currents introduce appreciable distortion, and there is usually a maximum recommended current. The power supplied to the triode is $p(t) = e_b(t)i_b(t) + e_c(t)i_c(t) \doteq e_b(t)i_b(t)$, and the hyperbola that forms the right-hand boundary of the operating region represents the maximum power-dissipation rating for the tube.

The graphical analysis of triode amplifiers is analogous to that for transistors, so the following discussion is more concise. For the simple amplifier shown in Fig. 12.1-13a, E_{cc} and E_{bb} are d-c bias supplies that are used to establish a suitable quiescent point, and $e_s(t)$ is a time-varying input signal. When $e_s(t) = 0$, and when capital letters are used to represent d-c currents and voltages,

$$I_b = -\frac{E_b}{R_2} + \frac{E_{bb}}{R_2}$$

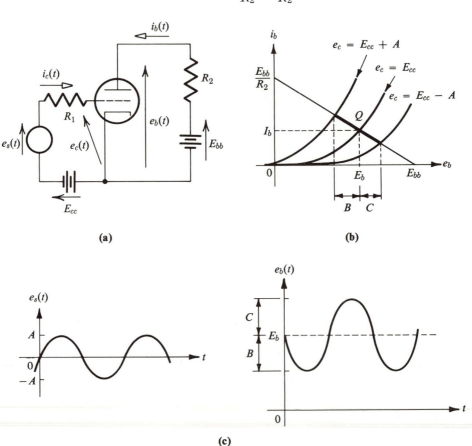

(a)

(b)

(c)

Figure 12.1-13

This equation describes a load line that has a slope $-1/R_2$ and that is shown in Fig. 12.1-13b. Since the grid voltage is normally negative, the grid current is negligible, and $E_c = E_{cc}$. The quiescent point Q is located at the intersection of the load line and the characteristic curve for $e_c(t) = E_{cc}$.

If an external signal $e_s(t) = A \sin \omega t$ is applied and if the grid voltage remains negative, $e_c(t) = E_{cc} + e_s(t)$. The path of operation is the heavy portion of the load line in Fig. 12.1-13b, and the waveshapes of the input and plate voltages are shown in Fig. 12.1-13c. The time-varying component of $e_b(t)$ is 180° out of phase with $e_s(t)$, but it may be 10 or 20 times as large as $e_s(t)$, so the input voltage is amplified.

The output waveshape will be severely distorted if the peak value of $e_s(t)$ is sufficiently large. If $e_s(t) + E_{cc}$ is positive for some values of t, grid current will flow and will limit the increase in grid voltage, since $e_c(t) = e_s(t) + E_{cc} - R_1 i_c(t)$. On the other hand, $i_b(t)$ is nearly zero for large negative values of grid voltage, and if $e_c(t)$ becomes still more negative there will not be a corresponding change in $e_b(t)$, and distortion will again occur.

The necessity for two separate bias supplies cannot be avoided by the use of a

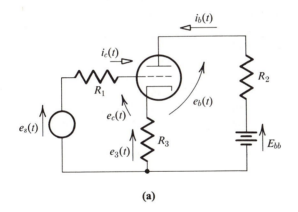

(a)

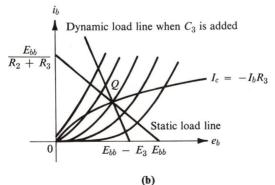

(b)

Figure 12.1-14

voltage divider (such as E_2, R_a, and R_b in Fig. 12.1-5b), because E_{cc} and E_{bb} have opposite polarities. However, E_{cc} may be eliminated if a resistance R_3 is added in series with the cathode, as in Fig. 12.1-14a. If $e_s(t) = 0$ and if no grid current flows, the quiescent plate current I_b flowing through R_3 will result in a negative cathode-to-grid voltage $E_c = -I_b R_3$. In order to locate the quiescent point, this relationship may be plotted on the i_b-e_b curves in Fig. 12.1-14b. The loop equation for the plate circuit is

$$R_3 I_b + E_b + R_2 I_b - E_{bb} = 0$$

or

$$I_b = \frac{-E_b}{R_2 + R_3} + \frac{E_{bb}}{R_2 + R_3}$$

which describes the static load line shown in the figure. The quiescent point Q is located at the intersection of the load line and the curve corresponding to $E_c = -I_b R_3$.

The addition of R_3 also tends to counteract any movement of the quiescent point that may result from aging of the tube or from other causes, although this problem is much less serious than for the transistor. Any increase in I_b will make the grid voltage $E_c = -I_b R_3$ more negative, which will in turn tend to decrease the plate current. This same action, however, reduces the amplification of an input signal $e_s(t)$, because $e_c(t) = e_s(t) - R_3 i_b(t)$. An increase in $e_s(t)$ results in an increase in $i_b(t)$, and the increase in grid voltage is limited by the $R_3 i_b(t)$ term.

To avoid this reduction in voltage amplification, a capacitance C_3 may be added in parallel with R_3. If the capacitance is sufficiently large to act like a short circuit to the time-varying component of $i_b(t)$, then $e_c(t) = e_s(t) - E_3$, where E_3 is the constant voltage $I_b R_3$. Hence, the full effect of any change in $e_s(t)$ is transmitted to $e_c(t)$. The equation

$$i_b(t) = -\frac{e_b(t)}{R_2} + \frac{E_{bb} - E_3}{R_2}$$

describes a load line (the dynamic load line) passing through the quiescent point

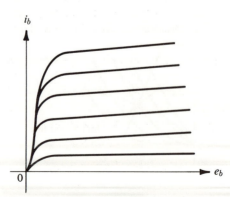

Figure 12.1-15

with a slope of $-1/R_2$. The response to $i_s(t) = A \sin \omega t$ can be found in the usual way by drawing the operating path along the dynamic load line.

The Pentode and the Field-Effect Transistor. The pentode is a vacuum tube to which a screen grid and a suppressor grid have been added. The suppressor grid is usually connected directly to the cathode, and the screen grid is maintained at a nearly constant voltage, even when a time-varying voltage is applied to the control grid. In most linear circuits, the control-grid current is negligibly small. Typical i_b-e_b curves are shown in Fig. 12.1-15. The curves are nearly horizontal in the linear operating range, so the device may be used to approximate the circuit in Fig. 12.1-3b. The analysis procedures for circuits containing pentodes are in general similar to those already discussed. The field-effect transistor is a semiconductor device having characteristics that are very similar to those for the pentode.

12.2
Incremental Models

In the graphical analysis of typical transistor and vacuum-tube circuits, the quiescent point is located by drawing a static load line, and the response to a time-varying input signal is found from the dynamic load line. In addition to the usual disadvantages associated with any graphical method, certain of the assumptions that were made seriously limit the techniques of the previous section. Characteristic curves can be used for a three-terminal device only if it can be described by *algebraic* equations, i.e., only if the applied signal varies slowly enough to allow the device to be treated as a resistive network. The external circuit also must be able to be described by algebraic equations. Any external capacitance or inductance must be approximated by a short or open circuit when constructing the dynamic load line. This was done, for example, in the analysis of Fig. 12.1-10a.

The principal interest in electronic circuits is in the time-varying component of the output. The d-c bias supplies are only used to establish a quiescent point in a linear region of operation and to supply the necessary power. In effect, they shift the origin for the characteristic curves from its original position in a nonlinear region to the center of a linear region.

Once the quiescent point has been found, an *incremental circuit model* may be used to calculate the time-varying component of the response, i.e., to find the deviation from the quiescent point as a function of time. The time-varying components of transistor voltages and currents are distinguished from the total instantaneous values by using lower-case subscripts. Thus,

$$i_B(t) = I_B + i_b(t)$$
$$e_{BE}(t) = E_{BE} + e_{be}(t)$$
$$i_C(t) = I_C + i_c(t)$$
$$e_{CE}(t) = E_{CE} + e_{ce}(t)$$

where, for example, I_B and $i_b(t)$ are the d-c and time-varying components, respectively, of the base current.

Suppose that the instantaneous base voltage and collector current for a transistor are expressed as functions of the other two variables.

$$e_{BE}(t) = f_1[i_B(t), e_{CE}(t)]$$

$$i_C(t) = f_2[i_B(t), e_{CE}(t)]$$

Let di_B and de_{CE} represent differential changes of $i_B(t)$ and $e_{CE}(t)$ from their quiescent values. The resulting differential changes in $e_{BE}(t)$ and $i_C(t)$ are given by

$$de_{BE} = \frac{\partial e_{BE}}{\partial i_B} di_B + \frac{\partial e_{BE}}{\partial e_{CE}} de_{CE}$$

$$(12.2\text{-}1)$$

$$di_C = \frac{\partial i_C}{\partial i_B} di_B + \frac{\partial i_C}{\partial e_{CE}} de_{CE}$$

The coefficients on the right side of these equations are denoted by the following symbols.

$$h_{11} = r_n = \frac{\partial e_{BE}}{\partial i_B} \doteq \frac{\Delta e_{BE}}{\Delta i_B}\Bigg]_{e_{CE}(t)=\text{constant}}$$

$$h_{12} = \mu = \frac{\partial e_{BE}}{\partial e_{CE}} \doteq \frac{\Delta e_{BE}}{\Delta e_{CE}}\Bigg]_{i_B(t)=\text{constant}}$$

$$(12.2\text{-}2)$$

$$h_{21} = \beta = \frac{\partial i_C}{\partial i_B} \doteq \frac{\Delta i_C}{\Delta i_B}\Bigg]_{e_{CE}(t)=\text{constant}}$$

$$h_{22} = g_o = \frac{\partial i_C}{\partial e_{CE}} \doteq \frac{\Delta i_C}{\Delta e_{CE}}\Bigg]_{i_B(t)=\text{constant}}$$

The h_{ij}'s are symbols often used for the description of any linear, three-terminal network, as discussed in Section 12.4, while r_n, μ, β, and g_o (or $1/r_o$) are usually

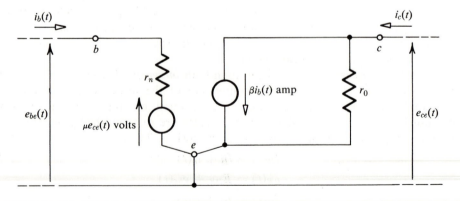

Figure 12.2-1

used only for a transistor. The quantities r_n and r_o have units of ohms, and μ and β are dimensionless.

Equations 12.2-1 are approximately valid even if the differential changes from the quiescent values are replaced by small incremental changes. Since the deviations from the quiescent values are simply the time-varying components,

$$e_b(t) = r_n i_b(t) + \mu e_{ce}(t)$$
$$i_c(t) = \beta i_b(t) + g_o e_{ce}(t)$$

$$(12.2\text{-}3)$$

These equations describe the incremental circuit model in Fig. 12.2-1, which is used extensively in the rest of this chapter.

Example 12.2-1. Show how the element values for the incremental model of a transistor may be found from the characteristic curves.

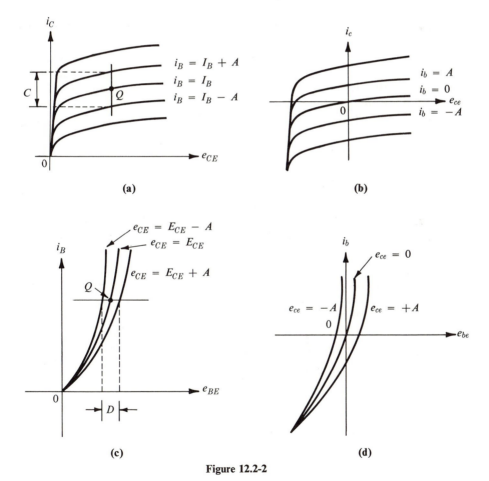

Figure 12.2-2

Solution.   To determine β from the i_C-e_{CE} characteristics in Fig. 12.2-2a, we draw a vertical line, along which $e_{CE}(t)$ is constant, through the quiescent point. Then, according to the definition in Eqs. 12.2-2, $\beta = C/2A$. The value of g_o equals the slope of the curve for $i_B(t) = I_B$ at the quiescent point. The same results are obtained by realizing that the variables in Eqs. 12.2-3 are deviations from the quiescent values, and that their use corresponds to shifting the origin to the quiescent point and to relabelling the curves of constant base current, as in Fig. 12.2-2b. If the curves are approximated by straight lines, $i_c(t) = g_o e_{ce}(t) + \beta i_b(t)$ where g_o is the slope and $\beta i_b(t)$ is the vertical intercept.

For the characteristic curves in Fig. 12.2-2c, $\mu = D/2A$, and $1/r_n$ is the slope of the curve for $e_{CB}(t) = E_{CB}$ at the quiescent point. If desired, the origin can again be relocated, and the curves can be labelled with the deviations from the quiescent values, as in Fig. 12.2-2d. As discussed in Section 12.1, the curves of constant collector voltage are very close together, so μ often can be considered to be zero. In this case, the incremental model reduces to the circuit in Fig. 12.1-3c. The slope of the i_B-e_{BE} curves is large near a typical quiescent point, so r_n is a small (sometimes negligible) resistance.

The use of an incremental model is equivalent to approximating the characteristic curves by a set of equally spaced, straight lines *in the vicinity of the quiescent point.* The values of the elements in the model depend upon the choice of the quiescent point, and they are sometimes called the *small signal parameters*, because they are valid only for small deviations from the quiescent point. Instead of determining the parameters from the characteristic curves, it is more common to measure them directly. One simple circuit for the measurement of β is shown in Fig. 12.2-3. The capacitances block the flow of any d-c current, but are large enough to act like short circuits to the time-varying components. Since

$$\beta = \left[\frac{\Delta i_C(t)}{\Delta i_B(t)}\right]_{e_{CE}(t)=\text{constant}} = \left[\frac{i_c(t)}{i_b(t)}\right]_{e_{ce}(t)=0}$$

the parameter β may be found by making two current measurements. Typical values for the parameters are given by the manufacturer.[†]

An incremental model for the triode can be developed in a manner similar to that used for the transistor. Let

$$i_c(t) = I_c + i_g(t), \qquad e_c(t) = E_c + e_g(t)$$

$$i_b(t) = I_b + i_p(t), \qquad e_b(t) = E_b + e_p(t)$$

where $i_g(t)$, $e_g(t)$, $i_p(t)$, and $e_p(t)$ are the time-varying components of the grid

---

[†] The h parameters for a transistor may be given for the "common-emitter connection" or for the "common-base connection." The difference between the two sets of parameters is discussed in one of the problems at the end of this chapter.

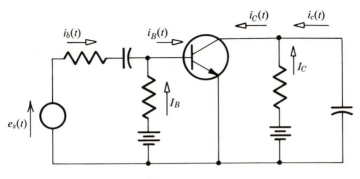

Figure 12.2-3

current, grid voltage, plate current, and plate voltage, respectively. If, for example, $i_c(t)$ and $e_b(t)$ are expressed as functions of $e_c(t)$ and $i_b(t)$,

$$di_c = \frac{\partial i_c}{\partial e_c} de_c + \frac{\partial i_c}{\partial i_b} di_b$$

$$de_b = \frac{\partial e_b}{\partial e_c} de_c + \frac{\partial e_b}{\partial i_b} di_b$$

$$(12.2\text{-}4)$$

The grid current remains negligible in most linear circuits, so the coefficients on the right side of the first equation are approximately zero. If

$$-\mu = \frac{\partial e_b}{\partial e_c} \doteq \left[\frac{\Delta e_b}{\Delta e_c}\right]_{i_b(t)=\text{constant}}$$

$$r_p = \frac{\partial e_b}{\partial i_b} \doteq \left[\frac{\Delta e_b}{\Delta i_b}\right]_{e_c(t)=\text{constant}}$$

$$(12.2\text{-}5)$$

then

$$e_p(t) = -\mu e_g(t) + r_p i_p(t) \qquad (12.2\text{-}6)$$

which describes the incremental model in Fig. 12.2-4a. Alternatively, if $i_b(t)$ is considered to be a function of $e_c(t)$ and $e_b(t)$,

$$di_b = \frac{\partial i_b}{\partial e_c} de_c + \frac{\partial i_b}{\partial e_b} de_b$$

With $\partial i_b/\partial e_c = g_m$ and $\partial i_b/\partial e_b = g_p$,

$$i_p(t) = g_m e_g(t) + g_p e_p(t) \qquad (12.2\text{-}7)$$

which describes the incremental model in Fig. 12.2-4b. In order for the circuits in parts (a) and (b) to be equivalent to each other, $g_m = \mu/r_p$ and $g_p = 1/r_p$. The symbols μ, r_p, and g_m are called the *amplification factor*, the *plate resistance*, and

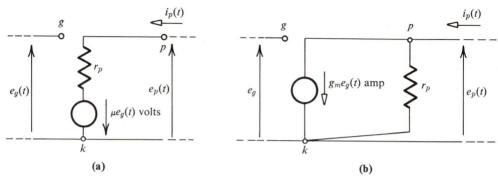

Figure 12.2-4

the *transconductance*, respectively. Their values may be found from the spacing and slope of the characteristic curves, or they may be measured directly.

The use of lower-case letters for the elements in the incremental models is intended to emphasize the fact that they are valid only for the calculation of the time-varying components. In order to use these models, the d-c voltage sources in the external circuit are replaced by short circuits, since there is no time-varying component of voltage across them. The complete waveshape of the instantaneous output, if it is desired, may be obtained by adding the d-c and time-varying components.

Example 12.2-2.   Find an expression for $e_2(t)$ for the circuit in Fig. 12.1-14a, where $e_2(t)$ is the time-varying component of the voltage across R_2.

Solution.   Replacing the triode by the incremental model in Fig. 12.2-4a, and re-placing the d-c source by a short circuit, we obtain the circuit in Fig. 12.2-5. This is essentially identical with the circuit solved in Example 2.3-6, and

$$e_2(t) = \frac{-\mu R_2}{R_3(1 + \mu) + r_p + R_2} e_s(t)$$

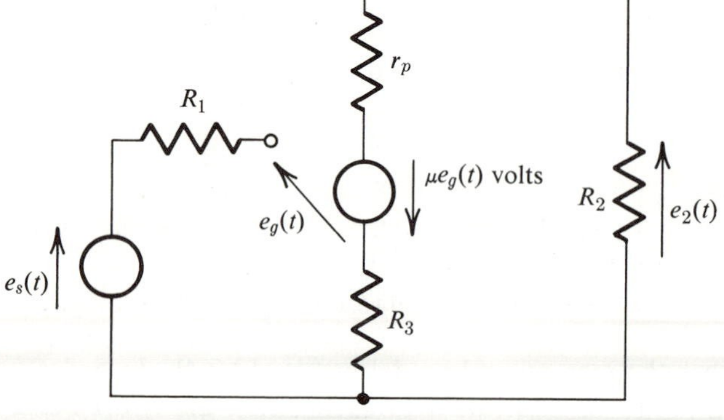

Figure 12.2-5

Notice that, since $\mu \gg 1$ for a typical triode, the presence of R_3 severely reduces the voltage gain, as was discussed in Section 12.1. If a large capacitance is placed in parallel with R_3, and if the capacitance acts like a short circuit to the time-varying signal,

$$e_2(t) = \frac{-\mu R_2}{r_p + R_2} \, e_s(t)$$

The voltage gain is increased if R_2 is increased, but it is always less than μ. The minus sign indicates that the output signal is 180° out of phase with the input, if $e_s(t)$ is sinusoidal.

Example 12.2-3. Find an expression for the current $i_o(t)$ for the common-base amplifier shown in Fig. 12.2-6a, if the capacitances act like short circuits for the time-varying signal. For the transistor, use the incremental model in Fig. 12.2-1 with $\mu = 0$.

Solution. The circuit model for the calculation of the time-varying currents and voltages, shown in Fig. 12.2-6b, can be described by two node equations. When $i_b(t)$ is replaced by $-e_1(t)/r_n$, the equations become

$$\left(G_1 + \frac{1}{r_n} + \frac{\beta}{r_n} + g_o\right) e_1(t) - g_o e_2(t) = i_s(t)$$

$$-\left(g_o + \frac{\beta}{r_n}\right) e_1(t) + (G_o + g_o)e_2(t) = 0$$

where $G_1 = 1/R_1$, etc. Solving simultaneously, we obtain

$$i_o(t) = G_o e_2(t) = \frac{G_o(g_o + \beta/r_n)}{G_o\left[G_1 + g_o + \frac{(\beta + 1)}{r_n}\right] + g_o(G_1 + 1/r_n)} \, i_s(t)$$

Notice that the current gain is less than unity. If R_1 and r_o are very large,

$$i_o(t) \doteq \frac{\beta}{\beta + 1} \, i_s(t)$$

The solution of Example 12.2-3 is very similar to the analysis of Fig. 2.5-4. Once the circuit model that is to be used for the calculation of the time-varying response has been drawn, all of the techniques of the first ten chapters may be applied. It is not unusual to find that the output power for a circuit containing a controlled source is greater than the power associated with the input signal, but this can be explained by referring to Example 12.1-1. In transistor and vacuum-tube circuits, the d-c bias supplies (which are not shown in the incremental model) not only establish a suitable quiescent point, but provide at least some of the power

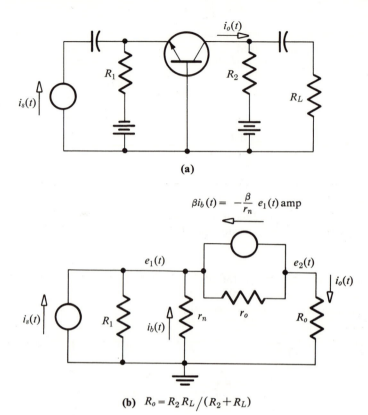

$$\beta i_b(t) = -\frac{\beta}{r_n} e_1(t) \text{ amp}$$

(b) $R_o = R_2 R_L / (R_2 + R_L)$

Figure 12.2-6

associated with the time-varying output. In any circuit with losses, the output power is always less than the input power when all the power sources are considered.

It is not necessary to assume that any inductances or capacitances act like open or short circuits to the time-varying signal in order to use an incremental model. In fact, the ability to include the effects of reactive elements is a principal advantage

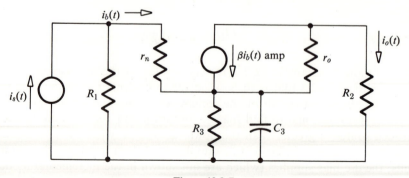

Figure 12.2-7

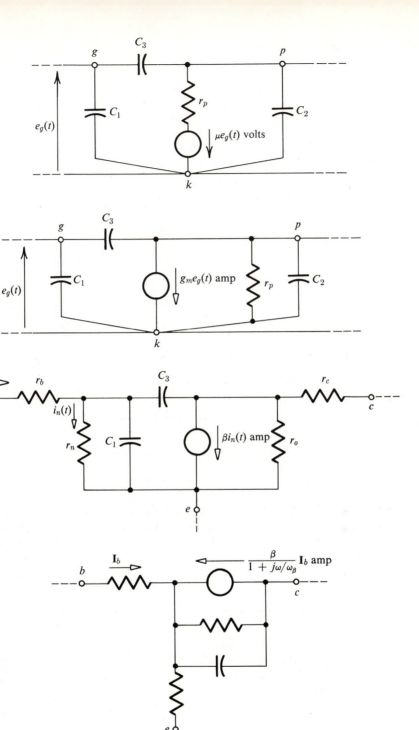

(a)

(b)

(c)

(d)

Figure 12.2-8

of the methods of this section, as compared to a graphical analysis. A complete incremental model for the circuit in Fig. 12.1-10a, for example, is given in Fig. 12.2-7. If r_o is very large (which is often true in practice), this circuit is similar to the one in Fig. 5.3-4a. The discussion in Example 5.3-3 about the variation of the output signal with changes in the angular frequency of a sinusoidal input is important, and the student is urged to review that example at this time.

The incremental models in Figs. 12.2-1 and 12.2-4 for the transistor and vacuum tube are composed of only controlled sources and resistances. For input signals that vary rapidly with time, these models may be too simple. One approach is to try to modify the models in a way that corresponds to the physical phenomena occurring within the device. In a vacuum tube, for example, the metallic electrodes act like the plates of a capacitance. This effect may be represented by the three interelectrode capacitances shown in Figs. 12.2-8a and 12.2-8b.

It is not easy to develop an incremental transistor model that is satisfactory for rapidly changing inputs. The choice of a model may depend upon the type of transistor that is used and upon the particular application. One model that is related to the physical phenomena within the device is shown in Fig. 12.2-8c. The letters C_1 and C_3 represent the capacitances that exist across the junctions of adjacent P- and N-type regions. Notice that the current source is controlled by the current through r_n and not by the current flowing into the base terminal. For a high-frequency, sinusoidal input, the impedance of the capacitances are small enough to seriously reduce the portion of the base current that passes through r_n, and thereby to reduce the current gain.

A somewhat different model that is sometimes used for a-c steady-state analysis is given in Fig. 12.2-8d. The value of the controlled source decreases as the angular frequency increases. For all these models, it may be possible to omit some of the elements in many applications, or it may be necessary to add additional elements. The test of the validity of a model is whether or not it gives reasonable agreement with experimental results. Some further comments will be found in Section 12.4.

12.3

Wide-Band Amplifiers

The input signal to an amplifier in a television system, or a similar system, may contain sinusoidal components that extend over a frequency range of several mega-cycles. In order for the input and output waveshapes to have the same shape, the a-c steady-state transfer function must have a constant magnitude and an angle proportional to ω over this frequency range. Wide-band amplifiers are frequently used to transmit pulses, which may contain information in digital form (as in pulse code modulation), or which may be used to initiate a specified reaction at a desired time. In the first part of this section, we examine the frequency response curves for simple transistor and vacuum-tube amplifiers, with particular emphasis on the factors that limit their low- and high-frequency performance. Later, an example of the relationship between the frequency response and the pulse response is considered.

Once an incremental circuit model that is sufficiently accurate over the entire frequency range of interest has been drawn, the a-c steady-state transfer function may be found and plotted by the methods in Chapters 5 and 6. However, this direct approach may require considerable labor, and it usually precludes a meaningful solution in literal form, since the resulting expression may be complicated and may involve high-order polynomials. A literal rather than a numerical solution is necessary in order to determine the effect of individual element values on the frequency response curves. Usually, the effect of certain elements is negligible except at low frequencies, and certain other elements can be neglected except at high frequencies. Then one circuit model may be used to study the behavior at low frequencies, and another model may be used for the high-frequency analysis. In order for this approach to be successful, both of these models must reduce to the same circuit at intermediate frequencies.

It is convenient to show the frequency response characteristics of an amplifier by drawing Bode diagrams, although most of the ensuing discussion is understandable even if the student has omitted Section 6.5. In many applications, several simple amplifiers are connected in cascade in order to obtain the necessary amplification. If the voltage or current gain of each stage is known, the gain for the cascaded combination may be found from Eqs. 6.5-2. When the gain of any one stage is to be calculated, the loading effect of subsequent stages must be included. (This matter was first discussed in connection with Example 3.5-5.) We shall account for this loading effect by adding one or more elements at the output terminals of the incremental model for the stage under consideration.

Vacuum-Tube Amplifiers.   The part of the circuit enclosed by the dashed lines in Fig. 12.3-1a represents one stage of a typical multistage, vacuum-tube amplifier. The quiescent plate voltage for each tube is a large positive quantity, while the grid voltage is negative. The *coupling capacitance* C_c isolates these two d-c voltages from each other, but is intended to act like a short circuit to a time-varying signal. As illustrated by Problem 1.6, the practical upper limit to the size of a capacitance depends upon the voltage that it must withstand. Since there is a large d-c voltage across C_c, it is expensive to use a capacitance that is larger than a few microfarads. If the input signal is a very low-frequency sinusoid, there will be a significant time-varying voltage across C_c. Because the d-c quiescent voltage across C_k is less than that across C_c, it may be possible to use a larger capacitance for C_k, and the effect of C_k may be less important in determining the lower limit of the bandwidth.

Under quiescent conditions, some of the electrons travelling from the cathode to the plate strike the grid, and if allowed to remain there they tend to make the grid voltage more negative. The presence of C_c blocks the d-c return path to the cathode that was formerly available, so a *grid leak resistance* R_g must be added. The d-c current through R_g does change the cathode-to-grid voltage, but since the grid current is extremely small R_g may be as large as 1 MΩ without substantially

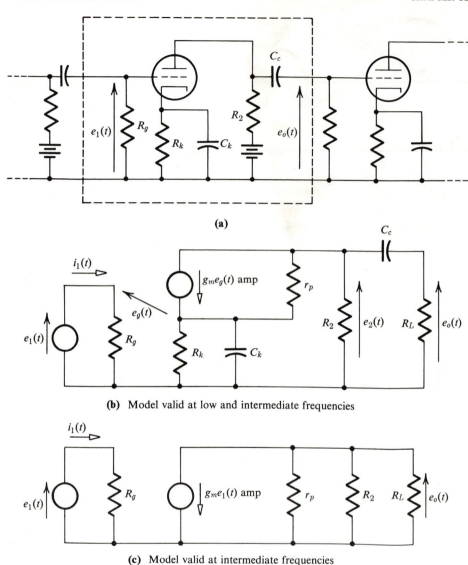

(a)

(b)   Model valid at low and intermediate frequencies

(c)   Model valid at intermediate frequencies

affecting the quiescent point. We shall assume that the loading effect of the sub-sequent stages or of any externally connected load can be represented by a resistance R_L at low and intermediate frequencies and by the parallel combination of a resistance R_L and a capacitance C_L at high frequencies.

The incremental models that apply at low, intermediate, and high frequencies in the a-c steady state are shown in Figs. 12.3-1b through 12.3-1d. These are identical with the models in Fig. 6.5-8, except for the addition of R_g, C_k, and R_k. The presence of R_g has no direct effect on the expression for the voltage gain $A(j\omega) = \mathbf{E}_o/\mathbf{E}_1$, although it does affect the input impedance $Z_{\text{in}}(j\omega) = \mathbf{E}_1/\mathbf{I}_1$. If we temporarily

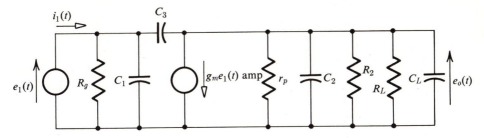

(d) Model valid at intermediate and high frequencies

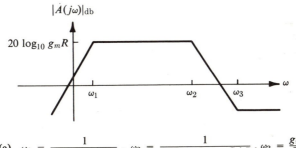

(e) $\omega_1 = \dfrac{1}{C_c(R_{p2} + R_L)}$, $\omega_2 = \dfrac{1}{(C_2 + C_3 + C_L)R}$, $\omega_3 = \dfrac{g_m}{C_3}$

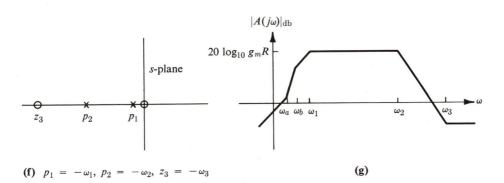

(f) $p_1 = -\omega_1$, $p_2 = -\omega_2$, $z_3 = -\omega_3$ **(g)**

Figure 12.3-1

assume that the capacitance C_k in the low-frequency model can be treated as a short circuit, then from Example 6.5-7

$$A(j\omega) = -g_m R \frac{j\omega/\omega_1}{1 + j\omega/\omega_1} \qquad \text{at low frequencies}$$

$$= -g_m R \qquad \text{at intermediate frequencies}$$

$$= -g_m R \frac{1 - j\omega(C_3/g_m)}{1 + j\omega/\omega_2} \qquad \text{at high frequencies}$$

where $\omega_1 = 1/C_c(R_{p2} + R_L)$ and $\omega_2 = 1/(C_2 + C_3 + C_L)R$ are the lower and upper limits to the bandwidth. R_{p2} is the resistance of the parallel combination of r_p and R_2, while R is the resistance of r_p, R_2, and R_L in parallel.

The Bode diagrams described by these equations were given in Fig. 6.5-8, and the decibel curve is repeated in Fig. 12.3-1e. The corresponding pole-zero plot is shown in Fig. 12.3-1f, but the drawing is not to scale. The pole p_1 is closer to the origin and the zero z_3 is more distant from the origin than is indicated in the figure. The bandwidth is $\omega_2 - \omega_1$ or approximately ω_2 radians per second. The product of the bandwidth and the magnitude of the intermediate-frequency gain is $g_m/(C_2 + C_3 + C_L)$, which is independent of R. Increasing R increases the intermediate-frequency gain, but reduces the bandwidth.

The input impedance $Z_{in}(j\omega)$ of one stage determines the values of R_L and C_L that should be used in the analysis of the previous stage. For this reason, the stage nearest the load is normally analyzed first. For the low- and intermediate-frequency models in Fig. 12.3-1, $Z_{in}(j\omega) = R_g$. At high frequencies,

$$\mathbf{I}_1 = \frac{1}{R_g}\,\mathbf{E}_1 + j\omega C_1\mathbf{E}_1 + j\omega C_3(\mathbf{E}_1 - \mathbf{E}_o)$$

or

$$\frac{\mathbf{I}_1}{\mathbf{E}_1} = \frac{1}{R_g} + j\omega C_1 + j\omega C_3[1 - A(j\omega)]$$

Since the voltage gain at high frequencies is a complex quantity, let $A(j\omega) = A_R + jA_I$. Then the input admittance is

$$Y_{in}(j\omega) = \frac{1}{R_g} + \omega C_3 A_I + j\omega\,[C_1 + C_3(1 - A_R)] = \frac{1}{R_{in}} + j\omega C_{in}$$

A_R is normally a large negative quantity, so the capacitance C_3 makes a much more substantial contribution to the input capacitance C_{in} than might have been anticipated. The multiplication of the value of C_3 by $(1 - A_R)$ is known as the *Miller effect*, and it can significantly decrease the value of ω_2 for the preceding stage.[†] Although A_R is a function of ω, the intermediate-frequency value is often used as a first approximation.

To complete the analysis of the low-frequency model in Fig. 12.3-1b, consider the effect of R_k and C_k. The expression obtained by a direct calculation of $A(j\omega)$ turns out to be the ratio of two quadratics, which are difficult to factor and to interpret in literal form. However, the resistance R_L is the grid leak resistance R_g for the following stage, which in practice is very large compared to R_2. Thus,

$$\frac{R_2(R_L + 1/j\omega C_c)}{R_2 + R_L + 1/j\omega C_c} \doteq R_2$$

---

† One advantage of the pentode over the triode is that for the former C_3 is negligibly small.

i.e., the loading effect of R_L and C_c on the calculation of $\mathbf{E_2}$ is negligible. Then, by the results of Example 12.2-2,

$$\frac{\mathbf{E_2}}{\mathbf{E_1}} = -\frac{\mu R_2}{Z_k(1 + \mu) + r_p + R_2}$$

where

$$Z_k = \frac{R_k/j\omega C_k}{R_k + 1/j\omega C_k} = \frac{R_k}{1 + j\omega R_k C_k}$$

Substituting this expression into the equation for $\mathbf{E_2}/\mathbf{E_1}$, and replacing μ by $g_m r_p$, we obtain

$$\frac{\mathbf{E_2}}{\mathbf{E_1}} = \frac{-g_m r_p R_2}{(r_p + R_2)(1 + K)}\left(\frac{1 + j\omega/\omega_a}{1 + j\omega/\omega_b}\right) = \frac{-g_m R}{1 + K}\left(\frac{1 + j\omega/\omega_a}{1 + j\omega/\omega_b}\right)$$

where $K = R_2(1 + \mu)/(r_p + R_2)$, $\omega_a = 1/R_k C_k$, $\omega_b = (1 + K)/R_k C_k$, and where $R = r_p R_2/(r_p + R_2)$ since R_L is large compared to R_2. Finally, $\mathbf{E_o}/\mathbf{E_2} = R_L/(R_L + 1/j\omega C_c)$, and

$$A(j\omega) = \frac{\mathbf{E_2}}{\mathbf{E_1}}\frac{\mathbf{E_o}}{\mathbf{E_2}} = -\frac{g_m R}{1 + K}\left[\frac{(1 + j\omega/\omega_a)(j\omega R_L C_c)}{(1 + j\omega/\omega_b)(1 + j\omega R_L C_c)}\right]$$

A typical decibel curve is shown in Fig. 12.3-1g. The presence of the $R_k C_k$ combination introduces two additional break points and decreases the low-frequency gain. The construction of a decibel curve for several identical stages in cascade was studied in Problem 6.30.

Transistor Amplifiers. Part of a typical multistage amplifier is shown in Fig. 12.3-2a. C_c is a large coupling capacitance that is used to isolate the d-c quiescent voltage on the collector of one stage from the d-c voltage on the base of the following stage or from the load resistance R_L. The coupling capacitance and the emitter bypass capacitance C_e act like short circuits to a sinusoidal input, except at low frequencies.

Consider first the final stage, which is enclosed by the dashed lines in Fig. 12.3-2a. The circuit models that will be used for the analysis of the low-, intermediate-, and high-frequency behavior are shown in parts (b) through (d) of the figure. R_1 is the resistance of the parallel combination of R_a and R_b. The high-frequency model for the transistor itself is similar to the one in Fig. 12.2-8c, except that r_b, r_c, and r_o have been omitted for simplicity. In the analysis, the algebraic details will be left for the student. At intermediate frequencies, the current gain is

$$\frac{\mathbf{I_o}}{\mathbf{I_1}} = \left(\frac{\beta R_1}{R_1 + r_n}\right)\left(\frac{R_2}{R_2 + R_L}\right)$$

and the input resistance is

$$R_{\text{in}} = \frac{R_1 r_n}{R_1 + r_n}$$

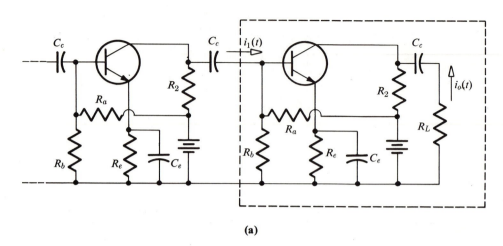

(a)

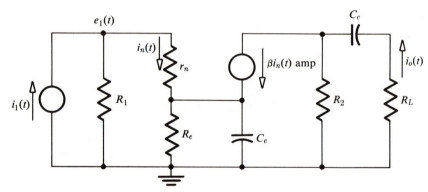

(b) Model valid at low and intermediate frequencies

The analysis of the low-frequency model in Fig. 12.3-2b was partly carried out in Example 5.3-3. From the results of that example,

$$\frac{\mathbf{I}_n}{\mathbf{I}_1} = \frac{R_1}{R_1 + r_n + (\beta + 1)Z_e}$$

where $Z_e = R_e/(1 + j\omega R_e C_e)$. Since

$$\frac{\mathbf{I}_o}{\beta \mathbf{I}_n} = \frac{R_2}{R_2 + R_L + 1/j\omega C_c}$$

the current gain is

$$\frac{\mathbf{I}_o}{\mathbf{I}_1} = \left[\frac{\beta R_1(1 + j\omega R_e C_e)}{(\beta + 1)R_e + (R_1 + r_n)(1 + j\omega R_e C_e)} \right] \left[\frac{j\omega R_2 C_c}{1 + j\omega(R_2 + R_L)C_c} \right]$$

$$= \left[\frac{\beta R_1 R_2}{(R_1 + r_n)(R_2 + R_L)(1 + K)} \right] \left[\frac{1 + j\omega/\omega_a}{1 + j\omega/\omega_b} \right] \left[\frac{j\omega/\omega_1}{1 + j\omega/\omega_1} \right]$$

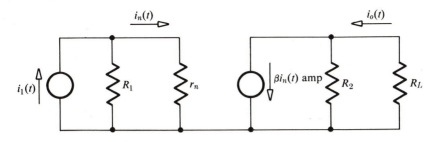

(c) Model valid at intermediate frequencies

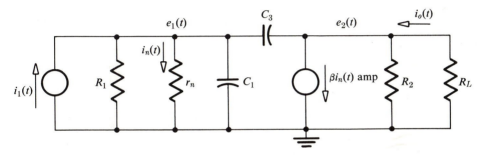

(d) Model valid at intermediate and high frequencies

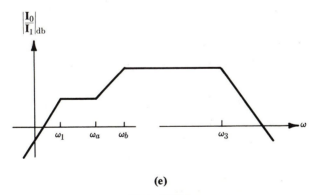

(e)

Figure 12.3-2

where $\omega_1 = 1/(R_2 + R_L)C_c$, $K = (\beta + 1)R_e/(R_1 + r_n)$, $\omega_a = 1/R_eC_e$, and $\omega_b = (1 + K)/R_eC_e$. The three break frequencies are shown on the typical Bode diagram in Fig. 12.3-2e, but the relative position of ω_1 with respect to ω_a and ω_b depends upon the element values. The R_eC_e combination reduces the amplification at low frequencies by the factor $1 + K$, while the coupling capacitance completely blocks the signal if ω approaches zero.

The expression for $\mathbf{I}_n/\mathbf{I}_1$ is in the form of a current divider rule and indicates that the input impedance consists of R_1 and $r_n + (\beta + 1)Z_e$ in parallel, so

$$Z_{\text{in}}(j\omega) = \frac{R_1[r_n + (\beta + 1)Z_e]}{R_1 + r_n + (\beta + 1)Z_e}$$

As discussed in Example 5.3-3, it is the increase in the resistance seen by the current $\mathbf{I}_n$ at low frequencies that is responsible for the factors

$$\frac{1 + j\omega/\omega_a}{(1 + K)(1 + j\omega/\omega_b)}$$

The direct calculation of the current gain $\mathbf{I}_o/\mathbf{I}_1$ for the high-frequency model in Fig. 12.3-2d leads to a quadratic that cannot be easily factored in literal form unless some approximations are made. One approach is to recognize that this circuit is essentially identical with the one in Fig. 12.3-1d, except that the input signal is a current rather than a voltage. In order to use the expression that was developed for the voltage gain of the triode circuit, let

$$R = \frac{R_2 R_L}{R_2 + R_L}, \qquad g_m = \frac{\beta}{r_n}, \qquad \omega_2 = \frac{1}{C_3 R}$$

Then

$$\frac{\mathbf{I}_o}{\mathbf{E}_1} = \frac{-\mathbf{E}_2/R_L}{\mathbf{E}_1} = \left(\frac{\beta}{r_n}\right)\left(\frac{R_2}{R_2 + R_L}\right)\left[\frac{1 - j\omega(r_n C_3/\beta)}{1 + j\omega/\omega_2}\right]$$

For a transistor amplifier, ω_2 and $\beta/r_n C_3$ are large compared to the upper limit of the bandwidth, so henceforth we shall omit the quantity in the brackets.

The current gain can be found by replacing $\mathbf{E}_1$ by $\mathbf{I}_1 Y_{\text{in}}$, as soon as a simple expression for the input admittance has been found. As a preliminary step, notice that for the circuit in Fig. 12.3-3

$$\mathbf{E}_1 = -Z_B\left(\frac{\beta}{r_n}\mathbf{E}_1 - \mathbf{I}_A\right) + Z_A \mathbf{I}_A$$

or

$$\frac{\mathbf{E}_1}{\mathbf{I}_A} = \frac{Z_A + Z_B}{1 + (\beta/r_n)Z_B}$$

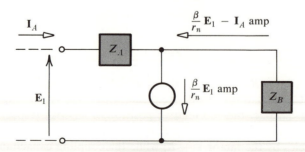

Figure 12.3-3

If the circuit in Fig. 12.3-3 represents the portion of Fig. 12.3-2d that lies to the right of C_1, it is found in practice that $|(\beta/r_n)Z_B| \gg 1$ and that $|Z_A| \gg |Z_B|$, so

$$\frac{\mathbf{E}_1}{\mathbf{I}_A} \doteq \frac{Z_A}{(\beta/r_n)Z_B} = \frac{1/j\omega C_3}{\beta R/r_n} = \frac{1}{j\omega(\beta R/r_n)C_3}$$

which describes a capacitance equal to $(\beta R/r_n)C_3$. This value is considerably larger than C_3 itself and is another example of the Miller effect. Then the total input admittance in the high-frequency model becomes

$$Y_{in} = \frac{1}{R_1} + \frac{1}{r_n} + j\omega\left(C_1 + \frac{\beta R}{r_n}C_3\right) = \frac{r_n + R_1 + j\omega R_1 r_n[C_1 + (\beta R/r_n)C_3]}{R_1 r_n}$$

so

$$\frac{\mathbf{I}_o}{\mathbf{I}_1} = \frac{1}{Y_{in}}\frac{\mathbf{I}_o}{\mathbf{E}_1} \doteq \left(\frac{R_1 r_n}{R_1 + r_n}\right)\left(\frac{1}{1 + j\omega/\omega_3}\right)\left(\frac{\beta}{r_n}\right)\left(\frac{R_2}{R_2 + R_L}\right)$$

$$= \frac{\beta R_1 R_2}{(R_1 + r_n)(R_2 + R_L)}\left(\frac{1}{1 + j\omega/\omega_3}\right)$$

where

$$\omega_3 = \frac{R_1 + r_n}{R_1 r_n[C_1 + (\beta R/r_n)\,C_3]}$$

The break frequency ω_3 is included in the Bode diagram in Fig. 12.3-2e. In the case of a transistor amplifier, the product of the bandwidth and the intermediate-frequency gain is not independent of the value of R_2.

The analysis of the *next* to the last stage in the amplifier in Fig. 12.3-2a is complicated by the fact that at low and high frequencies the loading effect of the subsequent stage cannot be adequately represented by a single resistance R_L. The resistance R_L in the circuit models in Figs. 12.3-2b and 12.3-2d must be replaced by the impedance Z_L, which equals the input impedance of the final stage. For intermediate frequencies, Z_L reduces to the resistance $R_L = R_1 r_n/(R_1 + r_n)$, and the current gain is again given by

$$\frac{\mathbf{I}_o}{\mathbf{I}_1} = \frac{\beta R_1 R_2}{(R_1 + r_n)(R_2 + R_L)}$$

At low frequencies, however,

$$Z_L = \frac{R_1[r_n + (\beta + 1)Z_e]}{R_1 + r_n + (\beta + 1)Z_e}$$

and, since $Z_e = R_e/(1 + j\omega R_e C_e)$, Z_L depends upon the angular frequency. The current gain at low frequencies is

$$\frac{\mathbf{I}_o}{\mathbf{I}_1} = \left[\frac{\beta R_1}{R_1 + r_n + (\beta + 1)Z_e}\right]\left[\frac{R_2}{R_2 + Z_L + 1/j\omega C_c}\right]$$

but, when the above expression for Z_L is used, the resulting equation becomes too complicated to be factored in literal form. In a numerical problem, of course, the factors and hence the break points in the Bode diagram could be obtained in a straightforward manner.

The high-frequency analysis of the next to the last stage is also complicated by the loading effect of the final stage. The input admittance to the final stage had the form $G_L + j\omega C_L$, where

$$G_L = \frac{1}{R_1} + \frac{1}{r_n}, \qquad C_L = C_1 + \frac{\beta R}{r_n}C_3$$

In Fig. 12.3-2d, R_L should be replaced by the parallel combination of R_L and C_L. As before,

$$\frac{\mathbf{I}_o}{\mathbf{I}_1} \doteq \frac{1}{Y_{\text{in}}}\left[\frac{\beta R_2}{r_n(R_2 + Z_L)}\right]$$

where Y_{in} is the input admittance of the stage under consideration. But, if R_L is replaced by R_L and C_L in parallel, and if the part of the circuit to the right of C_1 is again represented by the circuit in Fig. 12.3-3, then $1/Z_B = 1/R + j\omega C_L$, and

$$\frac{\mathbf{E}_1}{\mathbf{I}_A} \doteq \frac{Z_A}{(\beta/r_n)Z_B} = \frac{1/R + j\omega C_L}{j\omega(\beta/r_n)C_3} = \frac{1}{j\omega(\beta R/r_n)C_3} + \frac{C_L}{(\beta/r_n)C_3}$$

which describes the series combination of a capacitance *and a resistance*. The complete input admittance is

$$Y_{\text{in}} = \frac{1}{R_1} + \frac{1}{r_n} + j\omega C_1 + \frac{j\omega(\beta/r_n)C_3}{1/R + j\omega C_L}$$

which, when simplified, contains a quadratic factor in $j\omega$. The quadratic is again difficult to factor in literal form, and the effect of the various elements upon the upper limit of the bandwidth is no longer obvious.

Compensated Amplifiers.   Additional circuit elements are sometimes added to a simple amplifier in order to decrease the lower limit or increase the upper limit of the bandwidth. As one very simple example, consider the high-frequency vacuum-tube model in Fig. 12.3-1d when $C_3 = 0$ (which is a reasonable approximation for a pentode). R_L and C_L represent the loading effect of the following stage. The impedance seen by the controlled current source in the a-c steady state is the parallel combination of R and $1/j\omega C$, where R is the parallel combination of r_p, R_2, and R_L, and where $C = C_2 + C_L$. At high frequencies, the impedance and hence the output voltage falls off. In order to partially compensate for this effect, an inductance L can be placed in series with R_2, as in Fig. 12.3-4a, so that the impedance of this branch will *increase* as the frequency increases.

In order to be able to examine both the frequency response and the pulse response, we shall express the voltage gain as a function of the complex frequency s. From the discussion associated with Figs. 9.5-2 and 9.5-3, we also might expect that increasing the upper limit to the bandwidth will improve the pulse response. If $L = 0$, corresponding to an uncompensated circuit, the voltage gain is

$$A(s) = \frac{\mathbf{E}_o(s)}{\mathbf{E}_1(s)} = \frac{-g_m}{sC + 1/R} = \frac{-g_m/C}{s + 1/RC}$$

The frequency response curve, found by replacing s by $j\omega$, is given in Fig. 12.3-4b.

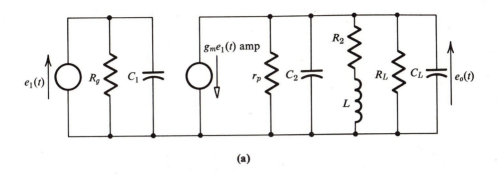

(a)

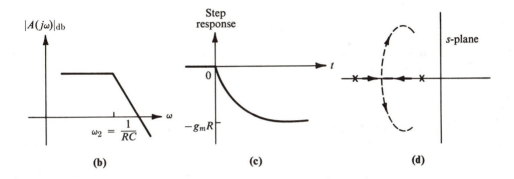

(b)   **(c)**   **(d)**

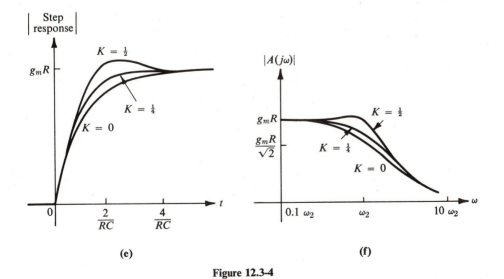

(e)   **(f)**

Figure 12.3-4

The unit step response, which is $\mathscr{L}^{-1}[A(s)/s]$, is shown in Fig. 12.3-4c.

If $L \neq 0$, it can be shown in a straightforward manner that

$$A(s) = \frac{E_o(s)}{E_1(s)} = \frac{-g_m}{C} \frac{s + \omega_2/K}{s^2 + (\omega_2/K + D)s + (\omega_2{}^2/K)}$$

where $K = \omega_2 L/R_2$, $\omega_2 = 1/RC$, and

$$D = \frac{\omega_2 R(R_L + r_p)}{R_L r_p}$$

K is called the *peaking factor* and is normally less than 1, while ω_2 is the upper limit of the bandwidth for the uncompensated circuit. In practice, R_L and r_p are large compared to R_2 and R, and $D \ll \omega_2/K$, so

$$A(s) \doteq \frac{-g_m}{C} \frac{s + \omega_2/K}{s^2 + (\omega_2/K)s + (\omega_2{}^2/K)}$$

The poles of this transfer function are located at

$$s = -\frac{\omega_2}{2K} (1 \pm \sqrt{1 - 4K})$$

and are complex if $K > \frac{1}{4}$. The locus of the poles as K is increased from a small value is shown in Fig. 12.3-4d. As K is increased, a zero moves along the negative real axis toward the origin, but this is not shown in the figure.

It should be expected from the discussion in Section 6.4 that as K increases a little beyond $\frac{1}{4}$ the upper limit of the bandwidth would also increase. Also for $K > \frac{1}{4}$, the complex poles will cause the free response of the circuit to contain oscillations, and the step response to have an overshoot. To verify these conclusions, careful plots of $|A(j\omega)|$ and of the unit step response $\mathscr{L}^{-1}[A(s)/s]$ can be made. Such curves appear in several journals and textbooks, and curves for $K = \frac{1}{4}$ and $K = \frac{1}{2}$ are sketched in Figs. 12.3-4e and 12.3-4f. In practice, values of K larger than $\frac{1}{2}$ are rarely used, because the peaks in the curves increase rapidly while the additional increase in the bandwidth is very slight. The response to a rectangular pulse can be found from the unit step response, as discussed in connection with Fig. 9.5-3.

Since the model used in Fig. 12.3-4a was the high-frequency model, the effect of the coupling and cathode bypass capacitances was not included in the calculations. These capacitances introduce additional zeros and an equal number of poles on the negative real axis very close to the origin of the s-plane. The additional poles and zeros are important only in the vicinity of $s = 0$, but they affect $A(j\omega)$ at very low frequencies and cause the unit step response to eventually fall off to zero instead of remaining at $-g_m R$ (a phenomenon known as "sag").

Other compensating networks can be used for both vacuum-tube and transistor amplifiers. Although most of these networks are more complicated than the one considered here, a similar analysis procedure may be used.

12.4
General Models for Two-Port Networks

Section 12.3 in effect consisted of three detailed examples of amplifier circuits. The analysis of other useful electronic circuits is considered in some of the problems at the end of this chapter. In many applications, the most difficult part of the solution is the selection of an appropriate incremental model and the use of any approximations that may be required to put the final answer in a tractable form. In this section, we discuss in general terms methods of characterizing any linear two-port network.

A *two-port* network is a network that has two pairs of externally accessible terminals, as shown in Fig. 12.4-1a, and that satisfies the following assumptions, which are nearly always satisfied in practice. The current entering one of the upper terminals must leave the corresponding lower terminal, so that there are only two independent external currents. It is also assumed that the voltage between the two bottom terminals is of no interest, and that the only external voltages that need be considered are $e_1(t)$ and $e_2(t)$. Finally, the network must contain no independent sources.

If the bottom terminals are directly connected within the two-port, as in Fig. 12.4-1b, the network is said to be an *unbalanced* or *common ground* two-port and is essentially the same as a three-terminal network. The majority of two-ports, such as the amplifiers discussed in Section 12.3, are unbalanced. Other examples of two-ports are filters used in communication systems and compensating networks used in control systems.

A discussion of the properties of two-ports and the effect of interconnecting them is helpful in the design of networks that must have a predetermined transfer function. Frequently, a complicated design problem can be decomposed into several simpler problems, each of which leads to a separate two-port. The final circuit is given by the proper interconnection of these two-ports. An example of this process is found in the problems at the end of this chapter.

The Characterization of Two-Ports.   The discussion concerning three-terminal networks in Sections 7.4 and 12.2 carries over to two-ports, where there are two

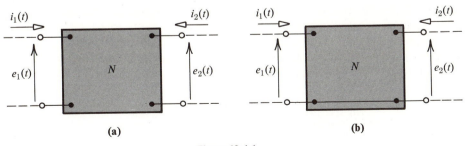

(a)   (b)

Figure 12.4-1

TABLE 12.4-1

Equations characterizing the two-port	Equation number
$E_1(s) = z_{11}(s)\,I_1(s) + z_{12}(s)\,I_2(s)$ $E_2(s) = z_{21}(s)\,I_1(s) + z_{22}(s)\,I_2(s)$	12.4-1
$I_1(s) = y_{11}(s)\,E_1(s) + y_{12}(s)\,E_2(s)$ $I_2(s) = y_{21}(s)\,E_1(s) + y_{22}(s)\,E_2(s)$	12.4-2
$E_1(s) = h_{11}(s)\,I_1(s) + h_{12}(s)\,E_2(s)$ $I_2(s) = h_{21}(s)\,I_1(s) + h_{22}(s)\,E_2(s)$	12.4-3
$I_1(s) = g_{11}(s)\,E_1(s) + g_{12}(s)\,I_2(s)$ $E_2(s) = g_{21}(s)\,E_1(s) + g_{22}(s)\,I_2(s)$	12.4-4
$E_1(s) = A(s)\,E_2(s) - B(s)\,I_2(s)$ $I_1(s) = C(s)\,E_2(s) - D(s)\,I_2(s)$	12.4-5

external currents and two external voltages of interest. Let the currents and voltages be represented by their Laplace transforms, and assume that the network contains no initial stored energy. Then any two of the transformed quantities $E_1(s)$, $E_2(s)$, $I_1(s)$, and $I_2(s)$ may be expressed in terms of the other two. One way, although not usually the best way, of doing this is to write and solve a set of transformed loop or node equations describing the network. Some of the ways in which a two-port can be characterized are summarized in Table 12.4-1.

The $A(s)$ in Eqs. 12.4-5 is not the same as the $A(s)$ used in Section 12.3 for the voltage gain. The minus signs in Eqs. 12.4-5 appear because these equations were first used in the analysis of transmission lines in power systems. Workers in this area use positive signs in front of all the terms, but choose the reference arrow for $I_2(s)$ in the opposite direction to that shown in Fig. 12.4-1. It is also possible to express $E_2(s)$ and $I_2(s)$ in terms of $E_1(s)$ and $I_1(s)$, but this is not often done.[†]

The coefficients on the right side of the equations in Table 12.4-1 are parameters that depend only upon the complex frequency s and upon the elements within the two-port. The equations can be used in a-c steady-state analysis if s is replaced by $j\omega$ in the parameters and if the transformed voltages and currents are replaced by

---

[†] Methods of *indirectly* specifying the relationships between the external voltages and currents also can be used. See, for example, S. Seshu and N. Balabanian, *Linear Network Analysis* (New York: Wiley, 1959), for a discussion of the *scattering parameters* that are used in microwave theory and the *image parameters* that are used in filter theory.

phasors. Notice that specifying four functions of s is sufficient to completely describe the external characteristics of a two-port. If the behavior at only one frequency in the a-c steady-state is required, the parameters reduce to four numbers, some of which may be complex.

For simplicity, the s in parentheses after the two-port parameters will be omitted in the rest of this chapter. Which set of parameters is the most convenient one to use depends upon the particular application, as illustrated later in this chapter. If the parameters for a given device are to be determined experimentally, a certain set may be particularly easy to measure or may be given by the manufacturer. For example, the h parameters are the ones usually given for a transistor.

It is important to be able to convert from one set of parameters to another, and this can be done by the algebraic manipulation of the equations in Table 12.4-1. If, for example, the h parameters are known, the $ABCD$ parameters can be found by solving Eqs. 12.4-3 for $E_1(s)$ and $I_1(s)$ in terms of $E_2(s)$ and $I_2(s)$.

$$I_1(s) = \frac{1}{h_{21}} I_2(s) - \frac{h_{22}}{h_{21}} E_2(s)$$

$$E_1(s) = \frac{h_{11}}{h_{21}} I_2(s) - \frac{h_{11}h_{22}}{h_{21}} E_2(s) + h_{12}E_2(s)$$

$$= \frac{h_{11}}{h_{21}} I_2(s) - \frac{h_{11}h_{22} - h_{12}h_{21}}{h_{21}} E_2(s)$$

By a comparison of these results to Eqs. 12.4-5,

$$A = \frac{-|h|}{h_{21}}, \qquad B = \frac{-h_{11}}{h_{21}}$$

$$C = \frac{-h_{22}}{h_{21}}, \qquad D = \frac{-1}{h_{21}}$$

where $|h|$ denotes the determinant

$$\begin{vmatrix} h_{11} & h_{12} \\ h_{21} & h_{22} \end{vmatrix} = h_{11}h_{22} - h_{12}h_{21}$$

The relationships between the different sets of parameters are summarized in Table 12.4-2, where the symbol η stands for the quantity $AD - BC$. In each of the rows, corresponding entries are equal.

It was pointed out in Section 7.4 that if a two-port contains only resistances, capacitances, inductances, and transformers then $z_{12} = z_{21}$. By the use of this fact and Table 12.4-2, it follows that for this subclass of two-port networks

$$z_{12} = z_{21}$$
$$y_{12} = y_{21}$$
$$h_{12} = -h_{21} \qquad \text{(12.4-6)}$$
$$g_{12} = -g_{21}$$
$$AD - BC = 1$$

Linear Electronic Circuits

12

TABLE 12.4-2

z		y		h		g		ABCD	
z_{11}	z_{12}	$\dfrac{y_{22}}{|y|}$	$\dfrac{-y_{12}}{|y|}$	$\dfrac{|h|}{h_{22}}$	$\dfrac{h_{12}}{h_{22}}$	$\dfrac{1}{g_{11}}$	$\dfrac{-g_{12}}{g_{11}}$	$\dfrac{A}{C}$	$\dfrac{\eta}{C}$
z_{21}	z_{22}	$\dfrac{-y_{21}}{|y|}$	$\dfrac{y_{11}}{|y|}$	$\dfrac{-h_{21}}{h_{22}}$	$\dfrac{1}{h_{22}}$	$\dfrac{g_{21}}{g_{11}}$	$\dfrac{|g|}{g_{11}}$	$\dfrac{1}{C}$	$\dfrac{D}{C}$
y_{11}	y_{12}	$\dfrac{z_{22}}{|z|}$	$\dfrac{-z_{12}}{|z|}$	$\dfrac{1}{h_{11}}$	$\dfrac{-h_{12}}{h_{11}}$	$\dfrac{|g|}{g_{22}}$	$\dfrac{g_{12}}{g_{22}}$	$\dfrac{D}{B}$	$\dfrac{-\eta}{B}$
y_{21}	y_{22}	$\dfrac{-z_{21}}{|z|}$	$\dfrac{z_{11}}{|z|}$	$\dfrac{h_{21}}{h_{11}}$	$\dfrac{|h|}{h_{11}}$	$\dfrac{-g_{21}}{g_{22}}$	$\dfrac{1}{g_{22}}$	$\dfrac{-1}{B}$	$\dfrac{A}{B}$
h_{11}	h_{12}	$\dfrac{|z|}{z_{22}}$	$\dfrac{z_{12}}{z_{22}}$	$\dfrac{1}{y_{11}}$	$\dfrac{-y_{12}}{y_{11}}$	$\dfrac{g_{22}}{|g|}$	$\dfrac{-g_{12}}{|g|}$	$\dfrac{B}{D}$	$\dfrac{\eta}{D}$
h_{21}	h_{22}	$\dfrac{-z_{21}}{z_{22}}$	$\dfrac{1}{z_{22}}$	$\dfrac{y_{21}}{y_{11}}$	$\dfrac{|y|}{y_{11}}$	$\dfrac{-g_{21}}{|g|}$	$\dfrac{g_{11}}{|g|}$	$\dfrac{-1}{D}$	$\dfrac{C}{D}$
g_{11}	g_{12}	$\dfrac{1}{z_{11}}$	$\dfrac{-z_{12}}{z_{11}}$	$\dfrac{|y|}{y_{22}}$	$\dfrac{y_{12}}{y_{22}}$	$\dfrac{h_{22}}{|h|}$	$\dfrac{-h_{12}}{|h|}$	$\dfrac{C}{A}$	$\dfrac{-\eta}{A}$
g_{21}	g_{22}	$\dfrac{z_{21}}{z_{11}}$	$\dfrac{|z|}{z_{11}}$	$\dfrac{-y_{21}}{y_{22}}$	$\dfrac{1}{y_{22}}$	$\dfrac{-h_{21}}{|h|}$	$\dfrac{h_{11}}{|h|}$	$\dfrac{1}{A}$	$\dfrac{B}{A}$
A	B	$\dfrac{z_{11}}{z_{21}}$	$\dfrac{|z|}{z_{21}}$	$\dfrac{-y_{22}}{y_{21}}$	$\dfrac{-1}{y_{21}}$	$\dfrac{-|h|}{h_{21}}$	$\dfrac{-h_{11}}{h_{21}}$	$\dfrac{1}{g_{21}}$	$\dfrac{g_{22}}{g_{21}}$
C	D	$\dfrac{1}{z_{21}}$	$\dfrac{z_{22}}{z_{21}}$	$\dfrac{-|y|}{y_{21}}$	$\dfrac{-y_{11}}{y_{21}}$	$\dfrac{-h_{22}}{h_{21}}$	$\dfrac{-1}{h_{21}}$	$\dfrac{g_{11}}{g_{21}}$	$\dfrac{|g|}{g_{21}}$

Thus, such networks can be described by three instead of four independent parameters.

Each of the sets of equations in Table 12.4-1 describes a general model that can be used to represent a given two-port. Equations 12.4-1 and 12.4-2 are similar to Eqs. 7.4-1 and 7.4-5, respectively, and the models based on the z and y parameters were given in Figs. 7.4-5a and 7.4-10. The models described by Eqs. 12.4-3 and 12.4-4 are shown in Fig. 12.4-2. For an unbalanced two-port, the two bottom terminals may be connected together. Notice that the model based on the h parameters is identical in form with the transistor model in Fig. 12.2-1. If $g_{11} = g_{12} = 0$, $g_{21} = -\mu$, and $g_{22} = r_p$, the model with the g parameters reduces to the incremental model in Fig. 12.2-4a for the vacuum tube. For those two-ports that satisfy Eqs. 12.4-6, a model containing no controlled sources can be found. This was done, for example, in the discussion associated with Figs. 7.4-8 and 7.4-11b.

Although the two-port parameters for a given network may be found by writing and solving a set of loop or node equations, this is usually not the easiest procedure. Furthermore, it is important to be able to determine the parameters experimentally by external measurements. This can be done by a set of open- and

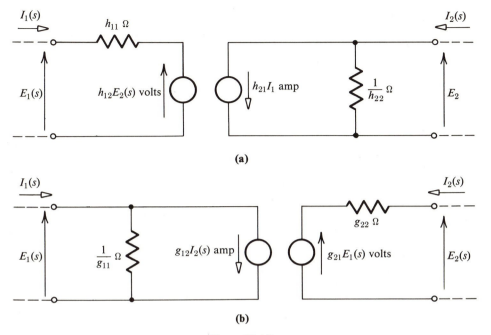

Figure 12.4-2

short-circuit tests, which were discussed in Section 7.4 for the z and y parameters. In terms of the h parameters, we can obtain the following relationships from the defining equations in Table 12.4-1 or from the model in Fig. 12.4-2a.

$$h_{11} = \left[\frac{E_1(s)}{I_1(s)}\right]_{E_2(s)=0}$$

$$h_{12} = \left[\frac{E_1(s)}{E_2(s)}\right]_{I_1(s)=0}$$

$$h_{21} = \left[\frac{I_2(s)}{I_1(s)}\right]_{E_2(s)=0} \tag{12.4-7}$$

$$h_{22} = \left[\frac{I_2(s)}{E_2(s)}\right]_{I_1(s)=0}$$

Replacing $E_2(s)$ by zero corresponds to placing a short circuit across the output port, while replacing $I_1(s)$ by zero corresponds to open-circuiting the left-hand port.

The h parameters are sometimes called the *hybrid parameters*, because they do not all have the same units. According to Eqs. 12.4-7, h_{11} is the *short-circuit input impedance*, h_{12} the *open-circuit reverse voltage gain*, h_{21} the *short-circuit current gain*, and h_{22} the *open-circuit output admittance*.

A circuit that can be used for measuring h_{21} for a transistor at a particular frequency in the a-c steady state was shown in Fig. 12.2-3. Remember that for the development of incremental models the voltages and currents in Eqs. 12.4-3 and 12.4-7 are deviations from the quiescent values. Thus, the requirement that $E_2(s) = 0$ is satisfied if the termination at the output port acts like a short circuit to the *time-varying* signal.

Example 12.4-1.   A simple incremental model for a common-base transistor, similar to one in Fig. 12.2-6, is shown in Fig. 12.4-3a. Find the *h* and the *g* parameters.

Solution.   In this problem the parameters are constants, because the given network does not contain inductance or capacitance. When $E_2(s) = 0$, as in Fig. 12.4-3b,

$$I_1(s) = \left(\frac{1}{r_n} + \frac{1}{r_o} + \frac{\beta}{r_n}\right) E_1(s)$$

and

$$I_2(s) = -\left(\frac{1}{r_o} + \frac{\beta}{r_n}\right) E_1(s) = -\frac{(1/r_o + \beta/r_n)I_1(s)}{1/r_n + 1/r_o + \beta/r_n}$$

so by Eqs. 12.4-7

$$h_{11} = \frac{1}{1/r_o + (\beta + 1)/r_n}$$

$$h_{21} = -\frac{1/r_o + \beta/r_n}{1/r_o + (\beta + 1)/r_n}$$

When $I_1(s) = 0$, as in Fig. 12.4-3c,

$$E_1(s) = r_n I_2(s)$$

$$E_2(s) = r_n I_2(s) + r_o\left[I_2(s) + \frac{\beta}{r_n} E_1(s)\right] = [r_n + (\beta + 1)r_o]I_2(s)$$

so

$$h_{12} = \frac{r_n}{r_n + (\beta + 1)r_o}$$

$$h_{22} = \frac{1}{r_n + (\beta + 1)r_o}$$

To calculate the *g* parameters, first let $I_2(s) = 0$, as in Fig. 12.4-3d.

$$I_1(s) = \frac{1}{r_n} E_1(s)$$

$$E_2(s) = E_1(s) + r_o \frac{\beta}{r_n} E_1(s)$$

so

$$g_{11} = \frac{1}{r_n}$$

$$g_{21} = 1 + \frac{\beta r_o}{r_n}$$

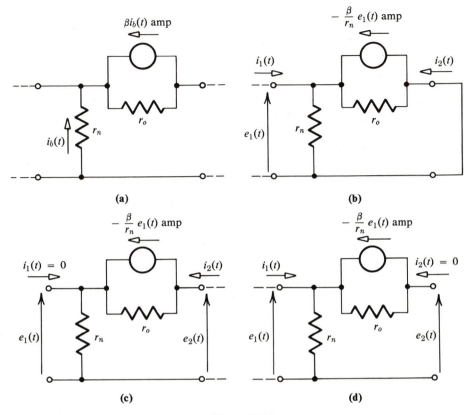

Figure 12.4-3

Finally, with $E_1(s) = 0$, $I_1(s) = -I_2(s)$ and $E_2(s) = r_o I_2(s)$, so

$$g_{12} = -1, \qquad g_{22} = r_o$$

As a check on the work, the h parameters may be obtained from the g parameters by using Table 12.4-2.

An appropriate set of four parameters completely characterizes a two-port regardless of the external connections, as long as these connections do not violate the assumptions stated in the second paragraph of Section 12.4. For a given set of external connections, the desired transfer function can be found by replacing the two-port by an appropriate model, or by the algebraic manipulation of the equations in Table 12.4-1.

Example 12.4-2. If the output port of the network N in Fig. 12.4-4 is terminated by a load impedance Z_L, derive expressions for the voltage gain $E_2(s)/E_1(s)$, for the current gain $I_2(s)/I_1(s)$, and for the input impedance $Z_{in}(s) = E_1(s)/I_1(s)$ in terms of the g parameters of the two-port.

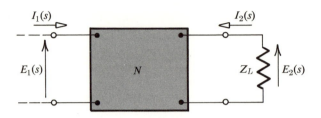

Figure 12.4-4

Solution. One approach is to replace the network N by the equivalent circuit in Fig. 12.4-2b. Another method is to replace $I_2(s)$ by $-E_2(s)/Z_L$ (with the minus sign needed because of the direction of the reference arrows) in the equations

$$I_1(s) = g_{11}E_1(s) + g_{12}I_2(s), \qquad E_2(s) = g_{21}E_1(s) + g_{22}I_2(s)$$

Then

$$E_2(s) = g_{21}E_1(s) - \frac{g_{22}}{Z_L} E_2(s)$$

or

$$\frac{E_2(s)}{E_1(s)} = \frac{g_{21}Z_L}{g_{22} + Z_L}$$

Also,

$$I_1(s) = g_{11}E_1(s) - \frac{g_{12}}{Z_L} E_2(s) = \left(g_{11} - \frac{g_{12}g_{21}}{g_{22} + Z_L} \right) E_1(s)$$

so

$$Z_{in}(s) = \frac{E_1(s)}{I_1(s)} = \frac{g_{22} + Z_L}{|g| + g_{11}Z_L}$$

where $|g| = g_{11}g_{22} - g_{12}g_{21}$.

$$\frac{I_2(s)}{I_1(s)} = \frac{-E_2(s)/Z_L}{I_1(s)} = \frac{-g_{21}}{|g| + g_{11}Z_L}$$

By the method used in the last example, the voltage and current gain and the

TABLE 12.4-3

$Z_{in}(s)$	$\dfrac{|z| + z_{11}Z_L}{z_{22} + Z_L}$	$\dfrac{y_{22} + Y_L}{|y| + y_{11}Y_L}$	$\dfrac{|h| + h_{11}Y_L}{h_{22} + Y_L}$	$\dfrac{g_{22} + Z_L}{|g| + g_{11}Z_L}$	$\dfrac{AZ_L + B}{CZ_L + D}$
$\dfrac{E_2(s)}{E_1(s)}$	$\dfrac{z_{21}Z_L}{|z| + z_{11}Z_L}$	$\dfrac{-y_{21}}{y_{22} + Y_L}$	$\dfrac{-h_{21}Z_L}{h_{11} + |h|Z_L}$	$\dfrac{g_{21}Z_L}{g_{22} + Z_L}$	$\dfrac{Z_L}{B + AZ_L}$
$\dfrac{I_2(s)}{I_1(s)}$	$\dfrac{-z_{21}}{z_{22} + Z_L}$	$\dfrac{y_{21}Y_L}{|y| + y_{11}Y_L}$	$\dfrac{h_{21}Y_L}{h_{22} + Y_L}$	$\dfrac{-g_{21}}{|g| + g_{11}Z_L}$	$\dfrac{-1}{D + CZ_L}$

input impedance for the circuit in Fig. 12.4-4 can be found in terms of other sets of parameters. The results are summarized in Table 12.4-3. This table is useful not only in circuit analysis but in the synthesis or design of circuits having a prescribed transfer function and a prescribed load impedance. One simple synthesis example is found in Problem 7.30.

Introduction to Matrix Algebra.　A very brief and elementary introduction to matrix algebra is included here, because it is convenient to state the general rules for the interconnection of two-ports in terms of matrices.[†] Matrices are also frequently used in advanced textbooks.

A matrix is a rectangular array of elements, such as

$$
\begin{bmatrix}
a_{11} & a_{12} & \ldots & a_{1n} \\
a_{21} & a_{22} & \ldots & a_{2n} \\
\multicolumn{4}{c}{\cdots\cdots\cdots\cdots\cdots} \\
a_{m1} & a_{m2} & \ldots & a_{mn}
\end{bmatrix}
\tag{12.4-8}
$$

The first subscript on an element denotes its row, and the second subscript its column. Unlike a determinant, a matrix cannot be reduced to a single element, and it is not necessarily a square array. It may, in fact, consist of a single row or a single column. The *order* of a matrix refers to the number of rows and columns and for the matrix written above is $m \times n$ (which is read "m by n"). The symbol A, $[a_{ij}]$, or $[a_{ij}]_{(m,n)}$ is often used as a shorthand notation for the above matrix. Care must be taken not to confuse capital letters representing matrices with those representing transformed quantities or phasors.

Two matrices are said to be equal if and only if they are of the same order and their corresponding elements are equal. Thus, equal matrices are identical matrices.

We next define the operations of addition, subtraction, and multiplication. Suppose that the matrix A is multiplied by the scalar k, which may be real or complex, but which is a single element and not a matrix. Then

$$
kA =
\begin{bmatrix}
ka_{11} & ka_{12} & \ldots & ka_{1n} \\
ka_{21} & ka_{22} & \ldots & ka_{2n} \\
\multicolumn{4}{c}{\cdots\cdots\cdots\cdots\cdots\cdots} \\
ka_{m1} & ka_{m2} & \ldots & ka_{mn}
\end{bmatrix}
\tag{12.4-9}
$$

Notice that, in contrast to the rule for multiplying a determinant by a scalar, every element in the matrix A is multiplied by k.

Two matrices may be added or subtracted only if they have the same order. The

---

[†] For a much more complete treatment of matrix theory, see Chapter 4 of P. M. DeRusso, R. J. Roy, and C. M. Close, *State Variables for Engineers* (New York: Wiley, 1965), or F. E. Hohn, *Elementary Matrix Algebra* (New York: Macmillan, 1958).

addition or subtraction is carried out by adding or subtracting corresponding elements. For example,

$$\begin{bmatrix} a_{11} & a_{12} \\ a_{21} & a_{22} \end{bmatrix} + \begin{bmatrix} b_{11} & b_{12} \\ b_{21} & b_{22} \end{bmatrix} = \begin{bmatrix} a_{11} + b_{11} & a_{12} + b_{12} \\ a_{21} + b_{21} & a_{22} + b_{22} \end{bmatrix} \tag{12.4-10}$$

The quantity $A - B$ can be interpreted as $A + (-B)$, where $-B$ denotes the matrix B multiplied by the scalar -1. The addition of matrices obeys the usual commutative and associative laws of ordinary algebra.

$$A + B = B + A$$

$$(A + B) + C = A + (B + C)$$

$$k_1 A + k_2 A = (k_1 + k_2)A$$

Two matrices can be multiplied only if the number of columns in the first matrix equals the number of rows in the second matrix. Thus, the product $AB = C$ is defined only if the order of A is $m \times n$ and the order of B is $n \times q$ (where m, n, and q may or may not be the same). The order of C is $m \times q$, i.e., the number of rows in C equals the number of rows in A, and the number of columns in C equals the number of columns in B. Matrix multiplication is defined in a way that allows a set of simultaneous equations to be replaced by a single matrix equation. If $A = [a_{rs}]_{(m,n)}$, $B = [b_{uv}]_{(n,q)}$, and $C = [c_{ij}]_{(m,q)}$, then

$$c_{ij} = \sum_{k=1}^{n} a_{ik} b_{kj} \tag{12.4-11}$$

The proper interpretation of this expression for c_{ij} is illustrated by the following example.

Example 12.4-3.   Find the matrix $C = AB$ if

$$A = \begin{bmatrix} 2 & -1 & 0 & 3 \\ 1 & 2 & 0 & -2 \\ 0 & 3 & -1 & 2 \end{bmatrix} \quad \text{and} \quad B = \begin{bmatrix} 1 & -1 \\ 2 & 3 \\ 4 & 0 \\ 1 & 2 \end{bmatrix}$$

Solution.   The matrix C must have three rows and two columns, so let

$$C = \begin{bmatrix} c_{11} & c_{12} \\ c_{21} & c_{22} \\ c_{31} & c_{32} \end{bmatrix}$$

To determine the element c_{12}, let $i = 1$, $j = 2$, and $n = 4$ in Eq. 12.4-11.

$$c_{12} = a_{11}b_{12} + a_{12}b_{22} + a_{13}b_{32} + a_{14}b_{42}$$

$$= (2)(-1) + (-1)(3) + (0)(0) + (3)(2) = 1$$

Thus, c_{12} is found by multiplying corresponding elements of the first row of matrix A and the second column of matrix B. Similarly,

$$c_{31} = a_{31}b_{11} + a_{32}b_{21} + a_{33}b_{31} + a_{34}b_{41}$$
$$= (0)(1) + (3)(2) + (-1)(4) + (2)(1) = 4$$

In summary, c_{ij} is formed by multiplying corresponding elements of the ith row of A and the jth row of B. The student should show that

$$C = \begin{bmatrix} 3 & 1 \\ 3 & 1 \\ 4 & 13 \end{bmatrix}$$

It can be shown that if A, B, and C represent three matrices, then

$$(AB)C = A(BC)$$
$$A(B + C) = AB + AC$$
$$(A + B)C = AC + BC$$

However, not all of the properties of ordinary algebra carry over to matrix multiplication. Except in special cases,

$$AB \neq BA$$

If, for example,

$$A = \begin{bmatrix} 1 & 2 & 3 \\ 4 & 5 & 6 \end{bmatrix} \quad \text{and} \quad B = \begin{bmatrix} 1 & 2 & 3 \\ 4 & 5 & 6 \\ 7 & 8 & 9 \end{bmatrix}$$

the product AB can be formed, but the product BA is undefined. If the matrix B is changed to

$$B = \begin{bmatrix} 1 & 2 \\ 3 & 4 \\ 8 & 9 \end{bmatrix}$$

both products exist, but AB is of order 2×2 while BA is of order 3×3. Even when AB and BA have the same order they are not necessarily equal. For example,

$$\begin{bmatrix} 1 & 2 \\ 3 & 4 \end{bmatrix} \begin{bmatrix} 1 & 0 \\ 0 & -1 \end{bmatrix} = \begin{bmatrix} 1 & -2 \\ 3 & -4 \end{bmatrix}$$

but

$$\begin{bmatrix} 1 & 0 \\ 0 & -1 \end{bmatrix} \begin{bmatrix} 1 & 2 \\ 3 & 4 \end{bmatrix} = \begin{bmatrix} 1 & 2 \\ -3 & -4 \end{bmatrix}$$

One application of matrices is associated with sets of simultaneous equations. Consider, therefore, the equations

$$a_{11}x_1 + a_{12}x_2 + \cdots + a_{1n}x_n = y_1$$
$$a_{21}x_1 + a_{22}x_2 + \cdots + a_{2n}x_n = y_2$$
$$\cdots\cdots\cdots\cdots\cdots\cdots\cdots\cdots\cdots\cdots$$
$$a_{m1}x_1 + a_{m2}x_2 + \cdots + a_{mn}x_n = y_m$$

(12.4-12)

and define the matrices

$$X = \begin{bmatrix} x_1 \\ x_2 \\ \cdots \\ x_n \end{bmatrix}, \qquad Y = \begin{bmatrix} y_1 \\ y_2 \\ \cdots \\ y_m \end{bmatrix}$$

$$A = \begin{bmatrix} a_{11} & a_{12} & \cdots & a_{1n} \\ a_{21} & a_{22} & \cdots & a_{2n} \\ \cdots\cdots\cdots\cdots\cdots\cdots\cdots \\ a_{m1} & a_{m2} & \cdots & a_{mn} \end{bmatrix}$$

Notice that the matrices X and Y consist of only one column. Then Eqs. 12.4-12 may be replaced by the matrix equation

$$\begin{bmatrix} a_{11} & a_{12} & \cdots & a_{1n} \\ a_{21} & a_{22} & \cdots & a_{2n} \\ \cdots\cdots\cdots\cdots\cdots\cdots\cdots \\ a_{m1} & a_{m2} & \cdots & a_{mn} \end{bmatrix} \begin{bmatrix} x_1 \\ x_2 \\ \cdots \\ x_n \end{bmatrix} = \begin{bmatrix} y_1 \\ y_2 \\ \cdots \\ y_m \end{bmatrix}$$

(12.4-13)

or

$$AX = Y$$

The fact that Eq. 12.4-13 is equivalent to Eqs. 12.4-12 can be verified by carrying out the indicated matrix multiplication and using the definition of matrix equality.

Each of the sets of equations in Table 12.4-1 may be replaced by a single matrix equation. For example, the equations

$$E_1(s) = z_{11}I_1(s) + z_{12}I_2(s)$$
$$E_2(s) = z_{21}I_1(s) + z_{22}I_2(s)$$

become

$$\begin{bmatrix} E_1(s) \\ E_2(s) \end{bmatrix} = \begin{bmatrix} z_{11} & z_{12} \\ z_{21} & z_{22} \end{bmatrix} \begin{bmatrix} I_1(s) \\ I_2(s) \end{bmatrix}$$

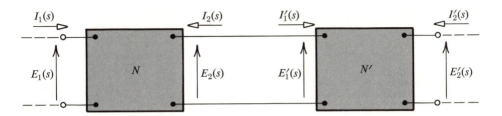

Figure 12.4-5

The Interconnection of Two-Ports. Consider first the cascade connection of two-ports shown in Fig. 12.4-5. The network N may be described in terms of its $ABCD$ parameters by the matrix equation

$$\begin{bmatrix} E_1(s) \\ I_1(s) \end{bmatrix} = \begin{bmatrix} A & B \\ C & D \end{bmatrix} \begin{bmatrix} E_2(s) \\ -I_2(s) \end{bmatrix}$$

Similarly, for network N'

$$\begin{bmatrix} E_1'(s) \\ I_1'(s) \end{bmatrix} = \begin{bmatrix} A' & B' \\ C' & D' \end{bmatrix} \begin{bmatrix} E_2'(s) \\ -I_2'(s) \end{bmatrix}$$

The interconnection of the two networks requires that

$$\begin{bmatrix} E_2(s) \\ -I_2(s) \end{bmatrix} = \begin{bmatrix} E_1'(s) \\ I_1'(s) \end{bmatrix}$$

so

$$\begin{bmatrix} E_1(s) \\ I_1(s) \end{bmatrix} = \begin{bmatrix} A & B \\ C & D \end{bmatrix} \begin{bmatrix} A' & B' \\ C' & D' \end{bmatrix} \begin{bmatrix} E_2'(s) \\ -I_2'(s) \end{bmatrix} \qquad (12.4\text{-}14)$$

Since $E_1(s)$, $I_1(s)$, $E_2'(s)$, and $I_2'(s)$ are the external voltages and currents for the cascaded combination, the $ABCD$ matrix for the combination is

$$\begin{bmatrix} A & B \\ C & D \end{bmatrix} \begin{bmatrix} A' & B' \\ C' & D' \end{bmatrix}$$

which is the product of the individual $ABCD$ matrices.

Example 12.4-4. The network N in Fig. 12.4-6 contains a transistor circuit for which the h parameters are $h_{11} = 10^3$ Ω, $h_{12} = 0$, $h_{21} = 10$, and $h_{22} = 10^{-4}$ mhos. (The parameters are assumed to be real constants.) Find $E_o(s)/E_1(s)$. Repeat the problem when the network N contains two identical transistor circuits in cascade, each of which has the h parameters given above.

Solution.   By Table 12.4-2, if the h matrix for network N is

$$\begin{bmatrix} 10^3 & 0 \\ 10 & 10^{-4} \end{bmatrix}$$

then the corresponding $ABCD$ matrix is

$$-\begin{bmatrix} 0.01 & 100 \\ 10^{-5} & 0.1 \end{bmatrix}$$

The $ABCD$ matrix for a two-port consisting of R_1 alone is

$$\begin{bmatrix} 1 & 100 \\ 0 & 1 \end{bmatrix}$$

Hence, for the cascaded combination of R_1 and N,

$$\begin{bmatrix} A & B \\ C & D \end{bmatrix} = -\begin{bmatrix} 1 & 100 \\ 0 & 1 \end{bmatrix}\begin{bmatrix} 0.01 & 100 \\ 10^{-5} & 0.1 \end{bmatrix} = -\begin{bmatrix} 0.011 & 110 \\ 10^{-5} & 0.1 \end{bmatrix}$$

Since this combination is terminated by a 10 kΩ load resistance, Table 12.4-3 may be used to find the voltage gain.

$$\frac{E_o(s)}{E_1(s)} = \frac{R_L}{B + AR_L} = \frac{10^4}{-110 - 110} = -45$$

If the network N consists of two identical transistor circuits, the $ABCD$ matrix for the combination of R_1 and N becomes

$$\begin{bmatrix} A & B \\ C & D \end{bmatrix} = \begin{bmatrix} 1 & 100 \\ 0 & 1 \end{bmatrix}\begin{bmatrix} 0.01 & 100 \\ 10^{-5} & 0.1 \end{bmatrix}\begin{bmatrix} 0.01 & 100 \\ 10^{-5} & 0.1 \end{bmatrix} = \begin{bmatrix} 1.2 \times 10^{-3} & 12 \\ 1.1 \times 10^{-6} & 0.011 \end{bmatrix}$$

so

$$\frac{E_o(s)}{E_1(s)} = \frac{10^4}{12 + 12} = 416$$

From the last example, notice that when individual networks are characterized by a set of two-port parameters the loading effect of one stage on the preceding

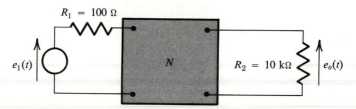

Figure 12.4-6

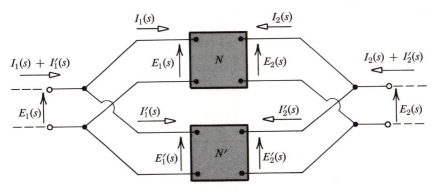

Figure 12.4-7

stage is automatically included in the matrix multiplication. In contrast to the procedure used in Section 12.3, the loading effect is not represented by separate elements.

Finally, consider the parallel connection shown in Fig. 12.4-7. The individual two-ports are described by the equations

$$\begin{bmatrix} I_1(s) \\ I_2(s) \end{bmatrix} = \begin{bmatrix} y_{11} & y_{12} \\ y_{21} & y_{22} \end{bmatrix} \begin{bmatrix} E_1(s) \\ E_2(s) \end{bmatrix}$$

$$\begin{bmatrix} I_1'(s) \\ I_2'(s) \end{bmatrix} = \begin{bmatrix} y_{11}' & y_{12}' \\ y_{21}' & y_{22}' \end{bmatrix} \begin{bmatrix} E_1'(s) \\ E_2'(s) \end{bmatrix}$$

For the entire combination, the matrix of the external currents is

$$\begin{bmatrix} I_1(s) + I_1'(s) \\ I_2(s) + I_2'(s) \end{bmatrix} = \begin{bmatrix} I_1(s) \\ I_2(s) \end{bmatrix} + \begin{bmatrix} I_1'(s) \\ I_2'(s) \end{bmatrix}$$

while the matrix of the external voltages is

$$\begin{bmatrix} E_1(s) \\ E_2(s) \end{bmatrix} = \begin{bmatrix} E_1'(s) \\ E_2'(s) \end{bmatrix}$$

Adding the equations for the individual two-ports, we obtain

$$\begin{bmatrix} I_1(s) + I_1'(s) \\ I_2(s) + I_2'(s) \end{bmatrix} = \left(\begin{bmatrix} y_{11} & y_{12} \\ y_{21} & y_{22} \end{bmatrix} + \begin{bmatrix} y_{11}' & y_{12}' \\ y_{21}' & y_{22}' \end{bmatrix} \right) \begin{bmatrix} E_1(s) \\ E_2(s) \end{bmatrix} \tag{12.4-15}$$

so the y matrix for the combination is the sum of the individual y matrices.

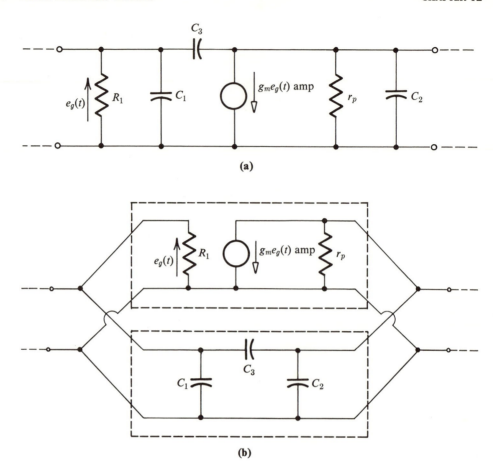

(a)

(b)

Figure 12.4-8

There is some question about the validity of Eqs. 12.4-15 in some cases because after the two networks in Fig. 12.4-7 have been connected in parallel the current leaving the lower left-hand terminal of network N does not necessarily equal $I_1(s)$. The only requirement is that the *sum* of the currents leaving the lower left-hand terminals of N and N' equal $I_1(s) + I_1'(s)$. Thus, the assumptions stated in the second paragraph of Section 12.4 may not be satisfied in all cases. It can be shown, however, that for the important case where both N and N' are unbalanced networks, Eqs. 12.4-15 are always valid. If there is any question about the validity of the equations, a specific test is available to resolve this question.[†]

[†] See, for example, Chapter 2 of A. J. Cote, Jr., and J. B. Oakes, *Linear Vacuum-tube and Transistor Circuits* (New York: McGraw-Hill, 1961), or Chapter 7, E. Peskin, *Transient and Steady-State Analysis of Electric Networks* (Princeton, N.J.: Van Nostrand, 1961).

Example 12.4-5. Find the y matrix for the incremental model of a vacuum-tube amplifier shown in Fig. 12.4-8a. C_1, C_2, and C_3 represent the interelectrode capacitances.

Solution. The circuit can be redrawn as the parallel combination of two two-port networks, as in Fig. 12.4-8b. The y matrix for the top two-port is

$$\begin{bmatrix} y_{11a} & y_{12a} \\ y_{21a} & y_{22a} \end{bmatrix} = \begin{bmatrix} G_1 & 0 \\ g_m & g_p \end{bmatrix}$$

where $G_1 = 1/R_1$ and $g_p = 1/r_p$. For the bottom two-port

$$\begin{bmatrix} y_{11b} & y_{12b} \\ y_{21b} & y_{22b} \end{bmatrix} = \begin{bmatrix} s(C_1 + C_3) & -sC_3 \\ -sC_3 & s(C_2 + C_3) \end{bmatrix}$$

Note that $y_{12a} \neq y_{21a}$ but that $y_{12b} = y_{21b}$. This should be expected, because Eqs. 12.4-6 apply to network b but not to a network with controlled sources. By Eq. 12.4-15, the y matrix for the circuit in Fig. 12.4-8a is

$$\begin{bmatrix} y_{11} & y_{12} \\ y_{21} & y_{22} \end{bmatrix} = \begin{bmatrix} G_1 + s(C_1 + C_3) & -sC_3 \\ g_m - sC_3 & g_p + s(C_2 + C_3) \end{bmatrix}$$

12.5

Summary

In transistor and vacuum-tube circuits, d-c bias voltages are used to establish a quiescent point in a linear region of operation and to supply the necessary power. The quiescent point may be found graphically from the characteristic curves for the electronic device and from a knowledge of the external circuit. The response to a time-varying input also may be found graphically provided that the circuit can be regarded as a resistive circuit for the given input. Normally, however, the time-varying components are found by making the d-c sources dead and by replacing the transistor or tube by an incremental model. Use of the incremental model is valid only when the input signal is small enough so that the characteristic curves may be approximated by equally spaced, straight lines in the operating region.

All the techniques of the first ten chapters may be applied in the analysis of the incremental model for a circuit. The frequency response curves for some typical wide-band amplifiers are examined in Section 12.3. If the individual stages are connected by a coupling capacitance (which provides d-c isolation), and if a parallel RC network is placed in series with the emitter or the cathode (to provide temperature stabilization or to eliminate the need for two different bias supplies), the capacitances limit the amplification at low frequencies. The amplification also falls off at high frequencies because of the wiring capacitance and the parasitic capacitances associated with the transistor or vacuum tube.

A general discussion of the characterization of two-port networks is found in Section 12.4. Any two of the four external voltages and currents may be expressed as functions of the other two, as in Table 12.4-1. If the two-port includes transistors or vacuum tubes, it is understood that the voltages and currents represent the deviations from the quiescent values. The external behavior can be characterized by four circuit parameters, which in general are functions of the complex frequency s (or of the angular frequency ω in the a-c steady state). The effect of connecting a load impedance or of connecting two networks in cascade or in parallel is considered. Other possible interconnections are examined in one of the problems at the end of this chapter.

PROBLEMS

12.1   For the circuit shown in Fig. P12.1a, $R_1 = 1\ \text{k}\Omega$, $R_2 = 2\ \text{k}\Omega$, $R_o = 4\ \text{k}\Omega$, $E_{bb} = 400$ volts, and $E_{cc} = -6$ volts. The characteristic curves for the vacuum tube are given in Fig. P12.1b. Assume that $i_c(t) = 0$ when $e_c(t) < 0$ and that the capacitance C acts like a short circuit to a time-varying signal.

    (a) Draw the static load line and find the quiescent plate current and plate voltage.

    (b) Draw the dynamic load line and find $e_o(t)$ when $e_s(t) = 2 \sin \omega t$. Calculate the average power supplied to each resistance and to the vacuum tube.

12.2   One method of making the operation of a transistor amplifier relatively insensitive to temperature changes was discussed in connection with Figs. 12.1-9a and 12.1-10a. Another method is to connect a resistance between the base and the collector, as in Fig. P12.2a. The characteristic curves for the transistor itself are given in Fig. P12.2b. Assume that the emitter-to-base voltage is negligibly small and that the capacitances act like short circuits at the angular frequency of the input signal.

    (a) If $E_2 = 25$ volts and $R_a = 5\ \text{k}\Omega$, and if the quiescent currents and voltages should be $I_B = 0.3$ ma, $I_C = 7$ ma, and $E_{CE} = 10$ volts, calculate the values of R_2 and R_b.

    (b) Draw the static and dynamic load lines. What is the maximum value of A (the peak value of the input signal) that can be used before serious distortion of the output waveform occurs as the result of nonlinearities in the transistor? What is $i_o(t)$ when this value of A is used?

    (c) Explain why the circuit provides temperature stabilization and why the capacitance C_a is needed. List any advantages or disadvantages of this circuit compared to the one in Fig. 12.1-10a.

12.3   The characteristic curves for the triode in Fig. P12.3 are given in Fig. P12.1b. Choose R_1 and R_2 so that the quiescent values are $I_b = 100$ ma and $E_b = 200$ volts. Draw the dynamic load line and sketch to scale $e_o(t)$ when $e_s(t) = 4 \sin 10^5 t$ volts. Justify any assumptions that need to be made.

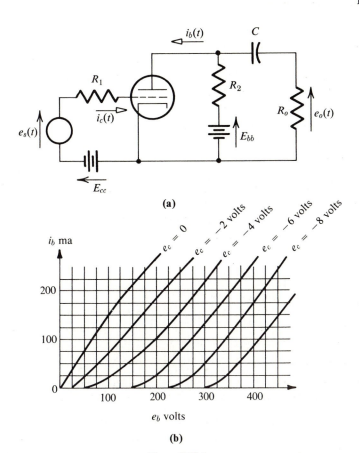

(a)

(b)

Figure P12.1

12.4 For each of the circuits in Fig. P12.4, draw the static and dynamic load lines on an i_C-e_{CE} plot. Assume that the quiescent currents are $I_B = 1$ ma and $I_C = 30$ ma and that any capacitances and inductances act like short or open circuits, respectively, to the time-varying signal.

12.5 For the circuit shown in Fig. P12.5, $R_1 = 15$ kΩ, $R_3 = \frac{1}{2}$ kΩ, $R_o = 100 \, \Omega$, $L_1 = 100$ mh, $L_2 = 4$ mh, and $M = 20$ mh. Assume that at the angular frequency ω the capacitance can be regarded as a short circuit and the transformer as an ideal transformer with a turns ratio $N = \sqrt{L_2/L_1}$. The characteristic curves for the transistor are given in Fig. P12.2b, and the quiescent point should correspond to $I_B = 0.3$ ma, $I_C = 7.5$ ma, and $E_{CE} = 15$ volts.

(a) Draw the static and dynamic load lines and calculate the values of E_1 and E_2.

(b) Graphically determine the output voltage $e_o(t)$ and calculate the collector efficiency.

12.6 The characteristic curves for the triode in Fig. P12.6 are given in Fig. P12.1b. Assume that at the angular frequency ω the capacitances can be regarded as short

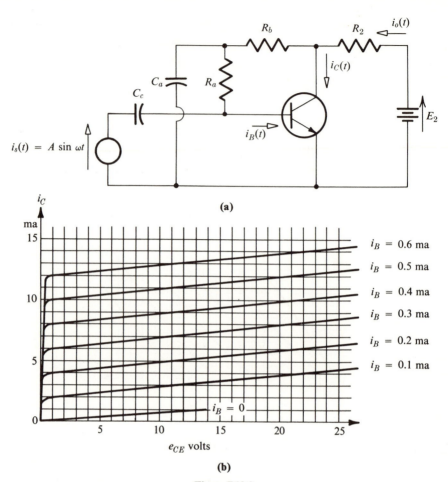

(a)

(b)

Figure P12.2

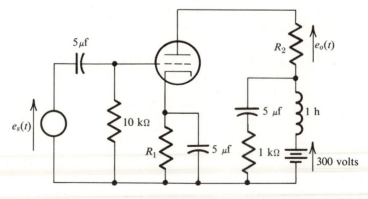

Figure P12.3

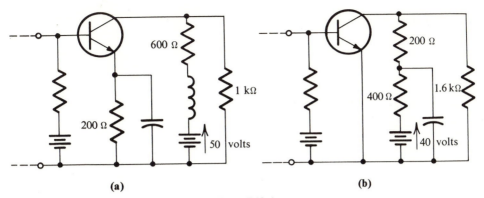

(a) (b)

Figure P12.4

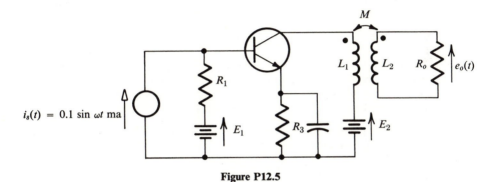

$i_s(t) = 0.1 \sin \omega t$ ma

Figure P12.5

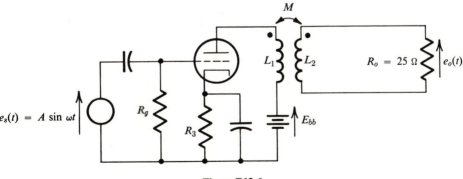

$e_s(t) = A \sin \omega t$

Figure P12.6

circuits and the transformer as an ideal transformer with a turns ratio $N = \sqrt{L_2/L_1}$. No grid current flows as long as the cathode-to-grid voltage is negative.

(a) If the quiescent plate current and plate voltage should be $I_b = 100$ ma and $E_b = 225$ volts, calculate the values of R_3 and E_{bb}.

(b) If the slope of the dynamic load line should be $-\frac{1}{5}$ ma per volt, find the value of the turns ratio N.

(c) What is the peak value of the largest sinusoidal input voltage that can be applied before serious nonlinear distortion occurs? What is $e_o(t)$ when such a signal is applied?

12.7 For the circuit in Fig. P12.7, assume that the quiescent point is at the middle of the dynamic operating path and that the operating path can extend nearly to the vertical and horizontal axes of the i_C-e_{CE} plot before the nonlinearities of the transistor become significant. Let E_{CE} and I_C denote the quiescent values of the collector voltage and current. Derive an expression for the maximum power that can be supplied to the resistance R_o, if $e_o(t)$ must be sinusoidal. Find the collector efficiency and compare the answer with the results of Example 12.1-1.

12.8 Find the values of μ, r_p, and g_m for a triode having the characteristic curves in Fig. P12.1b and a quiescent point at $E_b = 225$ volts and $E_c = -4$ volts. Draw circuits that might be used to measure these quantities experimentally.

12.9 The circuits in Fig. P12.9 contain two identical vacuum tubes or two identical transistors. For the vacuum tubes, use the incremental model in Fig. 12.2-4a; for the transistors, let $\mu = 0$ and $r_n \ll r_o$ in the incremental model in Fig. 12.2-1. For the time-varying components of voltage and current, find the voltage gain $e_o(t)/e_s(t)$ in Fig. P12.9a and the current gain $i_o(t)/i_s(t)$ in Fig. P12.9b. Assume that all capacitances can be regarded as short circuits to the time-varying signal.

12.10 A transistor circuit that is used as a negative impedance converter is given in Fig. P12.10. For simplicity, the bias circuits are not shown. In the incremental model for the transistors, assume that $\mu = r_n = g_o = 0$. Show that for the time-varying components of voltage and current $e_1(t) = e_2(t)$ and

$$i_1(t) = \left(\frac{\beta^2 + \beta + 2}{\beta^2 - \beta - 2}\right) i_2(t)$$

If $\beta \gg 1$, as is normally the case, $i_1(t) \doteq i_2(t)$. Compare the behavior of this circuit to that of the circuit in Fig. 1.4-10.

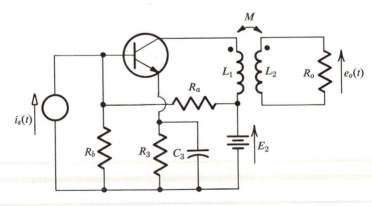

Figure P12.7

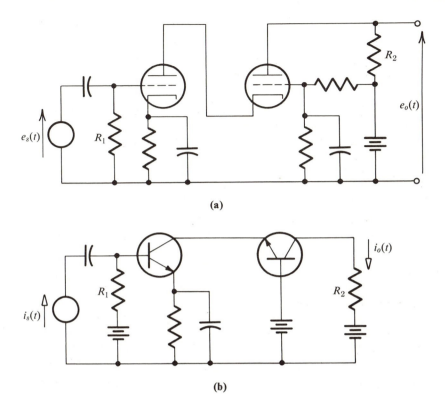

(a)

(b)

Figure P12.9

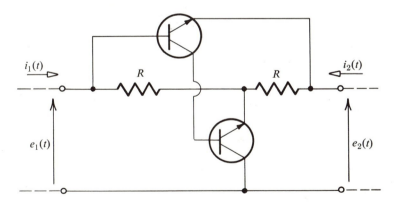

Figure P12.10

12.11 A "Darlington-connected" emitter follower is shown in Fig. P12.11. Draw the incremental circuit model if for each transistor μ and r_n are negligibly small and $r_0 \gg R_3$ and $r_0 \gg R_o$. Find the current gain $i_o(t)/i_s(t)$ and the input resistance $e_1(t)/i_s(t)$. Also determine the output resistance (the equivalent resistance that would be used in a Thévenin equivalent circuit).

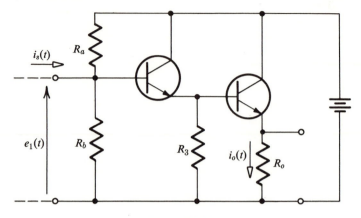

Figure P12.11

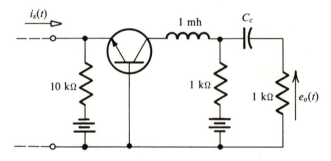

Figure P12.12

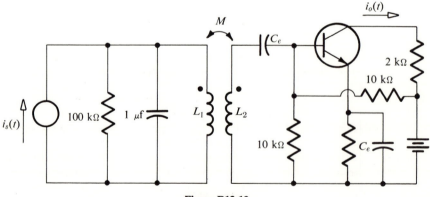

Figure P12.13

12.12 The incremental model for the transistor in Fig. P12.12 is given in Fig. 12.2-8c, with $r_b = r_c = g_o = 0$, $r_n = \frac{1}{2} k\Omega$, $\beta = 20$, $C_1 = 100$ pf, and $C_3 = 10$ pf (which includes the effect of any wiring capacitance). If C_c can be regarded as a short circuit to the time-varying input signal, find the transfer function $H(s) = E_o(s)/I_s(s)$.

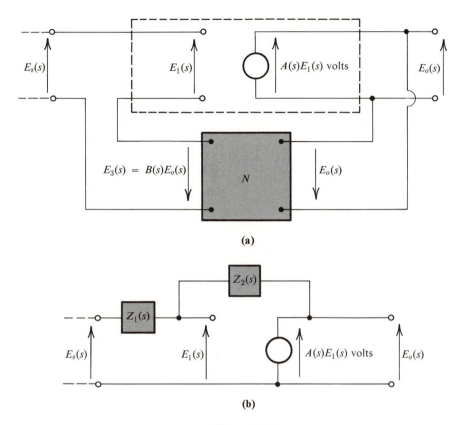

Figure P12.14

Plot the pole-zero pattern and sketch the frequency response curves for the a-c steady state. What is the resonant frequency and the bandwidth for the circuit?

12.13 A band-pass amplifier is shown in Fig. P12.13. For a time-varying signal, assume that C_e may be regarded as a short circuit and that the transformer may be replaced by the model in Fig. 7.4-15b, with $k = 1$, $L_1 = 1$ mh, and $N = \frac{1}{20}$. For the transistor itself, use the incremental model in Fig. 12.2-1, with $r_n = \frac{1}{2}$ kΩ, $\mu = 0$, $r_o = 50$ kΩ, and $\beta = 20$. Draw the complete incremental model for the circuit, and find the transfer function $I_o(s)/I_s(s)$. Sketch and fully label the frequency response curves for the a-c steady state. Why is a transformer often needed in this type of a band-pass amplifier?

12.14 A simple incremental model that may be used for a limited class of feedback amplifiers is shown in Fig. P12.14a. The network within the dashed lines is an ideal voltage amplifier, while $E_3(s) = B(s)E_o(s)$ for the network N. [The ideal voltage amplifier may be replaced by any network for which $E_o(s) = A(s)E_1(s)$ *provided* that the voltage gain $A(s)$ is calculated with the loading effect of the network N included.]

(a) Show that

$$A_f(s) = \frac{E_o(s)}{E_s(s)} = \frac{A(s)}{1 - A(s)B(s)}$$

and, by differentiating this equation, that

$$\frac{dA_f(s)}{A_f(s)} = \frac{1}{1 - A(s)B(s)} \frac{dA(s)}{A(s)}$$

$A(s)$ is the voltage gain without the feedback network N present, and $A_f(s)$ is the voltage gain with feedback.

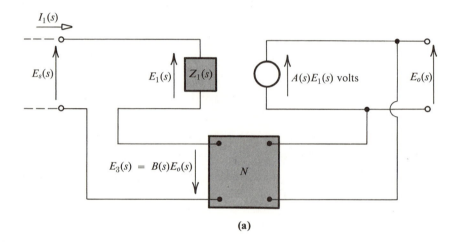

(a)

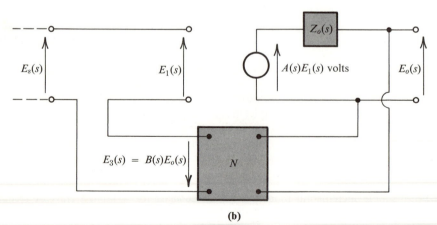

(b)

Figure P12.15

(b) The feedback is said to be positive when $|1 - A(s)B(s)| < 1$ and negative when $|1 - A(s)B(s)| > 1$. Discuss the effect of negative feedback upon the gain $A_f(s)$, and include the case where $|A(s)B(s)| \gg 1$. If the value of $A(s)$ changes by 5%, what is the corresponding change in $A_f(s)$? Discuss the effect of positive feedback and notice that if $A(j\omega)B(j\omega) = 1$ for some value of ω, then the free response will contain a sinusoidal oscillation.

(c) Find the voltage gain $E_o(s)/E_s(s)$ for the feedback circuit in Fig. P12.14b, and identify the quantity that is analogous to $B(s)$.

12.15 Some of the characteristics of positive and negative feedback were examined in Problem 12.14. The use of negative feedback to improve the temperature stability of a circuit was illustrated by the discussion associated with Fig. 12.1-9a. Two other effects are investigated in this problem.

(a) Find and discuss an expression for the input impedance $Z_{in}(s) = E_s(s)/I_1(s)$ for the circuit in Fig. P12.15a.

(b) Find and discuss an expression for the output impedance (the impedance that would be used in a Thévenin equivalent circuit) for the circuit in Fig. P12.15b.

12.16 Using the incremental model in Fig. 12.2-4a for the triode, redraw the cathode follower shown in Fig. P12.16 in the form of a feedback amplifier. Assume that $r_p \ll R_3$ and find the voltage gain.

12.17 Redraw the oscillator circuits in Figs. 7.2-4, P6.12, and 10.6-3 in the form of a feedback amplifier. In each case, what is $A(s)$ and $B(s)$?

12.18 Draw the low-, intermediate-, and high-frequency incremental models for the emitter follower circuit in Fig. P12.18. The coupling capacitance C_c acts like a short circuit except at low frequencies, and the high-frequency model for the transistor is given in Fig. 12.2-8c with $r_b = r_c = 0$.

(a) Find expressions for the a-c steady-state transfer function $H(j\omega) = E_o/E_1$ and draw a typical Bode diagram of decibel gain versus ω.

(b) Find expressions for the input impedance and output impedance (the equivalent impedance of the dead network in a Thévenin equivalent circuit) at intermediate frequencies.

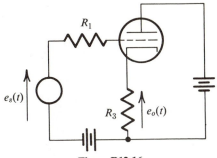

Figure P12.16

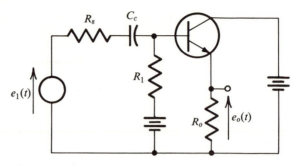

Figure P12.18

12.19 For the two-stage transistor amplifier shown in Fig. 12.3-2a, let $R_a = R_b = 20\ \text{k}\Omega$, $R_2 = R_L = 10\ \text{k}\Omega$, $R_s = 2\ \text{k}\Omega$, $C_s = 10\ \mu\text{f}$, and $C_c = 1\ \mu\text{f}$. Each transistor may be represented by the incremental model in Fig. 12.2-8c, with $r_b = r_c = 1/r_o = 0$, $r_n = 1\ \text{k}\Omega$, $C_1 = 100\ \text{pf}$, $C_3 = 5\ \text{pf}$, and $\beta = 50$. Find expressions for the a-c steady-state current gain $H(j\omega) = \mathbf{I}_o/\mathbf{I}_s$ at low, intermediate, and high frequencies, where the phasor $\mathbf{I}_s$ represents the current entering the upper left-hand terminal in Fig. 12.3-2a. Sketch the Bode diagram of decibel gain versus ω.

12.20 The incremental model of one stage of a pentode amplifier is given by the circuit in Fig. 12.3-1d with $C_3 = 0$. The a-c steady-state transfer function has the form

$$\frac{\mathbf{E}_o}{\mathbf{E}_1} = \frac{K}{1 + j\omega/\omega_2}$$

where the values of K and ω_2 include the loading effect of the following stage.

(a) If the amplifier consists of n identical stages in cascade, show that the upper limit to the bandwidth is at $\omega = \omega_2 \sqrt{2^{1/n} - 1}$.

(b) Suppose that in each stage of a five-stage amplifier $g_m = 0.01$ mhos, $C_2 + C_L = 15$ pf, $R_L = 100\ \text{k}\Omega$, and $r_p = 1\ \text{M}\Omega$. If the bandwidth of the amplifier should be 5 megacycles per sec, calculate the decibel voltage gain at intermediate frequencies. What should be the value of R_2?

12.21 Consider the a-c steady-state response of the feedback amplifier in Fig. P12.14a. Assume that $B(j\omega)$ is a constant but let

$$A(j\omega) = \frac{K}{1 + j\omega/\omega_2}$$

i.e., replace the circuit within the dashed lines by the high-frequency incremental model of a simple amplifier. The values of K and ω_2 must include any loading effect of the network N upon the amplifier within the dashed lines. Find an expression for $\mathbf{E}_o/\mathbf{E}_s$, and discuss the effect of negative feedback upon the bandwidth.

12.22 A transformer-coupled amplifier is shown in Fig. P12.22a, where R_L represents a load resistance or the input resistance to the following stage. The transformer provides the same d-c isolation between stages as a coupling capacitance, but its use in wide-band amplifiers often presents serious problems in size, cost, and low-frequency response. The incremental model to be used is given in Fig. P12.22b. The emitter bypass capacitance and the capacitances in the high-frequency model of a transistor are assumed to be less important than the transformer in limiting the frequency response and therefore are not included. The transformer has been replaced by the model in Fig. 7.4-15b, where the magnetizing inductance is $L_m = kL_1$ and where the leakage inductances are $L_a = (1 - k)L_1$ and $L_b = (1 - k)L_2$. In a wide-band amplifier the coefficient of coupling k is close to unity. Assume that L_a and L_b can be regarded as short circuits except at high frequencies and that L_m can be regarded as an open circuit except at low frequencies.

(a) Find expressions for the a-c steady-state current gain I_o/I_1 at low, intermediate, and high frequencies.

(b) Draw the Bode diagram of decibel gain versus ω, clearly labelling the break points.

(c) How would you expect the Bode diagram to be changed if two small capacitances, representing the distributed capacitance of the transformer windings, were added in parallel with r_o and R_L?

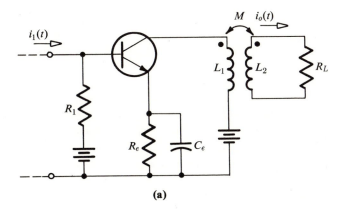

(a)

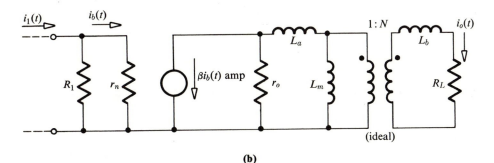

(b)

Figure P12.22

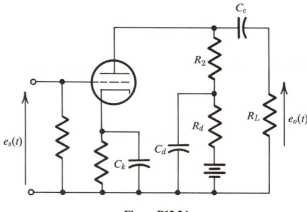

Figure P12.24

12.23 The frequency response of the compensated amplifier in Fig. 12.3-4a is shown in Fig. 12.3-4f for three values of the peaking factor K. Find the value of K for which the frequency response curve just fails to have a peak for $\omega > 0$.

12.24 One method for partly compensating for the effect of the coupling capacitance C_c on the low-frequency behavior of an amplifier is shown by the circuit in Fig. P12.24. Assume that the cathode bypass capacitance C_k can be regarded as a short circuit and use the incremental model in Fig. 12.2-4b for the vacuum tube.

(a) Explain in words why the addition of C_d should improve the low-frequency behavior of the amplifier.

(b) Find the a-c steady-state transfer function $H(j\omega) = \mathbf{E}_o/\mathbf{E}_s$ at low and intermediate frequencies if R_d and r_p approach infinity. Discuss the reasonableness of this assumption, and sketch the straight-line approximation to the decibel curve.

(c) Instead of the assumption used in part (b), assume that R_L and $1/\omega C_c$ are large compared to r_p and R_2. Again find the a-c steady-state transfer function and sketch the decibel curve.

12.25 One method of improving the high-frequency response of a wide-band amplifier is to add an inductance L in series with C_c in Fig. 12.3-1a. If $C_3 = 0$ and if R_g and R_L

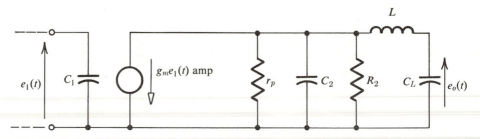

Figure P12.25

are very large (which are reasonable assumptions for a pentode amplifier), the incremental model in Fig. P12.25 can be used at intermediate and high frequencies.

(a) Show that the transfer function is

$$H(s) = \frac{E_o(s)}{E_1(s)} = \frac{-g_m/LC_2C_L}{s^3 + (1/RC_2)s^2 + (1/C_2 + 1/C_L)(s/L) + (1/LRC_2C_L)}$$

where $R = r_p R_2/(r_p + R_2)$.

(b) If $C_2 = 50$ pf, $C_L = 150$ pf, $R = 10$ kΩ, and $L = 13.3$ mh, find $|H(j\omega)|$ and sketch the decibel curve. Compare the result to the uncompensated amplifier with $L = 0$.

12.26 Find, if they exist, the g, h, and $ABCD$ parameters for the following circuits:

(a) The triode amplifier models in Fig. P7.23.

(b) The negative impedance converter that is enclosed by the dashed lines in Fig. 1.4-10.

(c) An ideal transformer.

12.27 Find the h parameters for the Darlington-connected two-port in Fig. P12.27 if each transistor may be replaced by the incremental model in Fig. 12.2-1 with $\mu = 0$. The bias circuits are not shown and may be disregarded.

12.28 Verify the entries in Table 12.4-3.

12.29 Three two-port networks that consist of a single transistor are given in Fig. P12.29. The bias supplies are not shown, since only incremental models are to be considered in this problem. Let the subscripts e, b, and c denote the parameters for the common-emitter, the common-base, and the common-collector connections, respectively. If

$$h_{11e} = 1 \text{ k}\Omega, \qquad h_{12e} = 10^{-3}$$

$$h_{21e} = 50, \qquad h_{22e} = 10^{-3} \text{ mho}$$

and if the same transistor is used in all three connections, find the h parameters for the common-base connection and for the common-collector connection.

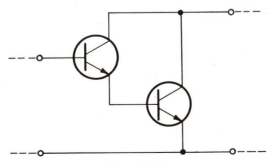

Figure P12.27

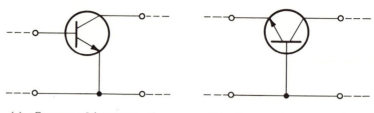

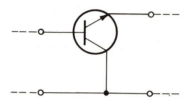

(a)   Common-emitter connection   **(b)**   Common-base connection

(c)   Common-collector connection

Figure P12.29

12.30 Two two-port networks are said to be equivalent if they cannot be distinguished from each other by measurements made at the input and output pairs of terminals, i.e., if they have identical matrices. Show, by finding the z matrices, that the symmetrical lattice and T networks in Fig. P12.30 are equivalent if

$$Z_A = Z_B = Z_a, \qquad Z_C = \tfrac{1}{2}(Z_b - Z_a)$$

If the networks do not contain controlled sources, can a symmetrical lattice always be replaced by an equivalent T network? Can a symmetrical T network always be replaced by a lattice?

12.31 The subnetwork N in Fig. P12.31 is described by the following parameters:

$$z_{11} = 3\ \Omega, \qquad z_{12} = 2\ \Omega, \qquad z_{21} = 1\ \Omega, \qquad z_{22} = 1\ \Omega$$

Find the $ABCD$ parameters for the entire circuit.

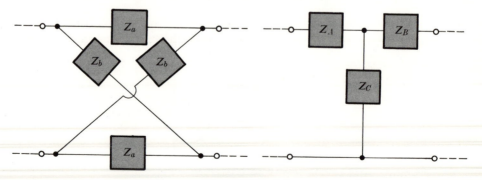

Figure P12.30

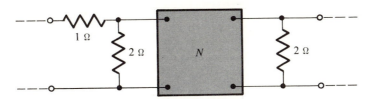

Figure P12.31

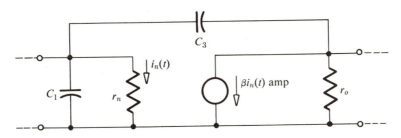

Figure P12.32

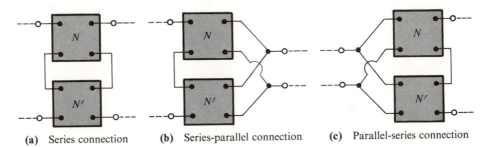

(a)   Series connection   **(b)**   Series-parallel connection   **(c)**   Parallel-series connection

Figure P12.33

12.32 Find the y and h parameters for the network shown in Fig. P12.32.

12.33 Three ways of interconnecting two-port networks are shown in Fig. P12.33. Assume that the interconnection does not change the parameters of the individual networks.

  (a) For the series connection show that the z matrix of the combination is the sum of the individual z matrices.

  (b) Find the h matrix for the series-parallel connection and the g matrix for the parallel-series connection in terms of the matrices for the individual networks.

12.34 For each of the configurations in Fig. P12.34, find expressions for z_{21} and for $[E_2(s)/E_1(s)]_{I_2(s)=0}$ in terms of the z parameters of the subnetworks N_a and N_b. The subnetwork labelled *NIC* is a negative impedance converter, which can be replaced by the model within the dashed lines in Fig. 1.4-10.

12.35 By use of the configuration in Fig. P12.34b, find a circuit for which

$$z_{21} = \frac{38s}{s^2 + 2s + 400} = \frac{38/(s + 20)}{(s^2 + 2s + 400)/[s(s + 20)]}$$

$$= \frac{\left[\dfrac{38}{s + 20}\right]}{\left(1 + \dfrac{20}{s}\right) - \dfrac{38}{s + 20}}$$

where the last form for the denominator follows by a partial fraction expansion. The subnetworks N_a and N_b correspond to special cases of the T network in Fig. 7.4-8.

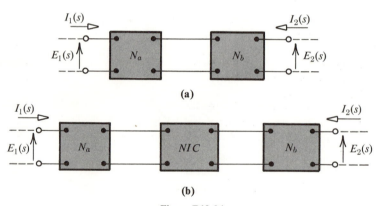

(a)

(b)

Figure P12.34

Answers to
Selected Problems

1

1.2 (a) Linear and time invariant; (b) linear and time invariant; (c) nonlinear and time invariant

1.4 $-\frac{2}{3}$ joule

1.6 For a 1 μf capacitance, $A \geq 38$ meters² for air and $A \geq 0.094$ meters² for mica.

1.12 $C = 4$ f and $R_2 = 1$ Ω

1.14 $de_o/dt + e_o/RC = de_1/dt$

1.17 If $A \to \infty$, $e_o(t) = -(R/L) \int e_1 \, dt$

1.19 Energy supplied by the source is $5/\epsilon = 1.84$ joules; energy stored in the inductance is 1.00 joule; energy dissipated in the resistance is 0.84 joule.

1.21 1.00 joule

1.22 $e(t) = 36$ volts

2

2.2 (b) 7 Ω; (d) -4 Ω

2.4 -4 volts

2.6 1.2 amp

2.9 15 volts

2.10 $R_1 R_4 = R_2 R_3$, $R_5 \to \infty$, or $e_1 = 0$

2.13 5 volts

2.16 $R_o = R_{eq}$

2.18 16 volts

2.20 $e_o/e_1 = -\frac{3}{10}$

2.23 $\frac{28}{9}$ volts

2.24 $\frac{6}{7}$ volt

2.26 (a) -37.5; (b) -7.1

2.28 $\frac{51}{22}$ Ω

2.30 (a) -1 volt; (b) $-\frac{5}{26}$ volt

3

3.2 $e(t) = 2(1 - \epsilon^{-t/2})$ for $0 < t < 1$, $e(t) = 2(\epsilon^{1/2} - 1)\epsilon^{-t/2}$ for $t > 1$

3.5 (a) $i_1(t) = 6$ amp, $e_o(t) = 6 - 12\epsilon^{-t}$ volts; (b) $i_1(t) = 3 + 3\epsilon^{-t}$, $e_o(t) = 6 - 12\epsilon^{-t}$; (c) $i_1(t) = 6 + 6\epsilon^{-t} - 6\epsilon^{-t/2}$, $e_o(t) = 6 - 6\epsilon^{-t/2} - 6\epsilon^{-t}$

3.8 $i_{sc}(t) = (\frac{1}{2} - \frac{1}{6}\epsilon^{-t}) U_{-1}(t)$

3.10 $r(t) = \frac{2}{3}(2 + \epsilon^{-2t}) U_{-1}(t) + \frac{2}{3} U_o(t)$

3.14 $4(1 + 2\epsilon^{-2t})$

3.16 $r(t) = -\frac{7}{10}(1 - \epsilon^{-10t/3}) U_{-1}(t)$,
unit ramp response is $[-\frac{7}{10}t + \frac{21}{100}(1 - \epsilon^{-10t/3})]U_{-1}(t)$.

3.18 $e(t) = (4KR^2C/a)(1 - \epsilon^{a/2RC})^2\epsilon^{-t/RC} \doteq Kah(t)$

3.19 (a) $10\epsilon^{-10} + 2\epsilon^{-5} - 2$; (b) $-4\epsilon^{-5}(4 + \epsilon^{-5})$

3.25 $4\epsilon^{-2t} - 2\epsilon^{-t}$

3.27 $\frac{1}{48}[3 - (4t^3 + 6t^2 + 6t + 3)\epsilon^{-2t}]U_{-1}(t)$

3.28 $\frac{80}{27}[(t/6) - 1]\epsilon^{-t/6}$

4

4.2 (a) $\frac{1}{8}(2t^2 - 2t + 9 - \epsilon^{-2t})$; (d) $\frac{1}{4}t \sin 2t$; (f) $\frac{1}{6}t^3\epsilon^{-t}$; (h) $\frac{1}{8}[(2t + 1)\epsilon^{-2t} - \cos 2t]$

4.3 The values when $e_1(t) = U_o(t)$ are $1/L$, $-R/L^2$, $1/RC$, $-1/R^2C^2$.

4.5 0, $-\frac{1}{2}$ volt per sec

4.9 $\epsilon^{-t}(\cos t + \sin t) - (2t + 1)\epsilon^{-2t}$

4.11 $r(t) = [1 - \frac{1}{3}(\epsilon^{-2t} + 2\epsilon^{-t/2})] U_{-1}(t)$, $h(t) = \frac{1}{3}(2\epsilon^{-2t} + \epsilon^{-t/2})U_{-1}(t)$

4.12 $e_o(t) = (2t + 1)\epsilon^{-t} - 1$ for $0 < t < 2$, $e_o(t) = (2t - 6.39)\epsilon^{-t}$ for $t > 2$

4.15 The circuit is unstable for $\beta < -2$, but the free response can never contain a constant-amplitude, sinusoidal oscillation.

4.17 20 volts, 2 volts per sec

4.19 $i(t) = (4/\sqrt{3})\epsilon^{-t/2} \cos [(\sqrt{3}/2)t - 30°]$. First minimum is -0.326 amp at $t = 2\pi/\sqrt{3}$ sec, and first maximum is 0.054 amp at $t = 4\pi/\sqrt{3}$ sec.

4.21 10 amp, $-\frac{5}{8}$ amp per sec

4.23 8 volts, -6 volts per sec

4.26 $26 + 87\epsilon^t$

5

5.2 (b) $-3\sqrt{3} + j3$; (d) $14.1(-1 + j)$; (f) $20\underline{/-37°}$; (h) $10\underline{/177°}$; (i) $20. 1\underline{/143°}$; (l) $82.2\underline{/158°}$; (m) $0.316\underline{/66.3°}$

5.5 $1.06 \cos (2t - 8.1°)$

5.7 $e_o(t) = [g_m/\sqrt{(g_m + 1/R)^2 + (\omega C)^2}] \cos (\omega t + \theta)$, where $\tan \theta = -\omega C/(g_m + 1/R)$ and $R = r_p R_k/(r_p + R_k)$

5.10 (a) $2\sqrt{2} \sin (2t - 45°)$; (b) $6.75 \cos (10^6 t - 88°)$

5.12 $20 \cos (2t + 30°)$

5.14 $10\sqrt{2}\underline{/-75°}$ Ω, $246 \cos (100t - 54°)$ volts, $22.3\underline{/34°}$ Ω

5.16 $R = 2.63$ Ω, $C = 0.088$ f; $R = \frac{17}{7}$ Ω, $L = \frac{17}{12}$ h

5.18 $\frac{4}{5}$ f

5.20 $\pm 5\epsilon^{-t/\sqrt{3}}$

5.23 $R_4 = R_2 R_3/R_1$, $L_4 = R_2 L_3/R_1$

5.27 $Z_a = 48.5 + j87.5 = 100\underline{/61°}\ \Omega$

5.33 (a) Semicircle of radius $\frac{1}{2}$; (b) semicircle of radius $\frac{1}{2}$; (c) line parallel to the vertical axis

5.36 The locus is not circular. The voltage and current are in phase when $\omega = 2/\sqrt{5}$ rad per sec.

5.37 (a) $5.83 \cos (3t - 149°)$; (b) $3.69 \cos (t + 67.5°)$; (c) $6.77 \cos (2t + 167°)$

6

6.2 (a) 0; (c) $3\sqrt{2}\ \epsilon^{-2t} \cos (2t + 45°)$; (e) $\frac{3}{2}$

6.4 (c) $(sCR)^3/[(sCR)^3 + 6(sCR)^2 + 5sCR + 1]$

6.6 $H(s) = [sC(R_2 - R_1) + 1]/[sC(R_2 + R_1) + 1]$

6.12 $H(s) = -\mu(sCR)^3/[(1 + \mu)(sCR)^3 + 6(sCR)^2 + 5sCR + 1]$. The free response will contain a constant-amplitude, sinusoidal oscillation at $1/\sqrt{6}\ CR$ rad per sec if $\mu = 29$.

6.15 $Z(s) = (6s + 2)/[(1 + 5g_m)s + 2g_m]$. When connected to an external current source, the circuit is unstable for $-\frac{1}{5} < g_m < 0$.

6.17 (a) $10\sqrt{2} \sin (t - 45°) + 10\epsilon^{-t} \cos t$; (b) $5(3t^2 - 4t + 1) + 5\epsilon^{-t}(3 \sin t - \cos t)$

6.18 (b) $(4 - 2t)\epsilon^{-t} - 4\epsilon^{-2t}$

6.20 (a) $H(s) = s/(s + 1)^2$; (c) $H(s) \doteq 9/(s^2 + s + 100)$

6.29 The multiplying constant should be changed to 0.57.

6.30 The bandwidth extends from 2 to 200 kiloradians per sec for one amplifier and extends from 3.1 to 129 kiloradians per sec for two amplifiers in cascade.

7

7.3 It is impossible to place three polarity dots so that they will be consistent with the dot convention.

7.4 (a) $(L_1 L_2 - M^2)/(L_1 + L_2 - 2M)$; (c) 84 h

7.6 $Z(s) = \frac{2}{3}(s^4 + 5s^2 + 4)/(s^3 + 2s)$

7.10 $1.02 \cos (t - 78.6°)$

7.14 (a) $2\epsilon^{-2t} + 2\epsilon^{-2t/3} - 4$; (b) $-3.95(\epsilon^{-0.46t} - \epsilon^{-6.54t})$

7.15 $R_o = 10\ \Omega$, $p(t) = \frac{5}{8} \cos^2 \omega t$, the average power is $\frac{5}{16}$ watt.

7.18 (a) $N^2 = \omega^2 LC$; (b) $N^2 = \omega C \sqrt{R^2 + (\omega L)^2}$

7.19 $1 - (1/5\sqrt{2}) \cos (200t - 45°)$

7.23 (b) $z_{11} = z_{22} = \frac{1}{2} (Z_B + Z_A)$, $z_{12} = z_{21} = \frac{1}{2} (Z_B - Z_A)$; (c) $y_{11} = 1/R_1$, $y_{12} = 0$, $y_{21} = \mu/r_p$, $y_{22} = 1/r_p + 1/R_2$; (e) $y_{11} = y_{12} = 0$, $y_{21} = -\mu/r_p$, $y_{22} = 1/R_2 + (1 + \mu)/r_p$

7.26 $I_a = (5/\sqrt{3})\underline{/-30°}$, $I_b = (5/\sqrt{3})\underline{/90°}$, $I_c = (5/\sqrt{3})\underline{/-150°}$ amp

7.28 $2/(7s + 4)$

8

8.2 (a) $\frac{4}{3}$ watts; (b) $\frac{13}{3}$ watts; (c) 25 watts; (d) 50 watts

8.4 (a) 2, $-\frac{2}{3}$, 2.22, and 1.48 amp; (c) 1.41, 1.27, 1.41, and 2.83 amp; (e) 1.63, $\frac{1}{2}$, 1.66, and 2.22 amp

8.7 6 watts

8.10 $-95.5 \ \Omega$

8.12 $425 \ \Omega$

8.15 (a) $\omega_0 L/R$ where $\omega_0 = \sqrt{1/LC}$; (c) 25

8.17 $N = 1.59$, $P = 0.102$ watt

8.18 (d) $-j3 \ \Omega$

8.20 $\frac{1}{4}$ f

8.26 Total power is 2.67 kw, total reactive power is 0, sum of the wattmeter readings is 2.42 kw.

8.29 $24.4 \underline{/183°}$ amp

8.30 37.3 amp

9

9.2 (a) $f_1(t) = \dfrac{1}{4} + \dfrac{2}{\pi^2}(\cos \omega_0 t + \dfrac{1}{9}\cos 3\omega_0 t + \dfrac{1}{25}\cos 5\omega_0 t + \cdots)$

$\qquad - \dfrac{1}{\pi}(\sin \omega_0 t + \dfrac{1}{2}\sin 2\omega_0 t + \dfrac{1}{3}\sin 3\omega_0 t + \cdots)$

$\qquad$ (b) $f_2(t) = \dfrac{1}{T} - \dfrac{2}{T}(\cos \omega_0 t - \cos 2\omega_0 t + \cos 3\omega_0 t - \cdots)$

9.6 (b) $f_2(t) = \dfrac{1}{4} + \dfrac{4}{\pi^2}(\sin \omega_0 t - \dfrac{1}{2}\cos 2\omega_0 t - \dfrac{1}{9}\sin 3\omega_0 t + \dfrac{1}{25}\sin 5\omega_0 t - \cdots)$

$\qquad$ (c) $f_3(t) = 2 + \dfrac{2}{\pi}(\cos \dfrac{\pi}{2} t + \sin \pi t - \dfrac{1}{3}\cos \dfrac{3\pi}{2} t + \dfrac{1}{5}\cos \dfrac{5\pi}{2} t + \cdots)$

9.10 $e_0(t) = 2.5 - 1.59 \sin (200\pi t - 64.3°) - 0.80 \sin (400\pi t - 103°) - \cdots$

9.12 The ratio of the third to the first harmonic in $e_o(t)$ is 1/6.7.

9.15 $C = 28 \ \mu f$, $R = 590 \ \Omega$, $Q = 31$

9.20 For the succession of impulses, $c_n = 0$ for n even and $c_n = 2/T$ for n odd.

9.21 (a) $\frac{1.5}{T}$; (b) $\frac{1.6}{T}$

9.24 $2a/(a^2 + \omega^2)$, $(\sqrt{\pi}/a)\,\epsilon^{-(\omega/2a)^2}$

9.26 $G(\omega) = \dfrac{\sin (a\omega - a)}{\omega - 1} + \dfrac{\sin (a\omega + a)}{\omega + 1}$

10

10.2 (b) $s/(s^2 - \beta^2)$; (d) $2\beta s/(s^2 + \beta^2)^2$; (f) $[1 - \epsilon^{-(1+s/a)}]/(s + a)$

10.3 (c) $1/[s(1 + \epsilon^{-s})]$; (e) $(1 - \epsilon^{-2\pi s})/(s^2 + 1)$

10.6 $\ln \dfrac{s+1}{s}$

10.9 (a) ϵ^{-t};   (c) $\frac{1}{2}(1 - \cos 2t)$

10.11 (c) $U_0(t) - 2\epsilon^{-t} \sin t$;   (e) $(1 - \cos \beta t)/\beta$;   (g) $t^2 + t - 2 + 2 \cos t - \sin t$;
(i) $8 - 4\epsilon^{-4t} \sin 4t$;   (k) $\frac{1}{9} \epsilon^{-t}[1 - t - 1.05 \cos (3t + 18.4°)]$

10.16 (a) $e_o(t) = 2t\epsilon^{-t}$;   (b) $e_o(t) = \frac{1}{2}(t - 1) \epsilon^{-t}$;   (c) $e_o(t) = \frac{3}{5}(1 + 4\epsilon^{-5t/3})$

10.18 (a) $i_o(t) = (2/\sqrt{3})\epsilon^{-t} \cos [(t/\sqrt{3}) - 30°]$;   (b) $i_o(t) = 9.4\epsilon^{-0.71t} - 4.4\epsilon^{-7.79t}$;
(c) $e_o(t) = \epsilon^{-3t/2}[2 \cos (\sqrt{15}t/2) + (2/\sqrt{15}) \sin (\sqrt{15}t/2)]$

10.20 (a) $e_K(t) = 2U_0(t) + (6 + \frac{8}{3} \epsilon^{-2t/3}) U_{-1}(t)$

10.22 (a) All finite values of s;   (b) none;   (c) none;   (d) $s \neq (2k - 1)(\pi/2)$, where k
is an integer.

10.23 (b) Residues are 1 and $-\epsilon^{-t}$.   (d) Residue is zero.

10.27 (b) $f(t) = t - (2/\pi)(\sin \pi t - \frac{1}{2} \sin 2\pi t + \frac{1}{3} \sin 3\pi t - \cdots)$
$= 2[U_{-1}(t - 1) + U_{-1}(t - 3) + U_{-1}(t - 5) + \cdots]$

10.28 (a) $h(t) \doteq 0.9\epsilon^{-t/2} \sin 10t$;   (b) $h(t) = (1/\sqrt{2}) \epsilon^{-t/2} \cos [(t/2) + 45°]$

11

11.7 $B = 20$ lb-sec per ft, maximum displacement $= 0.736$ ft

11.9 $M_2 = K/\omega_1^2$

11.10 (a) $x(t) = \frac{1}{4}(2t - 1 + \epsilon^{-2t})$;   (b) 0.42 ft per sec;
(c) $x(t) = \frac{1}{9} - 0.157\epsilon^{-t} \cos (\sqrt{8}t + 45°)$, if $x = 0$ and $t = 0$ correspond to the instant when the mass first hits the spring; (d) 0.715 sec, 0.146 ft-lb.

11.12 (a) $J = Ma^2/2$;   (b) $B_r = \pi \eta a^4/2d$

11.14 $\omega_2(t) = [B_{r1}/(B_{r1} + B_{r2})][1 - \epsilon^{-(B_{r1} + B_{r2})t/J}]$

11.18 (a) A/B_{r1}, 0, 0, A/J_3;   (b) opening the switch

11.19 The displacement of M_3 is 0.005 times the voltage across C_2.

11.22 It is not always possible to find a mechanical analog for a given electrical circuit.

11.24 $H(s) = (R_2/\rho g)/[R_1 R_2(A/\rho g)^2 s^2 + (2R_2 + R_1)(A/\rho g) s + 1]$

11.28 (b) $H(s) = (\mathcal{B}l/R)s/[Ms^2 + (B + \mathcal{B}^2 l^2/R)s + K]$, $\zeta = (B + \mathcal{B}^2 l^2/R)/(2\sqrt{MK})$

11.29 (a) $i(t) = (R/2)[\epsilon^{-t} + \sqrt{2} \cos (t - 135°)]$ for $t > 0$

11.36 (a) $\tau = (AJ/k_p)(d^2h/dt^2) + (AB_r/k_p)(dh/dt) + (\rho g k_p)h$;
(c) $\tau = (AJ/k_p)(d^2h/dt^2) + (AB_r/k_p + Ak_p R_1 + \rho g J/k_p R_2)(dh/dt)$
$+ (k_p + k_p R_1/R_2 + B_r/k_p R_2)\rho g h$

12

12.3 $R_1 = 40$ Ω, $R_2 = 960$ Ω, slope of dynamic load line $= -1/1960$ mho

12.5 (a) $E_1 = 8.4$ volts, $E_2 = 18.9$ volts; (b) $e_o(t) = -0.8 \sin \omega t$, $\eta_c = 2.3\%$

12.7 $P_{\max} = \frac{1}{2}E_{CB}I_C$, $\eta_c \doteq 50\%$

12.9 (a) $-\mu(\mu + 1)R_2/[(\mu + 2)r_p + R_2]$;   (b) $-\beta^2 R_1/(R_1 + r_n)[(\beta + 1) + (R_2/r_o)]$

12.12 $H(s) = 20 \times 10^3/[(10^{-14}s^2 + 5 \times 10^{-9}s + 1)(10^{-7}s + 42)]$, resonant angular frequency $= 10$ megaradians per sec, bandwidth $= \frac{1}{2}$ megaradian per sec

12.15 (a) $Z_{in}(s) = [1 - A(s)B(s)]Z_1(s)$;
(b) The output impedance is $Z_o(s)Z_B(s)/\{Z_o(s) + Z_B(s)[1 - A(s)B(s)]\}$, where $Z_B(s)$ is the impedance looking into the right-hand side of the network N.

12.18 (a) At intermediate frequencies, $H(j\omega) = [R_1/(R_1 + R_s)]\{(\beta + 1)R_2/[(\beta + 1)R_2 + R]\}$, where $R_2 = r_oR_o/(r_o + R_o)$, $R = r_n + R_{s1}$, $R_{s1} = R_sR_1/(R_s + R_1)$. The lower limit to the bandwidth is at $\omega_1 = [(\beta + 1)R_2 + r_n + R_1]/C_c(R_1 + R_s)[(\beta + 1)R_2 + r_n + R_{s1}]$, but the upper limit is not easily expressed in literal form.
(b) The impedance looking into the base of the transistor is $R_b = r_n + (\beta + 1)R_2$, the impedance seen by the voltage source is $R_s + R_1R_b/(R_1 + R_b)$, and the output impedance is $r_o(R_{s1} + r_n)/[R_{s1} + r_n + (\beta + 1)r_o]$.

12.21 $E_o/E_s = [K/(1 - KB)]/[1 + j\omega/\omega_2(1 - KB)]$

12.23 $K = 0.414$

12.26 For Fig. P7.23c, $g_{11} = 1/R_1$, $g_{12} = 0$, $g_{21} = -\mu R_2/(r_p + R_2)$, $g_{22} = r_pR_2/(r_p + R_2)$, $h_{11} = R_1$, $h_{12} = 0$, $h_{21} = \mu R_1/r_p$, $h_{22} = (r_p + R_2)/r_pR_2$. For Fig. P7.23d, $g_{11} = 1/R_1 + (\mu + 1)/(r_p + R_2)$, $g_{12} = -R_2/(r_p + R_2)$, $g_{21} = (\mu + 1)R_2/(r_p + R_2)$, $g_{22} = r_pR_2/(r_p + R_2)$, $h_{11} = r_p/(\mu + 1 + r_p/R_1)$, $h_{12} = 1/(\mu + 1 + r_p/R_1)$, $h_{21} = -(\mu + 1)/(\mu + 1 + r_p/R_1)$, $h_{22} = [\mu + 1 + (r_p + R_2)/R_1]/[(\mu + 1)R_2 + r_pR_2/R_1]$. For Fig. P7.23e, $g_{11} = g_{12} = 0$, $g_{21} = \mu R_2/[r_p + (\mu + 1)R_2]$, $g_{22} = r_pR_2/[r_p + (\mu + 1)R_2]$, and the h parameters do not exist.

12.29 $h_{11b} = 19.2\ \Omega$, $h_{12b} = 1.83 \times 10^{-2}$, $h_{21b} = -0.98$, $h_{22b} = 1.92 \times 10^{-5}$ mho, $h_{11c} = 10^3\ \Omega$, $h_{12c} = 1.00$, $h_{21c} = -51$, $h_{22c} = 10^{-3}$ mho

12.31 $A = \frac{27}{4}$, $B = \frac{5}{2}$, $C = \frac{13}{4}$, $D = \frac{3}{2}$

12.34 (a) $z_{21} = z_{21a}z_{21b}/(z_{22a} + z_{11b})$; (b) $z_{21} = -z_{21a}z_{21b}/(z_{22a} - z_{11b})$

Index

Page numbers in **boldface** refer to tables.